# 4"x4" D.O.T. LABELS—

 EXPLOSIVE A

 EXPLOSIVE B

 EXPLOSIVE C

 NON-FLAMMABLE GAS

 FLAMMABLE SOLID

 FLAMMABLE LIQUID

 FLAMMABLE GAS

 POISON

 POISON GAS

 ORGANIC PEROXIDE

 OXIDIZER

 IRRITANT

 CORROSIVE

 RADIOACTIVE I

 RADIOACTIVE II

 RADIOACTIVE III

 SPONTANEOUSLY COMBUSTIBLE

 IRRITANT

 DANGEROUS WHEN WET

 HARMFUL STOW AWAY FROM FOODSTUFFS

 OXYGEN

 CHLORINE

 BLASTING AGENT

EMPTY

 BIOMEDICAL MATERIAL

 MAGNETIZED MATERIAL

 DANGER

D1283017

# HANDBOOK OF CHEMICAL INDUSTRY LABELING

# HANDBOOK
# OF
# CHEMICAL INDUSTRY
# LABELING

Edited by

## Charles J. O'Connor

Labeling and Hazard Analysis Consultant
Greens Farms, Connecticut

## Sidney I. Lirtzman

Baruch College
City University of New York
New York, New York

**NOYES PUBLICATIONS**
Park Ridge, New Jersey, USA

Published in the United States of America by
Noyes Publications
Mill Road, Park Ridge, New Jersey 07656

10 9 8 7 6 5 4 3 2 1

Library of Congress Cataloging in Publication Data

Main entry under title:

Handbook of chemical industry labeling.

   Includes bibliographies and index.
   1. Chemicals--Labeling--Law and legislation--United
States. 2. Products liability--Chemical products--
United States. 3. Chemicals--Labeling--United States.
I. O'Connor, Charles J.
KF3958.H36 1984   344.73'0424   83-22108
ISBN 0-8155-0965-0   347.304424

# Preface

This handbook presents for the first time, in a single volume, a concise treatment of a long neglected subject—*Chemical Industry Labeling*. The necessity for such a volume has long existed and has been highlighted in recent months by news of newly legislated worker and public "Right to Know" obligations in Connecticut, California, New York, New Jersey, and Philadelphia as well as the recently published Federal OSHA Hazard Communication Standard.

The need for informative labeling in the workplace, transportation, distribution and disposal operations has been formally recognized in various city, state and Federal statutes and regulations. Society at large has demanded increased information on chemical products for many years; organized labor has long been in the forefront for an improved hazard label communication program; and the chemical industry published the first label guide for its members in 1938, and has updated its guide to precautionary labeling periodically, culminating in 1976 as the *American National Standard (Z129.1) for the Precautionary Labeling of Hazardous Industrial Chemicals.*

It was the need to serve these three major populations—labor, industry, and the public—that led to publication of this handbook. It is designed to provide an in-depth review of, and act as a source for the major elements of a *Hazard Label Communication Program.*

Today, labeling is defined as all written, printed and graphic material that accompanies or may accompany a product. In some cases it may include advertising and material data sheets. It is sometimes referred to as product communication, or product information.

Typically, a label provides information on what the product is, what it does, how to use it, how *not* to use it, and how to dispose of it, what to do if a fire, occurs, or a spill or leak, and first aid when appropriate.

A chemical label should include the name of the chemical in sufficient detail to permit (1) medical treatment in case of an accident and (2) appropriate and effective emergency response in case of fire, spill or leak. Medical instructions should be written by an occupational health physician. An additional prudent step would be to review such instructions with the National Poison Control Center in Pittsburgh.

This Center run by Richard Moriarity is probably the best single source of information for the emergency treatment of consumer and occupational chemical exposure. Operated by physicians, nurses and pharmacists 24 hours a day, every single day of the year, it is available instantly by phone. The Center will accept trade secret data on a confidential basis to be used only in case of a medical emergency. The true chemical identity of one's product will only be released to a physician who is treating a medical emergency. The cost for this service is very modest.

For the purpose of emergency response to a major spill, a fire or a leak, at the minimum the Department of Transportation (DOT) nomenclature should be followed; for those chemicals that meet the requirements of any one or more DOT classes a UN or NA number must also be assigned. Chemtrec, a 24 hour, every day, all year emergency response center, is available for emergencies by phone. Chemtrec will accept product information that would be useful in responding to fire, leak or spill emergency situations. This Center, run by the Chemical Manufacturers Association (CMA) as a public service is probably the best single source of emergency response information. The Chemtrec phone number with an appropriate statement should appear on the label. A closely related service is *The Emergency Response Guide* published by DOT. *The Emergency Response Guide,* produced by the applied Physics Laboratories of John Hopkins University and reviewed by an industrial chemical advisory group, was published by DOT in 1980. It is currently undergoing revision and a new edition should be available soon.

This DOT pamphlet is intended to accompany every transport vehicle that moves hazardous chemicals, be available in every fire house, on every fire truck, in every ambulance, police car, police department and emergency response center in the nation.

The Guide lists by proper shipping name and also by UN/NA number all chemical commodities regulated by DOT with cross-reference to specific instructions on what to do in case of fire, spill or leak. It is the best single, printed source of emergency response information available today.

There are a number of different ways of supplying sufficient information about a product, so that users, transporters and workers can be assured of adequate information in case of accidental exposure. There is evidence from the social sciences that simplified identification coupled with expert instruction is superior to reliance on chemical nomenclature to convey hazard information.

When one has selected some combination of chemical nomenclature, generic name, and first aid and emergency response instruction, one should next consider the selection of a signal word and a set of hazard statements. Historically, "Danger", "Warning", "Caution" have served as the three indicators of potential hazard. "Danger" indicates the highest level, "Caution" the lowest and "Warning" is intermediate. In many standards, these words are further defined in terms of $LC_{50}$, $LD_{50}$, flash point, skin corrosion and physical properties such as explosivity and radioactivity. Studies have shown that workers perceive a significant difference between "Danger", and "Warning" or "Caution". However, little if any difference is perceived between "Warning" and

"Caution." This seems to indicate that employers should instruct workers in the meaning of these terms.

The selection of a signal word for certain chronic effects, i.e. cancer, teratology and mutation is a controversial subject. When the chemical also possesses acute hazards, this fact may determine one's choice. If, for example, following the ANSI Appendix, the product is extremely flammable *and* a carcinogen, the signal word "Danger" is selected based on the flammability hazard.

The controversy centers about the fact that many practitioners feel the signal word "Danger" should be reserved for those hazards which are immediate and life-threatening. The probability of human harm is virtually certain with overexposure to acutely hazardous chemicals, but highly uncertain with overexposure to chemicals that possess chronic hazards. This concept is referred to as "Labeling Under Uncertainty".

This probabilistic property as applied to carcinogens, is further defined as follows: (1) known human carcinogen, (2) known animal and probably human carcinogen, (3) known animal carcinogen and (4) known mutagens. Much of our knowledge seems to indicate that some mutagens are probably carcinogens. Anthony Garro discusses this relationship in Part II. Some feel that action should be taken on this information and others prefer to at least obtain mammalian data before including any statement on a label.

An appendix to the CMA label proposal establishes a selection system which appears reasonable for most long-linked chronic hazards, i.e., cancer, mutagens and teratogens.

Although some effects as nephrotoxicity and hepatotoxicity are often treated as chronic effects, at times the onset of symptoms may be rapid, and when this is true, such effects should be handled as part of acute toxicity.

Statement of hazards should be simple, direct, and concise, but the stress should be on "simple". The use of what have become standard phrases is encouraged. A list of these phrases is contained in the LAPI/ANSI Guide discussed by Jay Young in Part IV.

This book is organized into four parts: Label Communication; Science and Labels; Product Liability, Regulations and Labels; and Industry Standards and Practice.

Part I is devoted to the perceptual and graphic elements of hazard label communication and the underlying science base which supports their practical use. In Chapter 1 of this section Sidney Lirtzman reports on a radical and revolutionary Hazard Label Communication Research Program conducted by O'Connor and Lirtzman, and the conclusions to which their research has led.

The research program utilized specially modified infra-red eye scan equipment. The basic equipment was provided by the Applied Science Laboratories as a working grant to the research team. Norma Skolnik, in the second chapter, provided a review and suggested program for utilizing both manual and machine based data services, as a source of label information. Harry Fund, in the last chapter of Part I deals with the graphic and production arts required to print and manufacture labels, placards and tags.

In Part II Adria Casey, Donald MacKellar, Anthony Garro and Richard Moriarity discuss the applied science that underlies much of labeling. Anthony Garro of Mt. Sinai School of Medicine and Donald MacKellar of Toxigenics,

Inc. explore the biological basis for chronic and acute toxicity, while Adria Casey explicates physical and chemical test parameters; and classification based on physical, chemical and biological data.

Labeling, Product Liability and Government Regulations form the major elements of Part III. David Zoll of the Chemical Manufacturers Association leads off with a discussion of product liability, the "prudent man" and case law. James Toupin of Covington & Burling follows with a chapter on trade secrets, patents and trademarks. Steven Jellinek discusses the label requirement for the sale and use of pesticides under FIFRA. TSCA and RCRA labeling regulations with specific case examples are explored by Robert Sussman and Jennifer Machlin. Robert Sussman also covers consumer product labeling in his chapter on the Consumer Product Safety Commission. John Gillick of the law firm of Kirby, Gillick, Schwartz and Tuohey reviews labeling in transportation. He details DOT label and placard regulations for air, water, road and rail movements for packages as well as for bulk containers. Flo Ryer, former Director of Health Standards for the Occupational Safety and Health Administration (OSHA), details the label requirements of OSHA's Health Standards and reviews OSHA's latest label standard proposal.

This controversial standard has been in development for more than eight years. OSHA's latest draft includes provisions for container and reactor labels, area placarding, a material safety data sheet in the workplace, worker training and a provision for maintaining trade secrets. Overall this standard is performanced-based, apparently permitting many existing systems to meet OSHA requirements.

In Part IV, the last section, Jay Young and Charles O'Connor discuss current and proposed industry standards. The Chemical Manufacturers Association (CMA) sponsored *ANSI Guide to Precautionary Labeling of Hazardous Chemicals* and the National Fire Protection Association's (NFPA) *Identification of Fire Hazards of Materials* are the two oldest standards. The ANSI Standard owes its beginnings to the original Labels and Precautionary Information Committee (LAPI) Guide published by CMA in 1938, while the NFPA System was first explicated in 1952.

Jay Young outlines the basic requirement of the CMA sponsored ANSI Standard and the specific elements required to compose an appropriate "Hazard Label". He brings special insight to this task, having served as the CMA executive responsible for the labeling activities of the association.

The two systems are complementary. NFPA uses a color keyed symbol system with high recognition value. This permits an observer to quickly assess the hazardous nature of a tank or area. ANSI, relying primarily upon text to convey its message, is better suited for container labeling. Intelligent application of both systems significantly improves label hazard communication. The current use and development of material safety data sheets are also discussed as a part of a hazard communication program.

An example of such a combined system is offered by the NIOSH Identification System for Occupationally Hazardous Materials. This identification system uses color-keyed symbols with numerical "degree of hazard" indicators for placards. The system adds precautionary text and hazard statements for labels, and requires the availability of a material safety data sheet in the workplace.

The authors also include a discussion of the current ASTM Z 535.2 proposal for Safety Signs. This system combines three elements: color, shape, and signal words to create three distinctive levels of hazard alert signs. ASTM uses the traditional signal words: "Danger", "Warning," "Caution". These words are not independent elements, but are always combined with specific colors and specific shapes.

"Danger" always appears with a combination of white, red and black on an oval shape; "Warning" is used with a combination of orange and black on a truncated diamond; and "Caution" always appears on a rounded-corner rectangle, colored yellow and black.

Jay Young and Charles O'Connor present the National Paint and Coatings Association (NCPA) Label Guide and in-plant Hazardous Materials Identification System (HMIS). HMIS is a complete hazard communication system. It utilizes labels, tags, wallet cards, wall posters, employee handouts, placards, symbols for personal protection, an audio visual program, and a rating system for health, reactivity, and flammability. This system will accommodate both acute and chronic health effects. As in the NFPA System, blue, red and yellow are used to highlight health, flammability, and reactivity. Rating or ranking for each hazard class runs from one (1) to five (5), with five (5) the most hazardous. The HMIS Manual also includes a glossary, information on how to assign hazard ratings, industrial hygiene, and raw material sheets. NPCA, under the guidance of Larry Thomas, Executive Director, has produced an integrated workable and highly valuable system.

I would like to express my thanks to Corrine Hessel for her help in preparing and reviewing the manuscript in development and through the galleys. This book would have been impossible without her professional help.

I hope that this handbook will provide a comprehensive library source, and be useful for the health, safety, and legal decisions which must be made by chemical manufacturers, attorneys, safety equipment producers, toxicologists, industrial safety engineers, waste disposal operators, health care professionals, and the many others who may have contact with or interest in the Chemical Industry due to their own or third party exposure.

Greens Farms, Connecticut                                Charles J. O'Connor
December, 1983

# Contributors

Adria C. Casey
Stauffer Chemical Company
Westport, CT

Harry Fund
Labelmaster
Chicago, IL

Anthony J. Garro
The City College of New York and
    the Mount Sinai School of
    Medicine
New York, New York

John E. Gillick
Kirby, Gillick, Schwartz and
    Tuohey, P.C.
Washington, D.C.

Steven D. Jellinek
Jellinek Associates, Inc.
Washington, D.C.

Sidney I. Lirtzman
Graduate School and University
    Center and Baruch College
City University of New York
New York, NY

Jennifer Machlin
Orrick, Herrington and Sutcliffe
San Francisco, CA

Donald G. MacKellar
Toxigenics, Inc.
Decatur, IL

Richard Moriarty
National Poison Center
Children's Hospital
Pittsburgh, PA

Charles J. O'Connor
Labeling and Hazard Analysis
    Consultant
Greens Farms, CT

Flo H. Ryer
U.S. Environmental Protection
    Agency
Washington, D.C.

Norma Skolnik
International Playtex, Inc.
Paramus, N.J.

Robert M. Sussman
Covington and Burling
Washington, DC

James Toupin
Covington and Burling
Washington, DC

**Jay A. Young**
Consultant
Silver Spring, MD

**David F. Zoll**
Chemical Manufacturers Association
Washington, D.C.

# Contents

## PART IV
## INDUSTRY STANDARDS AND PRACTICE

# Part I

# Label Communication

Underlying the practices associated with labeling in the chemical industry and in other industries, and for consumer product labeling as well, is the implied but usually unstated assumption that labeling is important because exposure to the label will cause changes to occur in the person who comes into contact with the product to which the label is affixed. These changes are construed to be either psychological in nature or changes in overt behavior associated with the problem to be avoided. From this perspective labeling is both a communication process and an information handling process as far as people are concerned. Thus, the evidence and data which behavioral science can bring to bear upon the labeling process is of vital importance to all those in the chemical industry who are charged with the design, evaluation, and production of labels to be used with or on products manufactured for distribution and sale and in the workplace.

This section presents three approaches to the behavioral science aspects of the labeling process. Chapter 1 reviews in detail behavioral considerations associated with labeling. In Chapter 2, there is a discussion of the information systems available to the labeler which are useful and necessary in decision-making activities with respect to chemical labeling and classification. Following in Chapter 3 comes a discussion of the technology of designing, developing and producing labels and placards.

Chapter 1 provides a discussion of the functions of labeling and the nature of labeling as a communications system whose function is to deliver a message or messages to a reader or consumer. The chapter provides a definition of labeling and discusses in some detail the purposes of labeling industrial products and products in general. Alternative perspectives of the importance of the labeling process in industry are presented as is an introduction to the problems involved in hazard labeling in the industry.

This is followed by a detailed summary of the major perceptual issues associated with labeling, including the processes of perception, attention, form, color, color preferences and emotional characteristics, and visibility, and a discussion of color and hazard labeling. The chapter also discusses the issues associated with legibility and summarizes the evidence and controversy involved with the use of symbols in the labeling process. The author also summarizes both the advantages and disadvantages of graphic symbols and provides some research evidence.

Also discussed in Chapter 1 is the issue of reading ability and reading levels in terms of the usefulness of the label. The chapter also presents data associated with the evaluation of label effectiveness and discusses three major criteria against which to evaluate results, including time, optical efficiency, and the communication value of labels, and relates these to the issues associated with hazard labeling. The reader is provided with a discussion of the methods now available and in use for testing the efficiency and effectiveness of labeling.

The chapter provides the reader with a summary of research and results of research conducted with respect to hazard labeling in the areas of foods, drugs, chemicals, children and hazard warnings. The author presents in some detail the results of research which has been conducted with respect to hazard warnings, hazard labeling, hazard communications over the past five years. He provides the reader with a series of recommendations for practically improving the effectiveness of the label and related guidelines.

A detailed reference list and suggested readings is provided for the reader.

Chapter 2 provides the reader with a wealth of useful information designed to improve the efficiency and speed of information access for those charged with the responsibility for making decisions about whether specific labels are required for a given product, and the nature of the information and warnings which may be necessary in designing a useful label for a chemical product.

The chapter is divided into two major sections: the first summarizing sources of regulatory information; the second sources of scientific information for use in the labeling process. The first section on regulatory information summarizes the nature of government documents and particularly discusses the Federal Register and the Code of Federal Regulations as they apply to the needs of the labeler. Other sources of Federal documents, including the *National Technical Information Service,* and the *Monthly Checklist of State Publications* are discussed. This section also summarizes the major computerized data systems for Federal information, including the Federal Index Data Base, the Legal Data System, including LEXIS and WESTLAW Systems. The chapter also discusses the easiest ways to obtain access to Government documents.

The second and more detailed section of the chapter discusses the scientific literature data bases relevant to the process of labeling in the chemical industry. It provides information with respect to services readily available to the labeler directly including libraries, chemical references and bibliographical sources for toxicological and medical information. The importance of the computerized literature retrieval system is discussed in detail because of its ability to increase the comprehensiveness and the efficiency of the search, especially for material involving hazardous chemical data in general and toxicological information in particular.

Chapter 2 provides a useful section on the issues associated with hazard evaluation and offers a compilation of sources for toxicological information which is very extensive and useful, especially to the novice in the field. The chapter then describes in detail government literature retrieval systems, especially the Medlar's System of the National Library of Medicine, which includes over fifteen relevant data bases, including the Toxicology Data Bank, MEDLINE, TOXLINE, RTECS (the Registry of Toxic Effects of Chemical Substances list), and CANCERLIT. A discussion on the Chemical Substance Information Network (CSIN) being developed by the Environmental Protection Agency is also included.

This chapter also discusses private on-line computerized abstracting services, including the Chemical Abstract Service, and CAS ONLINE which is the computerized system providing access to substance information from the chemical abstract system registry file. The Lockheed DIALOG Information Retrieval Service, BIOSIS Previews, CHEMSEARCH, CHEMSIS, ENVIROLINE, the Excerpta Medica, SCISEARCH, and related services are discussed. The Bibliographic Retrieval Services Data Base is also discussed in some detail. A particularly useful aspect of this chapter is a discussion of search services—their type and their costs—which enable the executive and decision-maker to do searches in-house rapidly and efficiently. In this regard both commercial and noncommercial search services are discussed and their advantages and disadvantages as well as costs are detailed.

Finally a very useful bibliography of information sources is provided to the reader.

Chapter 3 on labeling and placarding presents a very practical view of the issues, information, and recommendations which the person responsible for the actual production of product labels will find indispensable. In this chapter the author discusses the problems of production of labels and placards under various regulatory schemes. He also relates production problems to the national regulations and laws.

He then proceeds to a discussion of label technology, covering various types of labels, their advantages and disadvantages, as well as the problems associated with their use in different contexts, including transportation. He provides the reader with relevant excerpts from the laws and regulations governing the production and use of such labels. In addition, there is also a discussion of the history and use of placards which are functionally related but different from the label itself and have different production problems associated with them.

The author then discusses technical aspects of label production, covering in turn choice of labeling materials, different face and stocks, and the characteristics, advantages and disadvantages of the alternative materials available to the labeler. He provides a useful set of guidelines and recommendations with respect to the use of adhesives with certain types of labels and stocks. The chapter then discusses methods for printing labels, and provides a useful summary chart to the labeler, detailing the advantages and disadvantages of different methods of production for different types of labeling. The chapter then gives the labeler a series of guidelines for production of the label, and provides a bibliography of sources for further information on the functional labeling process.

# 1

# Labels, Perception and Psychometrics

**Sidney I. Lirtzman**
*Graduate School and University Center*
*and*
*Baruch College*
*City University of New York*
*New York, NY*

This chapter deals with the practical issues involved in the labeling of chemical products. Specifically. this chapter will cover (1) the functions of labeling, (2) the underlying perceptual issues associated with labeling, (3) the perceptual impact of labeling, (4) existing research information concerning hazard labeling, (5) procedures for evaluating labeling, and (6), suggestions to guide the development of labeling.

## FUNCTIONS OF LABELING

### Background

What is known about the factors contributing to the development and impact of labels for chemical products is largely a result of experience gathered over the past hundred years in the advertising and marketing of consumer products. As bulk shipment and packaging of products shifted toward unit packaging in response to a rise in general income, increased demand for utility and convenience by consumers as well as improved storage and transportation ability, the realization grew that packaging and labeling were important factors in attracting customers, informing them of product lines, and developing and maintaining product loyalty. The package and label gradually shifted away from almost purely functional roles to more sophisticated vehicles designed to communicate a variety of themes to potential and actual consumers.

A review of such sources as Dreyfus (1972), Kamekura (1965) and Humbert (1972) is instructive in seeing the radical changes in labeling over time.

Indeed, it has become very clear to all involved in labeling activity that through the package, the label(s) and related inserts, associated flyers and information, and the context of response, the consumer and the manufacturer have formed a communication system. The function of this system is to deliver a message or messages to the consumer—communication of brand name and product class, product function and instructions for use, and evocation or perception of product qualities felt likely to enhance trial purchase and/or repeat purchase.

Millions of dollars have been spent on the development and testing of labels and packages in the consumer and related industrial products area. Despite this, almost no published research or guides exist for use by people who are responsible for developing labeling in the chemical industry, particularly for chemical products falling under the general rubric of *hazardous products*. These include products whose manufacture and distribution are governed by statutes such as the Federal Hazardous Substances Act (FHSA), the Federal Insectiside, Fungicide, and Rodenticide Act (FIFRA), and the Toxic Substances Control Act (TSCA).

Much of the material covered in this chapter reflects information gathered and research conducted by the author during the course of his work in advertising, marketing research, and label and package design and evaluation. The remainder of the material is based on published materials and related research in psychology, behavioral science, economics, sociology, perception, and communications.

## Definition

While there a commonly accepted definition of the term *label,* the concept itself is surrounded by a considerable degree of ambiguity. In the chemical industry the term label or labeling incorporates the printed device(s) affixed directly or indirectly to a container surface, inserts found within the container, associated Material Safety Data Sheets, product descriptions, overpacks and wrappers, workshop signs and placards, most forms of advertising and the like. From this perspective, labeling of chemicals is a system involving a variety of elements.

However, for the purposes of this chapter, we will define a *label* as any element, affixed to or associated with a functional container or package, whose purpose is to communicate in language or other symbols, in color or form, specific information and emotional/perceptual stimuli designed to affect the perceptions and behavior of human beings who are prospective users of the products.

Although other factors such as the material safety data sheets (MSDS) are important parts of labeling we will not deal with them in this chapter. I will refer to them where appropriate in the context of discussion or research.

## Purpose of Labeling

Probably no aspect of a package is as frequently taken for granted as is the label. We all expect to see a label on a product but when called upon to describe the label we are often at a loss to do so with great detail. Yet, labels are functional; they are designed to change behavior of the people who are exposed to them.

In general, a label has at least eight common uses:

1. *Identifying the name of the product and manufacturer.* Most products are generally identified by a brand name or some common name. We call for a product by its name, and check the label to make sure we have the correct product.

2. *Marketing and promotional information.* The label can be used to promote the product by providing information stressing the *quality* and performance of the product (e.g., warranties, grades, seals of approval, image and status elements, advertising copy and the like).

3. *Identification of the function of the product.* The label tells a potential user what the product is supposed to do ("relief of sore muscle's ache," "contact adhesive," "oxidizer," etc.).

4. *Providing directions for use of the product.* The label informs the potential user of the correct or advised way of using the product. ("Take two teaspoonfuls every six hours;" "Apply adhesive to one surface of articles to be bonded").

5. *Education of the user.* The label provides information to the user which may be of potential value by listing components, nutritional values, identifying data such as presence of color additives, doses, weights and measures, etc.

6. *Providing hazard warnings.* The label gives warning of potential dangers related to use or misuse of the product and specifies the proper actions of the individual to avoid the danger.

7. *Provide remedial information.* The label will inform the user of actions to be taken or avoided if the warned against danger actually occurs (e.g., first aid information, notice to contact a physician, notice to avoid inducing vomiting, etc.).

8. *Idiosyncratic information.* The label provides information of potential importance to specific classes of prospective users with certain personal conditions, e.g., the presence of allergens, contra-indicated use for medical reasons, etc.

A review of the eight functions of labeling shows that the implicit if not explicit major assumptions of the labeling processes is that the label will cause changes to occur in the person who comes into contact with the product. These changes can be construed to be either psychological influences (changes in attitude, belief, emotion, comprehension, memory, risk evaluation mechanisms, etc.) or altered overt behavior (changes in actual handling of product, procedure of work, disposal of container or waste, reference to raw materials, questioning activity, etc.). Marketers, advertisers and regulators all share the unspoken assumption that the product *label itself,* irrespective of any other related elements of the product communication system *is capable of and does* cause changes in the overt and/or potential behavior of the person coming into contact with the label as consumer or worker. Two major corollaries to this implicit assumption are (1) that every element of the label is equally *capable of causing and likely to cause* a desired effect, (2) that every potential reader *values* the label, and therefore will read the label and all its elements. If these effects of labeling were not presumed, then there would be little concern on the part of manufacturers, government, consumers, or labor about what appears on labels.

In trying to assess the validity of these labeling assumptions it should be kept in mind that there is almost no empiric evidence publicly available which supports them. For the most part such proof as exists is proprietary and results from market, consumer and advertising research studies conducted by and for manufacturers, usually of consumer products, a portion of which involves packaging or labeling effectiveness or design issues. The remaining data are drawn from research in psychology and communications, or from a few relatively specific researchers into the effect of multifaceted programs designed to inform or warn consumers about ingredients or hazards associated with products or product use. In this regard, the editors have found no published research which clearly isolates the effect of a given label on a specific chemical product from the effects of other factors including inserts, training, general media information, advertising and promotion or consumerist activities.

Furthermore, almost nothing is known about what the consumer/reader/worker really *does* when exposed to a chemical product label, nor what this person may want to have appear in such labels.

### Views About Labeling

Since labeling was presumed to influence the behavior and psychological attributes of prospective readers, how to use and design product labels became important to the manufacturer. Labeling began to be evaluated as an important element in the execution of public policy with respect to consumer and worker protection (Food and Drug Administration [FDA], Federal Trade Commission [FTC], etc.), especially with respect to toxic and hazardous chemicals and substances (FIFRA, TSCA, FHSA). The *regulated* use of labels has proliferated to the point that chemical labeling is now an enterprise requiring professional attention. In deciding how and why to label, different interest areas bring different concerns to the process.

**Government:** Government approaches labeling from the perspective of public policy, asserting a police powers mandate *to protect* the citizen in his or her role as consumer or worker from non-negligible risks associated with the potential use or misuse of a product in commerce. The government perspective requires a decision to ban or label a product, and if labeling is the vehicle chosen, then government's perspective is directed toward label rule making which ostensibly will result in reduced risk to the citizen.

**Manufacturer/Seller:** In the absence of major constraints the manufacturer approaches labeling from the perspective of *sales and market advantage*. The label is perceived as a device to facilitate product use and selection, and labeling efforts are devoted toward improvement of the label's effectiveness as a sales support tool, and means of product image differentiation.

**Economist:** The economist approaches labeling as an element in the cost of sales. This perspective focuses on the decisions of the seller and not on the imputed effects on the behavior of the consumer. The economist will support labeling options which *reduce the cost of production*. Thus, good labeling decisions are those which will hold down seller costs or avoid negative consequences in the market activities of the seller.

**Behavioral Scientists:** The behavioral scientist views the label as an element in a complex communication and information-processing system whose general purpose is to affect consumer and worker behavior. To the extent that the label is effective, the benefits to the product user are enhanced. Behavioral science, therefore, *focuses on the impact of the label* so as to optimize the total amount of information potentially conveyed to the reader.

**Lawyer:** The attorney is generally concerned with the degree to which the law is used to control the decisions or presumed rights of the seller. The attorney tends to view the law or regulation as imposing unanticipated costs upon the seller or of restricting various rights and privileges previously assumed to be available to a manufacturer. The attorney, in general, tries to deal with labeling of products by *avoiding, controlling or minimizing the cost or restrictiveness* of labeling policy.

**Activist/Consumer:** Like the behavioral scientist, the activist approaches the label as a device for increasing the amount of information available to the consumer or worker. The underlying assumption is that information is inherently valuable, and more information is always in the public interest. The consumer/labor advocate wishes to label in the interest of *information transfer* to the user or worker, so as to enhance the "right to know." The reader may find it useful to review a related perspective on labeling functions and orientations provided in the Bambury Report #6 entitled "Product Labeling and Health Risks" edited by Morris et al. (1980)

### Hazard Labeling

A critical aspect of chemical industry product labeling is the fact of government regulation of many of the products which are considered to be potentially hazardous to consumers and/or workers. Such products are subject to various federal, state and local laws and ordinances which in many instances mandate the inclusion of specific warnings on product labels. These warnings describe the hazards and inform the user as to proper use procedures, actions to be taken in the event of exposure to hazard, first aid or medical advice, and disposal procedures, all in addition to whatever other material the seller might wish to include on the product label.

The need for hazard labeling obviously complicates the task of label design and placement in containers since products fall under various rules and regulations. Some require specific language and graphics to be incorporated in the label text; others state a "performance standard" and leave the specifics to the manufacturer. In some cases the law requires special additional labels to be affixed to packages, overwraps or shipping containers (e.g., Department of Transportation [DOT] rules), and label designers often try to incorporate such special labeling into an overall label, thus increasing label size, and complexity. In other cases, the hazard issue is met by adherence to voluntary standards such as ANSI (1976) which suggests a uniform approach to labeling certain chemical product hazards.

The hazard aspect of labeling probably produces the greatest increase in label design complexity and uncertainty because of the difficulty in making sure that the label does an effective job in projecting the nature and quality of the product, while at the same time meeting the requirements of law in an effective manner. A major contributor to the difficulties associated with hazard labeling is the fact that such labeling raises the issue of "perceived risk." Since hazards are *potential,* consumers or workers not only deal with label statements and information about hazards, they must assess the *risk* to themselves that use of the product might actually lead to the hazard described. Development of labels which successfully confront this issue is a major problem discussed later in this chapter.

## PERCEPTUAL ISSUES IN LABELING

### Perception

Policy formulation, label design and evaualtion of label effectiveness are all directly related to the psychological process of perception. Perception (an element of the process of cognition) involves the use of the senses either to (1) *obtain* information about the events, situations and states of our environment, or (2) to *maintain contact* with the real world and its constituent parts and states. The first part of the description deals with the active relations of people with the empiric world, while the second part covers the more or less automatic perceptual acts of people in their daily lives.

Labels are objects in our world, and we obtain information about labels and from labels via our senses, particularly but not limited to our visual sense. The *percept* of the label is the data we obtain from the process of perception as applied to the label: the meaningful individual experience resulting from sensory stimulation caused by the label.

Perception is thus the central process which mediates between exposure to the "objective" label as a stimulus and the resulting behavior of the person who is exposed to the label. (See Figure 1.1). Designing a label which has the greatest likelihood of having the effect you desire on a reader requires familiarity with a number of issues in perception. This section summarizes the factors in perception most important in labeling.

### Perceptual Representation

The first thing that someone actively involved in labeling policy, design, education or research must realize is that the "objective label" shown to consumers or workers so as to affect their behavior is not necessarily the label they *perceive.* The external sensory stimuli of the label—e.g., shape, color, design, size—are cognitively organized by the reader as *representations* of these exter-

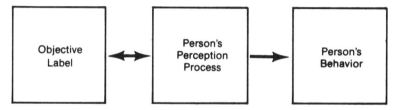

Figure 1.1: Basic perception model.

nal stimuli. The reader thus operates on the *perceived label,* a representation of the "objective label," and not the label itself. The effect of the "objective label" is literally unknown. The reader reacts to his or her perceived and cognitively represented label; a "label" which may be substantially different in terms of content, meaning and impact to the reader than the label you thought you were objectively presenting. This distinction is shown in Figure 1.2a. A related point also shown in Figure 1.2b is that no two people will perceive exactly the same "label." The objective label is perceived and cognitively repre-

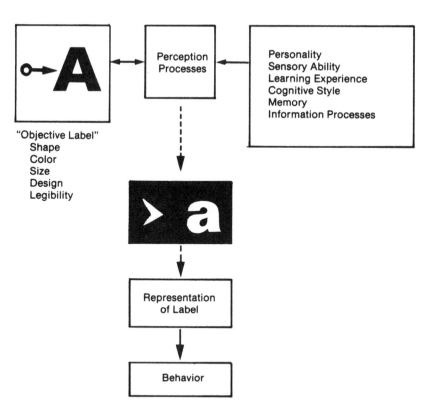

Figure 1.2a: Perception and representation of perception.

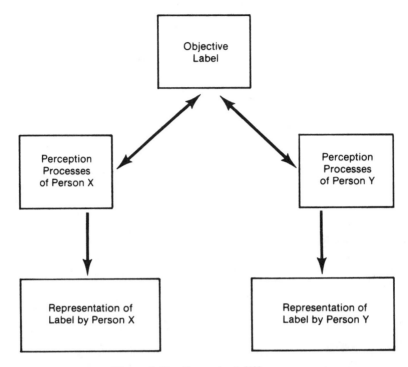

**Figure 1.2b**: Perceptual differences.

sented somewhat differently by different people. These differences between the objective label and the perceived label, or between the perceived labels of different people, occur because of differences in personality, sensory ability, learning experience, cognitive style, memory and information processing abilities among all people.

Obtaining reliable information about a label, the degree of identity between the "objective label" and the reader's perceptions and representation of the label, and the behavior elicited by the perceived label requires research and careful observation, and these will be discussed in later sections.

**Attention Processes**

There is a difference between *exposure* to a product label and *perception* of that label by the person exposed. Exposure sets up a potential for label perceptions and related behavior, whereas attention processes determine whether and to what extent the potential will become an actual perception. In general, attention deals with the selectivity of perception and cognition in people. People do not perceive or organize every possible stimulus to which they are exposed. They "attend" or focus on certain aspects of this potential. The process of focusing on a limited portion of the potential information surrounding a person, or, when perception occurs, on a fraction of this information is attention.

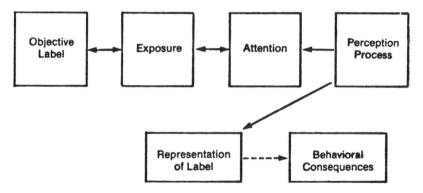

**Figure 1.2c:** Perception and attention.

See Figure 1.2c. Simply putting a label on the product (potential perception) does not necessarily mean that a given reader will attend it, or that having attended the label (actual perception) that the reader will have formed a representation of the label that is the same as the objective label itself.

### Form

While labels can and do occur in a variety of shapes most appear as regular, sharply angled forms such as rectangles and squares of various sizes. Research has shown that perception of forms is improved when the forms are regular and angular such as the rectangle. Ovals and circles tend to be less efficient perceptually, and irregular shapes tend to be least easily perceived, represented, memorized and recalled. In general, the label designer is probably best advised to select a rectangular shape for use with chemical products.

### Color

Whether or not to use color in the label is a major decision which every label designer must face. A large proportion of labels are printed in black and white, mainly for reasons of cost and simplification of design. But, once the decision is made to use color in the label, a number of perceptual issues arise.

First, although almost all sighted people can perceive shades of gray, e.g., black and white, with considerable acuity, this is not true of color. Visible color hues occur in a very limited portion of the electromagnetic radiation spectrum between ultraviolet and infrared radiation—from approximately 400 nm (violet) to 700 nm (red). A relatively large proportion of people have defective color vision not severe enough to preclude all color perceptions (color blind) but severe enough to interfere with proper perception of one or more colors (color weakness) or substantial enough to prevent perception of one or more colors (color deficient). Research has indicated that these color defective conditions are of genetic origin and are sex-linked, so that about 8 percent of all men and 1 percent of all women have color perception defects.

Complete color blindness in which a person sees only shades of gray is a very rare condition. Color deficiency, a partial color-blindness in which only yellows

Table 1.1: Problems in Color Perception

| Color Blindness | Color Deficiency | Color Weakness |
|---|---|---|
| Cannot perceive any colors, only black, white and shades of gray | Can perceive most, but not all colors | Can see all colors but have problems in accurate color matching |
| | Yellow confused with blue or gray or red confused with green or each with browns | Problem more obvious under low illumination |

and blues—or reds and greens—are seen, or color weakness, a deficiency in which people have difficulties in matching certain colors (red, green, yellow, or blue) especially under low levels of illumination are more common. The color deficient person may confuse red and green with each other or with browns and tans, or in some cases yellow or blue with grays. See Table 1.1.

Since most production workers in chemical plants still are men, color perception deficiency affects up to 8 percent of the employees and represents a potentially substantial problem for labelers of chemical products if color is to be used in the design of the label. This problem is compounded if the design of the label incorporates DOT symbols, or mandated symbols, such as the skull and crossbones which usually appear in red. In general, the labeler should be thoughtful about the voluntary use of red and green in a chemical product label, especially where the product is to be used in the workplace.

The second perceptual consideration involves the affective aspects of color. Color evokes emotional responses and people respond variably to different colors, and have preferences among the various colors. Color therefore, affects both attention and perceptual processes of the reader.

## Color Preferences

A considerable amount of perceptual research over the past 30 years has shown that the various primary hues (colors) are not equally preferred by people, and that relatively stable preference orders for colors characteristically develop. In the United States research has tended to indicate that when black and white are ignored, blues and greens tend to be the most preferred color, while yellows and reds are the least preferred.

A representative research was conducted with 90 college students having no color perception deficiency (Hopson et al., 1971), in which preference for colors was obtained for 10 colored Munsell papers with chroma values in 5/6 range. The colors included were red (R), yellow red (YR), Yellow (Y), green yellow (GY), green (G), blue green (BG), blue (B), purple blue (PB), purple (P) and red purple (RP). In various tests the colors were shown against white, gray and black backgrounds and under three different levels of illumination. The preference orders for the ten colors are shown for each background type, each illumination level and in total across all test conditions. As can be seen in Table 1.2, blue was most preferred under all conditions, with shades of blue usually next

Table 1.2:  Color Preferences Under Different Viewing Conditions

| Source | . . . . . . . . . . . . . Preference Order. . . . . . . . . . . . . . | | | | | | | | | |
|---|---|---|---|---|---|---|---|---|---|---|
| | ← - - - Higher | | | Preference | | | | Lower - - - → | | |
| White background | | | | | | | | | | |
| Order | B | PB | BG | GY | P | G | RP | R | Y | YR |
| Gray background | | | | | | | | | | |
| Order | B | PB | GY | BG | G | P | RP | R | Y | YR |
| Black background | | | | | | | | | | |
| Order | B | RP | BG | G | P | PB | GY | R | YR | Y |
| 40-W illumination | | | | | | | | | | |
| Order | B | PB | BG | P | GY | RP | G | Y | YR | R |
| 100-W illumination | | | | | | | | | | |
| Order | B | BG | PB | P | RP | G | GY | R | Y | YR |
| 200-W illumination | | | | | | | | | | |
| Order | B | GY | PB | G | BG | RP | P | R | YR | Y |
| Total | | | | | | | | | | |
| Order | B | PB | BG | GY | G | P | RP | R | Y | YR |

Note:  R = Red, YR = Yellow Red, Y = Yellow, GY = Green Yellow, G = Green, BG = Blue Green, B = Blue, PB = Purple Blue, P = Purple, RP = Red Purple

Source:  Hopson, Cogan and Batson (1971)

preferred, followed by greens. Purples and reds were usually less preferred, and yellows and yellow reds generally least preferred.

Another indicative study (Konz et al., 1972) tested both the legibility and color preferences among black, blue, green, purple, red, orange and yellow letters shown against either gray and brown backgrounds under constant illumination to samples of college students. As can be seen in Tables 1.3a and 1.3b, the blue and green hued letters tended to be both most preferred and most legible against either background, except that black was most legible and second most preferred against a brown background. Red, orange and yellow letters proved least legible (using distance methods) and least preferred.

Table 1.3a:  Legibility and Attractiveness for Colors
on Gray Cardboard in Experiment I

| Letter Color | Mean Legibility Distance, Meters | Letter Color | Mean Vote of Attractiveness* |
|---|---|---|---|
| Blue | 9.57 | Blue | 2.1 |
| Purple | 9.39 | Green | 3.3 |
| Green | 9.33 | Purple | 3.4 |
| Black | 9.27 | Black | 3.8 |
| Red | 8.72 | Red | 3.8 |
| Orange | 8.41 | Orange | 4.6 |
| Yellow | 6.31 | Yellow | 6.8 |

*1 = best; 7 = worst

Note:  Values connected by a vertical line are not significantly (p > 0.05) different

Source:  Konz, Chawla, Sothaye and Shah (1972)

Table 1.3b:  Legibility and Attractiveness for Colors
on Brown Cardboard in Experiment I

| Letter Color | Mean Legibility Distance, Meters | Letter Color | Mean Vote of Attractiveness* |
|---|---|---|---|
| Black | 9.39 | Blue | 2.4 |
| Blue | 9.33 | Black | 2.5 |
| Green | 9.17 | Green | 3.1 |
| Purple | 8.96 | Purple | 4.0 |
| Red | 8.66 | Red | 4.2 |
| Orange | 8.05 | Orange | 4.6 |
| Yellow | 6.55 | Yellow | 6.9 |

*1 = best; 7 = worst

Note:  Values connected by a vertical line are not significantly ($p > 0.05$)
        different

Source:  Konz, Chawla, Sothaye and Shaw (1972)

In general, color preferences appear to be relatively similar and stable across different national and cultural lines. Cross cultural color attitude research was conducted in 23 different countries throughout the world, including North and South America, Europe and Asia (Adams and Osgood, 1973). Their results show that over the 23 samples blue was the most highly evaluated (preferred) color of eight color concepts considered, followed by green and white. Red and yellow are evaluated near a neutral point, with gray and black generally low. If the shades of gray are ignored, the blue, green, red, yellow preference order is again seen.

### Color and Emotion

Although the evidence is somewhat less consistent than is the situation with preference or attractiveness, there is a substantial amount of evidence which relates different colors (hues) to differences in mood or emotional tone elicited. An early study (Wexner, 1954) showed that colors were variably associated with different mood conditions. Table 1.4 shows that red was almost universally associated with "exciting, stimulating," and to a lesser extent with "cheerful, defiance, hostility, powerful and strong." Blue and to a lesser extent green was associated with "sincere, comfortable, calm, peaceful and serene."

Orange was most often connected with the mood-tones "distressed, disturbed, upset," and "defiant, hostile." Yellow was most often connected with "cheerful, jovial, joyful," whereas black was associated with "distressed, disturbed, despondent, dejected, dignified, stately, powerful and masterful."

Other studies have shown similar findings. Red tended to be associated with "excitement, agitation, stimulation, hostility, aggression, activity and affection." Blue was generally felt to be indicative of "dignity, sadness, tenderness, cool, most pleasant, control, security, soothing." Green was usually perceived as "leisurely, control, youthful." Yellow was usually found to be "stimulating, unpleasant, exciting, cheerful, envious, hostile and aggressive." Orange was generally associated with "heat, stimulation, unpleasantness, warmth, de-

Table 1.4:  Emotions and Moods Most Often Associated with Different
            Colors in Various Researches

| Black | Sadness, anxiety, fear, depression, despondency, melancholy, unhappiness, dignity, strength, power, mastery, hostility |
| Black | Sadness, anxiety, fear, depression, despondency, melancholy, unhappiness, dignity, strength, power, mastery, hostility |
| White | Purity, solemnity, spirit, goodness, weakness |
| Red | Excitement, stimulation activity, aggression, intenseness, agitation, hostility, power, mastery, strength, heat, love, hate, protectiveness, defiance, potency |
| Orange | Stimulation, excitement, heat, emotion, unpleasantness, disturbing, distressed, warmth, happiness, upset, defiance, hostility |
| Yellow | Exciting, stimulation, unpleasantness, envy, hostility, aggression, cheer, joy, pleasantness |
| Green | Control, controlled emotion, calm, security, comfort, soothing, tenderness, youth, illness, goodness |
| Blue | Security, comfort, tenderness, calm, serenity, peacefulness, dignity, sadness, cool, pleasant, soothing, leisure, controlling, sadness, strength, depth, goodness |
| Purple | Dignity, stateliness, sadness, melancholy, unhappiness, depression, vigor, disagreeableness |

Sources:  Schaie (1966); Wexner (1954); Osgood and Adams (1973)

light, disturbance, depression and unpleasantness." Purple was found indicative of "depression, vigor, sadness, dignity, stateliness." Black, as might be expected was usually associated in research with "sadness, intense anxiety, fears, depression, erosion, power and mastery," whereas white generally was found associated with "solemnity and purity." Other lines of research have suggested that people tend to show less anxiety when exposed to blue or green light than with red or yellow (Jacobs and Sven, 1975).

Adams and Osgood's work (1973) also shows that the emotional aspects of color concepts are similar across the 23 cultures studied. When measured on "potency," e.g., the strength, weight or intensity of emotion or tone evoked by a color concept, black and red were cross-culturally perceived to be the most potent. All other colors were found to have negative average potency scores, with yellow, white and gray being weakest. Blue and green were somewhat more neutral.

When evaluated on the "activity" scale which measures the activity and excitement associated with color concepts red was noted most active and black and gray most passive.

In general, these cross-cultural studies found that black is perceived to be bad, strong and passive; gray—bad, weak and passive; white—good and weak; red—strong and active; yellow—weak; and blue and green are seen as good. Finally, a tabulation of emotion adjectives and nouns associated with colors found in the research literature has been made. As can be seen in Table 1.5, red is by far the most emotionally salient color with 1,199 associations, followed by blue with 377. Green appears to be the least salient with only 217 associations.

Table 1.5: Number of Emotional Word Associations Totaled Across
37 Major Research Sources for Each of Seven Colors

| Color | Number of Emotion Associations | (%) |
|-------|-------------------------------|-----|
| Red | 1,199 | 40 |
| Blue | 377 | 13 |
| Gray | 348 | 11 |
| Yellow | 345 | 11 |
| Black | 287 | 10 |
| White | 258 | 8 |
| Green | 217 | 7 |
| Total | 3,031 | 100 |

Source: Osgood and Adams (1973)

## Color Visibility

Ignoring color preferences or emotional associations, it has been found that different printed color combinations have varying reading efficiency potentials. Generally, black letters on white background is probably superior to all other color combinations, especially under difficult reading conditions. In descending order of efficiency, other combinations of print color and black and white, are blue on white, or yellow on black, yellow on blue, white on blue, yellow on purple, white on purple, blue on yellow, yellow on white, white on yellow, blue on black, red on yellow, green on white, orange on white, and red on green.

## COLOR AND HAZARD LABELING

The use of color in labeling and packaging is obviously widespread in consumer products and many industrial products. Color is used in such labels primarily because of its emotion-mood related characteristics which can, if effectively integrated into the design of label and package, contribute to the enhancement of attention processes and favorable image building qualities. Obviously, these tend to improve the probability of product trial and brand loyalty development.

However, the use of color with hazardous product labels presents some problems which should be considered: first, to determine whether or not the use of color is desirable in a given case, or if black and white design would be more appropriate; second, since substantial use of color can induce mood response in readers, the nature of mood or emotional reaction desired should be clarified. On balance it would seem desirable to use color only where color may enhance visibility of or attention to *required* or *desired* cautionary material related to the potential hazards associated with the product, yet not arouse or stimulate strong emotions or pleasant or happy moods.

Such a strategy would stress the use of blues, greens, browns and purples along with black and shades of gray, and avoidance of reds, yellows and oranges. In addition, the use of contrast coloring for visibility of specific areas of the label, such as hazard or safety data, must be considered if black on white is felt to be inadequate. In this regard blue on white or yellow on white might be preferred both for impact and distance legibility.

## LEGIBILITY

Reading requires the eye(s) to focus, not moving, on a particular aspect of the field to be read. The eye is seldom motionless, and usually is involved in rapid scanning shifts called saccades. Visual efficiency is a composite of a number of factors once attention has been focused; these include legibility, print size, use of color, length of word or sentence, spacing, type style, level of illumination and visual acuity.

Research in this area has tended to stress either *readability* (characteristics of printing or type which make it comfortable or easy to read—usually measured by rate of blinking) or *legibility* (characteristics that make it possible to read or increase quickness of recognition and perception). On balance legibility should be considered the primary criterion by labelers since one can have legibility without readability, but not readability without legibility.

The following relationships are useful in evaluating the legibility of a proposed label design, assuming minimally adequate levels of viewing illumination:

1.  For people with low reading levels, 10 to 14 point type tends to be more efficient (a point = 1/72 inch).

2.  Where the ability rapidly to find specific material in a printed stimulus is required, 12 point bold face type is more efficient.

3.  As the reading skill of people increases 11 point type is adequate.

4.  Use a standard type font where possible. Most standard type styles have about the same level of legibility. As a font becomes more exotic—a lot of curlicues, flourishes, graphic elements, etc.—legibility decreases and even simple information may be difficult to read or take longer to read.

5.  Space between printed lines (leading) should be at least two points.

6.  Printing in lower case or upper and lower case is almost always more legible than all upper case, except where it is important that a single word or short phrase be noted and read, upper case type is an advantage (e.g., DANGER, CAUTION: etc.).

7.  Avoid colored inks where possible. Black ink, especially on white, tends to be most legible. However, black ink on

colored paper loses considerable legibility. White ink on black background can lose up to 30 percent of legibility value.

When evaluating the readability of a proposed label design the following factors should be kept in mind:

1. Readability deals with factors involving ease of extended reading of given material and is affected by legibility, type face, size of type, word spacing, length of line, leading, page pattern, contrast of type and paper color, paper texture, and difficulty of content.

2. If your label will require sustained reading, increase the type size.

3. Avoid long lines of type where possible; 50 to 60 characters per line, or nine to ten average words tend to be a reasonable goal.

4. Readability tends to increase with the boldness of the type used.

5. As the level of reading difficulty (in grade equivalents) increases, readability tends to decrease. There is a complex relationship among the grade level of material, word length and sentence length and the resulting readability of material, especially for items such as labels. Figure 1.3 shows an example of some of these relationships. Careful attention to the level of difficulty of the content of the label is very critical in its design, especially where hazard information is involved.

6. Any factor which decreases readability of a given message increases the amount of work needed to see and comprehend a label. Increased readability is most easily measured by increase in eye blink rate.

## SYMBOLS AND LABELING

A major controversy in the field of labeling centers on the use of graphic symbols instead of, or as a complement to words in a chemical product label. Since words are themselves symbols, the real controversy is over *which* kind of symbol to use in a particular context.

When dealing with symbols and symbolic representation a commonly used concept is that of "coding." Psychologists believe that when an object or stimulus is perceived it is cognitively organized and placed in memory by a coding process which facilitates recall of the memory along with its referents and associated factors so that its meaning is available to a person.

A symbol is, however, arbitrarily related to the object or concept that it cognitively represents by convention; the symbol is coded along with its refer-

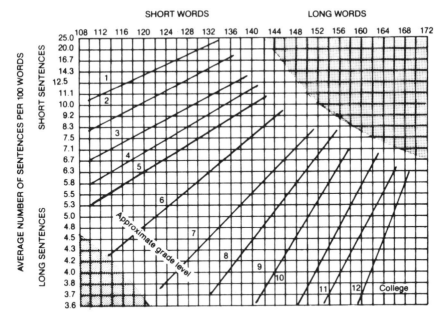

AVERAGE NUMBER OF SYLLABLES PER 100 WORDS

SHORT WORDS          LONG WORDS

**Directions:** Randomly select 3 one hundred word passages from a book or an article. Plot average number of syllables and average number of sentences per 100 words on graph to determine the grade level of the material. Choose more passages per book if great variability is observed and conclude that the book has uneven readability. Few books will fall in gray area but when they do grade level scores are invalid.

**Figure 1.3:** Relationships among grade level of reading difficulty, word length, sentence length and readability.

ent (that which the symbol represents). If the symbol is then a gain perceived, in order for the symbol's meaning to become available its referent must be *decoded*. The distinction can perhaps best be explained by comparison between a sign on a supermarket meat counter which shows the profile of a chicken, and the photograph of a chicken. The photograph does *not* require any coding or decoding for meaning. The photograph and its referent "a chicken" are directly related by projective correspondence of parts and elements. The outline sign in the supermarket generally means "chickens are sold here as food," and this referent must be coded to the percept of the sign, so that when next perceived the person will decode its symbolic meaning as "chickens are sold here for food." The more abstract the referent-symbol relationship the more difficult the coding process.

Specific symbols can be "public" or "private." The more widespread the cultural acceptance of a symbol's meaning, the more public it is since anyone who knows the coding process can understand the symbol. On the other hand, the more restricted the information about the symbol and its coding process, such as DOT symbols or electronic schematic symbols, the more "private" they are.

Those who oppose the use of symbols on labels do so because they feel that almost any graphic symbol used would be necessarily limited in its value to the extent that the coding for that symbol had been learned by the viewer. In the case of chemical labels this position reflects the fear that inexperienced workers are not adequately trained in symbol use and meaning. Indeed, a recent working paper prepared for the Interagency Regulatory Liaison Group on toxic chemicals labeling after first acknowledging that "symbols have quicker reaction time than written warnings, ability to transcend linguistic differences and communicate warning information to illiterates and children" (Murphy, 1980) backs itself into the position of opposing graphic symbols. This memorandum reports some work on symbol recognition and then states "many studies have shown that there is no positive correlation between what a person knows and what a person does. The central problem confronting government labelers, therefore, is not an informational one but a motivational one." It goes on to stress the point that younger inexperienced workers are the ones most likely to have accidents, creating real problems for symbolic warnings and indicating that OSHA has been wise to "abjure the use of warning symbols in its current rule-making effort."

Unfortunately, such a position in favor of *words* ignores the fact that words themselves are symbols; the referent of every word must be coded and learned by a person. Indeed, a dictionary is the code book for all of the words in a given language. The point to be considered is that although words are useful symbols, they are usually more difficult to perceive, attend, decode, comprehend and evoke response because with words two additional perceptual steps are usually needed. The person must first perceive the word stimuli as letters—themselves symbols—then perceive the cluster of letters as a word—another symbol—and finally decode this percept for referents. As Figure 1.4 shows, graphic symbols usually are perceived as a symbol directly and then decoded.

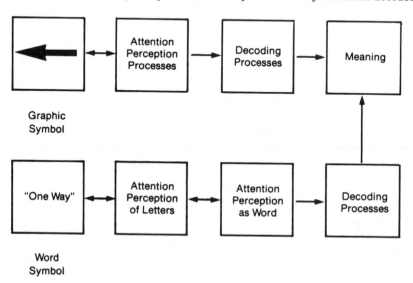

**Figure 1.4:** Difference in perception processes for word symbols and graphic symbols.

The inherent difficulty can perhaps best be appreciated with the example of acronyms. Suppose a worker sees "PCB" on a label. This acronym is really a very complex symbol. The person must perceive the letters PCB, presume them to be a word, perceive the whole word "PCB," and then in attempting to decode this "word" comprehend it is not a word but a different kind of symbol, and decode this new symbol. If the workers do not have the "code book" for PCB they are not much better off than with an "arbitrary" graphic symbol.

In this respect the recent Conservation Foundation report of the Belmont House Conference on labeling of hazardous and toxic substances in 1978 is instructive. The Labeling Committee report suggested the use of two types of graphic symbols; one, *generic,* designed to evoke a non-specific call for attention or "stop" on the part of the reader; and, two, a graphic symbol to inform the reader of the *specific* hazard. The report recommends a combination of text and symbols for the label.

Usually the situation is not all or none with respect to symbols in labeling. Rather, it is important to know about the strength and weakness of graphic symbols so that a labeler may prudently decide whether and how to use them.

### Advantages of Graphic Symbols in Labeling

Graphic symbols offer the following advantages:

1. Graphic symbols tend to be perceived more rapidly than words.

2. Graphic symbols can often be more easily transferred across cultures than languages. A label written in English has almost no meaning for someone who does not speak English. Symbol coding can be relatively easily taught in most cultures.

3. The referent meaning of graphic symbols can often be taught to and perceived by illiterates and children who cannot read.

4. Graphic symbols are often more legible than words.

5. Once the meaning of a graphic symbol is widely known its communication value is very substantial.

### Disadvantages of Graphic Symbols

The use of graphic symbols has the following disadvantages:

1. Where complexity or extent of the information to be conveyed is great, the effectiveness of graphic symbol as compared with words is greatly reduced.

2. Where new or "private" graphic symbols are to be used considerable effort in training in connecting the symbol to the desired referent or meaning is required since the symbol itself inherently usually is meaningless.

3. The use of pictographic symbols does not automatically guarantee widespread understanding of referent meaning.

For example, research on the European "public" railroad graphic symbols, and U.S. highway transportation symbols showed widespread lack of recognition and lack of correspondence between the pictograph and its supposed correspondence in the minds of the viewers.

An example of potential problems involved in the use of graphic symbols on labels was reported by Baldwin and Runkle in a 1967 article in *Science*. As it became clear that biologically hazardous materials were an increasing problem and potential danger, a group of scientists was assigned the task of developing a graphic symbol for labeling biological hazards or the containers in which they were placed. The symbol was to warn of the hazard involved and evoke appropriate safety behavior. The designers deliberately searched for a symbol that would be meaningless to people, thereby avoiding conflicting cultural or personal associations in the minds of viewers. Figure 1.5 shows the symbol that was developed; research conducted found it to be legible, memorable, meaningless. However, by most criteria this symbol is probably a failure. It is so devoid of meaning—so neutral—that the effort to attach (code) the appropriate referents—dread, fear, caution, possibility of death, etc.— would be enormous since the symbol has no possibility of "piggybacking" cultural meanings and associations to it. Suppose, on the other hand, some variation of the traditional skull and crossbones had been developed for this specific problem. There is little doubt that its effectiveness would have been enhanced significantly. The skull and crossbones has very widespread recognition and its implicit referents are culturally accepted by large numbers of people. Indeed, there is evidence that this symbol evokes fear reponses across most human cultures, even among young children. Further, there appears to be some evidence, reported by Hebb (1946), that this symbol may evoke fear responses in other species, especially primates.

Modley (1974) has proposed several criteria for good symbol design:

- Uniqueness of symbol
- Clarity of meaning (referent desired)

**Figure 1.5:** Symbol developed to warn of biohazards.
[Source: Baldwin and Runkle (1967)]

- Independence from language and cultural differences
- Visual directness of symbol (relative to referent)

Modley has indicated that symbol failure, at least for those symbols intended for general public use, usually involves one or more of the following:

- Conceptual failure: symbol does not express the referent.
- Poor draftmanship: symbol is poorly designed.
- Conflicting meanings: several symbols to convey one meaning or one to convey different meanings.
- Poor use of color: color selection ignores color blindness, cultural meanings of color.
- Failure to exploit available background shapes.
- Excessive symbol use: use of symbols even when they cannot explain the intended meaning.

Even a cursory review of the biological label using Modley's proposed criteria suggests that the design suffers very greatly from lack of *visual directness* of the symbol; it does not appear to connect to any culturally learned associations as referents. The symbol does not immediately suggest cognitive associations with danger, especially biologically selected hazards. It also probably suffers from too much *independence from language* and cultural differences, and also from difficulties associated with lack of clarity of meaning.

## READING LEVEL OF LABEL

As was indicated above, readability and legibility of a label decreases as level of reading difficulty (usually measured as grade level equivalent) increases. This imposes a very great constraint on label design since a label content of significantly high reading difficulty level could reduce the effectiveness of a chemical label by discouraging the viewer from complete reading of a label. The implication of such responses is doubly negative in the case of a label which requires that hazard warnings and information be present, read, and comprehended by a viewer.

The Belmont House Conservation Foundation Conference Report (1979) counsels that label warnings about risk, nature of hazard and remedies "should be readily understood by most people in the U.S. regardless of level of education and achievement." The report estimates that at least 35 million people in the U.S. over 14 years of age have reading skills at less than the eighth grade level, and that at lest half of these may have reading comprehension at the sixth grade level or less. These data suggest that many adults in the U.S. come very close to functional illiteracy, placing a great burden on the label designer to make certain that the labeling material is written at a very low level of required reading ability. This point largely has been lost on those setting chemical labeling regulatory standards.

**Estimating Reading Level**

Readability scores or formulae are usually applied to the assessment of printed materials although they have on occasion been used with verbal matter as well. These scores or formulae have gradually gained popularity because of their apparent objectivity; scores typically use counts of certain language variables to provide an index of probable difficulty for readers.

At least 30 readability formulae were developed prior to 1960, and more since then. However, only a small number (six to eight) have been extensively researched and cited in readability studies. Most of these popular formulae are used as regression equations involving two or more language factors—one measuring difficulty of words used (average number of syllables or number of words not on lists of "easy" words) and the second, length of the sentences used in text. Most yield an estimate of the grade level of the material being evaluated (or difficulty score). In most cases one of two validating criteria is used—comprehension measured by the *McCall-Crabbs Standard Test Lessons in Reading,* or the *Cloze Procedure* which involves omitting every $n^{th}$ word from a passage (usually every fifth word) and asking subjects to fill in the missing words.

The most frequently used and researched formulae are: the Flesch Formulae for Reading Ease and Human Interest (1948), the Dale-Chall Formula (Klare 1974), Gunnings FOG Index (1952), Coleman Formulae (Szalay, 1965), McGlaughlin's SMOG Formula (1969), and Fry's (1968) Readability Graph. (An example of the use of Fry is shown in Figure 1.3 above.)

Whenever significant hazards are associated with a chemical product, and where potential legal issues may be significant, the label designer is well advised to evaluate the level of reading difficulty of the label being considered. The lowest level of reading difficulty possible, consistent with regulatory requirements and other constraints should be the goal. The use of tests and consultants is advised whenever the label designer has minimal experience or expertise.

## EVALUATING THE IMPACT OF LABELING

Earlier it was indicated that implicit in all labeling activities is the assumption that labeling has effects on the behavior of those who perceive and comprehend the label. It is also implicit in this view that the resulting behavior will approximate that which the labeler intended. However, these assumptions are seldom verifiable, especially during the design phases except by evaluation research. The function of evaluation research is to tell the labeler and others interested in the label how and how well the label in fact functions.

Labeling and packaging research has generally used four procedures in developing data about the effectiveness of a given label design.

**Observation:** Procedures which permit the researcher to view—overtly or covertly—the reactions of people to the label when it is shown in various natural or laboratory situations.

**Interviews and Surveys:** As an extension of observation, by experimental or by survey procedures, the label viewer is asked to discuss what he or she remembers seeing and feeling when exposed to a given label or package system.

**Motivational Analysis:** An extension of interviewer procedures designed to uncover the conative and emotional content of response to label exposure. The primary interest is in determining what feelings or emotional impressions were evoked by a label and what behavior was motivated by these feelings.

**Instrumental Testing:** The use of sophisticated instruments in combination with interviews to evaluate the optical efficiency and communication value of a label. Such machines as the tachistoscope, the pupilometer, or the eye movement monitor or camera are most commonly used.

### Evaluation Criteria

Experience has shown that abstract testing of labels is for the most part of limited usefulness. Useful results require a fairly precise statement of what the labeler wants the proposed label to achieve, e.g., the goals of the specific labeling project. Once goals have been decided on there are usually three major criteria against which to evaluate results.

**Time:** The ability of a label to evoke behavior is related to the duration of exposure of the label to the viewer. A person exposed to a label may perceive, comprehend, and react to a label after latencies ranging from a few tenths of a second to minutes, depending on the complexity of the labeling system. For example, a label on a 55 gallon drum of industrial chemicals containing 30 lines of text and numerous symbols will normally require greater exposure time than one the same size showing only a skull and crossbones and the words "drink this and you will die."

In developing a label it is crucial that the labeler be able to state what the design must communicate in a given period of exposure time. Testing at various exposure periods will therefore yield desired effectiveness data.

**Optical Efficiency:** As discussed in above sections, the ability to perceive a label design varies with certain optical characteristics. After establishing the exposure time criterion the labeler must also define the distance at which viewing is to be tested and the level of illumination required and then state what elements of the label are to be perceived and remembered. Two criterion variables are then measured for the test label and compared to another version or to a competitive label.

*Legibility*—The number and type of label elements perceived and remembered by the viewer at a given distance, exposure duration and illumination level. (Respondent color weakness should be tested at this step).

*Readability*—The ease with which written materials are perceived, remembered and comprehended given the distance, duration of exposure and illumination levels set for the label.

**Communication Value:** Once perceived the label and its elements are organized cognitively by the viewer, its meaning is educed and the conative or emotional characteristics are registered. At this stage the respondent has decided what the label has told him or her, what it means, what feelings have been evoked, and latent behavior is motivated. In order to determine whether the label communicates what is desired the labeler should define in advance what information and meanings the proposed design is to communicate, what emotional tone or mood is to be evoked, and the behavioral intentions which are desired. The communication value of the design can then be tested against another version or competition.

### Implications for Hazard Labeling

A very important aspect of hazard label communication is the ability of the design to communicate risk and to do it in such a way as to evoke desired risk eliminating or reducing behavior. The problem is that the label describing acute hazards usually, and chronic hazards always, is asking people to deal with *possible* or *probable but not certain* consequences of exposure to or misuses of a hazardous chemical product. Risk is an exceptionally difficult concept to communicate accurately and with confidence of evoking desired behavior by means of labels.

Psychological research has shown that most people have considerable trouble in dealing with the notions of risk or uncertainty especially when related to rare events. This is especially true when the time to present risk information is limited (as is the message itself), the usual case with a hazard label. The labeler should keep the following facts in mind when designing a hazard label and testing it.

- People will tend to become anxious when faced with uncertainty or risk, and will immediately try to reduce anxiety by denying the risk and attempting to convert it to a negligible possibility which can be ignored, or to a certainty which the person must move to avoid. Either alternative may not be the correct response nor may it have anything to do with the label data per se.

- If confronted by a risk people often will reduce the level of the risk when considering themselves as opposed to ascribing great risk to others. This unrealistic optimism appears related to personal experience of the individual in dealing with certain well publicized perceived risks, for example, the smoker who continues to smoke knowing he has no lung cancer but knowing other smokers who have died from lung cancer. Such processes lead to distortions of label hazard data.

- People's perception of risk substantially increases for hazards which gain sudden public notoriety or publicity in the media. This distortion can occur regardless of labeling information or design.

- Statements about risks are tested against a person's beliefs. Beliefs tend to change slowly even when the person is shown or told evidence which is opposed to the initial belief. Evidence appears to be evaluated as favorable to the extent that it conforms with prior belief systems. The psychologist, Weinstein (1980) has indicated that people will accept information about a risk and reject reassuring information if they were interested in the topic. However, the fact that a person does not have risk information will not cause him to seek it.

The message to be used in the hazard label should therefore be very carefully evaluated and tested prior to inclusion to insure the ability of the message to communicate risk accurately.

## TESTING METHODS

### Tachistoscopic Procedures

An important procedure for testing optical efficiency and communication value of a proposed label design involves the use of tachistoscopes. A tachistoscope is a machine which permits precisely timed exposure of a set of visual stimuli. The tachistoscope can be built in a variety of forms permitting either direct viewing of the stimulus in a viewing box or similar device, or by projection on a screen. In some versions the actual object can be used; in others a photograph or slide. In tachistoscopic presentation the level of illumination is usually preset and kept constant, as is the actual or simulated eye to object distance. Precise exposure from hundredths of second, to full seconds or minutes can be programmed, and the scope can be used in conjunction with other equipment, such as sound generators, reaction time instruments, and the like. In using tachistoscopes for evaluating labels it is important to make sure that each respondent is tested for color weakness and for visual acuity before proceeding with label research.

A portable tachistoscopic instrument was developed by the author for use in labeling and package research. This tachistoscope consists of a binocular viewing chamber in which 35 mm slides are inserted, and a solid state timer control unit which permits selection of an exposure duration from 0.1 second to 10 seconds. When the experimenter pushes a button, an incandescent lamp lights in the viewer, allowing perception of the photograph. The advantages of this instrument lie in its lightness and portability, permitting optical impact and communication to be studied in the plant, the laboratory or the office.

A different version is the mass display tachistoscope which was designed by the author. This instrument consists of a viewing chamber 2 feet, by 2 feet, by 3 feet. The chamber contains variable shelving to permit the display of packages or containers of different sizes together with their labels, either individually or as massed displays of two or more containers. The interior perimeter of the display chamber is lined with miniature spot lamps to provide illumination levels variable between 20 and 300 watts. The face of the container is a one way mirror. When no illumination is on in the chamber the interior display is not visible to a respondent. The electronic control console controls the level of illumination in the chamber, the duration of exposure (from .01 seconds to continuous), and repetition sequence, e.g., a one second exposure at given illumination can be repeated x times every y seconds.

The mass display tachistoscope permits study of actual label designs on various size containers in a number of different contexts from stand-alone display to shelf facings. When combined with personal interview a wide range of perceptual and communication information about labels can be derived relatively easily and under highly controlled conditions.

### Eye Motion and Pupilometrics Equipment

Figure 1.6 shows the use of a form of equipment which the editors found exceptionally valuable in label design and evaluation research, the eye movement monitor. The equipment shown is the Model 200 Eye Trac manufactured by the Applied Science Laboratory of Gulf and Western Industries in Waltham, Massachusetts. The equipment consists of (a) a set of infra-red light sensors set into eyeglass frames, which are worn by a subject to monitor vertical and horizontal motion of the eye and (b) a microprocessor which converts light signals caused by the movement of the pupils of the eyes electronically to digital and analog information which shows the coordinate position of the pupils relative to focus on a piece of visual material. These signals can be sent directly to an XY plotter or to a video monitoring device which displays the point focused on by the eyes as they read and scan the material. The researcher is thus able to follow how a reader actually proceeds to look at and read a visual stimulus, and to record graphically or on videotape the pattern of perception.

Although this equipment requires accurate calibration of each subject individually and the use of a head/chin restraint to reduce head motion to a minimum, the editors have found the Model 200 Eye Trac of great value in label research. This equipment is particularly recommended for labelers who are re-

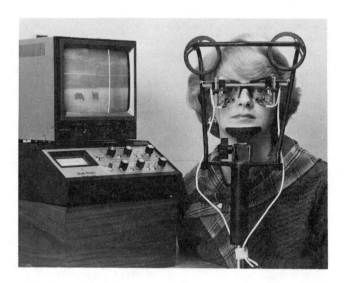

**Figure 1.6:** Model 200 Eye Trac. [Source:    Lirtzman and O'Connor (1980)]

sponsible for the design and evaluation of relatively large numbers of chemical product labels, and for research where litigation may be involved.

These perception instruments, when used in conjunction with observation, personal interviews and appropriate samples in both experimental and survey studies, are able to provide powerful tests of label design strengths and weaknesses and can lead to the development of highly effective chemical product labels.

## RESEARCH IN HAZARD LABELING

As has been noted there is a great discrepancy between the recognized significance of labeling of hazardous chemical substances and products, and the amount of available research findings which bear on labeling generally, and hazard labeling specifically.

The following overview is designed to acquaint the labeler with the current state of empirical research in the field of chemical and related product labeling. Although a reasonably large literature exists concerning the perceptual and cognitive factors related to labels and people's reaction to them, such as color, cognition, comprehension, fear, risk, attitudes and beliefs, almost no research can be cited about labels themselves. Most of such direct research has been done for commercial marketing or legal purposes and is not available to the public. The few studies which have been reported have dealt with the public's attitudes about hazard warnings and labels, food and drug labeling, pesticides, children's response to warning label elements, reaction to cancer labeling, and perception of similarity among certain label hazard signal words.

### Attitudes About Hazard Warnings on Labels

Although not a labeling study per se, a Shell Oil Company study (1978) about consumer and labor attitudes concerning risk of contracting cancer and carcinogens provided some data about what people wanted on hazard labels. Of 1500 people surveyed, two-thirds felt that proof that a substance really caused cancer before it was labeled as such was warranted. Further, over half of the respondents generally and nearly two-thirds of chemical workers questioned felt that the major responsibility for avoiding or reducing exposure to hazardous substances falls to the workers themselves. About two in three respondents also felt that there should be stricter regulations of hazardous substances where involuntary exposure would be involved as compared with voluntary exposure. The Shell survey results suggest that consumers may be more conservative about hazard labeling than is commonly assumed, and may want labeling information and warnings commensurate with objective risk.

## FOOD LABELING

### FDA and DHEW Research

The FDA conducted large scale consumer research to study the use of food and nutrition labels, and the attitudes of consumers about such labels, in 1974,

1975 and 1978. The findings of this research indicate that only about four in ten shoppers having primary responsibility for food shopping looked at ingredients on food labels, and only about one in four noted additives or preservatives, was willing to pay for nutrition data on labels, or even looked at nutritional information on the product label. By 1978 about half of the respondents reported using ingredient information to permit them to avoid certain specific ingredients which might be found in such products. A series of other studies by USDA and various private organizations has tended to report similar results.

## Drugs

FDA requires patient oriented labeling in the form of package inserts for certain drug products. A National Academy of Science report evaluated research on the use of these inserts for oral contraceptive and estrogen drugs. The various studies indicate that the inserts are "read" by a relatively large proportion of consumers, but with what thoroughness or effect is unknown. Use of inserts is apparently greatest at time of first use. However, data suggest that very little of the specific information on a labeling insert is recalled by the reader. Reading tests of such inserts indicate that they are usually at a very high level of reading difficulty.

## Pesticides

A series of researches was conducted by Salcedo et al. between 1971 and 1973 at the University of Illinois College of Agriculture to study a variety of factors associated with the readability of pesticide labels and those associated with the motivation to read these labels. They found that hazard data should be presented in 11 point type regardless of label size with two point leading. Data suggested that eight point type should be the minimum allowed. All caps printing should be avoided, with body copy in lower case. If color is used, brightness contrast—dark colored type on light color background is preferred. The requirement for printing the skull and crossbones and Poison-Danger in red and all caps on contrasting background is suggested. They also stress the need to test labels for reading level and reading ease.

## Children and Hazard Labels

Schneider (1977) investigated factors which might facilitate the design of packages and labels to control accidental childhood poisoning. Schneider studied the effects of label written warning, label pictorial warning, and packaging characteristics on the ability of a package to attract children. Labels differed only in terms of size of written warning and type of pictorial warning, and used three color printing. The two symbols used were a skull and crossbones, and Mr. Yuk. Subjects were 81 nursery school age children from the Minneapolis area. The criterion, "attraction" was defined as opening the test package while alone, in a free play situation for three minutes. Children tended to be *more* attracted to packages bearing some written warning than to those with none. Size or content of the warning had little effect on attraction, however. The presence of a pictorial symbol tended moderately to reduce the attractiveness of a package, with the skull and crossbones somewhat more effective than Mr. Yuk.

Although interesting, the nature of the sample, the procedures, and problems with some of the children limit the usefulness of this study.

## Hazard Warnings

The editors (O'Connor and Lirtzman) have been engaged in a program of perceptual and communication research designed to study the direct response of workers and consumers to various elements of labels for hazardous chemical products. These studies (Lirtzman and O'Connor, 1980) are believed to be the first reported direct research on hazard labeling other than pesticides. Early results reported in 1980 covered pilot studies involving carcinogen labeling, and perceived level of danger associated with signal words commonly used or mandated for use with hazardous products.

The first research studied the response for a group of chemical industry technicians to a set of labels for a fictitious metal cleaning product. A base label with no warnings served as a control. Other labels were created which warned that the product was a cancer hazard. In addition, the signal word was varied, showing either "caution," "warning" or "danger;" and the number of additional hazards listed varied from none to three. Where a hazard required a DOT label, this was added in reduced size in full color to the label. Labels were randomly exposed to subgroups for brief controlled periods totaling no more than 60 seconds at reading distance under standard illumination. Total exposure time was determined by the ability of the worker to report the name of product, product type and cancer hazard.

Results showed that cumulative exposure time needed for recognition of product name and type was lowest for the control label and increased with number of hazards on label. As Table 1.6 below shows, cumulative average exposure time needed for perception of cancer hazard increased with the number of hazards associated with a label, as did the average scale value on a five point work danger scale (5-high danger). The study also suggested that there is dif-

Table 1.6: Cumulative Exposure Time Needed to Perceive
Major Elements of Chemical Product Label
With Cancer Warning (Seconds)

| | | . . . . . . . Label Type . . . . . . . | | | . . . . .Signal Words. . . . . | | |
| | | Cancer Hazard Only | Cancer Hazard +1 Other | Cancer Hazard +3 Others | Caution | Warning | Danger |
| Criterion | Control | Cancer Hazard Only | Cancer Hazard +1 Other | Cancer Hazard +3 Others | Caution | Warning | Danger |
|---|---|---|---|---|---|---|---|
| Product Name | 5.3 | 8.2 | 15.0 | 12.4 | 5.9 | 7.2 | 19.3 |
| Product Type | 8.1 | 11.5 | 14.7 | 25.5 | 17.2 | 13.5 | 21.4 |
| Apprehension | 15.8 | 10.0 | 13.3 | 11.2 | 9.3 | 10.6 | 16.8 |
| Cancer Hazard | — | 5.4 | 7.9 | 11.8 | 9.1 | 8.1 | 9.6 |
| Hazard Scale* | 2.3 | 3.9 | 4.2 | 4.5 | 4.8 | 3.8 | 4.5 |

*5 point scale; 1 = low, 5 = high

Source:  Lirtzman and O'Connor (1980).

ferential meaning associated to signal words. Mean exposure time needed for perception of product name, type and level of anxiety reported was uniformly highest for labels with the word "danger." Very little difference was seen in terms of "caution" or "warning."

A second study investigated the perception of danger associated with the three key words in combination with four standard hazard warnings (poison, may be fatal if swallowed; corrosive, causes severe eye and skin burns; cancer hazard; and extremely flammable). A control anchor, high voltage electric current, was also used. In one experiment samples of chemical workers and consumers/graduate students were shown all pairs of these signal words and hazard warnings and asked to estimate their similarity as dangers using a danger similarity scale. After multidimensional scaling analysis perceptual maps of the perceived similarity in danger were created for the worker and consumer respondent data. These are shown in Figures 1.7 and 1.8.

The results suggest that people appear to use two dimensions in evaluating danger associated with hazard warnings: a short/long range hazard (time), and degree of danger perceived inherent in the warning.

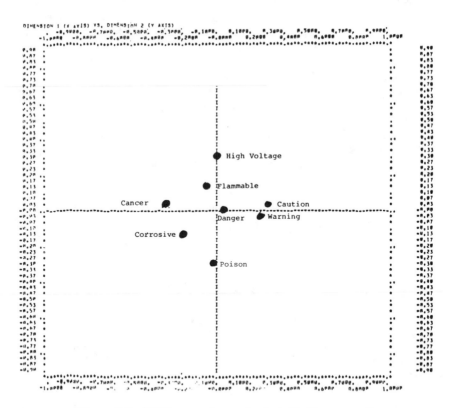

**Figure 1.7:** MDS map of perceived similarity among 8 hazard warnings—chemical workers. [Source: Lirtzman and O'Connor (1980)]

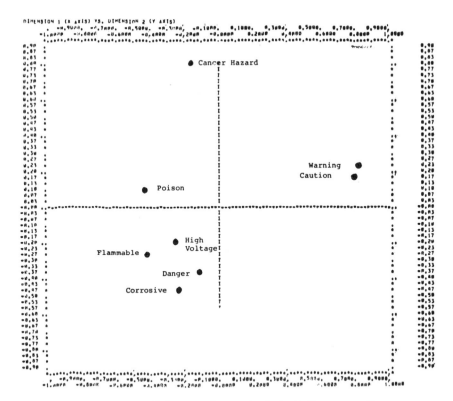

**Figure 1.8**: MDS map of perceived similarity among 8 hazard warnings—students/consumers. [Source: Lirtzman and O'Connor (1980)]

Student consumers apparently perceive cancer as a very long range hazard of moderately high danger. Caution and warning are perceived to be almost identical cognitively and seem to reflect low danger and longer range hazards. Flammability, high voltage, corrosiveness and *danger* signal word are perceived to be reasonably similar—short range, moderate to high hazards. Poison is perceived as a somewhat longer range hazard.

The picture seems to change for chemical workers. Although the dimensions probably are similar, degree of perceived danger and immediacy (short/long range), the perceptions differ. Cancer hazard is at the high pole of the danger axis, while "caution" and "warning" signals are at the other, low danger pole. "Danger" signal is about at the mid-point of both danger and immediacy axes. Cancer is perceived as a moderate range hazard, while high voltage and flammability are immediate dangers. Corrosiveness and poison are seen as long range dangers. In general, among workers all eight hazard stimuli are perceived to be closer together, e.g., more similar as dangers than is the case among student/consumers.

These findings suggest that it is probably quite naive to make too many assumptions about people's response to labeling especially hazard labeling. Certain hypotheses can be raised:

- The more one puts into a chemical label the harder it is for a worker to perceive its elements and evolve meaning. This suggests that clutter may have a significant negative effect on the value of a chemical hazard label.

- The more hazards one puts on a label the more anxiety the label arouses but probably at the cost of clear understanding of the individual hazards.

- It is probably a great mistake to assume all label readers are alike. Different audiences for example, workers and non-workers, probably approach a label differently and with different perceptual and cognitive habits and attitudes. A careful decision about who the major audience is to be is warranted and careful research should precede the design of the label.

- The labeler's assumptions about what words mean should be critically evaluated before casting convention into labeling concrete. Most people probably perceive the word danger to mean danger, whereas they probably do not differentiate the meanings of caution and warning, neither of which apparently are perceived to imply much of a hazard.

- No one knows much about what a label reader does when reading a hazard label, or what the reader wants and expects in labeling of chemical products—especially hazard labeling.

## IMPROVING LABEL EFFECTIVENESS

This chapter has provided an overview of the behavioral and perceptual issues involved in the design of labels for chemical products. It should by now be obvious that labeling is one of the few relatively low cost avenues open to the manufacturer for marketing support of a chemical product in terms of shelf visibility and communication of mood and product image, as well as for transmission of product use information and warnings. The time and money cost of improving or optimizing a product label is usually little in comparison to the sales and legal advantages which can accrue to the seller. The following suggestions will be helpful when reviewing a label with an eye towards improvement:

1. Always make a conscious effort to define in detail what the major audience(s) of the label is. Design the label for that audience.

2. Avoid unusual or gimmicky shapes and forms in the label.

3. If color is to be used avoid color combinations which can cause problems with color deficient readers.

4. Consider the context in which the label is most likely to be read, and design for that context. A label which will appear on a drum amid other labels and required symbols will pose a different set of design constraints than will a label to appear alone on a one gallon tin of adhesive to be used directly by a worker.

5. Do not expect a label to deliver more than it can. Extensive information, hazard warnings, safety information, imagery, use instructions, etc. cannot effectively be squeezed into a small label. If a small label is all that's possible, then perceptual elements must predominate— size and variety of print, color and color contrast, isolation and emphasis of hazard and warning areas, layout factors. At the very least, the label must communicate the product name, what it does, legally required warnings, and necessary use information rapidly and with minimal difficulty.

6. Keep label text as brief as is possible without being incomprehensible to a reader or ambiguous in meanings. Where warning information is involved, err on the side of completeness and concreteness, especially where specific behavior is desired. For example, "Wear protective clothing when using this product" is not as good as "Before you use (this product) always put on a pair of safety goggles, a rubber apron and rubber gloves, and your safety shoes." Also, "Use only with adequate ventilation and avoid open flames or sparks" is not as good as "Use only with adequate ventilation, the same as if it seemed you were out of doors. Do not use anywhere near any flame, spark or pilot light, or if there is air-conditioning machinery, electric fans or any other equipment in operation because these and equipment like them can cause a spark to occur."

7. Organize the text carefully in sections, and identify each section appropriately and clearly so that the reader can go to a needed portion easily. Within a label and within a section always put general statements first, and specific statements later. Material which must be remembered should come later in a passage, and be emphasized in some way by type style, use of color, etc.

8. Test the text for reading difficulty. Never use text which is more difficult than it has to be or which is inappropriately difficult for the intended audience. Recent studies of health related materials have, for example, shown that only 16 percent of them would be understood by 75

percent of the people in this country, and that as many as two-thirds of them could not be understood by at least half of those reading the material.

Use simple words and short sentences of five to ten words wherever possible.

9. Where one has a choice use the active voice and active verbs in the label text. Avoid non-important text (e.g., "Doctors often say," "That is to say," etc.). Be consistent in the use of words and terms.

10. The more hazardous the product the more subdued and sober should be the tone of the label.

11. Review labels at least yearly, and certainly whenever a change to the product is made.

The design of labels should be an active and continuing process. If the labeler is not experienced or wants objective evaluation of a label, the labeler should not hesitate to make use of laboratories and consultants, especially when litigation may be a factor.

The time and care spent in the design and testing of labels will be recouped in increased sales, product satisfaction, and safe use of the product and reduced litigation—clear bottom line motives for attention to the behavioral and perceptual aspects of labeling.

## REFERENCES

Adams, F. M. and Osgood, C. E., A cross cultural study of the affective meanings of color. *Journal of Cross Cultural Psychology* 4:2 (1973).

*ANSI Standard Z129.1-1976,* American National Standards Institute, Washington, D.C.

Baldwin, C. L. and Runkle, R. S., Biohazards symbol: Development of a biological hazards warning signal. *Science* 158 (1967).

Cahill, M., Interpretability of graphic symbols as a function of context and experience factors. *Journal of Applied Psychology* 60:3 (1975).

Department of Health, Education and Welfare and Food and Drug Administration: Consumer Nutrition Knowledge Survey, Report 1: 1973-74. Government Printing Office, Washington, D.C.; *Consumer Food Labeling Survey,* Government Printing Office, Washington, D.C.

Dreyfus, H., *Symbol Sourcebook,* New York: McGraw-Hill (1972).

Flesch, R., À new readability yardstick. *Journal of Applied Psychology* 32 (1948).

Fry, E., A readability formula that saves time. *Journal of Reading* 12 (1963).

Gunning, R., The FOG Index after twenty years. *Journal of Business Communications* 6 (Winter 1968).

Gusman, S. and Irwin, F. (Eds.), *Chemical Hazard Warnings: Labeling for Effective Communication,* The Conservation Foundation, Washington, D.C., 1979.

Hebb, D. O., On the nature of fear, *Psychological Review* 53 (1946).

Hopson, J., Cogar, P. and Batson, C., Color preference as a function of background and illumination. *Perceptual and Motor Skills* 33 (1971).

Humbert, C., *Label Design Evolution Function and Structure of Label,* New York: Watson-Guptill Publications (1972).

Institute of Medicine, *Evaluating Patient Package Inserts.* 1979, August (79-05), National Academy of Sciences, Washington, D.C.

Kamekura, Y., *Trademark and Symbols of the World,* New York: Reinhold Publishing Corporation (1965).

Klare, G. R., Assessing readability. *Reading Research Quarterly* 10 (1974).

Konz, S., Chawla, S., Sothaye, S. and Shah, P., Attractiveness and legibility of various colors when presented on cardboard. *Ergonomics* 15:2 (1972).

Lirtzman, S. I., and O'Connor, C. J., A Pilot Study of the Potential Effect of Label Characteristics and of Reader Attitudes About Hazard Warnings, Working Paper, Baruch College, CUNY (1980).

McGlaughlin, G. H., Clearing of SMOG. *Journal of Reading* 13 (December 1969).

Modley, R., World language without words. *Journal of Communication* 24 (1974).

Morris, A. L. et al., eds, Product Labeling and Health Risks, Banbury Report 6, Cold Spring Harbor Laboratories (1980).

Murphy, D. C., Toxic Substances and Hazardous Materials Labeling Regulations and Labeling Efficacy Studies: The State of the Art, Interagency Regulatory Liaison Group (August 1980).

Salcedo, R. N., Read, H., Evans, J. F. and Kony, A. C., *Improving User Attitude Toward and Readership of Pesticide Labels,* Agricultural Communications Research Report #26, Office of Agricultural Communication, University of Illinois (1973).

Schaie, K. W., On the relation of color and personality. *Journal of Projective Techniques and Personality Assessment* 30 (1966).

Schneider, K. C., Prevention of accidental poisoning through package and label design. *Journal of Consumer Research* 4 (1977).

Shell Oil Company, *Public and Worker Attitudes Toward Carcinogens and Cancer Risk,* Houston, Texas (1978).

Szaley, T. G., Validation of the coleman readability formulae. *Psychological Reports* 17 (1965).

Weinstein, N. D., Unrealistic optimism about future life events. *Journal of Personality and Social Psychology* 29 (5): 806-820 (1980).

Wexner, L. B., The Degree to which colors (hues) are associated with Modd-Tones. *Journal of Applied Psychology* 38: 6 (1954).

## ADDITIONAL SUGGESTED READINGS

Combs, B. and Slovic, P., Newspaper coverages of causes of death. *Journalism Quarterly* 56 (1979).

Geldard, F. A., *The Human Senses,* Second Ed. New York: John Wiley and Sons, Inc. (1972).

Lee, M., *Bookmaking: The Illustrated Guide to Design and Production,* New York: R. R. Bowker & Company (1965).

Leventhal, H., Singer, R. P. and Jones, S., Effects of fear and specifying of recommendations upon attitudes and behavior. *Journal of Personality and Social Psychology* 2 (1965).

Ley, P., Memory for medical information. *British Journal of Social and Clinical Psychology* 18 (1979).

Ley, P., Pike, L. A., Whitworth, M. A. and Woodward, R., Effects of source, context of communication, and difficulty level on the success of health educational communications. *Health Education Journal* 38 (1979).

Lichtenstein, S., Slovic, P., Fischoff, B., Layman, M., and Combs, B., Judged frequency of lethal events. *Journal of Experimental Psychology: Human Learning and Memory* 4 (1978).

Luckiesh, M. and Moss, F., *Reading as A Visual Task,* New York: Van Nostrand Company (1942).

McCall, W. A. and Crabbs, L. M., *Standard Test Lessons in Reading,* 1961 Edition, New York, Bureau of Publications, Teachers College, Columbia University.

Rosenberg, J., A question of ethics: The DNA controversy. *American Educator* 2 (1978).

Schiff, W., *Perception: An Applied Approach,* Boston: Houghton Mifflin Company (1980).

Taylor, W. L., Cloze Procedure: A new tool for measuring readability. *Journalism Quarterly* 30 (Fall 1953).

Thorndyke, P. W., Cognitive structures in comprehension and memory of narrative discourse. *Cognitive Psychology 9 (1977).*

Twerski, A. D., Weinstein, A. S., Donaher, W. A. and Piehler, H. P., The use and abuse of warnings in product liability: Design and defect litigation comes of age. *Cornell Law Review* 61 (1976).

# 2

# Information Sources and Systems for Labeling

**Norma Skolnik**
*International Playtex, Inc.*
*Paramus, N.J.*

## INTRODUCTION

Information is the key to decision-making with respect to chemical labeling and classification. Obtaining data of various types—regulatory, legal, toxicological—is the first step in the labeling process. The purpose of this chapter is to discuss sources of data and efficient ways one can obtain information for hazard evaluation and regulatory decision-making. Without efficient access to this data, the labeling process becomes more complicated and costly than it needs to be.

## REGULATORY INFORMATION

### Government Documents

The first step in information management for regulatory compliance is obtaining the necessary documents. Obtaining essential government documents for chemical industry needs requires a knowledge of how the U.S. Government handles agency publications. *Federal Register* and *Codes of Federal Regulations* (once one has determined which titles are relevant to ones company's needs) are available from the Government Printing Office (GPO) in Washington. The *Federal Register* is the daily record of Government agency publications that is published Monday through Friday except for legal holidays. All Notices of Rulemaking, Advance Notices of Rulemaking and Final Rules issued by any Federal agency must be published in the *Federal Register,* an important document to review to maintain one's awareness of regulatory happenings. *Codes of Federal Regulations* are codifications of the rules published in the *Federal Register* by Executive departments and Federal agencies. There are 50 titles in the Code which represent broad areas subject to federal regulation. Each title is divided into chapters which usually bear the name of the issuing agency. Following is a complete list of CFR titles:

Title   1—General Provisions

Title   2—Reserved

Title   3—The President (Includes Presidential proclamations, Executive orders, etc.)

Title   4—Accounts

Title   5—Administrative Personnel

Title   6—Economic Stabilization

Title   7—Agriculture

Title   8—Aliens & Nationality

Title   9—Animals & Animal Products

Title 10—Energy

Title 11—Federal Elections

Title 12—Banks & Banking

Title 13—Business Credit & Assistance

Title 14—Aeronautics & Space

Title 15—Commerce & Foreign Trade

Title 16—Commercial Practices (This title includes Consumer Product Safety Regulations, such as the Federal Hazardous Substances Act, which are important to labeling considerations for chemical products)

Title 17—Commodity & Securities Exchanges

Title 18—Reserved

Title 19—Customs Duties

Title 20—Employees' Benefits

Title 21—Foods & Drugs (This title includes FDA Regulations as well as those of the Drug Enforcement Administration, Dept. of Justice. This is an important title for chemical industry needs, particularly for those companies who either manufacture or supply to pharmaceutical, medical device, or food or cosmetic product manufacturers)

Title 22—Foreign Relations

Title 23—Highways

Title 24—Housing & Urban Development

Title 25—Indians

Title 26—Internal Revenue

Title 27—Alcohol, Tobacco Products & Firearms

Title 28—Judicial Administration

Title 29—Labor (This important title includes Dept. of Labor Regulations, Equal Employment Opportunity Commission rules, and Occupational Safety & Health Administration rules (OSHA))

Title 30—Mineral Resources

Title 31—Money & Finance: Treasury

Title 32—National Defense

Title 33—Navigation

Title 34—Government Management

Title 35—Panama Canal

Title 36—Parks, Forestry & Public Property

Title 37—Reserved

Title 38—Pensions, Bonuses & Veterans' Relief

Title 39—Postal Service

Title 40—Protection of Environment (This title is of paramount importance to the chemical industry. It includes such areas of EPA regulation as Air Pollution, Water Pollution & Hazardous Waste rules. Council on Environmental Quality standards are also included)

Title 41—Public Contracts and Property Management

Title 42—Public Health

Title 43—Public Lands

Title 44—Reserved

Title 45—Public Welfare

Title 46—Shipping

Title 47—Telecommunication

Title 48—Reserved

Title 49—Transportation (An essential title for chemical industry needs, this contains regulations issued by the Federal Railroad Administration, the Federal Highway Administration, the Coast Guard, the National Transportation Safety Board and the Interstate Commerce Commission. DOT's Hazard-

ous Materials Table, as well as other essential
chemical industry rules are included in this title).

Title 50—Wildlife and Fisheries

Many federal documents, particularly those on environmental subjects, are obtainable from the National Technical Information Service (NTIS). NTIS, an agency of the U. S. Department of Commerce, is the central source for the public sale of U. S. sponsored research, development and technical reports. The *Monthly Catalog* lists by agency and is available from GPO. Many publications on safety and toxicology that relate to chemical industry needs, however, must be obtained from the National Institute of Occupational Safety and Health (NIOSH) in Cincinnati. Securing state documents can be equally frustrating.

State publications are listed in the Monthly Checklist of State Publications. These are usually available from one's own Department of State. State statutes must be purchased from West Publishing Company on Long Island. State Administrative Codes are obtained from your Department of State, Division of Administrative Procedure, located in the state capitol. In many states, one must buy the entire code and cannot obtain individual titles, thus defeating one of the main principles of efficient information management—not to be burdened with data that is not relevant to one's operational needs.

## Computerized Data Systems

There is a serious problem with many data bases in use today. Some contain data which is either unsupported, or at times simply in error. There is a growing concern among information scientists that a review of data content is long overdue. One such program has been undertaken by the United Nations. It is a time consuming activity, requiring review by highly specialized scientists. Attention must be given to data quality, validity and origin. A lack of a critical evaluation of data can have important effects both in the marketplace and in the laboratory.

One may lose the use of a valuable commercial product or one may waste precious resources in needless testing. In any event, a competent scientist should review computer generated information before it is used as a basis for Hazard Communication.

**Federal Index:** This data base covers virtually all federal documents issued between October 1976 and the present. Available from Capitol Services, Inc., Washington, D.C., the Federal Index covers proposed rules, regulations, speeches, bill introductions, congressional hearings, court decisions, executive orders, *Washington Post* articles, as well as all *Federal Register* documents. The cost for one hour of on-line connect time is $90.00 for subscribers to Lockheed's Dialog Information Retrieval Service.

**Legal Data System:** A major area of concern to those involved in regulatory compliance is obtaining legal information. The area of regulatory law can be intimidating to laymen—but one need not be a lawyer to obtain legal information pertinent to chemical industry needs. Such information is available through two data bases for legal research: LEXIS and WESTLAW. These are computerized retrieval systems for information on such areas as environmental law and product liability. These systems make it possible to search statutes, regulations, decisions, and cases which may have a bearing on the legal problem being researched. Of course, automated legal research only puts the information in our hands; it does not provide the legal interpretation which is the function of the lawyer. However, it does give one access to legal information and provide for fast case finding; and if one is interested in re-searching product liability as it might pertain to a product, then one could search either of these legal data bases.

Suppose one wanted to search for cases of product liability regarding haz-ards of chemical products. Once one enters the request, e.g. "Product Liability and Chemicals" on the terminal and the computer has done its search, the number of cases found with the search request appears on the screen. Since this number may be large, one may want to narrow the request further (haz-ards of corrosive chemical products; flammable chemical products, toxic chem-ical products, etc.). The computer indicates how many headnotes and/or cases have been retrieved and will be displayed. Once promising headnotes are found in a search, the searcher may ask to have the full text displayed. If it is not available on line, it's possible to find the case in its bound form.

### Access to Government Documents

The easiest way to obtain Government documents is to set up a deposit ac-count with the Government Printing Office. This is done by sending a letter with $50.00 or more to:

> Superintendent of Documents
> Government Printing Office
> Washington, D.C. 20402

Once a deposit account has been established, one simply phones in an order by calling (202) 783-3238 and charges it to one's deposit account.

To obtain technical documents published by NTIS it is also necessary to send a letter stating the company name and enclosing a $25.00 mini-deposit. This letter would be sent to:

> Deposit Accounts Section—NTIS
> 5288 Port Royal Road
> Springfield, Virginia 22161
> (703) 487-4770

Information about NTIS information service is included in an excellent free pamphlet published by the Department of Commerce and known as NTIS. General Catalog No. 7a.

National Institute for Occupational Safety and Health documents relevent to chemical industry needs can be identified in the *Catalog of NIOSH Publica-tions*. This free catalog can be obtained from:

> Publications Department—NIOSH
> Mailstop R6
> 4676 Columbia Parkway
> Cincinnati, Ohio 45226

Once identified, NIOSH documents can be ordered from this office free of charge.

Information on the LEXIS Data System can be obtained from:

> Mead Data Central
> 9333 Springboro Pike
> Miamisburg, Ohio 45342
>
> or
>
> 200 Park Avenue
> New York, New York 10017
> (212) 883-8560

Information on the WESTLAW system is available from:

> West Publishing Company
> Mineola, New York
> (516) 248-1900

## SCIENTIFIC LITERATURE

### Self Services

Another of this chapter's objectives is demonstrating how the labeling process can be aided by efficient information management. Access to a good collection of chemical reference books is very helpful. The hazards of many chemical substances can be determined by using published information sources. A bibliography of sources for toxicological and medical information appears at the end of this section. These are standard toxicology reference books which provide toxicological data assembled by authorities in the field. As resources available for hazard labeling, these works are of great assistance.

Toxicology books alone are neither the only or the most efficient sources of information about chemical hazards. They are definitely the least expensive source of such information. In many ways they are the easiest to use. A far more comprehensive and efficient method of searching for hazardous chemical data in general, and toxicology information in particular, is computerized literature retrieval. This can be done by companies who own or lease their own computer terminals. The initial costs for companies who want to do their own searching involve purchasing a computer terminal at an approximate cost of about $2,000.00 each or leasing terminals at a cost of approximately $136.00 per month, plus the cost of training employees or hiring employees trained in on-line searching of chemical and toxicological data. These costs may sound high, but in the long run they are lower than the cost of hiring a search service to perform frequent searches.

If one's company only requires a few searches per year, then it makes sense to use a search service. If, however, the company will require many searches, an "in-house" self-service system of searching is more cost effective. For those companies that do not have an available computer terminal, there are information retrieval services, libraries, university research centers, medical centers, etc. that will perform computerized searches for a fee. NIOSH will perform appropriate searches free provided one can wait the 4 to 5 weeks that NIOSH often takes because of current backlog.

## Hazard Evaluation
## A Compilation of
## Sources for Toxicological Information

Publications—refer to most recent editions.

1. American Conference of Governmental Industrial Hygienists, Inc: *Documentation of Threshold Limit Values.* Cincinnati, Ohio. (P.O. Box 1937, Cincinnati, Ohio 45201)

   This important guide to recommended Threshold Limit Values goes beyond O.S.H.A.'s chemical exposure ceiling limits. It is a valuable reference work for safety managers and toxicologists, as well as industrial hygienists.

2. American Mutual Insurance Alliance: *Handbook of Hazardous Materials.* Chicago, Illinois (20 N. Wacker Drive, Chicago, Ill.)

   This volume and the following are both written with a view to product liability considerations.

3. American Mutual Insurance Alliance: *Handbook of Organic Industrial Solvents.* Chicago, Illinois (20 N. Wacher Drive, Chicago, Illinois)

4. Browning, E.: *Toxicity and Metabolism of Industrial Solvents.* New York: American Elsevier, 1965 (52 Vanderbilt Ave., New York, New York 10017)

5. Browning, E.: *Toxicity of Industrial Metals.* 2nd ed., New York: Appleton-Century-Crofts, 1969. (292 Madison Avenue, New York, New York 10017)

6. Casarett, Louis and Doull, John: *Toxicology: The Basic Science of Poisons.* New York: Macmillan Publishing Co., Inc. (866 Third Avenue., New York, N.Y. 10022)

7. *CHRIS (Chemical Hazards Response Information System) Hazardous Chemical Data.* U.S. Department of Transportation, U.S. Coast Guard, Washington, D.C. (U.S. Government Printing Office, Dvn. of Public Documents, Washington, D. C. 20402)

8. Clayton, George D. & F. E. Clayton: *Patty's Industrial Hygiene and Toxicology.* New York: Wiley-Interscience. (605 Third Avenue, New York, N.Y. 10016)

A classic in the field of toxicological information, this extremely thorough reference text has recently been expanded to a four volume set (Volumes I, IIa and IIb, Volume III).

9. *Condensed Chemical Dictionary.* Edited by G. Hawley, New York: Van Nostrand Reinhold (135 West 50th Street, New York, N.Y. 10020)

10. *Fenaroli's Handbook of Flavor Ingredients,* Volume 1 & 2. Edited, translated, and revised by T. E. Furia and N. Bellanca, 2nd ed. 1975, Boca Raton: CRC Press, (2000 N.W. 24th Street, Boca Raton, Fla.)

11. Friberg, L., G. R. Nordberg, and V. B. Vouk: *Handbook on the Toxicology of Metals.* New York: Elsevier North Holland, 1979 (52 Vanderbilt Ave., New York, N.Y. 10017)

12. *Goodman & Gilman's The Pharmacological Basis of Therapeutics.* Edited by A. G. Gilman, L. S. Goodman, and A. Gilman, 6th ed. New York: Macmillan Publishing Co., Inc. 1980 (866 Third Ave., New York, N.Y. 10022)

This classic in the field of pharmacology provides detailed information on anatomy and functions of the autonomic nervous system. Although primarily concerned with drugs and how they act upon the body, there are also excellent sections on the toxicity of gases and vapors and heavy metals. Goodman and Gilman thoroughly evaluate many of the more widely used compounds utilized by the pharmaceutical industry.

13. Gosselin, Robert et al: *Clinical Toxicology of Commercial Products; Acute Poisoning.* Baltimore: Williams and Wilkins, (428 E. Preston Street, Baltimore, Md. 21202)

14. Grant, W. Morton: *Toxicology of the Eye.* Springfield: Charles C. Thomas (327 E. Lawrence Ave., Springfield, Ill. 62717)

15. Haley, T. J., and C. Thienes: *Clinical Toxicology* 5th ed. Philadelphia: Lea and Febiger, 1972. (600 S. Washington Sq., Philadelphia, Pa. 19106)

16. Hamilton, Alice and Hardy, Harriet: *Industrial Toxicology.* Publishing Sciences Group, Inc. Acton, Mass.

17. Hayes, Wayland J., Jr. *Toxicology of Pesticides.* Baltimore: Williams and Wilkins, 1975. (428 E. Preston St., Baltimore, Md. 21202)

18. Lefaux, Rene: *Practical Toxicology of Plastics*. Boca Raton: CRC Press, 1968 (2000 N.W. 24th Street, Boca Raton, Fla. 33431)

19 *Merck Index*. Edited by Martha Windholz, Rahway, New Jersey, Merck and Company (Box 2000, Rahway, N.J. 07065)

20. National Fire Protection Association: *Hazardous Materials Data*. Boston, Massachusetts, NFPA Pub. No. 49 (470 Atlantic Avenue, Boston, Massachusetts 02110)

    This book is particularly instructive with regard to flammability hazards and the control of fires that involve hazardous chemicals.

21. NIOSH/OSHA *Pocket Guide to Chemical Hazards*. U.S. Department of Health, Education, and Welfare, NIOSH DHEW Pub. No. 78-210 (U.S. Government Printing Office, Washington, D.C. 20402).

22. *Registry of Toxic Effects of Chemical Substances*. U.S. Department of HEW-NIOSH, DHEW, Pub. No. 79-100 (U.S. Government Printing Office, Washington, D.C. 20402)

    Although this is an important reference text, it does contain some unvalidated information. References to international journal articles on toxicology of specific chemicals are particularly helpful.

23. Sax, Irving: *Dangerous Properties of Industrial Materials*. New York: Van Nostrand Reinhold (135 W. 50th Street, New York, N.Y. 10020)

    An easy-to-use guide to the hazardous properties of chemicals. Alphabetically arranged, this is a ready reference aid that would be helpful for the novice who wants simple toxicological data.

24. Sittig, Marshall: *Handbook of Toxic and Hazardous Chemicals*. Park Ridge: Noyes Publications (Mill Road at Grand Avenue, Park Ridge, N.J. 07656)

25. Sunshine, Irving (ed.) *Methodology for Analytical Toxicology*. Boca Raton: CRC Press, Inc., 1975 (2000 N.W. 24th Street, Boca Raton, Fla.)

26. Weiss, G.: *Hazardous Chemicals Data Book*. Park Ridge: Noyes Data Corporation (Mill Road at Grand Avenue, Park Ridge, N.J. 07656)

### Private On-Line Services

**Chemical Abstracts Service:** One major source of scientific and technical information for the chemical industry is Chemical Abstracts Service (CAS) of Columbus, Ohio. CAS is a self-sustaining division of the American Chemical Society (ACS) and monitors new developments published in the world's chemical literature. In a recent five year period, CAS monitored over 15,000 journals, periodicals, conferences and technical reports gathered from over 150 nations and published in over 50 languages. CAS also monitors the patents from 26 major countries of the world for new chemical developments. CAS reviews this literature for new information published related to chemical engineering and over 75 different fields of chemistry. Each week for the past 74 years CAS has published the abstracts or summaries pertaining to the new developments cited in the world's chemical literature. On the average, over 500,000 citations are being added to the CAS information base each year.

Information in the CAS data base is selected, combined and organized automatically to produce a wide range of publications, indexes, computer and microform information files, and special services. Together they provide a variety of approaches to locating, retrieving, and keeping up with chemical information.

The most comprehensive and best known of CAS's publications is *Chemical Abstracts* recognized the world over as the key to the world's chemical literature. Abstracts are published in weekly issues that are about half the size of the Manhattan Telephone Directory and contain 8,000 to 9,000 abstracts each.

CAS ONLINE is Chemical Abstracts Service's computerized system that provides access to substance information from the CAS registry file. CAS ONLINE allows users to search for substances on the basis of structural units, registry number, chemical name, or molecular formula. Portions of the CAS information base are licensed by several information vendors.

CA Selects is a series of current awareness publications derived from the Chemical Abstracts data base. For each topic in the series, a special profile assures the retrieval of pertinent abstracts. The *CA Selects* series topic that is most relevant to chemical labeling needs is Chemical Hazards, Health and Safety. This current-awareness service covers documents dealing with safety in the chemical industry, as well as the health and safety of personnel working with hazardous substances. Coverage includes effects of human exposure to hazardous substances and hazardous properties of chemical substances and reactions. The annual cost for this service is $75.00.

For information on Chemical Abstracts Service, contact Brian Cannan, P.O. Box 3012 Columbus, Ohio 43210.

**System Development Corporation (SDC):** System Development Corporation of Santa Monica, California has a large collection of chemical information on-line and quite a few data bases, many of them exclusives, that address the needs of the chemical community.

Some of the highlights of SDC's file offerings include the following:

| | |
|---|---|
| CHEMDEX | —a chemical dictionary file |
| CA Search | —includes 3 Chemical Abstracts literature files |

WPI/WPIL       —Derwent's World Patents Index is an SDC exclusive.

APILIT/APIPAT—also an exclusive. This file gives full coverage of petroleum and petroleum-related literature and patents. This file allows retrieval of chemicals by the specific role that they play in reactions, i.e., reactant, product, etc.

One SDC data base that would be relevant to hazard evaluation needs is called "Safety". It provides broad, interdisciplinary coverage of literature related to safety and is particularly concerned with hazard identification and control. Costs for SDC computer-connect time run on average between $60.00-$100.00 per hour.

For further information on SDC files, contact Kathleen Shenton (213) 820-4111.

**Lockheed:** The DIALOG Information Retrieval Service, from Lockheed Information Systems, has been serving users since 1972. The DIALOG service has over 100 data bases, several of which are relevant to chemical industry needs. These include those which follow.

On average, costs for DIALOG data bases run between $65.00-$120.00 for an hour of on-line connect time.

For more information about the Lockheed DIALOG service call their toll free number (800) 227-1927.

*1. BIOSIS Previews,* 1969-present, 3,049,540 records, monthly updates (Biosciences Information Service, Philadelphia, PA).

BIOSIS Previews contains citations from both *Biological Abstracts* and *Biological Abstracts/RRM* (formerly entitled Bioresearch Index), the major publications of BioSciences Information Service of *Biological Abstracts.* Together, these publications constitute the major English language service providing comprehensive worldwide coverage of research in the life sciences. Nearly 8,000 primary journals as well as symposia, reviews, preliminary reports, semipopular journals, selected institutional and government reports, research communications, and other secondary sources provide citations on all aspects of the biosciences and medical research. File 5 contains all the citations from 1972 through the present. The citations for the years from 1969 through 1971 are available in File 55.

*2. CHEMSEARCH* 183,876 chemical substances, derived from CA Search (File 4), biweekly updates, (DIALOG Information Retrieval Service, Palo Alto, CA and Chemical Abstracts Service, Columbus, OH).

CHEMSEARCH is a dictionary listing of the most recently cited substances in CA Search (File 4) and is a companion file to CHEMNAME (File 31). For each substance listed, the following information is provided: CAS Registry Number, molecular formula, and CA Substance Index Names. CHEMSEARCH includes all new chemical substances cited in the latest six issues of *Chemical Abstracts.* Also included are additional chemical substance search terms generated specifically for all DIALOG chemical name dictionaries. Chemical substances already in CHEMNAME are not included. Only the most recent six issues of *Chemical Abstracts* are covered by the data base at any one time.

3. *CHEMSIS,* 1972-1976, 1,300,000 records, closed file; 1977-present, 1,320,880 records, irregular updates. (DIALOG Information Retrieval Service, Palo Alto, CA and Chemical Abstracts Service, Columbus, OH)

CHEMSIS (CHEM Singly Indexed Substances) is a dictionary, non-bibliographic file containing those chemical substances cited once during a Collective Index period of *Chemical Abstracts.* CHEMSIS is an important access point to chemical information as 75% of the chemicals cited appear only once. For each substance listed, the CAS Registry Number, molecular formula, CA Substance Index Name for the Ninth Collective Index Period, available synonyms, ring data, and other chemical substance data are included. As with CHEMNAME the file purpose is to support specific substance searching and substructure searching via nomenclature in the DIALOG Chemical Information System (consisting of CA Search, CHEMNAME, CHEMSEARCH, and CHEMSIS).

4. *ENVIROLINE,* 1971-present, 86,126 citations, monthly updates (Environment Information Center, Inc., New York, NY)

ENVIROLINE, produced by the Environment Information Center, covers the world's environmental information. Its comprehensive, interdisciplinary approach provides indexing and abstracting coverage of more than 5,000 international primary and secondary source publications reporting on all aspects of the environment. Included are such fields as: management, technology, planning, law, political science, economics, geology, biology, and chemistry as they relate to environmental issues. Literature covered includes periodicals, government documents, industry reports, proceedings of meetings, newspaper articles, films and monographs. Also included are excerpts from the *Federal Register* and patents from the *Official Gazette.*

5. *Excerpta Medica,* June 1974-present, 1,346,913 records, monthly updates (Excerpta Medica, Amsterdam, The Netherlands)

*Excerpta Medica* is one of the leading sources for searching the biomedical literature. It consists of abstracts and citations of articles from over 3,500 biomedical journals published throughout the world. It covers the entire field of human medicine and related disciplines. The on-line file corresponds to the 43 separate specialty abstract journals and 2 literature indexes which make up the printed *Excerpta Medica* plus an additional 100,000 records annually that do not appear in the printed journals. *Excerpta Medica* is widely used by physicians, medical researchers, medical libraries, hospitals, medical schools, health organizations, and chemical and pharmaceutical companies. In addition to providing abstracts in all fields of medicine, *Excerpta Medica* provides extensive coverage of the drug and pharmaceutical literature and of other health related sciences such as environmental health and pollution control, forensic science, health economics and hospital management, and public health. In-process records are placed initially in File 73. As processing is completed by *Excerpta Medica,* records are periodically transferred to File 72. File 172 contains records from 1974-1979.

6. *SCISEARCH* January 1974-present, 3,096,000 citations, monthly updates (Institute for Scientific Information, Philadelphia, PA)

SCISEARCH is a multidisciplinary index to the literature of science and technology prepared by the Institute for Scientific Information (ISI). It contains all the records published in *Science Citation Index* (SCI) and additional records from the *Current Contents* series of publications that are not included in the printed version of SCI. SCISEARCH is distinguished by two important and unique characteristics. First, journals indexed are carefully selected on the basis of several criteria, including citation analysis, resulting in the inclusion of 90 percent of the world's significant scientific and technical literature. Second, citation indexing is provided, which allows retrieval of newly published articles through the subject relationships established by an author's reference to prior articles. SCISEARCH covers every area of pure and applied sciences.

The ISI staff indexes all significant items (articles, reports of meetings, letters, editorials, correction notices, etc.) from about 2600 major scientific and technical journals. In addition, the SCISEARCH file for 1974-75 includes approximately 38,000 items from *Current Contents*-Clinical Practice. Beginning January 1, 1976, all items from *Current Contents*-Engineering, Technology, and Applied Science and *Current Contents*-Agriculture, Biology, and Environmental Sciences that are not presently covered in the printed SCI are included each month. This expanded coverage adds approximately 58,000 items per year to the SCISEARCH file.

Records from 1974-1977 are in File 94. File 34 contains records from 1978 through the present.

7. *SSIE Current Research,* last two years, e.g., 1978-present, 162,288 citations, monthly updates (Smithsonian Science Information Exchange, Washington, D.C.)

SSIE (Smithsonian Science Information Exchange) Current Research is a data base containing reports of both government and privately funded scientific research projects, either currently in progress or initiated and completed during the most recent two years. SSIE data are collected from the funding organizations at the inception of a research project and provide a source for information on current research long before first or progress reports appear in the published literature. SSIE Current Research encompasses all fields of basic and applied research in the life, physical, social, and engineering sciences.

Project descriptions are received from over 1,300 organizations that fund research including federal, state, and local government agencies; nonprofit associations and foundations; and colleges and universities. A small amount of material is provided from private industry and foreign research organizations, while 90% of the information in the data base is provided by agencies of the federal government.

Research projects in SSIE Current Research include work in progress in the agricultural, behavioral, and biological sciences; chemistry and chemical engineering; electronics, physics, materials science; engineering; mathematics; medical sciences; and the social sciences and economics.

8. *TSCA Initial Inventory,* 1979, 43,278 records, irregular updates (DIALOG Information Retrieval Service, Palo Alto, CA and Environmental Protection Agency, Office of Toxic Substances, Washington, D.C.)

TSCA Initial Inventory (Derived from the Initial Inventory of the Toxic Substances Control Act Chemical Substance Inventory) is a non-bibliographic dictionary listing chemical substances in commercial use in the U.S. as of June 1, 1979. TSCA Initial Inventory is not a list of toxic chemicals since toxicity is not a criterion for inclusion in the list. For each substance the following are provided: CAS Registry Number, preferred name, synonyms, and molecular formula. Confidential substances, definitions of complex substances with no appropriate molecular formulas, and additional synonyms, however, are not included in the on-line file.

**Bibliographic Retrieval Services (BRS):** The Bibliographic Retrieval Services, Inc. of Scotia, New York also offer efficient access to several data bases that are relevant to chemical industry needs. BRS offers an on-line version of Chemical Abstracts CA search under the file label CHEM. Often a CHEM search can be supplemental with other BRS data bases. One can use the CROS file, which provides an index to all data bases on-line at BRS. BRS data bases include NTIS, Science SCISEARCH Citation Index, the Smithsonian Information Exchange (SMIE), and BIOSIS Previews. All are described in the previous section. A few of the MEDLARS data bases are available from BRS, as well as a data base called Pharmaceutical News Index that covers drug industry news.

BRS seems to offer comparatively lower prices for their data bases. Data bases identical to those offered by Lockheed (BIOSIS, SSIE, etc.) are cheaper from BRS. On average, costs for BRS data bases run between $46.00-$70.00 for an hour of on-line connect time. Only one of their data bases (Science Citation Index) is fairly costly ($120.00, 1 hour connect time).

For further information about Bibliographic Retrieval Services, Inc. contact the BRS Customer Service Department, Scotia, New York 12302, (518) 374-5011.

### Search Services—Types and Costs

**In-House Direct Access:** For those who own or lease their own terminals, and have already had a staff member (e.g. a librarian or information scientist) trained in on-line searching by the National Library of Medicine (NLM), there are several advantages. One benefit of having an in-house terminal is having immediate access to the data base and being able to obtain the literature results that much more quickly. With one's own terminal one can even do after hours searching when emergency information needs mandate completing a search in a hurry. Another advantage is the security factor of knowing that one's company alone has knowledge of the data for which one is searching. No one else will find out what information areas are being explored.

*Costs*—Recent charges for MEDLARS data bases have been as follows:

| | |
|---|---|
| MEDLINE & RTECS | $22.00 each, per hour |
| Off-line print out | .18 per page |
| CHEMLINE | $54.00 per hour |
| Off-line print out | .41 per page |
| TOXLINE | $55.00 per hour |
| Off-line print out | .41 per page |

Since these charges are frequently raised, it is best to consult NLM about current costs for data bases and off-line print outs.

**Non-Commercial Search Services:** Among the noncommercial purveyors of search services are institutional and university-run research centers. The fees charged by these institutional search services vary from vendor to vendor, but are usually reasonable. These charges are *in addition* to the above-listed costs for MEDLARS on-line. All of these search services access MEDLARS data bases. Some of them also access other commercial data bases (Lockheed, SDC, BRS, etc.) and may provide other services. The Franklin Institute in Philadelphia, Pa., for example, provides annotated search print outs, obtains all relevant articles cited in the search, and also provides transactions where necessary.

### Examples of Non-Commercial Search Services

Since rates are always subject to change, check with vendors to verify current charges.

1. Downstate Medical Center
   State University of New York
   450 Clarkson Avenue
   Brooklyn, New York

   $25.00 surcharge for each data base used.

2. Rutgers University
   Research Information Service
   New Brunswick, New Jersey 08901

   $50.00 per hour surcharge for search analyst's time
   $ 5.00 charge for billing and handling

3. St. Joseph's Hospital and Medical Center
   703 Main Street
   Paterson, New Jersey 07503

   $20.00 surcharge for each data base used.

4. Indiana University
   Department of Chemistry
   Chemical Information Center (CIC)
   Bloomington, Indiana 47401

   $35.00 surcharge for each data base used

5. The Franklin Institute in Philadelphia will not perform any literature search for less than $250.00.

6. University of Cincinnati
   Science Retrieval Center
   Cincinnati, Ohio 45221

   $5.00 charge for billing and handling, plus 10% surcharge for each data base used. (Not charged to non-profit users.)

7. U.C.L.A.
   Medical Sciences Library
   Los Angeles, California 90024

   $13.00 surcharge for Toxline

8. University of California, Davis
   Health Sciences Library
   Davis, California 95616

   $13.00 surcharge for Toxline

9. Northwestern University Medical School
   303 East Chicago Avenue
   Chicago, Illinois 60611

   $15.00 surcharge per data base.

**Commercial Services:** A growing number of commercial search services provide on-line access to many data bases in addition to MEDLARS. Some of these commercial services can obtain almost any type of information for you. The costs tend to be quite high, but the vendors will remind you that "you get what you pay for". Since the competition for clients is great among commercial service vendors, their marketing is often aggressive. A client can, however, often obtain extensive personalized service from a commercial service. Space does not permit a description of all the commerical search services available. As an example, however, one or two commercial services can be examined.

"FIND, S.V.P." located at 500 Fifth Avenue in New York, New York is probably the best-known search service in the U.S. They will do an individual search using three MEDLARS data bases for approximately $300.00-$350.00 (depending upon the length of the search). For this amount, the customer receives off-line print outs that have been reviewed and highlighted by a FIND staff member, as well as copies of up to five articles cited in the search. They search other type of data bases and will answer all types of reference questions.

Another commercial search service is Information Specialists, Inc., in Cleveland, Ohio (216) 321-7500. This service also performs various types of searches and will answer all types of reference questions for their clients. Business intelligence information and third party inquiries are areas where services like Information Specialists and FIND can be of great assistance.

For a guide to commercial and non-commercial search services, one can consult the *Encyclopedia of Information Systems and Services.* This book provides extensive listings of Search Services, Data Base Producers and Publishers, On-line Vendors, Computer Service Companies, Computerized Retrieval Systems and Research Consultants located throughout the United States. Edited by Anthony Kruzas, the *Encyclopedia of Information Systems and Services* is published by the Gale Research Company, Detroit, Michigan.

### Government Literature Retrieval Systems

**The MEDLARS System:** An astoundingly complete scientific data retrieval system, MEDLARS, is the computerized literature retrieval service of the National Library of Medicine (NLM). This overall system contains over 5,000,000 references to journal articles and books. The MEDLARS network includes many specialized scientific data bases. Only those relevant to the chemical industry will be discussed here. There are 15 other data bases available through MEDLARS.

*MEDLARS Data Bases*—For further information on the data files cited and on NLM on-line training programs contact: National Library of Medicine, 8600 Rockville Pike, Bethesda, Maryland 20209, Telephone: (301)-496-6193.

*Toxicology Data Bank:* Toxicology Data Bank (TDB) contains chemical, pharmacological, and toxicological information. Use of National Library of Medicine's Toxicology Data Bank is highly recommended for the 2,600 substances for which it provides animal and human toxicology data, chemical and physical properties, formulas, laboratory methods, poisoning potential and antidote information. Data for the TDB are extracted from handbooks and textbooks and reviewed by a peer review group of subject specialists.

*MEDLINE:* This data base contains approximately 600,000 references to biomedical journal articles published in the current as well as the two preceding years. MEDLINE can also be used to update a search periodically. Coverage of previous periods (back to 1966) is provided by back files. An English abstract, if available, is frequently included. The articles are from 3,000 journals published in the U.S. and 70 foreign countries. MEDLINE also includes a limited number of chapters and articles from selected monographs. Coverage of previous periods (back to 1966) is provided by back files that total some 2,700,000 references. MEDLINE is updated monthly and is used to publish *Index Medicus* and other recurring bibliographies.

*TOXLINE:* (Toxicology Information On-Line) contains a collection of about 800,000 toxicologically related references published since 1976 together with earlier references concerning mutagens and teratogens. Coverage of previous periods (in some cases, antedating 1965) is provided by TOXBACK. Almost all references have abstracts or indexing terms and most chemical compounds mentioned in TOXLINE are further identified with Chemical Abstracts Service Registry Numbers. The references are from five major published secondary sources and five special literature collections.

*RTECS:* Also available on microfiche and in book form (Registry of Toxic Effects of Chemical Substances List) is an annual compilation of unvalidated toxicity data prepared by the National Institute for Occupational Safety and Health (NIOSH). There is no back file. RTECS contains toxicity data for approximately 40,000 substances. Threshold limit values, recommended standards in air, and aquatic toxicity data are also included in this file.

*CANCERLIT:* (formerly called Cancerline) is sponsored by National Institute of Health's National Cancer Institute and contains more than 240,000 references dealing with cancer research. All references have English abstracts. Over 3,500 U.S. and foreign journals, as well as selected monographs, meeting papers, reports and dissertations are abstracted for inclusion in CANCERLIT. Information on carcinogenic substances is included in CANCERLIT in greater depth than in other files.

**Chemical Substances Information Network (CSIN):** The Toxic Substances Control Act (TSCA), upon passage in 1976, became part of a complex array of at least 12 pieces of federal legislation concerned with chemicals and their mixtures. This legislation requires the private and public sectors to collect and analyze data and information relevant to the development, use and disposal of chemical substances. An analysis conducted by EPA identified several hundred information resources that these sectors were using when responding to these needs. It became obvious that a new approach was needed to reduce the burden on organizations for the identification, acquisition and processing of material relevant to chemical substances.

CSIN was developed to provide a methodology to efficiently and effectively identify, access and use data and information in diverse information resources. Subsequent to its initial development CSIN has further evolved to serve the needs of managers, scientists, science administrators, and engineers in industry, academe, and federal and state government whose responsibilities include the development, production, use, environmental fate and regulation, of chemical substances. CSIN can provide such data as nomenclature, molecular structure, physical-chemical properties, toxicology, production, control technology, economics, uses, as well as the development and/or interpretation of regulations and guidelines.

CSIN has been criticized for going beyond what some see as its legislative mandate. Legislation history seems to indicate that Congress intended the agencies to develop a system to access and organize data submitted under TSCA and other federal statutes and regulations concerned with health and the environment. The intent of the program was to make available all federal data in a form that would be useful, to EPA and other federal agencies, in their regulatory decisions.

The CSIN network as developed accesses government *and* private data systems. These private data systems are, naturally, beyond the agencies' control. While EPA can insist upon quality indicators for data entered by government agencies or departments, it cannot control the private data systems.

The question of data quality has not been solved, either by government or industry. CMA has recognized the importance of developing data quality indicators and has begun a cooperative program with government, academe and industry to accomplish this goal.

CSIN is supported by a technology which is the state of the computer science art and is described as distributed data base management. It provides users with the ability to access and process data in a large number of information resources that are geographically scattered. These information sources are independent and autonomous, with differing requirements for hardware, software, record format and data and information content. CSIN enables users to manage data search and report generation without requiring them to directly interact with each system individually.

The prototype CSIN is presently operational. CSIN administration has developed a group of organizational users, from the public and private sectors, to test the systems utility and process capabilities. This new technology, distributed data base management, holds great promise for enabling organizations to efficiently and effectively manage their own data and information systems. Inquiries regarding participation in this effort are welcome. Please contact Dr. Sidney Siegel, Administrator, Chemical Substances Information Network, Office of Toxics Integration, United States Environmental Protection Agency, Washington, DC 20460.

The following figures further describe the Chemical Substances Information System.

### Definition
- A network of coordinated online information systems concerning chemical substances

### Purpose
- Satisfy information requirements of toxic substances legislation and a broad spectrum of related activities

### Information Content
- Provides access to information on
  - Nomenclature and Composition
  - Properties
  - Production and Commerce
  - Products and Uses
  - Exposure
  - Effects
  - Studies and Research
  - Regulations and Controls

of Chemical Substances

Figure 2.1: CSIN concept.

| Without CSIN | With CSIN |
|---|---|
| User must know the languages and interfaces of many systems | User must know only one interface; queries and processing of replies done automatically |
| User must frequently reenter long lists of chemical identification and query terms | Chemical identification and query lists transmitted automatically |
| User must reformat lists for each system | Lists automatically reformatted for each system |
| User must be capable of selecting the appropriate data base and then must repeatedly connect to and log into systems | Connection and log-in can be done to systems and data bases selected by the user |
| Many chances for error | Few chances for error |

Figure 2.2: CSIN scripts function.

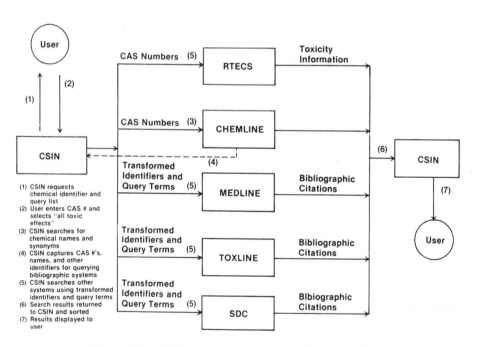

(1) CSIN requests chemical identifier and query list
(2) User enters CAS # and selects "all toxic effects"
(3) CSIN searches for chemical names and synonyms
(4) CSIN captures CAS #'s, names, and other identifiers for querying bibliographic systems
(5) CSIN searches other systems using transformed identifiers and query terms
(6) Search results returned to CSIN and sorted
(7) Results displayed to user

Figure 2.3: CSIN script operation, sample toxic effects script.

| Organization \ Data Base | AGLI | AGRICOLA | APILIT | APIPAT | BIOSIS | CA | CA PATENT | CANCERLIT | CANCERPROJ | CATLINE | CHEMDEX | CHEMNAME | DRVGINFO | EISIP | ENVIROLINE | FEDREG | NEISS | NOHS | PNI | PTS FOS | PTS FI | PTS IATS | PTS ISA | PTS PROMT | PTS USATS | PTS USSA | RTECS | SSIE | TDB | TOXLINE | TSCA | ··· Etc. |
|---|---|---|---|---|---|---|---|---|---|---|---|---|---|---|---|---|---|---|---|---|---|---|---|---|---|---|---|---|---|---|---|---|
| NCI | o | o | o | o | o | o | o | o | o | o | o | • | • | o | • | o | o | o |  | o | • | • | • | • | • | • | • | • | o | • | o | • |
| CDC |  | o | o | o | o | o | o | o | o | o | o | • | • |  |  | • | o | • |  | o | • | • | • | • | • | • | • | • | o | • | o | • |
| DOD | o | o | o | o | o | o | o | o | o | o | o | • | • | o | • |  | o | • |  | o | • | • | • | • | • | • | • | • |  | • |  | • |
| EPA/ORD | o | o | o | o | o | o | o | o | o | o | o | • | • | o | • | o | o | • |  | o | • | • | • | • | • | • | • | • | o | • | o | • |
| FDA/BD | o | o |  |  | o | o | o | o | o |  | • |  | • | o |  | • | o |  |  | • | • | • | • | • | • | • | • | • | • | • | o |  |  | o | • |
| SRI | o | o | o | o | o | o | o | o | o | o | o | • | • |  | • |  | • |  |  | • | • | • | • | • | • | • | • | • |  | • |  | • |
| OSHA | o | o | o | o | o | o | o | o | o | o | o | • | • |  | • |  | • |  |  | • | • | • | • | • | • | • | • | • |  | • |  | • |
| . . . Etc. |  |  |  |  |  |  |  |  |  |  |  |  |  |  |  |  |  |  |  |  |  |  |  |  |  |  |  |  |  |  |  |  |

• Primary Sources
o Secondary Sources

**Figure 2.4:** Chemical data bases used by surveyed organizations.

## REFERENCES

American Conference of Government Industrial Hygienists, Inc., *Documentation of Threshold Limit Values,* Cincinnati, Ohio (1971).

American Mutual Insurance Alliance, *Handbook of Hazardous Materials,* Chicago, Ill. (1974).

American Mutual Insurance Alliance, *Handbook of Organic Industrial Solvents,* Chicago, Ill. (1980).

Browning, E., *Toxicity and Metabolism of Industrial Solvents,* New York: Appleton-Century Crofts (1969).

Casarett, L. and Doull, J., *Toxicology: The Basic Science of Poisons,* New York: Macmillan (1980).

CHRIS Hazardous Chemical Data, U.S. Department of Transportation, U.S. Coast Guard, Washington, D.C. (1978).

Clayton, G. and Clayton, F.E., *Patty's Industrial Hygiene and Toxicology,* New York: Wiley-Interscience (1981).

*Fenaroli's Handbook of Flavor Ingredients,* Volume 1 and 2 (T.E. Furia and N. Bellancy, eds.) Second Ed., Cleveland: The Chemical Rubber Co. (1975).

Friberg, L., Nordberg, G. and Vouk, V., *Handbook on the Toxicology of Metals,* New York: Elsevier, North Holland (1979).

Goodman and Gilman's *Pharmacological Basis of Therapeutics* (A.G. Gilman, L.S. Goodman & L. Gilman, eds.) Sixth Ed., New York: Macmillan Publishing Co., Inc. (1980).

Gosselin, et al., *Clinical Toxicology of Commercial Products: Acute Poisoning,* Baltimore, Md. (1976).

Grant, W. Morton, *Toxicology of the Eye,* Springfield, Mo.: Charles C. Thomas (1974).

Haley, T.J. and Thienes, C., *Clinical Toxicology,* Fifth Ed., Philadelphia: Lea and Febiger (1972).

Hamilton, A. and Hardy, H., *Industrial Toxicology,* Acton, Mass.: Publishing Sciences Group, Inc.

Hawley, G., *Condensed Chemical Dictionary,* New York: Van Nostrand Reinhold (1981).

Hayes, W., *Toxicology of Pesticides,* Baltimore: Williams and Wilkins (1975).

Kruzas, *Encyclopedia of Information Systems and Services,* Gale Research Co., Detroit (1979).

Lefaux, R., *Practical Toxicology of Plastics,* Cleveland: CRC Press (1969).

Merck Index (Windholz, M., ed.) Rahway, N.J.: Merck & Co.

NIOSH, Catalog of NIOSH Publications, Cincinnati, Ohio (1980).

NIOSH/OSHA, *Pocket Guide to Chemical Hazards,* U.S. Dept. of Health, Education & Welfare, NIOSH DHEW Pub. No. 78-210.

Registry of Toxic Effects of Chemical Substances, U.S. Dept. of HEW-NIOSH, DHEW Pub. No. 79-100.

Sax, I., *Dangerous Properties of Industrial Materials,* Fifth Ed., New York: Van Nostrand Reinhold.

Sittig, M., *Handbook of Toxic and Hazardous Chemicals,* Park Ridge, N.J.: Noyes Data Corporation (1981).

Sunshine, I. (ed.) *Methodology for Analytical Toxicology,* Cleveland: CRC Press, Inc. (1975).

U.S. Dept. of Commerce, NTIS Current Published Searches, Springfield, Va. (1980).

U.S. Dept. of Commerce, NTIS General Catalog 7a, Springfield, Va. (1981).

U.S. Government Printing Office, Monthly Catalog, Washington, D.C.

U.S. Government Printing Office, Codes of Federal Regulations, Titles 1-50, Washington, D.C.

Weiss, G., *Hazardous Chemicals Data Book,* Park Ridge, N.J.: Noyes Data Corporation (1980).

# 3

# Transportation Labels and Placards: Technology

**Harry Fund**
*Labelmaster*
*Chicago, IL*

## INTRODUCTION

It is difficult to believe the use of labels to identify packaging containing hazardous materials was required by the railroads as far back as 1910. Then, they were prescribed by the Bureau of Explosives. Ultimately, these labels were adopted by the Interstate Commerce Commission (ICC).

The labels known today date back to April 1967 with the establishment of the Department of Transportation (DOT) and its Office of Hazardous Materials Operations as the regulator for the transportation of hazardous materials.

While the original diamond shape of the label was retained, the content was changed dramatically. The old ICC labels were heavily worded, consisting of phrases such as "Do Not Drop" or "Don't Shake." In addition, there was a Certification on the label so each package had to be signed. These ICC labels bear very little resemblance to the comparatively simplified "symbol" label now in use.

This chapter will concern itself with all of these labels and placards. It will examine the state of the art today and attempt to predict label and placard appearance in the future.

Most readers are aware of the international relationships which influence, not only the specification of such labels and placards, but of all markings, packaging and documentation as well. In fact, the impact of these international regulations has been so noticeable during the past few years it is probably safe to predict the regulations of any one nation will soon be a carbon copy of those of all other nations. Accordingly, an examination of the origin of these regulations would be helpful.

This internationalization of hazardous materials regulations, or at least the impact of its importance, is of recent origin. Up until the last dozen years each country went its own way; there are still some major developed nations without regulatory statutes in place. However, as in all other areas of international trade, the amount of chemicals shipped became so great, and thus the potential hazards so significant, that the inherent requirements for standardized markings could no longer be denied.

It should be stated here, that there are many enterprises within the United States strongly resentful of the imposition of international standards on U.S. regulations which they consider to be purely domestic matters. They argue, not without some justification, that international trade represents just ten percent of the movement of hazardous materials and, therefore, should not be permitted to influence U.S. regulations so strongly.

The response from our government appears to be that this ten percent represents tens of thousands of shipments; and, to have different sets of regulations for international and domestic shipments is completely impractical and apparently, unacceptable as further actions are proposed. (The most apparent aberration resulting from this controversy is the existence of two commodity lists [49CFR 172.101 and 172.102]. DOT's recent attempt to authorize either table for use of the proper shipping name and United Nations ID Number was strongly objected to by the railroads. As a result, and at the date of this writing, 172.102 is to be used for international shipments only.)

Whichever side of this volatile argument with which one identifies, it appears a single set of regulations is preferable from the point of view of the chemical manufacturer who ships both domestically and internationally. With regard to the emergency response group requiring the information contained on a label, a placard, or a shipping paper, an accident is an accident regardless of place of origin of the material. Consider their problem in not knowing whether the UN Identification Number shown has been derived from 172.101 or 172.102!

In approaching a discussion of the international mechanism, one must only remember it was not too long ago each of the U.S. transportation modes (Federal Railroad Administration, Coast Guard, Federal Aviation Administration and Federal Highway Administration) issued their own regulations. It is a matter of some pride that our nation now has a single set of statutes regarding the matter of markings. Considering the great difficulties encountered in achieving intermodal agreement, one can imagine the obstacles standing in the way of international agreements. Commenting objectively, the progress has been remarkable. While still imperfect in many ways, a full set of reference material does exist.

As the international body is now constituted, the senior representatives of the world's major nations meet biannually in Geneva under the auspices of the United Nations Economic and Social Council (ECOSOC). In addition, there is ongoing work by a number of standing subcommittees, such as those on packaging, explosives, etc. (See Figure 3.1). Known in the past as the UN Group of Rapporteurs, these representatives continually examine, amend and expand their recommendations having to do with every aspect of the carriage of hazardous materials. These are contained in a book known as the "Trans-

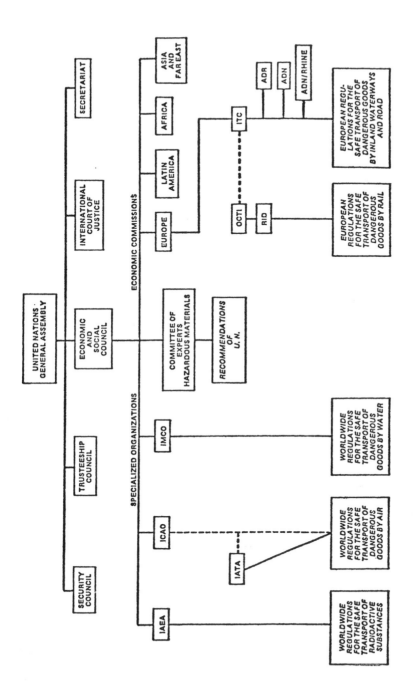

Figure 3.1: Major organizations involved in international regulations for transporting hazardous materials.

65

port of Dangerous Goods: Recommendations Prepared by the United Nations Committee of Experts." More popularly known as the UN Orange Source Book, this work is the seed material from which all nations seek their guidance in proposing their own regulations. It is also the reference for the two international regulators by sea and air respectively, IMO and ICAO.

IMO, acronym for the International Maritime Organization, has for many years published a full set of definitive regulations for the shipment of Dangerous Goods by Sea. ICAO, acronym for the International Civil Aviation Organization, has just issued its third complete edition of Technical Instructions for the Safe Transport of Dangerous Goods by Air.

To summarize, all governments, IMO and ICAO issue their regulations based on the recommendations of the UN Committee of Experts. In this manner, it is hoped there will emerge a standardized format for labels and placards in all matters having to do with the transport of dangerous goods.

## LABEL TECHNOLOGY

The first labels used were of the gummed adhesive type which had to be wetted prior to use. More recently, the pressure sensitive label has gained the most prominence. Two types of labels are most frequently offered. The first is a 60 lb paper using an acrylic type pressure sensitive adhesive backing. These are usually produced by the flexographic roll printing method, and shipped in rolls of 500. Such labels are convenient and easily applied. The second type is a vinyl label, usually 4 mil, which has been screen processed using acrylic type paints or via the new UV screen printing technology. These latter labels, while more expensive than the paper variety, may be relied upon to withstand several years of outdoor exposure. Other label durability enhancements exist. Special chemically resistant clear finishes or film laminates may be used when such additional specifications are required.

The UV technology alluded to above holds out the promise of even more superior labels. In this process, ultraviolet cured inks are used as the printing medium. These are cured by exposure to industrial type UV mercury vapor lamps in either a non-inert system (air environment) or in an inert (oxygen-free) system. The result is excellent adhesion to vinyls, polyesters, polycarbonates, butyrates and acetates; and a cured elongation of between 20 and 40% to avoid embrittlement of the substrate. Initial testing has exhibited little sign of degradation and excellent chemical resistance after three years of exterior exposure. Such UV labels as well as other techniques should be considered whenever use demands extended label performance.

In this context, there have been a number of legal actions involving the use of paper labels on 55 gallon drums. In each instance, labels had been subjected to prolonged outdoor storage and had completely faded. The subsequent misuse of the drums resulted in accidents. Accidents resulting from the absence of such warning labels obviously become a greater possibility. The diversion of at least some of the millions of chemical drums into other uses such as trash cans or make-shift barbecue grills, or their reuse with other chemicals without purging, is well known. In such circumstances, it always seems that the only iden-

## § 172.407  Label specifications.

(a) Each label, affixed to or printed on a package must be durable and weather resistant. Black and any color on a label must be able to withstand, without substantial change—

(1) A 72-hour fadeometer test (for a description of equipment designed for this purpose, see ASTM G 23-69 (1975), or ASTM G 26-70); and

(2) A 30-day exposure to conditions incident to transportation that reasonably could be expected to be encountered by the labeled package.

(b) Each diamond (square-on-point) label prescribed in this part must be at least 4 inches (101 mm.) on each side with each side having a black solid line border ¼-inch (6.3 mm.) from the edge.

(c) Except for size and color, the printing, inner border, and symbol on each label must be as shown for each label.

(d) A color on a label, upon visual examination, must fall within the color tolerances displayed on the appropriate Office of Hazardous Materials Label and Placard Color Tolerance Chart.

(1) A set of six charts, dated January 1973, for comparison with labels and placards surfaced with paint, lacquer, enamel, plastic or other opaque coatings, or ink, may be purchased from the Office of Hazardous Materials Regulation, U.S. Department of Transportation, Washington, D.C. 20590, for $5.50.

(2) A set of six charts, dated January 1974, for comparison with labels and placards surfaced with ink, may be similarly purchased for $12.50.

(3) Both sets of charts may be inspected in Room 8426, Nassif Building, 400 7th Street S.W., Washington, D.C. 20590, or any of the offices of the Federal Highway Administration listed at 49 CFR 390.40.

(4) The technical specifications for each chart are set forth in Appendix A to this Part.

(5) The requirements of paragraph (d) of this section do not apply to labels printed directly onto the surface of a packaging before July 1, 1979: *Provided,* The colors of such labels comply with the appropriate colors described in §§ 172.411 through 172.450. Such labels printed on or after July 1, 1979, must comply with color specifications in one of the appropriate tables in Appendix A to this part.

(e) The specified label color must extend to the edge of the label in the area designated on each label except the CORROSIVE, MAGNETIZED MATERIAL, RADIOACTIVE YELLOW-II, and RADIOACTIVE YELLOW-III labels.

(f) A label may contain form identification information, including the name of its maker, provided that information is printed outside of the solid line inner border in no larger than 10-point type.

(g) A label may contain the UN and (IMCO) hazard class number and, when appropriate, the division number. The number must be—

(1) Black, unless it is on a CORROSIVE label when it must be white, or unless other colors are authorized by this Part.

(2) Located in the lower corner of the label, and

(3) One-half inch (12.7 mm.) or less in height.

(h) For import shipments only, a label conforming to the requirements of IMCO or the United Nations Recommendations affixed to a package in another country may contain inscriptions required by the country of origin.

(i) The dotted line border shown on each label is not part of the label specification, except when used as an alternative for the solid line outer border to meet the requirements of § 172.406(d).

(j) EXPLOSIVE A, EXPLOSIVE B, and EXPLOSIVE C labels may bear inscriptions in addition to those prescribed in this subpart, if required for import or export purposes.

(49 U.S.C. 1808)

[Amdt. 172-29, 41 FR 15996, Apr. 15, 1976, as amended by Amdt. 172-29A, 41 FR 40679, Sept. 20, 1976]

the color tolerances specified in Table 3. Central colors and tolerances described in Table 2 approximate those described in Table 1 while allowing for differences in production methods and materials used to manufacture labels and placards surfaced with printing inks. Primarily, the color charts based on Table 1 are for label or placard colors applied as opaque coatings such as paint, enamel or plastic, whereas color charts based on Table 2 are intended for use with labels and placards surfaced only with inks.

For labels printed directly on packaging surfaces, Table 3 may be used, although compliance with either Table 1 or Table 2 is sufficient. However, if visual reference indicates that the colors of labels printed directly on package surfaces are outside the Table 1 or 2 tolerances, a spectrophotometer or other instrumentation may be required to insure compliance with Table 3.

APPENDIX A—OFFICE OF HAZARDOUS MATERIALS REGULATION COLOR TOLERANCE CHARTS AND TABLES

The following are Munsell notations and Commission Internationale de L'Eclariage (CIE) coordinates which describe the Office of Hazardous Materials Regulation Label and Placard Color Tolerance Charts in Tables 1 and 2, and the CIE coordinates for

TABLE 1—SPECIFICATIONS FOR COLOR TOLERANCE CHARTS FOR USE WITH LABELS AND PLACARDS SURFACED WITH PAINT, LACQUER, ENAMEL, PLASTIC, OTHER OPAQUE COATINGS, OR INK [1]

| Color | Munsell notations | CIE data for source C | | |
|---|---|---|---|---|
| | | Y | x | y |
| **Red:** | | | | |
| Central color | 7.5R 4.0/14 | 12.00 | .5959 | 3269 |
| Orange | 8.5R 4.0/14 | 12.00 | .6037 | 3389 |
| Purple and vivid | 6.5R 4.0/14 | 12.00 | .5869 | .3184 |
| Grayish | 7.5R 4.0/12 | 12.00 | .5603 | .3321 |
| Vivid | 7.5R 4.0/16 | 12.00 | .6260 | .3192 |
| Light | 7.5R 4.5/14 | 15.57 | .5775 | .3320 |
| Dark | 7. 5R 3.5/14 | 09.00 | .6226 | .3141 |
| **Orange:** | | | | |
| Central color | 5.OYR 6.0/15 | 30.05 | .5510 | .4214 |
| Yellow and Grayish | 6.25YR 6.0/15 | 30.05 | .5452 | .4329 |
| Red and vivid | 3.75YR 6.0/15 | 30.05 | .5552 | .4091 |
| Grayish | 5.OYR 6.0/13 | 30.05 | .5311 | .4154 |
| Vivid | 5.OYR 6.0/16 | 30.05 | .5597 | .4239 |
| Light | 5.OYR 6.5/15 | 36.20 | .5427 | .4206 |
| Dark | 5.OYR 5.5/15 | 24.58 | .5606 | .4218 |
| **Yellow:** | | | | |
| Central color | 5.OY 8.0/12 | 59.10 | .4562 | .4788 |
| Green | 6.5Y 8.0/12 | 59.10 | .4498 | .4865 |
| Orange and vivid | 3.5Y 8.0/12 | 59.10 | .4632 | .4669 |
| Grayish | 5.OY 8.0/10 | 59.10 | .4376 | .4601 |
| Vivid | 5.OY 8.0/14 | 59.10 | .4699 | .4920 |
| Light | 5.OY 8.5/12 | 68.40 | .4508 | .4754 |
| Dark | 5.OY 7.5/12 | 50.68 | .4620 | .4823 |
| **Green:** | | | | |
| Central color | 7.5G 4.0/9 | 12.00 | .2111 | .4121 |
| Bluish | 0.5BG 4.0/9 | 12.00 | .1974 | .3809 |
| Green-yellow | 5.0G 4.0/9 | 12.00 | 2237 | .4399 |
| Grayish A | 7.5G 4.0/7 | 12.00 | .2350 | .3922 |
| Grayish B [2] | 7.5G 4.0/6 | 12.00 | .2467 | .3822 |
| Vivid | 7.5G 4.0/11 | 12.00 | .1848 | .4319 |
| Light | 7.5G 4.5/9 | 15.57 | .2204 | .4060 |
| Dark | 7.5G 3.5/9 | 09.00 | .2027 | .4163 |
| **Blue:** | | | | |
| Central color | 2.5PB 3.5/10 | 09.00 | .1691 | .1744 |
| Purple | 4.5PB 3.5/10 | 09.00 | .1796 | .1711 |
| Green and vivid | 10.0B 3.5/10 | 09.00 | .1557 | .1815 |
| Grayish | 2.5PB 3.5/8 | 09.00 | .1888 | .1964 |
| Vivid | 2.5PB 3.5/12 | 09.00 | .1516 | .1547 |
| Light | 2.5PB 4.0/10 | 12.00 | .1805 | 1888 |
| Dark | 2.5PB 3.0/10 | 06.55 | .1576 | .1600 |

TABLE 1—SPECIFICATIONS FOR COLOR TOLERANCE CHARTS FOR USE WITH LABELS AND PLACARDS SURFACED WITH PAINT, LACQUER, ENAMEL, PLASTIC, OTHER OPAQUE COATINGS, OR INK [1]—Continued

| Color | Munsell notations | CIE data for source C | | |
|---|---|---|---|---|
| | | Y | x | y |
| Purple: | | | | |
| Central color | 10.0P 4.5/10 | 15.57 | .3307 | .2245 |
| Reddish purple | 2.5RP 4.5/10 | 15.57 | .3584 | .2377 |
| Blue purple | 7.5P 4.5/10 | 15.57 | .3068 | .2145 |
| Reddish gray | 10.0P 4.5/8 | 15.57 | .3280 | .2391 |
| Gray [2] | 10.0P 4.5/6.5 | 15.57 | .3254 | .2519 |
| Vivid | 10.0P 4.5/12 | 15.57 | .3333 | .2101 |
| Light | 10.0P 5.0/10 | 19.77 | .3308 | .2328 |
| Dark | 10.0P 4.0/10 | 12.00 | .3306 | .2162 |

[1] Maximum chroma is not limited.
[2] For the colors green and purple, the minimum saturation (chroma) limits for porcelain enamel on metal are lower than for most other surface coatings. Therefore, the minimum chroma limits of these two colors as displayed on the Charts for comparison to porcelain enamel on metal is low, as shown for green (grayish B) and purple (gray).

NOTE: CIE = Commission Internationale de L'Eclairage.

TABLE 2—SPECIFICATIONS FOR COLOR TOLERANCE CHARTS FOR USE WITH LABELS AND PLACARDS SURFACED WITH INK

| Color/series | Munsell notation | CIE data for source C | | |
|---|---|---|---|---|
| | | Y | x | y |
| Red: | | | | |
| Central series: | | | | |
| Central color | 6.8R 4.47/12.8 | 15.34 | .5510 | .3286 |
| Grayish | 7.2R 4.72/12.2 | 17.37 | .5368 | .3348 |
| Purple | 6.4R 4.49/12.7 | 15.52 | .5442 | .3258 |
| Purple and vivid | 6.1R 4.33/13.1 | 14.25 | .5529 | .3209 |
| Vivid | 6.7R 4.29/13.2 | 13.99 | .5617 | .3253 |
| Orange | 7.3R 4.47/12.8 | 15.34 | .5572 | .3331 |
| Orange and grayish | 7.65R 4.70/12.4 | 17.20 | .5438 | .3382 |
| Light series: | | | | |
| Light | 7.0R 4.72/13.2 | 17.32 | .5511 | .3322 |
| Light and orange | 7.4R 4.96/12.6 | 19.38 | .5365 | .3382 |
| Light and purple | 6.6R 4.79/12.9 | 17.94 | .5397 | .3289 |
| Dark series: | | | | |
| Dark A | 6.7R 4.19/12.5 | 13.30 | .5566 | 3265 |
| Dark B | 7.0R 4.25/12.35 | 13.72 | .5522 | .3294 |
| Dark and purple | 7.5R 4.23/12.4 | 13.58 | .5577 | .3329 |
| Orange: | | | | |
| Central series: | | | | |
| Central color | 5.0YR 6.10/12.15 | 31.27 | .5193 | .4117 |
| Yellow and grayish A | 5.8YR 6.22/11.7 | 32.69 | .5114 | .4155 |
| Yellow and grayish B | 6.1YR 6.26/11.85 | 33.20 | .5109 | .4190 |
| Vivid | 5.1YR 6.07/12.3 | 30.86 | .5226 | .4134 |
| Red and vivid A | 3.9YR 5.87/12.75 | 28.53 | .5318 | .4038 |
| Red and vivid B | 3.6YR 5.91/12.6 | 29.05 | .5291 | 4021 |
| Grayish | 4.9YR 6.10/11.9 | 31.22 | .5170 | .4089 |
| Light series: | | | | |
| Light and vivid A | 5.8YR 6.78/12.7 | 39.94 | .5120 | .4177 |
| Light and yellow | 6.0YR 6.80/12.8 | 40.20 | .5135 | .4198 |
| Light and vivid B | 4.9YR 6.60/12.9 | 37.47 | .5216 | .4126 |
| Dark series: | | | | |
| Dark and yellow | 5.8YR 5.98/11.0 | 29.87 | .5052 | .4132 |
| Dark A | 5.1YR 5.80/11.1 | 27.80 | .5127 | .4094 |
| Dark B | 5.0YR 5.80/11.0 | 27.67 | .5109 | .4068 |
| Yellow: | | | | |
| Central series: | | | | |
| Central color | 4.3Y 7.87/10.3 | 56.81 | .4445 | .4589 |
| Vivid A | 4.5Y 7.82/10.8 | 55.92 | .4503 | .4658 |
| Vivid B | 3.3Y 7.72/11.35 | 54.24 | .4612 | .4624 |
| Vivid and orange | 3.2Y 7.72/10.8 | 54.25 | .4576 | .4572 |
| Grayish A | 4.1Y 7.95/9.7 | 58.18 | .4380 | .4516 |
| Grayish B | 5.1Y 8.06/9.05 | 60.12 | .4272 | .4508 |
| Green-yellow | 5.2Y 7.97/9.9 | 58.53 | .4356 | .4605 |

TABLE 2—SPECIFICATIONS FOR COLOR TOLERANCE CHARTS FOR USE WITH LABELS AND
PLACARDS SURFACED WITH INK—Continued

| Color/series | Munsell notation | CIE data for source C | | |
|---|---|---|---|---|
| | | Y | x | y |
| Light series: | | | | |
|   Light | 5.4Y 8.59/10.5 | 70.19 | .4351 | .4628 |
|   Light and green-yellow | 5.4Y 8.56/11.2 | 69.59 | .4414 | .4692 |
|   Light and vivid | 4.4Y 8.45/11.4 | 67.42 | .4490 | .4662 |
| Dark series: | | | | |
|   Dark and green-yellow | 4.4Y 7.57/9.7 | 51.82 | .4423 | .4562 |
|   Dark and orange A | 3.4Y 7.39/10.4 | 48.86 | .4584 | .4590 |
|   Dark and orange B | 3.5Y 7.41/10.0 | 49.20 | .4517 | .4544 |
| Green: | | | | |
|   Central series: | | | | |
|     Central color | 9.75G 4.26/7.75 | 13.80 | .2214 | .3791 |
|     Grayish | 10G 4.46/7.5 | 15.25 | .2263 | .3742 |
|     Blue A | 1.4BG 4.20/7.4 | 13.36 | .2151 | .3625 |
|     Blue B | 1.0BG 4.09/7.75 | 12.60 | .2109 | .3685 |
|     Vivid | 8.4G 4.09/8.05 | 12.59 | .2183 | .3954 |
|     Vivid green-yellow | 7.0G 4.23/9.4 | 13.54 | .2292 | .4045 |
|     Green-yellow | 7.85G 4.46/7.7 | 15.23 | .2313 | .3914 |
|   Light series: | | | | |
|     Light and vivid | 9.5G 4.45/8.8 | 15.21 | .2141 | .3863 |
|     Light and blue | 0.2BG 4.31/8.8 | 14.12 | .2069 | .3814 |
|     Light and green-yellow | 8.3G 4.29/9.05 | 14.01 | .2119 | .4006 |
|   Dark series: | | | | |
|     Dark and green-yellow | 7.1G 4.08/7.1 | 12.55 | .2354 | .3972 |
|     Dark and grayish | 9.5G 4.11/6.9 | 12.70 | .2282 | .3764 |
|     Dark | 8.5G 3.97/7.2 | 11.78 | .2269 | .3874 |
| Blue: | | | | |
|   Central series: | | | | |
|     Central color | 3.5PB 3.94/9.7 | 11.58 | .1885 | .1911 |
|     Green and grayish A | 2.0PB 4.35/8.7 | 14.41 | .1962 | .2099 |
|     Green and grayish B | 1.7PB 4.22/9.0 | 13.50 | .1898 | .2053 |
|     Vivid | 2.9PB 3.81/9.7 | 10.78 | .1814 | .1852 |
|     Purple and vivid A | 4.7PB 3.53/10.0 | 9.15 | .1817 | .1727 |
|     Purple and vivid B | 5.0PB 3.71/9.9 | 10.20 | .1888 | .1788 |
|     Grayish | 3.75PB 4.03/9.1 | 12.17 | .1943 | .1961 |
|   Light series: | | | | |
|     Light and green A | 1.7PB 4.32/9.2 | 14.22 | .1904 | .2056 |
|     Light and green B | 1.5PB 4.11/9.6 | 12.72 | .1815 | .1971 |
|     Light and vivid | 3.2PB 3.95/10.05 | 11.70 | .1831 | .1868 |
|   Dark series: | | | | |
|     Dark and grayish | 3.9PB 4.01/8.7 | 12.04 | .1982 | .1992 |
|     Dark and purple A | 4.8PB 3.67/9.3 | 9.95 | .1918 | .1831 |
|     Dark and purple B | 5.2PB 3.80/9.05 | 10.76 | .1985 | .1885 |
| Purple: | | | | |
|   Central series: | | | | |
|     Central color | 9.5P 4.71/11.3 | 17.25 | .3274 | .2165 |
|     Red | 1.0RP 5.31/10.8 | 22.70 | .3404 | .2354 |
|     Red and vivid A | 1.4RP 5.00/11.9 | 19.78 | .3500 | .2274 |
|     Red and vivid B | 0.2RP 4.39/12.5 | 14.70 | .3365 | .2059 |
|     Vivid | 8.0P 4.04/12.0 | 12.23 | .3098 | .1916 |
|     Blue | 7.0P 4.39/10.8 | 14.71 | .3007 | .2037 |
|     Grayish | 8.8P 5.00/10.3 | 19.73 | .3191 | .2251 |
|   Light series: | | | | |
|     Light and red A | 0.85RP 5.56/11.1 | 25.18 | .3387 | .2356 |
|     Light and red B | 1.1RP 5.27/12.3 | 22.27 | .3460 | .2276 |
|     Light and vivid | 9.2P 4.94/11.95 | 19.24 | .3247 | .2163 |
|   Dark series: | | | | |
|     Dark and grayish | 9.6P 4.70/10.9 | 17.19 | .3283 | .2204 |
|     Dark and vivid | 8.4P 4.05/11.6 | 12.35 | .3144 | .1970 |
|     Dark and blue | 7.5P 4.32/10.5 | 14.19 | .3059 | .2078 |

TABLE 3—SPECIFICATION FOR COLORS FOR USE WITH LABELS PRINTED ON PACKAGINGS SURFACES

| CIE data for source C | Red | Orange | Yellow | Green | Blue | Purple |
|---|---|---|---|---|---|---|
| x | .424 | .460 | .417 | .228 | .200 | .377 |
| y | .306 | .370 | .392 | .354 | .175 | .205 |
| x | .571 | .543 | .490 | .310 | .255 | .377 |
| y | .306 | .400 | .442 | .354 | .250 | .284 |
| x | .424 | .445 | .390 | .228 | .177 | .342 |
| y | .350 | .395 | .430 | .403 | .194 | .205 |
| x | .571 | .504 | .440 | .310 | .230 | .342 |
| y | .350 | .430 | .492 | .403 | .267 | .284 |
| Y (high) | 23.0 | 41.6 | 72.6 | 20.6 | 15.9 | 21.2 |
| Y (low) | 7.7 | 19.5 | 29.1 | 7.4 | 6.5 | 8.2 |

(49 U.S.C. 1803,1804,1808, 49 CFR 1.53)

[Amdt. 172-50, 44 FR 9757, Feb. 15, 1979; Amdt. 172-50, 44 FR 10984, Feb. 26, 1979, as amended by Amdt. 172-50, 44 FR 22467, Apr. 16, 1979]

tification remaining on the label is the corporate name responsible for the drum's original use. It is to that corporation the lawyers look for damages, even though the cause of the accident was four or five times removed from the original use.

This is perhaps one of the most serious problems facing both industry and government as regulators. The durability requirements set by government are many times established at the lowest level. This is, no doubt, due to the pressures exerted by "economic impact." On the other hand, the consequences of a civil or criminal suit far exceed the economic impact of the cost difference between a paper and vinyl label. The same may be said for placards.

Some recognition of the need to discriminate as regards durability may be found in a new IMO requirement, 7.2.2. from amendment 16-78:

> The method of marking the label and correct technical name
> on packages containing dangerous substances should be such
> that this information will still be identifiable on packages
> surviving *at least three months' immersion in the sea.* In con-
> sidering suitable marking methods, account should be taken
> of the durability of the materials used and of the surface of
> the package. (our italics)

Perhaps this accounts for the success of vinyl *Empty Drum* warning labels introduced two years ago. The message appears to be clear. Design the labels for the durability required. There is no reason a paper label should not be used on corrugated when one knows it will never be used outside. However, more careful consideration should be given to drums, and even five gallon containers, where the possibility for exterior exposure or label-destroying spills exists.

In these instances, one should not wait for government specifications to improve. Rather, a strict course of determining use requirements should be followed and acted upon.

## Labels Other than DOT

In May of 1980, DOT and EPA issued combined regulations for the marking of hazardous waste and hazardous substances. Generators of such waste should be particularly careful in the type of label they specify. 49CFR 172.304 states in part " . . . must be durable." An interpretation from both DOT and EPA is that the marking be durable and permanent when exposed to weather extremes during extended on-site storage, transportation and disposal. Paper labels are obviously not recommended, but the term "must be durable" requires further explanation & specification.

## Markings

DOT, IMO and ICAO regulations now require each container of dangerous goods to be marked with the proper shipping name and its UN number. One may wish to give some consideration to incorporating these requirements in the same labels containing the class symbols. Such combinations eliminate the use of separate label and marking.

### § 172.304   Marking requirements.

(a) The marking required in this subpart—(1) Must be durable, in English and printed on or affixed to the surface of a package or on a label, tag, or sign.

(2) Must be displayed on a background of sharply contrasting color;

(3) Must be unobscured by labels or attachments; and

(4) Must be located away from any other marking (such as advertising) that could substantially reduce its effectiveness.

[Amdt. 172-29, 41 FR 15996, Apr. 15, 1976, as amended by Amdt. 172-29B, 41 FR 57067, Dec. 30, 1976]

### § 172.308   Authorized abbreviations.

(a) Abbreviations may not be used in a proper shipping name marking except in the following instances—

(1) For marking descriptions of ammunition, such as Ammunition for cannon without projectile, etc., the words "with" or "without" may be abbreviated as "W" or "W/O". For example: "Ammunition for cannon W/O projectile."

(2) The abbreviation "ORM" may be used in place of the words "Other Regulated Material."

[Amdt. 172-101, 45 FR 74666, Nov. 10, 1980]

## PLACARDS

The history of placards, while more recent, parallels that of labels. However, the changes have been far more dynamic. It is most assuredly an area where additional amendments may be expected. The concept of identifying vehicles transporting dangerous goods is an obvious one . . . it is the method which can most quickly communicate the contents to an emergency response group.

Again, the latest change requiring the display of the four-digit identification number began with the United Nations Committee of Experts. It was their determination that while a worded placard might be sufficient for dry freight shipments, full loads contained in cargo tanks and tank cars represented a much greater hazard. Therefore, a solution which would more specifically identify that hazard was required. Several years were spent in examination and comparison of many different systems. The final selection was the use of the four-digit UN identification number. It must be assumed this system was chosen primarily because this number is common to all regulations; and secondly, because it acts as a "fingerprint" identification of that specific commodity or group of commodities.

Except for *Explosives, Poison Gases* and *Radioactives* in large quantities, the shipper has been given the option of displaying this number on the diamond-shaped placard in place of words, or on a rectangular orange panel alongside the worded placard. As of this time orders seem to indicate that the diamond-shaped placard is the overwhelming winner. As of this writing, such placards are now mandatory in the U.S. for tank shipments, and as of May 1, 1982 in Canada for rail tank shipments. They have also been proposed for IMO shipments but not yet regulated. The Identification Number system has resulted in the issuance of an Emergency Response Guidebook by the Department of Transportation, a revised edition of which will be available in late fall of 1983.

## Placard Technology

The *minimum* specification required by DOT for placard construction is 125 pounds per ream of 24″ × 36″ white tagboard with the ability to pass a 60 psi Mullen test, and the ability to withstand open weather exposure for 30 days without a substantial reduction in effectiveness. Again, some shippers and transporters have determined this specification to be much too risky for their longer durability requirements and have sought out more durable materials for this purpose. Indeed, a number of alternatives do exist.

The most common device for displaying tagboard placards, particularly on tank cars, is a slide holder. The holder is manufactured of either aluminum or heavy duty steel, and permits the insertion and removal of a tagboard placard. More recently, a 20 mil rigid vinyl has been introduced which is far more durable than tagboard and may be reused for dozens of trips.

One of the most popular substitutes for tagboard has been a 4 mil flexible vinyl with a removable pressure sensitive adhesive. This material has the virtue of complete exterior durability, yet it can be removed easily, when required, without leaving a residue. A slide holder is not required. Instead, these placards are affixed directly on the tank.

With regard to dry freight, the majority of trucks and trailers have been equipped with what has come to be known as a permanent flip-placard system. Provided in aluminum for highway use, and in heavy duty steel for rail use, these assemblies are made up of all the placards, and simply flip from one to the other as required.

### § 172.519 General specifications for placards.

(a) A placard may be made of any plastic, metal, or other material that is equal to or better in strength and durability than the tagboard specified in paragraph (b) of this section. Also, reflective or retroreflective materials may be used on a placard providing the prescribed colors, strength and durability are maintained.

(b) A placard made of tagboard must be of material that has—

(1) A quality at least equal to that designated commercially as white tagboard;

(2) A weight of 125 pounds per ream of 24 by 36-inch sheets;

(3) The ability to pass a 60 p.s.i. Mullen test; and

(4) The ability to withstand open weather exposure for 30 days without a substantial reduction in effectiveness.

(c) A placard may contain form identification information, including the name of its maker if that information is printed in the outer ½-inch (12.7 mm.) border in no larger than 10-point type.

(d) The hazard class and division number prescribed for dangerous goods in the UN Recommendations titled "Transport of Dangerous Goods" may be entered in the lower corner of the diamond on each placard. If a placard is used to display identification numbers as authorized by § 172.332, the class number must be entered in a numeral approximately 1¾ inches (45 mm.) in height (numeral height may be between 1⅝ inches (41 mm.) and 1¾ inches (45 mm.)). It must be black on each placard except when on a NON-FLAMMABLE GAS, FLAMMABLE GAS, FLAMMABLE, COMBUSTIBLE or CORROSIVE placard. The class number on a NON-FLAMMABLE GAS, FLAMMABLE GAS, FLAMMABLE and COMBUSTIBLE pacard may be white or black.

The class number on a CORROSIVE placard must be white, and on a COMBUSTIBLE placard with a white bottom as prescribed by § 172.332(c)(4), the class number must be red or black.

(e) Surface pigmentation on a placard must meet the following requirements:

(1) Black and any color must be able to withstand, without substantial change—

(i) A 72-hour fadeometer test (for a description of equipment designed for this purpose, see ASTM G 23-69 (1975), or ASTM G 26-70); and

(ii) A 30-day exposure to open weather conditions.

(2) A color on a placard, upon visual examination, must fall within the color tolerances displayed on the appropriate Office of Hazardous Materials Label and Placard Color Tolerance Chart (see § 172.407(d)).

(f) Except as provided in § 172.334, placards shall be as described in this section and as prescribed in Appendix B to this Part.

(g) The dotted line at the outside of the ½-inch (12.7 mm.) white border on each placard is not part of the placard specification. However, a dotted or solid line outer border may be used when needed to indicate the full size of a placard that is part of a larger format or is on a background the color of which does not contrast with the placard color.

(49 U.S.C. 1803, 1804, 1808; 49 CFR 1.53(e))

[Amdt. 172-29, 41 FR 15996, Apr. 15, 1976, as amended by Amdt. 172-29A, 41 FR 40680, Sept. 20, 1976; Admt. 172-37, 42 FR 34285, July 5, 1977; Amdt. 172-58, 45 FR 34702, May 22, 1980; Amdt. 172-58, 45 FR 74668, Nov. 10, 1980]

APPENDIX B—DIMENSIONAL
SPECIFICATION FOR PLACARDS

1. *Placard specifications.* (a) The print
type on each placard must be Franklin
Gothic Condensed.

(b) Each square-on-point placard must
measure 10¾ inches (273.0 mm.) on each
side, the outer, ½-inch (12.7 mm.) of which
must be white.

NOTE: The measurements in these specifications may be rounded to the nearest ¹⁄₃₂ of
an inch and to the nearest whole millimeter.

(c) Specifications for each placard to aug-
ment those in paragraphs (1)(a) and (b) of
this Appendix and those contained in Subpart F, Part 172 of this subchapter are as
follows:

(1) *DANGEROUS placard.* The word
"DANGEROUS" must be across the center
of the placard and made with letters ²⁷⁄₃₂
inches (56.4 mm.) high with a ⅜-inch (9.5
mm.) stroke. The white section of the placard must be centered across the placard and
5 inches (127 mm.) wide. The two ends of
the white area must have an ⅛-inch (3.2
mm.) red solid line border to indicate the
outer ½-inch (12.7 mm.) white placard
border. The placard color must be red,
white, and black.

(2) *EXPLOSIVES A placard.* The word
"EXPLOSIVES" must be across the center
area of the placard and made with letters
1⅞ inches (47.6 mm.) high with a ⁵⁄₁₆-inch
(7.9 mm.) stroke. The top of the letters in
the words "EXPLOSIVES" must be 1½
inches (38.1 mm.) above the placard horizontal center line. The top of the letter "A"
must be ½-inch (12.7 mm.) below the horizontal center line. The letter "A" must be
approximately 2 inches (50.8 mm.) high.
The base of the symbol must be 2¹⁄₁₆ inches
(52.4 mm.) above the placard horizontal

center line, and must be 4½ inch (114.3
mm.) high when measured from a horizontal line touching the lowest extremity. The
width of the symbol must be 7⅞ inches
(200.0 mm.) when measured between two
lines perpendicular to the base line and
touching the widest extremities on each
side. The radius of the bomb must be ¾-
inch (19.1 mm.). The placard color must be
orange, black, and white.

(3) Except for the letter "B", the EXPLO-
SIVES B placard specifications are the same
as those for the EXPLOSIVES A placard.
The location, height, and stroke for the
letter "B" are the same as those prescribed
for the letter "A".

(4) The word "NON-FLAMMABLE" must
be across the center area with the word
"GAS" centered beneath the word "NON-
FLAMMABLE." The letters in both words
must be 1⁹⁄₁₆ inches (39.6 mm.) high and
made with a ⁹⁄₃₂-inch (7.1 mm.) stroke. The
top of the letters in the words "NON-
FLAMMABLE" must be 1⅖ inches (35.9
mm.) above the placard horizontal center
line, and the top of the letters in the word
"GAS" must be ⁹⁄₁₆-inch (14.3 mm.) below
the placard horizontal center line. The base
of the symbol must be 3⅛ inches (79.3 mm.)
above the horizontal centerline with the top
4¹⁵⁄₁₆ inches (125.4 mm.) above the placard
horizontal center line. The lower portion of
the cylinder (symbol) must be ¹⁷⁄₃₂-inch
(13.5 mm.) wide with the neck ¼-inch (6.3
mm.) wide. The symbol must be 3⁹⁄₁₆ inches
(90.4 mm.) long. The placard color must be
green and white.

(5) The word "OXYGEN" must be centered on the placard horizontal center line
in letters 2½ inches (63.5 mm.) high and
made with a ⁷⁄₁₆-inch (11.1 mm.) stroke. The
base of the bar in the symbol must be 2¹⁄₁₆
inches (52.4 mm.) above the placard horizontal center line. The overall height of the

symbol must be $4\frac{5}{16}$ (109.5 mm.) with the bar measuring $\frac{1}{8}$-inch (3.2 mm.) wide and, $3\frac{3}{16}$ inches (55.5 mm.) long.

The symbol must be $2\frac{3}{8}$ inches (60.3 mm.) across the widest part.

The outer $\frac{1}{2}$-inch (12.7 mm.) of the $10\frac{3}{4}$ inches (273.0 mm.) square on-point placard must be white. The placard color must be yellow, black, and white.

(6) *FLAMMABLE GAS placard.* The word "FLAMMABLE" must be across the placard center area with the word "GAS" centered beneath the word "FLAMMABLE". The letters in both words must be 2 inches (50.8 mm.) high and made with a $\frac{3}{8}$-inch (9.5 mm.) stroke. The top of the letters in the word "FLAMMABLE" must be $1\frac{5}{8}$ inches (41.3 mm.) above the placard horizontal center line and the top of the word "GAS" must be $\frac{5}{8}$-inch (15.9 mm.) below the placard horizontal center line. The base of the symbol bar must be $2\frac{1}{4}$ inches (57.1 mm.) above the placard horizontal center line, and must be $4\frac{17}{32}$ inches (115.1 mm.) high and $3\frac{5}{16}$ inches (84.1 mm.) wide. The bar must be $\frac{5}{32}$-inch (4.0 mm.) wide and $3\frac{5}{16}$ inches (84.1 mm.) long. The outer $\frac{1}{2}$ (12.7 mm.) of the $10\frac{3}{4}$ inches (273.0 mm.) square-on-point placard must be white. The placard color must be red and white.

(7) *CHLORINE placard.* The specifications for the CHLORINE placard are the same as those for the POISON GAS placard except for the word "CHLORINE" and the symbol. The word "CHLORINE" must be centered on the placard horizontal center line in letters $2\frac{1}{2}$ inches (63.5 mm.) high and made with a $\frac{7}{16}$-inch (11.1 mm.) stroke. The lowest part of the symbol must be $1\frac{3}{4}$ inches (44.5 mm.) above the placard horizontal center line. The symbol must be $3\frac{11}{16}$ inches (93.6 mm.) high and $5\frac{1}{8}$ inches (130.2 mm.) across the widest extremities.

(8) *POISON GAS placard.* The word "POISON" must be across the center area of the placard with the word "GAS" centered beneath the word "POISON." The letters in both words must be $2\frac{3}{16}$ inches (55.5 mm.) high and made with a $\frac{13}{32}$-inch (10.3 mm.) stroke. The top of the letters in the word "POISON" must be $2\frac{1}{4}$ inches (57.1 mm.) above the horizontal center line. The lowest part of the symbol must be $2\frac{3}{4}$ inches (69.8 mm.) above the horizontal center line and must be $3\frac{1}{4}$ inches (82.5 mm.) high and $4\frac{5}{16}$ inches (109.5 mm.) across the widest extremeties. The $\frac{1}{8}$-inch (3.2 mm.) black border must be $\frac{1}{2}$-inch (12.7 mm.) in from the placard edge. The placard color must be black and white.

(9) *FLAMMABLE placard.* The word "FLAMMABLE" must be centered on the placard horizontal center line. The letters in the word "FLAMMABLE" must be 2 inches (50.8 mm.) high and made with an $\frac{11}{32}$-inch (8.7 mm.) stroke. The base of the symbol bar must be $2\frac{1}{4}$ inches (57.1 mm.) above the placard horizontal center line. The symbol must be $4\frac{9}{16}$ inches (115.9 mm.) high and $3\frac{5}{16}$ inches (84.1 mm.) wide. The bar must be $\frac{1}{8}$-inch (3.2 mm.) wide and $3\frac{5}{16}$ inches (84.1 mm.) long. The outer $\frac{1}{2}$-inch (12.7 mm.) of the $10\frac{3}{4}$ inches (273.0 mm.) square-on-point placard must be white.

(10) *EMPTY placard.* The specifications for the FLAMMABLE-EMPTY placard is representative of the requirements for the following EMPTY placards: NON-FLAMMABLE GAS; POISON GAS; CHLORINE; OXYGEN; FLAMMABLE GAS; FLAMMABLE; FLAMMABLE SOLID; FLAMMABLE SOLID W; OXIDIZER; ORGANIC PEROXIDE; POISON; and CORROSIVE. The specification for each EMPTY placard must be the same as those prescribed for each placard except for the top triangle in the placard.

The base of the black triangle must be $\frac{1}{4}$-inch (6.3 mm.) above the top of the letters in the placard name. The base of the letters in the word EMPTY must be $3\frac{1}{4}$ inches (82.5 mm.) above the placard horizontal center line.

The letters in the word "EMPTY" must be 1-inch (25.4 mm.) high and made with a $\frac{7}{32}$-inch (5.5 mm.) stroke.

The EMPTY placards may be made in any of the three ways cited in § 172.525(c), Subpart F of Part 172.

(11) *COMBUSTIBLE placard.* The specification for the COMBUSTIBLE placard are the same as those prescribed for the FLAMMABLE placard except the letters in the word "COMBUSTIBLE" must be $1\frac{7}{8}$ inches (47.6 mm.) high and made with an $\frac{11}{32}$-inch (8.7 mm.) stroke.

(12) *FLAMMABLE SOLID placard.* The word "FLAMMABLE" must be across the center of the placard with the word "SOLID" centered beneath the word "FLAMMABLE." The letters in the word "FLAMMABLE" must be 2 inches (50.8 mm.) high and made with a $\frac{3}{8}$-inch (9.5 mm.) stroke. The letters in the word "SOLID" must be $1\frac{1}{2}$ inches (38.1 mm.) high and made with a $\frac{1}{4}$-inch (6.3 mm.) stroke. The top of the letters in the word "FLAMMABLE" must be $1\frac{3}{16}$ inches (30.1 mm.) above the placard horizontal center line, and the top of the word "SOLID" must be 1-inch (25.5 mm.) below the placard horizontal center line. The base of the symbol bar must be $2\frac{1}{4}$ inches (57.1 mm.) above the placard horizontal center line. The symbol must be $4\frac{9}{16}$ inches (115.9 mm.) high and $3\frac{5}{16}$ inches (84.1 mm.) wide. The outer $\frac{1}{2}$-inch (12.7 mm.) of the $10\frac{3}{4}$ inches (273.0 mm.) square-on-point placard must be white. Each red and white stripe must be approximately 1 inch (25.4 mm.) wide. The placard must have seven red stripes and six white stripes. One red stripe must be approximately centered on the vertical center

line of the placard. The placard color must be black, white, and red.

(13) *FLAMMABLE SOLID W placard.* The specifications for the FLAMMABLE SOLID W are the same as the specifications for the FLAMMABLE SOLID placard except for the top triangle. The base of the blue triangle must be 2 inches (50.8 mm.) above the placard horizontal center line with the base of the symbol 2⅜ inches (60.3 mm.) above the placard horizontal center line. The symbol must be 2¼ inches (57.1 mm.) high; 2¾ inches (69.8 mm.) across the top 1¾ inches (44.4 mm.) across the base, and made with a ⁵⁄₁₆-inch (7.9 mm.) stroke. The white stripe in the symbol must be ⁷⁄₃₂-inch (5.5 mm.) wide and 3½ inches (88.9 mm.) long. The white stripe must slant upward from right to left at an angle of approximately 21 degrees from the horizontal. This placard may be made in any of the three ways cited in § 172.548, Subpart F of Part 172.

App B *OIDIZER placard.* The word "OXI-DIZER" must be centered on the placard horizontal center line in letters 2½ inches (63.5 mm.) high with a ¹⁵⁄₃₂-inch (11.9 mm.) stroke. The base of the bar of the symbol must be 2¹⁄₁₆ inches (52.4 mm.) above the placard horizontal center line. The overall height of the symbol must be 4⁵⁄₁₆ inches (109.5 mm.) with the bar measuring ⅛-inch (3.2 mm.) wide and 2³⁄₁₆ inches (55.6 mm.) long. The symbol must be 2⅜ inches (60.3 mm.) across the widest part. The outer ½-inch (12.7 mm.) of the 10¾ inches (273.0 mm.) placard must be white. The placard color must be yellow, black, and white.

(15) *ORGANIC PEROXIDE placard.* The word "ORGANIC" must be across the center line of the placard with the word "PEROXIDE" centered beneath the word "ORGANIC." The letters in both words must be 2 inches (50.8 mm.) high and made with an ¹¹⁄₃₂-inch (8.7 mm.) stroke. The top of the letters in the word "ORGANIC" must be 2⅛ inches (54.0 mm.) above the placard horizontal center line, and the top of the letters in the words "PEROXIDE" must be ⁵⁄₁₆-inch (7.9 mm.) below the placard horizontal center line. The base of the symbol bar must be 2⅞ inches (73.0 mm.) above the horizontal center line. The symbol must be 3¹¹⁄₁₆ inch (93.6 mm.) high and 2¹⁄₁₆ inches (52.3 mm.) wide with the bar ³⁄₁₆-inch (4.8 mm.) wide and 1⅞ inches (47.6 mm.) long. The outer ½-inch (12.7 mm.) of the 10¾ inches (273.0 mm.) square-on-point placard must be white. The placard color must be yellow, black, and white.

(16) *POISON placard.* The word "POISON" must be centered on the placard horizontal center line in letters 3¹⁄₁₆ inches (77.8 mm.) high and made with a ⁹⁄₁₆-inch (14.3 mm.) stroke. The lowest point on the symbol must be 2⅛ inches (54.0 mm.) above the placard horizontal center line. The symbol must be 3¹¹⁄₁₆ inches (93.6 mm.) high

and 4¹⁵⁄₁₆ inches (125.4 mm.) across the widest extremities. The ⅛-inch (3.2 mm.) black border must be ½-inch (12.7 mm.) in from the placard edge. The placard color must be black and white.

(17) *RADIOACTIVE placard.* The word "RADIOACTIVE" must be centered on the placard horizontal center line in letters 2 inches (50.8 mm.) with an ¹¹⁄₃₂-inch (8.7 mm.) stroke. The lower edge of the yellow triangle must be 1⅛ inches (28.6 mm.) above the placard horizontal center line. The lower edge of the symbol must be 1¼ inches (31.7 mm.) above the placard horizontal center line. The symbol must be made as shown with the following dimensions:

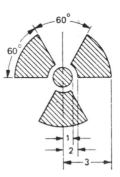

1 = Radius of Circle —
1/2 inch (12.7 mm.)

2 = 1 1/2 Radii

3 = 4 1/2 Radii

The lower white area must have a ⅛-inch (3.2 mm.) black solid line border extended from the edge of the yellow area to indicate the outer ½-inch (12.7 mm.) white placard border. The placard color must be yellow, black, and white.

(18) *CORROSIVE placard.* The word "CORROSIVE" must be across the center of the placard and made with letters 2¹⁄₁₆ inches (52.4 mm.) high with a ¹¹⁄₃₂-inch (8.7 mm.) stroke. The base of the top white triangle must be 1½ inches (38.1 mm.) above the placard horizontal center line. The lowest part of the symbol must be 1⅝ inches (41.3 mm.) above the placard horizontal center line. The height of the symbol measured from a horizontal line extended from the lowest part of the symbol must be 3¼ inches (82.5 mm.) and the width across the widest part must be 7⅜ inches (187.3 mm.). The upper white area must have a ⅛-inch (3.2 mm.) black solid line border as an extension from the edge of the black area to

indicate the outer ½-inch (12.7 mm.) white placard border. The placard color must be black and white.

(19) BLASTING AGENTS placard. The words BLASTING AGENTS must be across the center area of the placard and made with letters 1⅞ inches (47.6 mm) high with a ⁵⁄₁₆-inch (7.9 mm) stroke.

(49 U.S.C. 1803, 1804, 1808; 49 CFR 1.53 and App. A to Part 1)

[Amdt. 172-29, 41 FR 15996, Apr. 15, 1976, as amended by Amdt. 172-29A, 41 FR 40680, Sept. 20, 1976]

NOTE: For amendments to Appendix B of Part 172 see the List of CFR Sections Affected appearing in the Finding Aids section of this volume.

## APPENDIX C—DIMENSIONAL SPECIFICATIONS FOR RECOMMENDED PLACARD HOLDER

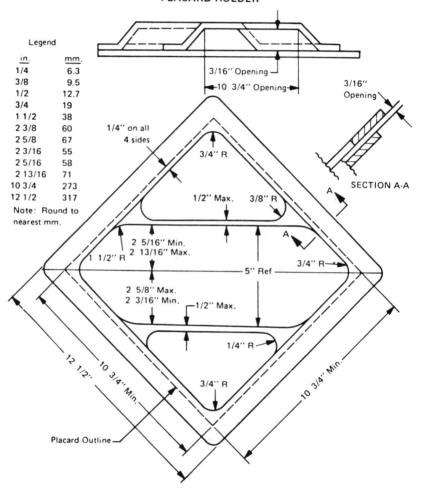

Legend

| in. | mm. |
|-----|-----|
| 1/4 | 6.3 |
| 3/8 | 9.5 |
| 1/2 | 12.7 |
| 3/4 | 19 |
| 1 1/2 | 38 |
| 2 3/8 | 60 |
| 2 5/8 | 67 |
| 2 3/16 | 55 |
| 2 5/16 | 58 |
| 2 13/16 | 71 |
| 10 3/4 | 273 |
| 12 1/2 | 317 |

Note: Round to nearest mm.

## THE TECHNICAL ASPECTS OF LABEL PRODUCTION

The two major factors involved in producing labels are first, the material to be used, and second, the production method for printing the stock, including the type of ink and clearcoat.

### Choosing the Label Material

Typical pressure sensitive label construction consists of three layers as shown in Figure 3.2.

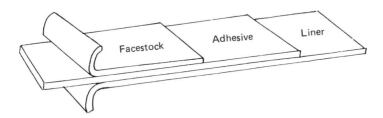

**Figure 3.2:** Pressure sensitive label construction.

**Facestock:** The printable facestock imparts the basic appearance and properties of a pressure sensitive label. A broad variety of facestocks may be specified according to the required use. The selection which follows suggests this variety, but is in no way limited to this list.

*Standard Paper Products*

Uncoated Litho—40 lb, 45 lb and 60 lb uncoated litho stock for low cost labels.

Semigloss Litho—60 lb machine finished semigloss paper featuring good quality print characteristics.

Gloss Litho—60 lb highly flexible medium gloss stock with good printability.

Gloss—60 lb cast coated one side premium gloss paper featuring high internal strength and excellent ink receptivity.

Mirror Finish—Cast coated paper for high quality printing. Use for demanding pharmaceutical applications.

Matte Litho—60 lb matte finish premium litho stock with a smudgeproof coating.

Fluorescent—Matte fluorescent coated litho stock in a number of colors for attention getting labels.

Laminated Foils—0.00035 foil laminated to a 55 lb base paper. Choice of bright and dull silver or gold.

### Latex Impregnated

Texoprint—80 lb Texoprint features a high degree of flexibility and conformability. Latex impregnated on both sides for high moisture resistance.

Kimdura—3.2 mil biaxially oriented polypropylene with good durability and printing surface.

Tyvek—8 mil spunbonded white olefin with high tear resistance, good opacity and dimensional stability.

### Vinyls

Soft or Hard 4 mil.—Durable semirigid or flexible vinyl suitable for outdoor use.

### Polyesters

Clear Polyesters—1 and 2 mil clear durable polyester film suitable for outdoor use

Silver Polyesters—1 and 2 mil bright and matte chrome polyester film suitable for outdoor use

### Overlaminates

Polypropylene—1 mil clear, conformable film for a brilliant, solvent and abrasion resistant finish for labels

Polyester—0.0005 and 1 mil clear film for brillant solvent and abrasion resistance finish for labels

Acetate—2 mil clear cast acetate film. Recommended for indoor applications and overlaminating.

**The Adhesive:** The choice of the proper adhesive for one's label is quite obviously of major importance. The two types of adhesives generally used for pressure sensitive label construction are rubberbase and acrylic emulsion. In addition, most converters of label stock offer a variety of specialty adhesives to meet particularly demanding applications. Some of the terminology associated with pressure sensitive adhesives is as follows.

Permanent Adhesives—Provides high tack, ultraviolet resistance and strong ultimate adhesion.

Removable Adhesive—Provides medium tack and low ultimate adhesion.

Tack—(Also called initial adhesion, grab, quick-stock) Ability of a pressure sensitive to instantly adhere to a surface.

Adhesion—The bond ultimately established between the adhesive and the surface, sometimes requires a buildup time of 24 to 48 hours.

Holding Power—The characteristic of a pressure sensitive adhesive to resist removal.

Peel Adhesion—The force required to pull a gasket from a surface at a specified angle and rate of speed.

Shear Adhesion—(Also called creep resistance) The force required to pull a gasket from a standard surface in a direction parallel to that surface.

The following adhesive selector chart provides some preliminary information. Other decisions will have to do with application temperatures and ultimate adhesion requirements.

### Table 3.1: Adhesive Selector Chart

|  | Acrylic | Rubberbase |
|---|---|---|
| Acetate film | X | X |
| Aluminum | X | X |
| Cork | X | T |
| Corrugated | X | X |
| Cotton | X | NR |
| Dacron | X | NR |
| Felt | X | T |
| Glass | X | X |
| Leather | X | T |
| Neoprene | T | NR |
| Nylon | X | NR |
| Paint | X | X |
| Paper kraft | X | X |
| Plexiglass | X | X |
| Polyethylene | T | T |
| Polypropylene |  |  |
|   Treated | X | X |
|   Untreated | T | X |
| Polystyrene | X | X |
| Polyurethane foam | T | T |
| Rayon | X | NR |
| Satin film | X | X |
| Skin | T | NR |
| Styrofoam | T | T |
| Tedlar |  |  |
|   Treated | X | X |
| Vinyl | X | NR |
| Wood | X | X |

Note:   X  =  Excellent adhesion.
       NR  =  Not recommended.
       T  =  Test prior to using.

**The Liner:** Liners serve two basic functions. They protect the adhesive from damage during processing, and provide stay-flat body to the free-standing label. Liners also act as a carrier for the die-cut roll label. Liners are either 40 lb, 50 lb or 90 lb paper weight. The lighter weights are normally used for roll labels, the heavier weight for sheeted or individual labels to help maintain their flatness.

## Methods for Printing Labels

Generally speaking, chemical identification labels are printed by any one of five different methods:

1. Letterpress (Flatbed)
2. Offset
3. Flexographic
4. Rotary (Continuous) Letterpress
5. Screen Process

Each of the above possesses both virtues and limitations as regards their value to the chemical industry. An examination of each in turn will reveal their desirability for one's own requirements.

**Letterpress, Flatbed, Sheet Fed:** This is one of the oldest printing methods wherein direct contact is made between raised type, usually metal, and the facestock. The quality of the printing would be rated as fair to good since control of the ink is not as positive as some of the other methods. It should be considered for very short run paper stock where copy is fairly simple. The label stock usually used is a standard Crack/Peel type with very little selection of the adhesive. This type of factory-cut liner is more difficult to remove than other types, particularly for a large label. Not recommended for vinyls or other exterior grade use.

**Offset Printing, Sheet and/or Web:** This is a highly popular process for printing sheet-fed labels. Printing is accomplished via the indirect method through a transfer of the inks from aluminum plates. Its desirability is the relatively low cost of the plates and good control of the inks. Further economies are possible because of the availability of multiple color presses. Four-color process labels would, of course, also be possible on such presses. The disadvantages of sheet fed process are the same as those for letterpress. Web offset by its very nature is recommended only for very large runs.

**Flexography:** This is presently the most prevalent method for producing labels, particularly in roll form. Its virtues are a relatively low plate cost (rubber plates are most popular though a greater use of polymer plates is anticipated) and the ability to print in multiple colors, die-cut, and laminate all in line. Both solvent and water based inks are heat cured. The pressure sensitive roll label has continued to gain in popularity. It is mandatory for automatic applications, and also desirable for its ease of handling in manual use. The disadvantages of flexo are its relatively poor control of the inks and the tendency to show 'gear marks' in large solid areas of color. Similarly it is not recommended for printing small type such as might be required on a pharmaceutical label or for printing four color process. Quality of the printing is rated as fair to good. Outdoor exposure is not recommended.

**Rotary Letterpress:** The rotary letterpress is state of the art equipment for producing either roll or sheeted pressure sensitive labels. The printing medium is ultraviolet curable inks cured by exposure to UV mercury vapor lamps. The quality of printing is rated as good to excellent, coming very close to the quality of offset. It is highly recommended for large solids, small type and four-color process. These new presses also incorporate flat bed, steel-rule die cutting making small runs practical. Chemically resistant clear coats are available in line. Rotary letterpress should also be considered for printing exterior labels since the equipment prints equally well on papers, vinyls and polyesters.

**Screen Process:** This printing method has been previously discussed. Whether the drying of the inks is by heat or by UV curing, the screen process permits heavier deposits of ink laydown, and thus provides the greatest durability for outdoor use.

Table 3.2 illustrates the properties of each of the above printing processes.

## PLANNING A LABEL PROGRAM

Specifying one's label and/or placard is a fairly simple matter once the proper shipping name has been determined. If the label manufacturer has any difficulty at all in day to day operations, it is with those companies who have not properly identified the hazard class for their particular compound. In the waste area, it often appears that the proper shipping name, *Hazardous Waste, Liquid, n.o.s.* has become a convenient catch-all for materials where the hazards are not known. Proper testing of the material is essential for it is only in that manner one may be certain what the hazard class really is.

Once the proper shipping name has been determined, the Commodity List (49CFR 172.101) will indicate the correct label. If testing has indicated the presence of a substantial secondary hazard, then you should double label the container. You should also ascertain the ways in which the product is to be packaged. It is entirely possible one will want to specify paper labels for five gallon pails, and vinyl labels for the same product when packaged in drums. It must be decided whether to incorporate the proper shipping name and UN number on the hazard label or as a separate marking. As an alternative, it might be best to incorporate all of the information on one larger product marking, complete with corporate identification. You also need to take into consideration the mode by which the product is being shipped. If by air, perhaps a "Cargo Aircraft Only" label is also required. A special orientation or fragile label may help get the package to its destination with greater safety.

These are some of the decisions one needs to make in establishing label procedures. Within the pages of 49CFR one should find everything one needs to know with regard to labeling a product, including methods for testing and packaging. The same may be said for placarding. The tables shown in 49CFR 172.504 will determine very quickly whether or not placarding is required for the hazard class being shipped.

### Reference Guides

One should obtain as much information as possible with regard to the shipment of hazardous materials. One may often find oneself double checking the proper shipping name in two or three separate source books. Not only does this increase one's knowledge, but many times it results in the discovery of subtle differences which greatly assist in arriving at the final conclusion. Such efforts will allow one to document findings, and to benefit from the use of this information for a very long time. Maintaining the currency of one's guides will assure a continued awareness of the regulations, and permit updating of docu-

Table 3.2: Properties of Printing Processes

| Print Method | Color Solids | Small Type | 4-Color Process | Produce Rolls | Produce Sheets | Die-Cut In-Line | Laminate In-Line | UV Clear In-Line | Prints, Vinyls, Polyesters | Outdoor Labels | Economical Short Run | Economical Long Run |
|---|---|---|---|---|---|---|---|---|---|---|---|---|
| Letterpress | F-G | F-G | P-F | N | Y | N | N | N | F | P | Y | N |
| Offset, sheet | G-E | G-E | G-E | N | Y | N | N | N | F | P | M | Y |
| Offset, web | G-E | G-E | G-E | Y | Y | N | N | M | F | P | N | Y |
| Flexography | P-F | P-F | P-F | Y | Y | Y | Y | N | F | P | Y | Y |
| Rotary letterpress | G-E | G-E | G-E | Y | Y | Y | Y | Y | E | G-E | Y | Y |
| Screen process | G-E | G-E | F-G | Y/N | Y | N | N | N | E | E | M | M |

Note: P = Poor
F = Fair
G = Good
E = Excellent
Y = Yes
N = No
M = Medium

mentation to maintain a posture of compliance. With continued use of reference books one will find a growing familiarity with their form and technique. There is no greater assest to planning one's program than the time spent in gaining this familiarity.

## THE FUTURE

It is hoped that the information presented in this chapter will make the trend for the near future somewhat discernible. A first glimpse may come from a new set of regulations soon to be issued by Transport Canada. Canada has the unique opportunity of writing on a clean slate. While from a practical point of view, Canadian regulations must interact with those of the U.S., their regulators have the advantage of a perspective of the past coupled with their awareness of present planning. Their initial drafts have introduced the concept of wordless labels and placards. That is, only the symbol and U.N. hazard class number would be shown on the prescribed color background. (The regulation will also permit, however, the option of using words such as Flammable Liquid, Corrosive, etc.) While this idea is undoubtedly presented in response to Canada's bilingual condition, the concept is one which should and will have the attention of the UN Committee of Experts. Since Spanish is the language of Mexico and most of Central and South America, there has been a consistent chorus of requests that U.S. labels and placards be bilingual for these nations—English and Spanish. Wordless labels would make a contribution in this area as well.

A language barrier should not stand in the way of hazard communication. In some ways, a precedent has already been set vis-a-vis the four-digit placard. This placard, without the use of words, communicates the symbol, color and class number. The four-digit UN number simply identifies the specific commodity. Is the white flame on a red background with Class Number 3 any less a *Flammable Liquid* label than with the use of those words? Or, is the green label with the white cylinder and Class Number 2 any less a *Non-Flammable Gas?* There will be those who will object to the additional training necessary in a world without words. It is suggested that such a step will not be as difficult as it may appear, and further, it will promote safety and economic benefits in both international and domestic trade.

The safe transportation of dangerous goods benefits all. Labels, placards, and markings contribute to that end. A well planned program designed to meet the conditions of transport, storage and use of the package assures not only compliance with the regulations, but anticipates any field problems which may occur. Setting parameters which accommodate those possible problems is the best guarantee of a successful program.

## REFERENCES

1. *Transport of Dangerous Goods:* Recommendations Prepared by the United Nations Committee of Experts (U.N. Orange Book), Second Revised Ed., Chicago: Intereg (1982).

2. IMO *International Maritime Dangerous Goods Code,* Five Volume Edition, London: International Maritime Organization (1981) Amendment 19-80 and Amendment 20-82 for the above volumes. *Medical First Aid Guide for Use in Accidents Involving Dangerous Goods* (MFAG) Revised 1982 *Emergency Procedures for Ships Carrying Dangerous Goods (1981)*

3. ICAO *Technical Instructions for the Safe Transport of Dangerous Goods by Air,* International Civil Aviation Organization (1984) Available in English, Spanish, French and Russian. Chicago: Interleg

4. IATA *24th Edition Dangerous Goods Regulations, Consolidated Version* Montreal: International Civil Aviation Organization.

5. *Title 49 Code of Federal Regulations* Parts 100-199 Transportation of Hazardous Materials, Washington, DC: General Services Administration (1982).

6. Title 40 Code of Federal Regulations Parts 100-399 Hazardous Waste Regulations, Washington, DC: General Services Administration (1982).

7. *Emergency Response Guidebook,* Department of Transportation (1983) Materials Transportaion Bureau U. S. Department of Transportation, Washinton DC.

8. *Emergency Response Guide for Dangerous Goods,* Copp Clark Pitman in cooperation with Transport Canada, Transport of Dangerous Goods Branch, and the Canadian Government Publishing Centre, Ottowa, Second Edition (1982)

9. *Emergency Handling of Hazardous Materials in Surface Transportation,* Washington DC: Association of American Railroads, Bureau of Explosives (1981)

10. UPS *Guide for Shipping Hazardous Materials via UPS,* United Parcel Service of America, Inc., Greenwich, CT. (1981; Reprinted 1982)

All of the above publications are also available from Labelmaster, 5724 N. Pulaski Road, Chicago, IL 60646

# Part II

# Science and Labels

All labeling is based upon a thorough understanding of the basic properties of the subject product. These properties fall into three groups: (1) physical-chemical; (2) human-biological; and (3) environmental. In this section groups one and two are examined.

Chapter 4 explores in some detail the major subclasses of physical-chemical importance in classification and labeling. The author uses a combined DOT-RCRA scheme as a convenient classification device. Flammability, corrosivity, explosivity, oxidizers, peroxides, gases, heavy metal toxicity (EP toxicity), and radioactivity, among other properties are discussed. As pointed out, classification (or categorization as EPA refers to it) of a chemical is the first step in labeling. The inherent chemical, physical and biological properties of a chemical must be understood before judgments can be rendered as to appropriate label statements. The interaction of such variables as vapor pressure, physical state, melting point, boiling point and freezing point will dictate how to evaluate the importance of $LC_{50}$ and $LD_{50}$ values.

Chapters 6 and 7 split the field of toxicology and human effects into acute and chronic effect areas. While this is artificial to some extent, it is useful for label classification and development. Typically, acute data is obtained first, and, at relatively modest cost. Although hepatotoxicity, neurotoxicity, and nephrotoxicity are usually thought of as chronic effects, they may in fact be acute effects. The real division between the two fields is in immediacy of effect following exposure. Thus, acute effects follow directly and immediately upon exposure, while chronic effects may be delayed for many years, e.g. cancer.

The results obtained from a battery of acute toxicity tests often indicate whether additional studies are needed and help guide the choice and design of such additional studies. The current state of acute effect testing is well-developed and although some refinement still occurs, there is good general agreement on the interpretation of acute test results.

The situation is very different with chronic effect testing. The scientific community is split on both what tests should be performed and how test results should be interpreted. While a very rough consensus does appear to be developing recently, much work remains to be done in correlating bacterial and cellular tests to ninety day and lifetime effect testing.

Today, however, many scientists claim the only reliable results are obtained from lifetime studies. Mammalian cell tests, bacterial tests and ninety day tests are, however, important indicators of the biologic activity of a chemical.

In Chapter 5 the author addresses medical treatment and first aid directions for labels. Appropriate and rapid first aid will often greatly reduce the severity of the results of accidental exposure. At its most basic, first aid is designed to prolong life until medical treatment is available. Every facility that handles hazardous chemicals must have a well trained first aid squad, with on-going training programs that will ensure skill and knowledge maintenance.

An occupational health physician or the National Poison Control Center Network, located in Pittsburgh, should either prepare or review medical treatment instructions for labeling (labels and material safety data sheets). The Center's charge for this service is quite modest.

## 4

# Hazard Assessment and Classification for Labeling

**Adria C. Casey**
*Stauffer Chemical Company*
*Westport, CT*

## INTRODUCTION

Classification or categorization of a product according to the hazards it presents is the first step in determining the precautionary text, hazard warning labels, etc., that must be used as part of the hazard communication system for the product. For transportation and for disposal as wastes, the classification of products or other materials is dictated by federal and state regulations. From a regulatory standpoint, hazard classification is perhaps the most important and complex evaluation of the characteristics of a material. It is important because once it has been determined that a material is hazardous, a series of regulatory requirements must be observed. It is complex because hazard determination must combine the technical assessment of the physical, chemical and biological properties of the material with a thorough understanding of the regulatory framework.

Several federal agencies have rules on the categorization of hazardous materials and the applicability of these rules is clearly related to the mandate of the agency and the place, conditions and group of people the regulations are designed to protect. Thus, the Department of Transportation hazardous materials regulations are directed to providing protection from materials posing an unreasonable risk to *health, safety and property* when *transported* in commerce. Under Subtitle C of the Solid Waste Disposal Act, as amended by the Resource Conservation and Recovery Act of 1976, (RCRA), the Environmental Protection Agency (EPA) is directed to promulgate regulations to protect *hu-*

*man health* and *the environment* from the improper *management of hazardous wastes*. Thus, during its lifetime, an industrial product could be regulated under different sets of rules that may classify the material as hazardous or not, or even assign to it different hazard classes depending on the philosophy of the regulatory framework.

In this chapter, the classification of industrial products for transportation under present DOT regulations and disposal classifications under RCRA will be discussed. A brief discussion of pesticide classification under FIFRA is also presented. It must be kept in mind that regulations are a living set of rules which are frequently modified by amendments, deletions and new definitions. Thus, while an attempt has been made to present the subject of product classification in a manner that will be affected as little as possible by the process of regulatory change, this is clearly not entirely possible. The reader must be aware of this fact and of his responsibility for keeping abreast of any new developments in this field.

## CLASSIFICATION UNDER DOT REGULATIONS

To the uninitiated, it would seem that all materials offered for transportation are considered hazardous. Obviously, this is not the case, since not all materials transported pose an unreasonable risk to health, safety and property, the basic criteria for regulating a material for transportation. Only those materials classified as hazardous under the Department of Transportation criteria are regulated and the degree of restriction which the regulation imposes depends on the kind of hazard or hazards for which the product is classified.

The regulations applicable to the transportation of hazardous materials are found in the *Code of Federal Regulations* (CFR) Title 49. The section of 49CFR concerned with the classification of hazardous materials is Chapter I, Parts 100–199 (See Chapter 14 of this Handbook for a full discussion of 49CFR and *Labeling in Transportation*). Reference to applicable paragraphs of 49CFR will be made here. It is highly recommended to read this chapter in conjunction with the latest printing of the DOT regulations.

A hazardous material is a material listed by DOT in the Hazardous Materials Table (§172.101) by technical name, or, if not listed by technical name, a material that meets the specific criteria of a hazard class. A plus sign (+) on Column I of the Hazardous Materials Table fixes the hazard class and proper shipping name for that material without regard as to whether that material meets the definition of that hazard class. If petitioned, an alternate hazard class, in that case, may be authorized by the Associate Director, Office of Hazardous Materials Regulation, MTB (§172.101(b)(1)). If a material listed by technical name on the Hazardous Materials Table meets the definition of a hazard class other than the class shown in association with the technical name, the material must be classified in accordance with the appropriate hazard class. It must then be described by the shipping name that best describes the material listed in association with the correct hazard classification. If the material is not listed by technical name, then its hazard class must be determined and an appropriate shipping name selected from the general descriptions or n.o.s. (not otherwise specified) entries corresponding to the specific hazard class of the material being shipped.

At present, DOT regulates as hazardous materials those which meet the criteria of one or more of the following hazard classes.

A. Forbidden Materials

B. Explosives or Blasting Agents

C. Flammable or Combustible Liquids

D. Flammable Solids, Oxidizers or Organic Peroxides

E. Corrosive Materials

F. Non-flammable and Flammable Gas

G. Poison or Irritating Materials

H. Etiological Agents

I. Radioactive Materials

J. Other Regulated Materials (ORM): ORM-A, ORM-B, ORM-C, ORM-D or ORM-E.

**Forbidden Materials**

Forbidden materials are those considered too hazardous to be permitted in commercial transportation. Although "FORBIDDEN" is not specifically designated as a hazard class, the decision as to whether a material is forbidden for transportation is the first evaluation to be made when classifying the material. The chemical names of many such chemicals are found in the Hazardous Materials Table, Section 172.101. However, others must be evaluated against the criteria given in Section 173.21. This criteria covers:

1. A material which is liable to decompose or polymerize at a temperature of 130°F (54.4° C) or less *with* an evolution of a dangerous quantity of heat or gas unless stabilized or inhibited in a manner that will preclude such evolution.

The determination of whether a material is forbidden according to the above criteria may be made by one of the following methods: Standard Method of Test for Constant Temperature Stability of Chemical Materials (ASTM E-487-74)[1] or the Self Accelerating Decomposition Temperature Test published by the Organic Peroxide Producers' Safety Division.[2]

2. Materials (other than explosives) which *when packaged* will detonate in a fire. Detonation is a type of explosion in which the shock wave travels through the material at a speed greater than the speed of sound in the

undecomposed material. Testing for this type of forbidden material must be done or approved by: (a) the Bureau of Explosives; or (b) some government agencies under whose direction, supervision or request these materials have been manufactured. There are three tests specified in the regulations for determining if a packaged material detonates as a result of thermal stimulus. These are referred to in Sections 173.88(g), Note 2; 173.86(b)(1) and (2); and 173.114a(b)(6). However, this type of testing should be done only by pesonnel who are well versed in the testing of explosives and this fact is cleary stated in the regulations.

As with other areas of the transportation regulations, it is the responsibility of the person offering the material for transportation to ascertain that the classification of the material and all other requirements stemming from such classification are complied with. Thus, the development of familiarity with the "forbidden" material area is essential.

Examples of forbidden materials are ethyl hydroperoxide which explodes above 100° C, benzoyl azide, hexanitroethane, etc.

### Explosives and Blasting Agents

An explosive material is defined as "any chemical compound, mixture or device, the primary or common purpose of which is to function by explosion" (49CFR §173.50(a)). Defined in this manner, a material must have been manufactured with the express intent to be an explosive, i.e., to cause substantial instantaneous release of energy, usually in the form of gas or heat, to be considered in this hazard class. Explosives are further divided as Class A, B or C explosives, depending on the magnitude of the hazard they possess, with Class A presenting the highest danger.

A blasting agent is a material designed for blasting which has been tested by a series of tests provided in §173.114a(b) and found so insensitive that there is very little probability of accidental initiation to explosion or of transition from deflagration to detonation.

New explosives or blasting agents are either materials not previously produced or if previously produced, materials to which changes have been made in the formulation, design, process or production (49CFR 173.86(a)) and (49CFR 173.114a(d)). Classification for new explosives must be provided by the Associate Director of Operations and Enforcement (OE) of the Materials Transportation Bureau after the Bureau of Explosives has examined the material. Classification of a blasting agent must be provided, after examination, by the Bureau of Explosives, with final approval given by the Associate Director for OE. Thus, no further discussion of the classification of Explosives and Blasting Agents will be advanced here.

It must be noted that some explosives are forbidden for transportation. There are several forbidden explosives listed in 49CFR 173.51. The underlying criteria is given in §173.51(a) as follows:

"Explosive compounds, mixtures or devices which ignite spontaneously or undergo marked decomposition when subjected to a temperature of 167°F (75°C) for 48 consecutive hours" are forbidden for transportation.

Examples of explosive materials are trinitrotoluene, TNT (Class A explosive), and common fireworks (Class C explosive).

## Flammable and Combustible Liquids

Depending on the ignitability of their vapors, liquids are classified as flammable or combustible. These two classes differ in their flash point range.[3] A liquid that ignites spontaneously when in contact with dry or moist air, a pyrophoric liquid, is classified as a flammable liquid. The criteria for these hazard classifications (49CFR 173.115) are as follows:

(a) *Flammable liquid.*

(1) . . . a flammable liquid means any liquid having a flash point below 100°F (37.8°C), with the following exceptions:

(i) Any liquid meeting one of the definitions specified in §173.300 (definition of a compressed gas);

(ii) Any mixture having one component or more with a flash point of 100°F (37.8°C), or higher, that makes up at least 99 percent of the total volume of the mixture.

(2) For the purposes of the regulation, a distilled spirit of 140 proof or lower is considered to have a flash point no lower than 73°F.

(b) *Combustible liquid.*

(1) For the purposes of the regulation a combustible liquid is defined as any liquid that does not meet the definition of any other classification specified in the regulation and has a flash point at or above 100°F (37.8°C) and below 200°F (93.3°C) except any mixture having one component or more with a flash point at 200°F (93.3°C) or higher, that makes up at least 99 percent of the total volume of the mixture.

(2) For the purposes of the regulation an aqueous solution containing 24 percent or less alcohol by volume is considered to have a flash point no less than 100°F (37.8°C) if the remainder of the solution does not meet the definition of a hazardous material as defined in this subchapter.

(3) 200°F (93.3°C) is a limitation of the application of the regulations and should not be construed as indicating that liquids with higher flash points will not burn. Markings such as "NONFLAMMABLE" or "NONCOMBUSTIBLE" should

not be used on a vehicle containing a material that has a flash point of 200°F (93.3°C) or higher.

(c) *Pyrophoric liquids.*

(1) For the purposes of the regulation a pyrophoric liquid is any liquid that ignites spontaneously in dry or moist air at or below 130°F (54.4°C).

As discussed earlier, the extent of the regulatory requirements depends on the degree of hazard assigned to a material. Thus, flammable liquids are almost always regulated, while combustible liquids are only regulated when shipped in large quantities (greater than 110 gallons). Combustible liquids, however, are regulated in all quantities and modes of transportation when they also qualify for an ORM-E hazard class (see ORM-E definition).

Gasoline, ether and lighter fluid are examples of flammable liquids, while fuel oil and many alcohols are combustible liquids. Metal alkyl solutions are, if of relatively high concentration, pyrophoric liquids.

## Flammable Solids, Oxidizers and Organic Peroxides

For the purpose of the DOT regulation, "Flammable solid" is any solid material, other than one classed as an explosive, which under conditions normally incident to transportation is liable to cause fires through friction, retained heat from manufacturing or processing, or which can be ignited readily and persistently as to create a serious transportation hazard. Included in this class are spontaneously combustible and water-reactive materials. (49CFR 173.150)

An oxidizer is a substance, liquid or solid "such as a chlorate, permangate, inorganic peroxide, or a nitrate, that yields oxygen readily to stimulate the combustion of organic matter." (49CFR 173.151)

An organic peroxide is "an organic compound containing the bivalent -O-O- structure and which may be considered a derivative of hydrogen peroxide where one or more of the hydrogen atoms have been replaced by organic radicals . . ." (49CFR 173.151a).

Because of the descriptive nature of the flammable solid and oxidizer definitions it is difficult to determine if a material falls within the definition of either hazard class. In efforts to make both definitions more specific and to provide tests which would allow appropriate classification, the Materials Transportation Bureau has requested comments on new definitions and proposed test methods for the flammable solid and oxidizer hazard classes.[4] The reader should become aware of these test methods definitions since they could become part of the regulatory framework. In addition, the test methods could serve as guidance, even at present, for decisions involved in classifying a material as a flammable solid or as an oxidizer.

## Corrosive Materials

"A corrosive material is a liquid or solid that causes visible destruction or irreversible alterations in human skin tissue at the site of contact or in the case of a leakage from its packaging a liquid that has a severe corrosion rate on steel (49CFR 173.240(a))". Defined in this manner, a liquid must be evaluated by prescribed animal testing, as well as by metal corrosion test before being categorized as a corrosive material. A solid, on the other hand, needs only to be evaluated by prescribed animal tests.

The criteria used for skin and metal corrosion are as follows:

(1) A material is considered to be destructive or to cause irreversible alteration in human skin tissue if when tested on the intact skin of the albino rabbit by the technique prescribed[5], the structure of the tissue at the site of contact is destroyed or changed irreversibly after an exposure period of 4 hours or less.

(2) A liquid is considered to have a severe corrosion rate if its corrosion rate exceeds 0.250 inch per year (IPY) on steel (SAE 1020) at a test temperature of 130°F. An acceptable test is described in NACE Standard TM-02-69.[6]

As in many other instances in the DOT regulations, if human experience or other data indicates that the hazard of a material is greater or less than indicated by the results of the tests, DOT may revise the material's classification or make the material subject to the regulations.

Examples of corrosive liquids are concentrated acids and bases, acid chlorides, etc. Titanium tetrachloride, a solid, is also a corrosive material.

Some common chemicals have been tested for the skin corrosion under DOT protocols at a variety of concentrations. The results of this study can be used as a guide when classifying some of these materials.[7] It must be remembered that the responsibility of determining if a material belongs to a hazard class rests with the shipper and that the use of published data should only serve as a guideline.

### Non-flammable and Flammable Gases

The term compressed gas is used to describe a material or mixture having in the container an absolute pressure exceeding 40 pounds per square inch (psi) at 70°F, or, regardless of the pressure at 70°F, having an absolute pressure exceeding 104 psi at 130°F; or any liquid flammable material having a vapor pressure exceeding 40 psi absolute at 100°F as determined by ASTM Test D-323 (49CFR 173.300(a)).

Absolute pressure is the total pressure exerted by the gas plus atmospheric pressure (14.7 psi).

A compressed gas is a "flammable compressed gas" if it meets any of the following criteria:

(1) Either a mixture of 13% or less (by volume) with air forms a flammable mixture or the flammable range with air is wider than 12% regardless of the lower limit. These limits shall be determined at atmospheric temperature and pressure. The method of sampling and test procedure shall be acceptable to the Bureau of Explosives and approved by the Associate Director of OE.

(2) Using the Bureau of Explosives' Flame Projection Apparatus[8] the flame projects more than 18 inches beyond the ignition source with valve opened fully, or the flame flashes back and burns at the valve with any degree of valve opening.

(3) Using the Bureau of Explosives' Open Drum Apparatus[8] there is any significant propagation of flame away from the ignition source.

(4) Using the Bureau of Explosives' Closed Drum Apparatus[8] there is any explosion of the vapor-air mixture in the drum.

If a compressed gas does not meet the criteria of flammability, it is obviously a non-flammable gas. Other definitions involved are:

A "non-liquified compressed gas" is a gas, other than gas in solution, which under the charged pressure is entirely gaseous at a temperature of 70°F.

A "liquified compressed gas" is a gas which under the charged pressure, is partially liquid at a temperature of 70°F.

A "compressed gas in solution" is a non-liquified compressed gas which is dissolved in a solvent.

### Poisonous and Irritating Materials

Poisonous materials are divided into three groups according to the degree of hazard presented in transportation. In decreasing order of hazard they are:

(1) Poison A

(2) Poison B

(3) Irritating Materials

Class A Poisons are extremely dangerous poisonous gases or liquids of such nature that a very small amount of the gas, or vapor of the liquid, mixed with air is dangerous to life. In general, these materials appear listed by technical name in the Hazardous Materials Table.

Class B Poisons are dangerous liquids or solids which are known to be so toxic to man as to afford a hazard to health during transportation, or which, in the absence of human toxicity, are presumed to be toxic to man because they meet one of the following criteria when tested on laboratory animals:

(1) Oral toxicity. Those liquids or solids which produce death within 48 hours in half or more than half of a group of 10 or more white laboratory rats weighing 200 to 300 grams, at a single dose of 50 milligrams or less per kilogram of body weight, when administered orally.

(2) Toxicity on inhalation. Those liquids or solids which produce death within 48 hours in half or more than half of a

group of 10 or more white laboratory rats weighing 200 to 300 grams, when inhaled continuously for a period of one hour or less at a concentration of 2 milligrams or less per liter of vapor, mist, or dust, provided such concentration is likely to be encountered by man when the chemical product is used in any reasonable foreseeable manner.

(3) Toxicity by skin absorption. Those liquids or solids which produce death within 48 hours in half or more than half of a group of 10 or more rabbits tested at a dosage of 200 milligrams or less per kilogram body weight, when administered by continuous contact with the bare skin for 24 hours or less.

If the physical characteristics or the probable hazards to humans shown by experience indicate that the substances will not cause serious sickness or death, positive results in the above animal tests would not be considered a sufficient reason for any material to be classified as a Poison B. This is so despite the display of danger or warning labels pertaining to the use of the material. These danger and warning labels are often required by a different regulatory agency or standard (49CFR 173.343(a)(1), (2) and (3) and 173.343(b)).

Irritating materials are liquids or solids which on contact with fire or when exposed to air give off dangerous or intensely irritating fumes. Examples of this hazard class are tear gas agents. The Irritating Material hazard designation should not be confused with the use of the term irritant made by other agencies or standards. In the DOT sense, the materials classified as irritants must impair a person's ability to function in case of accident during transportation.

## Etiologic Agents

Materials classified as etiologic agents are living microorganisms or the toxins they produce that cause or may cause human disease. Only those agents listed by the Department of Health and Human Services in 42CFR 72.3 are included in this classification. These are:[9]

### Bacterial Agents

*Acinetobacter calcoaceticus.*
*Actinobacillus*—all species.
*Actinomycetaceae*—all members.
*Aeromonas hydrophila.*
*Arachnia propionica.*
*Arizona hinshawii*—all serotypes.
*Bacillus anthracis.*
*Bacteroides* spp.
*Bartonella*—all species.
*Bordetella*—all species.
*Borrelia recurrentis, B. vincenti.*
*Brucella*—all species.
*Campylobacter (Vibrio) foetus, C. (Vibrio) jejuni.*

*Chlamydia psittaci, C. trachomatis.*
*Clostridium botulinum, Cl. chauvoei, Cl. haemolyticum, Cl. histolyticum, Cl. novyi, Cl. septicum, Cl. tetani*
*Corynebacterium diphtheriae, C. equi, C. haemolyticum, C. pseudotuberculosis, C. pyogenes, C. renale.*
*Edwarsiella tarda.*
*Erysipelothrix insidiosa.*
*Escherichia coli,* all enteropathogenic serotypes.
*Francisella (Pasteurella) Tularensis.*
*Haemophilus ducreyi, H. influenzae.*
*Klebsiella*—all species and all serotypes.
*Legionella*—all species and all Legionella-like organisms.
*Leptospira interrogans*—all serovars.
*Listeria*—all species.
*Mimae polymorpha.*
*Moraxella*—all species
*Mycobacterium*-all species.
*Mycoplasma*—all species.
*Neisseria gonorrhoeae, N. meningitidis.*
*Nocardia asteroides.*
*Pasteurella*—all species.
*Plesiomonas shigelloides.*
*Proteus*—all species.
*Pseudomonas mallei.*
*Pseudomonas pseudomallei.*
*Salmonella*—all species and all serotypes.
*Shigella*—all species and all serotypes.
*Sphaerophorus necrophorus.*
*Staphylococcus aureus.*
*Streptobacillus moniliformis.*
*Streptococcus pneumoniae.*
*Streptococcus pyogenes.*
*Treponema careteum, T. pallidum,* and *T. pertenue.*
*Vibrio cholerae, V. parahemolyticus.*
*Yersinia (Pasteurella) pestis, Y. enterocolitica.*

## Fungal Agents

*Blastomyces dermatitidis.*
*Coccidioides immitis.*
*Cryptococcus neoformans.*
*Histoplasma capsulatum.*
*Paracoccidioides brasiliensis.*

## Viral and Rickettsial Agents

Adenoviruses—human—all types.
Arboviruses—all types.
*Coxiella burnetii.*
Coxsackie A and B viruses—all types.

Creutzfeldt—Jacob agent

Cytomegaloviruses.

Dengue viruses—all types.

Ebola virus.

Echoviruses—all types.

Encephalomyocarditis virus.

Hemorrhagic fever agents including, but not limited to, Crimean hermorrhagic fever (Congo), Junin, Machupo viruses, and Korean hemorrhagic fever viruses.

Hepatitis associated materials (hepatitis A, hepatitis B, hepatitis nonA-nonB).

Herpesvirus—all members.

Infectious bronchitis-like virus.

Influenza viruses—all types.

Kuru agent.

Lassa virus.

Lymphocytic choriomeningitis virus.

Marburg virus.

Measles virus.

Mumps virus.

Parainfluenza viruses—all types.

Polioviruses—all types.

Poxviruses—all members.

Rabies virus—all strains.

Reoviruses—all types.

*Rickettsia*—all species.

*Rochalimaea quintana.*

Rotaviruses—all types.

Rubella virus.

Simian virus 40.

Tick-borne encephalitis virus complex, including Russian spring-summer encephalitis, Kyasanur forest disease, Omsk hemorrhagic fever, and Central European encephalitis viruses.

Vaccinia virus.

Varicella virus.

Variola major and Variola minor viruses.

Vesicular stomatis viruses—all types.

White pox viruses.

Yellow fever virus.

## Radioactive Materials

This hazard class covers any material or combination of materials that spontaneously emit ionizing radiation. Radioactive emissions are generally of one of three primary forms: gamma rays, alpha particles or beta particles. Gamma rays are not particulate but are short wavelength electromagnetic radiation from the nucleus of radioactive atoms. They are the most penetrating form of radiation and travel great distances in air. Alpha particles are positively charged particles emitted from the nucleus of atoms having a mass and charge equal to the nucleus of a helium atom (2 protons plus 2 neutrons). They have a very little penetrating ability and travel very short distances in air. Beta particles are negatively charged particles emitted from the nucleus of an atom and have a mass and charge equal to that of an electron. They have an intermediate penetrating ability and travel greater distances in air than alpha particles. These radioactive emissions are capable, because of their high energy, to ionize any atoms in their path. Exposure to ionizing radiation can be highly dangerous and even lethal. For this reason, the primary consideration in shipping radioactive materials is the use of proper packaging for the specific radioactive materials to be transported.

## Other Regulated Materials (ORM)

In general, a material corresponding to the Other Regulated Material (ORM) hazard class is a material that may pose an unreasonable risk to health and safety or property when transported in commerce and that does not meet the definition of any other hazard class. Consumer commodities, as will be seen later, are an exception.

There are five classes of ORM materials, A through E, as follows:

(1) An ORM-A material is a material which has an anesthetic, irritating, noxious, toxic or other similar property and which can cause extreme annoyance or discomfort to passengers and crew in the event of leaking during transportation. Chloroform and carbon tetrachloride belong to this hazard class.

Materials classified as ORM-A are commonly regulated for air shipment only, unless they are also hazardous substances or hazardous wastes (see ORM-E).

(2) An ORM-B material is a material (including a solid when wet with water) capable of causing significant damage to a transport vehicle or vessel from leakage during transportation. Materials meeting one or both of the following criteria are ORM-B materials:

(i) A liquid substance that has a corrosion rate exceeding 0.250 inch per year (IPY) on aluminum (nonclad 7075-T6) at a test temperature of 130°F. An acceptable test is described in NACE Standard TM-01-69.

(ii) Specifically designated by name in the Hazardous Materials Table.

Dilute hypochlorite solutions, stannous chloride solid, etc., are ORM-B materials.

(3) An ORM-C material is a material which has other inherent characteristics not described as an ORM-A or ORM-B but which make it unsuitable for shipment, unless properly identified and prepared for transportation. ORM-C materials are specifically named in the Hazardous Materials Table. Sulfur, solid, when shipped by water, is classified as an ORM-C material.

(4) An ORM-D material is a material such as a consumer commodity which, although a hazardous material, presents a limited hazard during transportation due to its form, quantity and packaging. They must be materials for which exceptions are provided in the Hazardous Materials Table.

(5) An ORM-E is a material that is not included in any other hazard class, but is subject to regulations because it is a:

   (i) Hazardous waste, or a

   (ii) Hazardous substance.

A hazardous waste, as defined by DOT, means "any material subject to the hazardous waste manifest requirements of the EPA specified in 40CFR Part 262" (49CFR 172.8 and *Classification Under RCRA* later in this chapter).

A hazardous substance, as defined by DOT, is a material and its mixture or solutions identified by the letter E in Column 1 of the Hazardous Materials Table (§172.101). Only those chemicals listed under 40CFR Part 116 of the Clean Water Act have an E on the Hazardous Materials Table. Those chemicals are regulated for transportation by DOT as hazardous substances *only* if the quantity of the material contained in one package or in one transport vehicle when not packaged, equals or exceeds the reportable quantity (RQ) for that material. This quantity is shown for each regulated material next to its proper shipping name on the Hazardous Materials Table. In the case of a mixture or solution containing a hazardous substance, it will only be regulated if its concentration is equal *or* greater than that shown on Table 4.1 *and* if the quantity of the hazardous substance contained in the package is equal to or greater than the reportable quantity for that substance.

### Table 4.1: DOT Reportable Quantity-Concentration

| RQ (lb) | RQ (kg) | . Concentration by Weight . (%) | (ppm) |
|---|---|---|---|
| 5,000 | 2,270 | 10 | 100,000 |
| 1,000 | 454 | 2 | 20,000 |
| 100 | 45.4 | 0.2 | 2,000 |
| 10 | 4.54 | 0.02 | 200 |
| 1 | 0.45 | 0.002 | 20 |

The applicability of the ORM-E hazard class as related to hazardous substances is best illustrated by the use of examples.

Dinitrotoluene, liquid, is listed in the Hazardous Materials Table with the letter E on Column 1, indicating it is a hazardous substance. In addition, the RQ follows the proper shipping name, in pounds and kilograms as shown:

Dinitrotoluene, liquid (RQ-1000/454)

When dinitrotoluene, liquid, is shipped in less than 1,000 pound quantity in a container, the chemical is not regulated for transportation. If, however, it is shipped in a quantity equal or greater than 1,000 pounds in a package, the chemical is regulated as an ORM-E material.

Carbon tetrachloride is regulated for shipments by air or water as an ORM-A material, as indicated by the letters A and W on Column 1 of the Hazardous Materials Table. This chemical also has on the table the letter E on column 1. As with all hazardous substances, the reportable quantity (RQ) for that material follows the name of the substance, 5,000 pounds in this case. Thus, if more than 5,000 pounds of carbon tetrachloride are shipped in one package, or in one transport vehicle if not packaged, carbon tetrachloride is regulated, as a *hazardous substance,* by all modes of transportation. However, the hazard class remains ORM-A and the regulatory requirements are met by the addition of the letter RQ to the proper shipping description.

The next example illustrates the use of the Concentration Table. Product "Formula XYZ" contains 20% by weight of carbon tetrachloride. Because this concentration exceeds the 10% limit shown on Table 4.1 for hazardous substances having an RQ of 5,000 pounds, "Formula XYZ" will be regulated as a hazardous substance when shipped in one package in quantities equal or greater than 25,000 pounds. This quantity contains the 5,000 pounds reportable quantity for carbon tetrachloride.

If the concentration of carbon tetrachloride in "Formula XYZ" were 5%, the product would be exempted from the regulation as a hazardous substance regardless of the quantity shipped in one package or in a transport vehicle if not packaged.

## Materials Meeting the Criteria of More Than One Hazard Class

If a material to be classified meets the definition of more than one hazard class, its hazard class must be assigned according to the following order of precedence (§173.2(a)):

> Radioactive material.
> Poison A.
> Flammable gas.
> Non-flammable gas.
> Flammable liquid.
> Oxidizer.
> Flammable solid.
> Corrosive material (liquid).
> Poison B.
> Corrosive material (solid).

> Combustible liquid (in containers having capacities exceed-
> ing 100 gallons or less).
> ORM-B
> ORM-A
> Combustible liquid (in containers having capacities of 110
> gallons or less).
> ORM-E.

Thus, if a liquid material meets the definition of flammability and corrosivity, its hazard class must be Flammable liquid. However, as discussed in Chapter 14, for shipment this material will require both labels: *Flammable Liquid* and *Corrosive*.

### Selection of a Proper Shipping Name

Once it has been determined that a material is a hazardous material and the appropriate hazard class has been assigned, a proper shipping name must be selected. The Hazardous Materials Table lists a number of general entries from which the selection must be made. Proper shipping name assignment is an important step in the preparation of a material for transportation, since a number of proper shipping names corresponding to the same hazard class require different specification packagings, with other requirements such as allowable quantity in one package for air shipment and exemptions and exceptions being different also. In addition, the identification number which leads to emergency response action may also be different. For these reasons, selection of the proper shipping name that best describes the material must be made by regulatory requirement.

For example, an alcohol not listed by name in the Hazardous Materials Table, which meets the definition of a Flammable liquid can be assigned the proper shipping names Alcohol, n.o.s., or Flammable liquid, n.o.s. The identification numbers and specific packing requirements for these entries are different. The regulation requires the selection of Alcohol, n.o.s, as the proper shipping name, since this name best describes the material to be shipped.

When the material meets more than one hazard class, the proper shipping name must reflect this fact. For example, a liquid that is flammable and corrosive must be assigned the proper shipping name Flammable liquid, corrosive, n.o.s., if its technical name does not appear on the Hazardous Materials Table.

## CLASSIFICATION UNDER RCRA REGULATION

The regulations promulgated by the EPA under RCRA are a complex and extensive set of rules designed to create a "cradle-to-grave" system of accounting for hazardous wastes. The regulations published in the *Federal Register* of May 19, 1980, have been amended, interpreted and expanded several times before and after their effective date, November 19, 1980. Thus, correct classification of a waste under RCRA requires constant monitoring of new regulatory developments. The identification and listing of hazardous wastes can be found in Title 40 of the Code of Federal Regulations Part 261.

In contrast with hazard classification of a material under DOT regulations, classification under RCRA does not take into account degree of hazard. A waste is either hazardous and thus fully regulated or non-hazardous and not covered by the regulation. Once it has been established that a waste meets the criteria of being a hazardous waste under the Act, full implementation of the regulation must follow, for the storage, treatment or disposal of the material. Hazard determinations must be applied to both wastes that are disposed of off-site as well as wastes that remain on-site. Transportation of hazardous wastes disposed off-site is regulated under DOT and RCRA with requirements that were coordinated by both agencies.

### Definition of Hazardous Wastes

The starting point for the hazardous waste classification system is the definition of a solid waste, since no material can be a hazardous waste without first being a solid waste. The definition of a solid waste is as follows (§261.2):[13]

> (a) A solid waste is any garbage, refuse, sludge or any other waste material which is not excluded under §261.4(a).

> (b) An "other waste material" is any solid, semi-solid or contained gaseous material, resulting from industrial, commercial, mining or agricultural operations, or from community activities which:

>> (1) Is discarded or is being accumulated, stored or physically, chemically or biologically treated prior to being discarded; or

>> (2) Has served its original intended use and sometimes is discarded; or

>> (3) Is a manufacturing or mining by-product and sometimes is discarded.

> (c) A material is "discarded" if it is abandoned (and not used, reused, reclaimed or recycled) by being:

>> (1) Disposed of; or

>> (2) Burned or incinerated, except where the material is being burned as a fuel for the purpose of recovering usable energy; or

>> (3) Physically, chemically, or biologically treated (other than burned or incinerated) in lieu of or prior to being disposed of.

> (d) A material is "disposed of" if it is discharged, deposited, injected, dumped, spilled, leaked or placed into or on any

land or water so that such material or any constituent thereof may enter the environment or be emitted into the air or discharged into ground or surface waters.

(e) A "manufacturing or mining by-product" is a material that is not one of the primary products of a particular manufacturing or mining operation, is a secondary and incidental product of the particular operation and would not be solely and separately manufactured or mined by the particular manufacturing or mining operation. The term does not include an intermediate manufacturing or mining product which results from one of the steps in a manufacturing or mining process and is typically processed through the next step of the process within a short time.

This definition of solid waste has caused great confusion. The expression "and sometimes is discarded" has led to ambiguities in deciding if a by-product which is reused by a segment of the regulated community and discarded by others is to be considered a solid waste.

Section 261.4(a) addresses materials which are not solid waste for the purpose of the regulation. These are:

(1) (i) Domestic sewage; and
(ii) Any mixture of domestic sewage and other wastes that passes through a sewer system to a publicly-owned treatment works for treatment.

"Domestic sewage" means untreated sanitary wastes that pass through a sewer system.

(2) Industrial wastewater discharges that are point source discharges subject to regulation under Section 402 of the Clean Water Act, as amended.
(Comment: This exclusion applies only to the actual point source discharge. It does not exclude industrial wastewaters while they are being collected, stored or treated before discharge, nor does it exclude sludges that are generated by industrial wastewater treatment).

(3) Irrigation return flows.

(4) Source, special nuclear or by-product material as defined by the Atomic Energy Act of 1954, as amended 42 U.S.C. 2011 et seq.

(5) Materials subjected to in-situ mining techniques which are not removed from the ground as part of the extraction process.

A solid waste is a hazardous waste if:

1) It is listed; or

2) It is a mixture of solid waste and one or more hazardous waste listed; or

3) It exhibits any of the characteristics of ignitability, corrosivity, reactivity or toxicity as defined (see below).

In Subparts D of the regulations a number of wastes from non-specific as well as from specific processes are listed which are considered hazardous unless proven otherwise by the process of delisting (Sections 261.31 and 261.32 of the regulation respectively). These lists are shown on Table 4.2 and 4.3. The basis for listing a waste as hazardous in §261.31 (Table 4.2) and 261.32 (Table 4.3) is indicated alongside the name or description of each waste. In initiating a delisting petition for a listed waste, it must be proven that the waste does not possess the hazard indicated or any of the characteristics enumerated in three above.

A number of commercial products, their off-specification species and spill residues and debris thereof are listed as materials that when discarded or intended to be discarded, would become hazardous wastes. Two lists of substances are given under Sections 261.33(e) and 261.33(f) of the regulation (Tables 4.4 and 4.5). It should be understood that these lists include not only the commercially pure grade of the chemical, but also any technical grades of the chemical that are produced or marketed and all formulations in which the chemical is the sole active ingredient.

The chemical products listed on Table 4.4 are identified as acute hazardous wastes and a container or inner liner removed from a container that was used to hold any of these products is also considered a hazardous waste unless it has been triply rinsed.

It is important to note that solid wastes which contain one or more of the chemicals listed in Tables 4.4 and 4.5 are not considered hazardous merely by the presence of these materials. These solid wastes would be hazardous if listed in Sections 261.31 or 261.32 (Tables 4.2 and 4.3) or if they meet one or more of the hazard characteristics. However, when a solid waste is mixed with one of these discarded materials, the resulting mixture is a hazardous waste.

If the waste is not listed or if it is not a mixture of a listed waste and a solid waste, then it must be determined if the waste exhibits any of the four hazard characteristics identified by RCRA: ignitability, corrosivity, reactivity or toxicity (as demonstrated by the extraction procedure).

**Characteristic of Ignitability:** A solid waste exhibits the characteristic of ignitability if a representative sample has any of the following properties:

(1) It is a liquid, other than an aqueous solution containing less than 24 percent alcohol by volume and has flash point less than 60°C (140°F), as determined by a Pensky-Martens Closed Cup Tester, using the test method specified in ASTM Standard D-93-79 or D-93-80 or a Setaflash Closed Cup Tester, using the test method specified in ASTM Standard D-3278-78, or as determined by an equivalent test method approved by the Administrator.[3]

## Table 4.2: RCRA Listed Hazardous Wastes from Nonspecific Sources

| Industry and EPA hazardous waste number | Hazardous waste | Hazard Code* |
|---|---|---|
| Generic: | | |
| F001 . . | The following spent halogenated solvents used in degreasing: tetrachloroethylene, trichloroethylene, methylene chloride, 1,1,1-trichloroethane, carbon tetrachloride, and chlorinated fluorocarbons; and sludges from the recovery of the solvents in degreasing operations. | (T) |
| F002 . . | The following spent halogenated solvents: tetrachloroethylene, methylene chloride, trichloroethylene, 1,1,1-trichloroethane, chlorobenzene, 1,1,2-trichloro-1,2,2-trifluoroethane, ortho-dichlorobenzene, and trichlorofluoromethane; and the still bottoms from the recovery of these solvents. | (T) |
| F003 . . | The following spent non-halogenated solvents: xylene, acetone, ethyl acetate, ethyl benzene, ethyl ether, methyl isobutyl ketone, n-butyl alcohol, cyclohexanone, and methanol; and the still bottoms from the recovery of these solvents. | (I) |
| F004 . . | The following spent non-halogenated solvents: cresols and cresylic acid, and nitrobenzene; and the still bottoms from the recovery of the solvents. | (T) |
| F005 . . | The following spent non-halogenated solvents: toluene, methyl ethyl ketone, carbon disulfide, isobutanol, and pyridine; and the still bottoms from the recovery of these solvents. | (I,T) |
| F006 . . | Wastewater treatment sludges from electroplating operations except from the following processes: (1) sulfuric acid anodizing of aluminum (2) tin plating of carbon steel; (3) zinc plating (segregated basis) on carbon steel; (4) aluminum or zinc-aluminum plating on carbon steel; (5) cleaning/stripping associated with tin, zinc and aluminum plating on carbon steel; and (6) chemical etching and milling of aluminum. | (T) |
| F019 . . | Wastewater treatment sludges from the chemical conversion coating of aluminum. | (T) |
| F007 . . | Spent cyanide plating bath solutions from electroplating operations (except for precious metals electroplating spent cyanide plating bath solutions). | (R,T) |
| F008 . . | Plating bath sludges from the bottom of plating baths from electroplating operations where cyanides are used in the process (except for precious metals electroplating plating bath sludges). | (R,T) |
| F009 . . | Spent stripping and cleaning bath solutions from electroplating operations where cyanides are used in the process (except for precious metals electroplating spent stripping and cleaning bath solutions). | (R,T) |
| F010 . . | Quenching bath sludge from oil baths from metal heat treating operations where cyanides are used in the process (except for precious metals heat-treating quenching bath sludges). | (R,T) |
| F011 . . | Spent cyanide solutions from salt bath pot cleaning from metal heat treating operations (except for precious metals heat treating spent cyanide solutions from salt bath pot cleaning). | (R,T) |
| F012 . . | Quenching wastewater treatment sludges from metal heat treating operations where cyanides are used in the process (except for precious metals heat treating quenching wastewater treatment sludges). | (T) |

```
* Ignitable Waste  . . . . . . . . . .  (I)
  Corrosive Waste  . . . . . . . . . .  (C)
  Reactive Waste   . . . . . . . . . .  (R)
  EP Toxic Waste   . . . . . . . . . .  (E)
  Acute Hazardous Waste  . . . . . . .  (H)
  Toxic Waste  . . . . . . . . . . . .  (T)
```

## Table 4.3: RCRA Listed Hazardous Wastes from Specific Sources

| Industry and EPA hazardous waste number | Hazardous waste | Hazard Code* |
|---|---|---|
| **Wood Preservation:** | | |
| K001 . . | Bottom sediment sludge from the treatment of wastewaters from wood preserving processes that use creosote and/or pentachlorophenol. | (T) |
| **Inorganic Pigments:** | | |
| K002 . . | Wastewater treatment sludge from the production of chrome yellow and orange pigments. | (T) |
| K003 . . | Wastewater treatment sludge from the production of molybdate orange pigments | (T) |
| K004 . . | Wastewater treatment sludge from the production of zinc yellow pigments | (T) |
| K005 . . | Wastewater treatment sludge from the production of chrome green pigments | (T) |
| K006 . . | Wastewater treatment sludge from the production of chrome oxide green pigments (anhydrous and hydrated). | (T) |
| K007 . . | Wastewater treatment sludge from the production of iron blue pigments | (T) |
| K008 . . | Oven residue from the production of chrome oxide green pigments | (T) |
| **Organic Chemicals:** | | |
| K009 . . | Distillation bottoms from the production of acetaldehyde from ethylene | (T) |
| K010 . . | Distillation side cuts from the production of acetaldehyde from ethylene | (T) |
| K011 . . | Bottom stream from the wastewater stripper in the production of acrylonitrile | (R,T) |
| K013 . . | Bottom stream from the acetonitrile column in the production of acrylonitrile | (R,T) |
| K014 . . | Bottoms from the purification column in the production of acrylonitrile | (T) |
| K015 . . | Still bottoms from the distillation of benzyl chloride | (T) |
| K016 . . | Heavy ends or distillation residues from the production of carbon tetrachloride | (T) |
| K017 . . | Heavy ends (still bottoms) from the purification column in the production of epichlorohydrin | (T) |
| K018 . . | Heavy ends from the fractionation column in ethyl chloride production | (T) |
| K019 . . | Heavy ends from the distillation of ethylene dichloride in ethylene dichloride production | (T) |
| K020 . . | Heavy ends from the distillation of vinyl chloride in vinyl chloride monomer production | (T) |
| K021 . . | Aqueous spent antimony catalyst waste from fluoromethanes production | (T) |
| K022 . . | Distillation bottom tars from the production of phenol/acetone from cumene | (T) |
| K023 . . | Distillation lights ends from the production of phthalic anhydride from naphthlene | (T) |
| K024 . . | Distillation bottoms from the production of phthalic anhydride from naphthalene | (T) |
| K093 . . | Distillation light ends from the production of phthalic anhydride from ortho-xylene | (T) |
| K094 . . | Distillation bottoms from the production of phthalic anhydride from ortho-xylene | (T) |
| K025 . . | Distillation bottoms from the production of nitrobenzene by the nitration of of benzene | (T) |
| K026 . . | Stripping still tails from the production of methyl ethyl pyridines | (T) |
| K027 . . | Centrifuge and distillation residues from toluene diisocyanate production | (R,T) |
| K028 . . | Spent catalyst from the hydrochlorinator reactor in the production of 1,1,1-trichloroethane | (T) |
| K029 . . | Waste from the product steam stripper in the production of 1,1,1-trichloroethane | (T) |
| K095 . . | Distillation bottoms from the production of 1,1,1-trichloroethane | (T) |
| K096 . . | Heavy ends from the heavy ends column from the production of 1,1,1-trichloroethane | (T) |
| K030 . . | Column bottoms or heavy ends from the combined production of trichloroethylene and perchloroethylene | (T) |
| K083 . . | Distillation bottoms from aniline production | (T) |

**(continued)**

## Table 4.3: (continued)

| Industry and EPA hazardous waste number | Hazardous waste | Hazard Code* |
|---|---|---|

**Organic Chemicals (continued):**

K103 . . Process residues from aniline extraction from the production of aniline     (T)

K104 . . Combined wastewater streams generated from nitrobenzene/aniline production     (T)

K085 . . Distillation or fractionation column bottoms from the production of chloro-benzenes     (T)

K105 . . Separated aqueous stream from the reactor product washing step in the production of chlorobenzenes     (T)

**Inorganic Chemicals:**

K071 . . Brine purification muds from the mercury cell process in chlorine production where separately prepurified brine is not used.     (T)

K073 . . Chlorinated hydrocarbon waste from the purification step of the diaphragm cell process using graphite anodes in chlorine production.     (T)

K106 . . Wastewater treatment sludge from the mercury cell process in chlorine production     (T)

**Pesticides:**

K031 . . By-product salts generated in the production of MSMA and cacodylic acid     (T)

K032 . . Wastewater treatment sludge from the production of chlordane     (T)

K033 . . Wastewater and scrub water from the chlorination of cyclopentadiene in the production of chlordane     (T)

K034 . . Filter solids from the filtration of hexachlorocyclopentadiene in the production of chlordane     (T)

K097 . . Vacuum stripper discharge from the chlordane chlorinator in the production of chlordane     (T)

K035 . . Wastewater treatment sludges generated in the production of creosote     (T)

K036 . . Still bottoms from toluene reclamation distillation in the production of disulfotone     (T)

K037 . . Wastewater treatment sludges from the production of disulfoton     (T)

K038 . . Wastewater from the washing and stripping of phorate production     (T)

K039 . . Filter cake from the filtration of diethylphosphorodithioic acid in the production of phorate     (T)

K040 . . Wastewater treatment sludge from the production of phorate     (T)

K041 . . Wastewater treatment sludge from the production of toxaphene     (T)

K098 . . Untreated process wastewater from the production of toxaphene     (T)

K042 . . Heavy ends or distillation residues from the distillation of tetrachlorobenzene in the production of 2,4,5-T     (T)

K043 . . 2,6-Dichlorophenol waste from the production of 2,4-D     (T)

K099 . . Untreated wastewater from the production of 2,4-D     (T)

**Explosives:**

K044 . . Wastewater treatment sludges from the manufacturing and processing of explosives     (R)

K045 . . Spent carbon from the treatment of wastewater containing explosives     (R)

K046 . . Wastewater treatment sludges from the manufacturing, formulation and loading of lead-based initiating compounds     (T)

K047 . . Pink/red water from TNT operations     (R)

**Petroleum Refining:**

K048 . . Dissolved air flotation (DAF) float from the petroleum refining industry     (T)

K049 . . Slop oil emulsion solids from the petroleum refining industry     (T)

K050 . . Heat exchanger bundle cleaning sludge from the petroleum refining industry     (T)

K051 . . API separator sludge from the petroleum refining industry     (T)

**(continued)**

## Table 4.3: (continued)

| Industry and EPA hazardous waste number | Hazardous waste | Hazard Code* |
|---|---|---|
| | **Petroleum Refining (continued):** | |
| K052 . . | Tank bottoms (leaded) from the petroleum refining industry | (T) |
| | **Iron and Steel:** | |
| K061 . . | Emission control dust/sludge from the primary production of steel in electric furnaces | (T) |
| K062 . . | Spent pickle liquor from steel finishing operations | (C,T) |
| | **Secondary Lead:** | |
| K069 . . | Emission control dust/sludge from secondary lead smelting | (T) |
| K100 . . | Waste leaching solution from acid leaching of emission control dust/sludge from secondary lead smelting | (T) |
| | **Veterinary Pharmaceuticals:** | |
| K084 . . | Wastewater treatment sludges generated during the production of veterinary pharmaceuticals from arsenic or organo-arsenic compounds | (T) |
| K101 . . | Distillation tar residues from the distillation of aniline-based compounds in the production of veterinary pharmaceuticals from arsenic or organo-arsenic compounds. | (T) |
| K102 . . | Residue from the use of activated carbon for decolorization in the production of veterinary pharmaceuticals from arsenic or organo-arsenic compounds | (T) |
| | **Ink Formulation:** | |
| K086 . . | Solvent washes and sludges, caustic washes and sludges, or water washes and sludges from cleaning tubs and equipment used in the formulation of ink from pigments, driers, soaps, and stabilizers containing chromium and lead. | (T) |
| | **Coking:** | |
| K060 . . | Ammonia still-lime sludge from coking operations | (T) |
| K087 . . | Decanter tank tar sludge from coking operations | (T) |

## Table 4.4: RCRA 261.33(e) Table

| Hazardous waste No. | Substance | Hazardous waste No. | Substance |
|---|---|---|---|
| P023 . . | Acetaldehyde, chloro- | P031 . . | Cyanogen |
| P002 . . | Acetamide, N-(aminothioxomethyl)- | P033 . . | Cyanogen chloride |
| P057 . . | Acetamide, 2-fluoro- | P036 . . | Dichlorophenylarsine |
| P058 . . | Acetic acid, fluoro-, sodium salt | P037 . . | Dieldrin |
| P066 . . | Acetimidic acid, N((methylcarbamoyl) oxy)thio-, methyl ester | P038 . . | Diethylarsine |
| | | P039 . . | O,O-Diethyl S-(2-(ethylthio)ethyl) phosphorodithioate |
| P001 . . | 3-(alpha-Acetonyl-benzyl)-4-hydroxycoumarin and salts | P041 . . | Diethyl-p-nitrophenyl phosphate |
| P002 . . | 1-Acetyl-2-thiourea | P040 . . | O,O-Diethyl O-pyrazinyl phosphorothioate |
| P003 . . | Acrolein | | |
| P070 . . | Aldicarb | P043 . . | Diisopropyl fluorophosphate |
| P004 . . | Aldrin | P044 . . | Dimethoate |
| P005 . . | Allyl alcohol | P045 . . | 3,3-Dimethyl-1-(methylthio)-2-butanone, O-((methylamino)carbonyl) oxime |
| P006 . . | Aluminum phosphide (R,T) | | |
| P007 . . | 5-(Aminomethyl)-3-isoxazolol) | | |
| P008 . . | 4-Aminopyridine | P071 . . | O,O-Dimethyl O-p-nitrophenyl phosphorothioate |
| P009 . . | Ammonium picrate (R) | | |
| P119 . . | Ammonium vanadate | P082 . . | Dimethylnitrosamine |
| P010 . . | Arsenic Acid | P046 . . | Alpha, alpha-Dimethylphenethylamine |
| P012 . . | Arsenic (III) oxide | P047 . . | 4,6-Dinitro-o-cresol and salts |
| P011 . . | Arsenic (V) oxide | P034 . . | 4,6-Dinitro-o-cyclohexylphenol |
| P011 . . | Arsenic pentoxide | P048 . . | 2,4-Dinitrophenol |
| P012 . . | Arsenic trioxide | P020 . . | Dinoseb |
| P038 . . | Arsine, diethyl- | P035 . . | Diphosphoramide, octamethyl- |
| P054 . . | Aziridine | P039 . . | Disulfoton |
| P013 . . | Barium cyanide | P049 . . | 2,4-Dithiobiuret |
| P024 . . | Benzenamine, 4-chloro- | P084 . . | Ethenamine, N-methyl-N-nitroso- |
| P077 . . | Benzenmine, 4-nitro- | P101 . . | Ethyl cyanide |
| P028 . . | Benzene, (chloromethyl)- | P054 . . | Ethylenimine |
| P042 . . | 1,2-Benzenediol, 4-(1-hydroxy-2-methyl-amino)ethyl)- | P097 . . | Famphur |
| | | P056 . . | Fluorine |
| P014 . . | Benzenethiol | P057 . . | Fluoroacetamide |
| P028 . . | Benzyl chloride | P058 . . | Fluoroacetic acid, sodium salt |
| P015 . . | Beryllium dust | P109 . . | Dithiopyrophosphoric acid, tetraethyl ester |
| P016 . . | Bis(chloromethyl)ether | | |
| P017 . . | Bromoacetone | P050 . . | Endosulfan |
| P018 . . | Brucine | P088 . . | Endothall |
| P021 . . | Calcium cyanide | P051 . . | Endrin |
| P123 . . | Camphene, octachloro- | P042 . . | Epinephrine |
| P103 . . | Carbamimidoselenoic acid | P046 . . | Ethanamine, 1,1-dimethyl-2-phenyl- |
| P022 . . | Carbon bisulfide | P065 . . | Fulminic acid, mercury(II) salt (R,T) |
| P022 . . | Carbon disulfide | P059 . . | Heptachlor |
| P095 . . | Carbonyl chloride | P051 . . | 1,2,3,4,10,10-Hexachloro-6,7-epoxy-1,4,4a,5,6,7,8,8a-octahydro-endo, endo-1,4:5,8-dimethanonaphthalene |
| P033 . . | Chlorine cyanide | | |
| P023 . . | Chloroacetaldehyde | | |
| P024 . . | p-Chloroaniline | P037 . . | 1,2,3,4,10,10-Hexachloro-6,7-epoxy-1,4,4a,5,6,7,8,8a-octahydro-endo, exo-1,4,5,8-demethanonaphthalene |
| P026 . . | 1-(o-Chlorophenyl)thiourea | | |
| P027 . . | 3-Chloropropionitrile | | |
| P029 . . | Copper cyanides | P060 . . | 1,2,3,4,10,10-Hexachloro-1,4,4a,5,8,8a-hexahydro-1,4;5,8-endo,endo-dimethanonaphthalene |
| P030 . . | Cyanides (soluble cyanide salts), | | |

(continued)

## Table 4.4: (continued)

| Hazardous waste No. | Substance | Hazardous waste No. | Substance |
|---|---|---|---|
| P004 . . | 1,2,3,4,10,10-Hexachloro-1,4,4a,5, 8,8a-hexahydro-1,4,5,8-endo,exodi-methanonaphthalene | P048 . . | Phenol, 2,4-dinitro |
| | | P047 . . | Phenol, 2,4-dinitro-6-methyl- and salts |
| P060 . . | Hexachlorohexahydro-endo,endo-dimethanonaphthalene | P020 . . | Phenol, 2,4-dinitro-6-(1-methylpropyl) |
| P062 . . | Hexaethyl tetraphosphate | P009 . . | Phenol, 2,4,6,-trinitro-, ammonium salt (R) |
| P116 . . | Hydrazinecarbothioamide | | |
| P068 . . | Hydrazine methyl- | P036 . . | Phenyl dichloroarsine |
| P063 . . | Hydrocyanic acid | P092 . . | Phenylmercuric acetate |
| P063 . . | Hydrogen cyanide | P093 . . | N-Phenylthiourea |
| P096 . . | Hydrogen phosphide | P094 . . | Phorate |
| P064 . . | Isocyanic acid, methyl ester | P095 . . | Phosgene |
| P007 . . | 3(2H)-Isoxazolone, 5-(aminomethyl)- | P096 . . | Phosphine |
| P092 . . | Mercury, (acetato-O)phenyl- | P041 . . | Phosphoric acid, diethyl p-nitrophenyl ester |
| P065 . . | Mercury fulminate (R,T) | | |
| P016 . . | Methane, oxybis(chloro- | P044 . . | Phosphorodithioic acid, O,O-dimethyl S-(2-(methylamino-2-oxoethyl)ester |
| P112 . . | Methane, tetranitro-(R) | | |
| P118 . . | Methanethiol, trichloro- | P043 . . | Phosphorofluoridic acid, bis(1-methyl-ethyl)-ester |
| P059 . . | 4,7-Methano-1H-indene, 1,4,5,6,7,8, 8-heptachloro-3a,4,7,7a-tetrahydro- | P094 . . | Phosphorothioic acid, O,O-diethyl S-(ethylthio)methyl ester |
| P066 . . | Methomyl | P089 . . | Phosphorothioic acid, O,O-diethyl O-(p-nitrophenyl)ester |
| P067 . . | 2-Methylaziridine | | |
| P068 . . | Methyl hydrazine | P040 . . | Phosphorothioic acid, O,O-diethyl O-pyrazinyl ester |
| P064 . . | Methyl isocyanate | | |
| P069 . . | 2-Methyllactonitrile | P097 . . | Phosphorothioic acid, O,O-dimethyl O-(p-((dimethylamino)-sulfonyl)phenyl) ester |
| P071 . . | Methyl parathion | | |
| P072 . . | alpha-Naphthylthiourea | | |
| P073 . . | Nickel carbonyl | P110 . . | Plumbane, tetraethyl- |
| P074 . . | Nickel cyanide | P098 . . | Potassium cyanide |
| P074 . . | Nickel (II) cyanide | P099 . . | Potassium silver cyanide |
| P073 . . | Nickel tetracarbonyl | P070 . . | Propanal, 2-methyl-2-(methylthio)-, O-((methylamino)carbonyl)oxime |
| P075 . . | Nicotine and salts | | |
| P076 . . | Nitric oxide | P101 . . | Propanenitrile |
| P077 . . | p-Nitroaniline | P027 . . | Propanenitrile, 3-chloro- |
| P078 . . | Nitrogen dioxide | P069 . . | Propanenitrile, 2-hydroxy-2-methyl- |
| P076 . . | Nitrogen (II) oxide | P081 . . | 1,2,3-Propanetriol, trinitrate-(R) |
| P078 . . | Nitrogen (IV) oxide | P017 . . | 2-Propanone, 1-bromo- |
| P081 . . | Nitroglycerine (R) | P102 . . | Propargyl alcohol |
| P082 . . | N-Nitrosodimethylamine | P003 . . | 2-Propenal |
| P084 . . | N-Nitrosomethylvinylamine | P005 . . | 2-Propen-1-ol |
| P050 . . | 5-Norbonene-2,3-dimethanol, 1,4,5,6,7, 7-hexachloro,cyclic sulfite | P067 . . | 1,2-Propylenimine |
| | | P102 . . | 2-Propyn-1-ol |
| P085 . . | Octamethylpyrophosphoramide | P008 . . | 4-Pyridinamine |
| P087 . . | Osmium oxide | P075 . . | Pyridine, (S)-3-(1-methyl-2-pyroli-dinyl)-, and salts |
| P087 . . | Osmium tetroxide | | |
| P088 . . | 7-Oxabicyclo(2.2.1)heptane-2,3-dicarboxylic acid | P111 . . | Pyrophosphoric acid, tetraethyl ester |
| | | P103 . . | Selenourea |
| P089 . . | Parathion | P104 . . | Silver cyanide |
| P034 . . | Phenol, 2-cyclohexyl-4,6-dinitro- | P105 . . | Sodium azide |

(continued)

## Table 4.4: (continued)

| Hazardous waste No. | Substance | Hazardous waste No. | Substance |
|---|---|---|---|
| P106 . . | Sodium cyanide | P045 . . | Thiofanox |
| P107 . . | Strontium sulfide | P049 . . | Thioimidodicarbonic diamide |
| P108 . . | Strychnidin-10-one, and salts | P014 . . | Thiophenol |
| P018 . . | Strychnidin-10-one,2,3-dimethoxy- | P116 . . | Thiosemicarbazide |
| P108 . . | Strychnine and salts | P026 . . | Thiourea, (2-chlorophenyl)- |
| P115 . . | Sulfuric acid, thallium(I) salt | P072 . . | Thiourea, 1-naphthalenyl- |
| P109 . . | Tetraethyldithiopyrophosphate | P093 . . | Thiourea, phenyl- |
| P110 . . | Tetraethyl lead | P123 . . | Toxaphene |
| P111 . . | Tetraethylpyrophosphate | P118 . . | Trichloromethanethiol |
| P112 . . | Tetranitromethane (R) | P119 . . | Vanadic acid, ammonium salt |
| P062 . . | Tetraphosphoric acid, hexaethyl ester | P120 . . | Vanadium pentoxide |
| P113 . . | Thallic oxide | P120 . . | Vanadium (V) oxide |
| P113 . . | Thallium (III) oxide | P001 . . | Warfarin |
| P114 . . | Thallium (I) selenide | P121 . . | Zinc cyanide |
| P115 . . | Thallium (I) sulfate | P122 . . | Zinc phosphide (R,T) |

(2) It is not a liquid and is capable, under standard temperature and pressure, of causing fire through friction, absorption of moisture or spontaneous chemical changes and, when ignited, burns so vigorously and persistently that it creates a hazard.

(3) It is an ignitable compressed gas as defined by DOT, as determined by the test methods provided by that agency or equivalent test methods approved by the Administrator.

(4) It is an oxidizer as defined by DOT (see *Flammable Solids, Oxidizers and Organic Peroxides* earlier in the chapter).

**Characteristic of Corrosivity:** A solid waste exhibits the characteristic of corrosivity if a representative sample of the waste has either of the following properties:

(1) It is aqueous and has a pH less than or equal to 2 or greater than or equal to 12.5, as determined by a pH meter using either an EPA test method or an equivalent test method approved by the Administrator. The EPA test method for pH is specified as Method 5.2 in "Test Methods for the Evaluation of Solid Waste, Physical/Chemical Methods."[10]

(2) It is a liquid and corrodes steel (SAE 1020) at a rate greater than 6.35 mm (0.250 inch) per year at a test temperature of 55°C (130°F) as determined by the test method specified in NACE (National Association of Corrosion Engineers) Standard TM-01-69[6] or an equivalent test method approved by the Administrator.

## Table 4.5: RCRA 261.33(f) Table

| Hazardous waste No. | Substance | Hazardous waste No. | Substance |
|---|---|---|---|
| U001 . . | Acetaldehyde (I) | U028 . . | 1,2-Benzenedicarboxylic acid, (bis(2-ethylhexyl))ester |
| U034 . . | Acetaldehyde, trichloro- | | |
| U187 . . | Acetamide, N-(4-ethoxyphenyl)- | U069 . . | 1,2-Benzenedicarboxylic acid, dibutyl ester |
| U005 . . | Acetamide, N-9H-fluoren-2-yl- | | |
| U112 . . | Acetic acid, ethyl ester (I) | U088 . . | 1,2-Benzenedicarboxylic acid, diethyl ester |
| U144 . . | Acetic acid, lead salt | | |
| U214 . . | Acetic acid, thallium (I) salt | U102 . . | 1,2-Benzenedicarboxylic acid, dimethyl ester |
| U002 . . | Acetone (I) | | |
| U003 . . | Acetonitrile (I,T) | U107 . . | 1,2-Benzenedicarboxylic acid, di-n-octyl ester |
| U004 . . | Acetophenone | | |
| U008 . . | 2-Acetylaminofluorene | U070 . . | Benzene, 1,2-dichloro- |
| U006 . . | Acetyl chloride (C,R,T) | U071 . . | Benzene, 1,3-dichloro- |
| U007 . . | Acrylamide | U072 . . | Benzene, 1,4-dichloro |
| U008 . . | Acrylic acid (I) | U017 . . | Benzene, (dichloromethyl) |
| U009 . . | Acrylonitrile | U223 . . | Benzene, 1,3-diisocyanatomethyl-(R,T) |
| U150 . . | Alanine, 3-(p-bis(2-chloroethyl)amino)phenyl-,L- | U239 . . | Benzene, dimethyl-(I,T) |
| | | U201 . . | 1,3-Benzenediol |
| U011 . . | Amitrole | U127 . . | Benzene, hexachloro- |
| U012 . . | Aniline (I,T) | U056 . . | Benzene, hexahydro-(I) |
| U014 . . | Auramine | U188 . . | Benzene, hydroxy- |
| U015 . . | Azaserine | U220 . . | Benzene, methyl- |
| U010 . . | Azirino(2',3':3,4)pyrrolo(1,2-a)indole-4,7-dione, 6-amino-8-((aminocarbonyl)oxy)methyl)-1,1a,2,8,8a,8b-hexahydro-8a-methoxy-5-methyl- | U105 . . | Benzene, 1-methyl 1-2,4-dinitro- |
| | | U106 . . | Benzene, 1-methyl-2,6-dinitro |
| | | U203 . . | Benzene, 1,2-methylenedioxy-4-allyl |
| U157 . . | Benz(j)aceanthrylene, 1,2-dihydro-3-methyl- | U141 . . | Benzene, 1,2-methylenedioxy-4-propenyl- |
| | | U090 . . | Benzene, 1,2-methylenedioxy-4-propyl- |
| U016 . . | Benz(c)acridine | U055 . . | Benzene, (1-methylethyl)- (I) |
| U016 . . | 3,4-Benzacridine | U169 . . | Benzene, nitro- (I,T) |
| U017 . . | Benzal chloride | U183 . . | Benzene, pentachloro |
| U018 . . | Benz(a)anthracene | U185 . . | Benzene, pentachloronitro- |
| U018 . . | 1,2-Benzanthracene | U020 . . | Benzenesulfonic acid chloride (C,R) |
| U094 . . | 1,2-Benzanthracene,7,12-dimethyl- | U020 . . | Benzenesulfonyl chloride (C,R) |
| U012 . . | Benzenamine (I,T) | U207 . . | Benzene, 1,2,4,5-tetrachloro- |
| U014 . . | Benzenamine, 4,4'-carbonimidoylbis(N,N-dimethyl- | U023 . . | Benzene, (trichloromethyl)-(C,R,T) |
| | | U234 . . | Benzene, 1,3,5-trinitro- (R,T) |
| U049 . . | Benzenamine, 4-chloro-2-methyl- | U021 . . | Benzidine |
| U093 . . | Benzenamine, N,N'-dimethyl-4-(phenylazo) | U202 . . | 1,2-Benzisothiazolin-3-one, 1,1-dioxide, and salts |
| U158 . . | Benzenamine 4,4' methylenebis(2-chloro- | | |
| U222 . . | Benzenamine, 2-methyl-, hydrochloride | U120 . . | Benzo(j,k)fluorene |
| U181 . . | Benzenamine, 2-methyl-5-nitro | U022 . . | Benzol(a)pyrene |
| U019 . . | Benzene (I,T) | U022 . . | 3,4-Benzopyrene |
| U038 . . | Benzeneacetic acid, 4-chloro-alpha-(4-chlorophenyl)-alpha-hydroxy, ethyl ester | U197 . . | p-Benzoquinone |
| | | U023 . . | Benzotrichloride (C,R,T) |
| | | U050 . . | 1,2-Benzphenanthrene |
| U030 . . | Benzene, 1-bromo-4-phenoxy- | U085 . . | 2,2-Bioxirane (I,T) |
| U037 . . | Benzene, chloro- | U021 . . | (1,1'-Biphenyl)-4,4'-diamine |
| U190 . . | 1,2-Benzenedicarboxylic acid anhydride | U073 . . | (1,1'-Biphenyl)-4,4'-diamine,3,3'-dichloro- |

(continued)

## Table 4.5: (continued)

| Hazardous waste No. | Substance | Hazardous waste No. | Substance |
|---|---|---|---|
| U091 . . | (1,1'-Biphenyl)-4,4'-diamine, 3,3',-dimethoxy- | U049 . . | 4-Chloro-o-toluidine, hydrochloride |
| U095 . . | (1,1'-Biphenyl)-4,4'-diamine, 3,3'-dimethyl | U032 . . | Chromic acid, calcium salt |
| U024 . . | Bis(2-chloroethoxy)methane | U050 . . | Chrysene |
| U027 . . | Bis(2-chloroisopropyl) ether | U051 . . | Creosote |
| U244 . . | Bis(dimethylthiocarbamoyl) disulfide | U052 . . | Cresols |
| U028 . . | Bis(2-ethylhexyl) phthalate | U052 . . | Cresylic acid |
| U246 . . | Bromine cyanide | U053 . . | Crotonaldehyde |
| U225 . . | Bromoform | U055 . . | Cumene (I) |
| U030 . . | 4-Bromophenyl phenyl ether | U246 . . | Cyanogen bromide |
| U128 . . | 1,3-Butadiene, 1,1,2,3,4,4-hexachloro- | U197 . . | 1,4-Cyclohexadienedione |
| U172 . . | 1-Butanimine, N-butyl-N-nitroso- | U056 . . | Cyclohexane (I) |
| U035 . . | Butanoic acid, 4-(bis(2-chloroethyl) amino) benzene- | U057 . . | Cyclohexanone (I) |
| U031 . . | 1-Butanol (I) | U130 . . | 1,3-Cyclopentadiene, 1,2,3,4,5,5-hexachloro- |
| U159 . . | 2-Butanone (I,T) | U058 . . | Cyclophosphamide |
| U160 . . | 2-Butanone peroxide (R,T) | U240 . . | 2,4-D, salts and esters |
| U053 . . | 2-Butenal | U059 . . | Daunomycin |
| U074 . . | 2-Butene, 1,4-dichloro- (I,T) | U060 . . | DDD |
| U031 . . | n-Butyl alcohol (I) | U061 . . | DDT |
| U136 . . | Cacodylic acid | U142 . . | Decachlorooctahydro-1,3,4-methano-2H-cyclobuta(c,d)pentalen-2-one |
| U032 . . | Calcium chromate | U062 . . | Diallate |
| U238 . . | Carbamic acid, ethyl ester | U133 . . | Diamine (R,T) |
| U178 . . | Carbamic acid, methylnitroso-ethyl ester | U221 . . | Diaminotoluene |
| U176 . . | Carbamide, N-ethyl-N-nitroso- | U063 . . | Dibenz(a,h)anthracene |
| U177 . . | Carbamide, N-methyl-N-nitroso- | U063 . . | 1,2:5,6-Dibenzanthracene |
| U219 . . | Carbamide, thio- | U064 . . | 1,2:7,8-Dibenzopyrene |
| U097 . . | Carbamoyl chloride, dimethyl- | U064 . . | Dibenz(a,i)pyrene |
| U215 . . | Carbonic acid, dithallium (I) salt | U066 . . | 1,2-Dibromo-3-chloropropane |
| U156 . . | Carbonochloridic acid, methyl ester (I,T) | U069 . . | Dibutyl phthalate |
| | | U062 . . | S-(2,3-Dichloroallyl) diisopropylthiocarbamate |
| U033 . . | Carbon oxyfluoride (R,T) | U070 . . | o-Dichlorobenzene |
| U211 . . | Carbon tetrachloride | U071 . . | m-Dichlorobenzene |
| U033 . . | Carbonyl fluoride (R,T) | U072 . . | p-Dichlorobenzene |
| U034 . . | Chloral | U073 . . | 3,3'-Dichlorobenzidine |
| U035 . . | Chlorambucil | U074 . . | 1,4-Dichloro-2-butene (I,T) |
| U036 . . | Chlordane, technical | U075 . . | Dichlorodifluoromethane |
| U026 . . | Chlornaphazine | U192 . . | 3,5-Dichloro-N-(1,1-dimethyl-2-propynyl) benzamide |
| U037 . . | Chlorobenzene | U060 . . | Dichloro diphenyl dichloroethane |
| U039 . . | 4-Chloro-m-cresol | U061 . . | Dichloro diphenyl trichloroethane |
| U041 . . | 1-Chloro-2,3-epoxypropane | U078 . . | 1,1-Dichloroethylene |
| U042 . . | 2-Chloroethyl vinyl ether | U079 . . | 1,2-Dichloroethylene |
| U044 . . | Chloroform | U025 . . | Dichloroethyl ether |
| U046 . . | Chloromethyl methyl ether | U081 . . | 2,4-Dichlorophenol |
| U047 . . | beta-Chloronaphthalene | U082 . . | 2,6-Dichlorophenol |
| U048 . . | o-Chlorophenol | U240 . . | 2,4-Dichlorophenoxyacetic acid, salts and esters |

(continued)

## Table 4.5: (continued)

| Hazardous waste No. | Substance | Hazardous waste No. | Substance |
|---|---|---|---|
| U083 . . | 1,2-Dichloropropane | U043 . . | Ethene, chloro- |
| U084 . . | 1,3-Dichloropropene | U042 . . | Ethene, 2-chloroethoxy- |
| U085 . . | 1,2:3,4-Diepoxybutane (I,T) | U078 . . | Ethene, 1,1-dichloro- |
| U108 . . | 1,4-Diethylene dioxide | U079 . . | Ethene, trans-1,2-dichloro- |
| U086 . . | N,N-Diethylhydrazine | U210 . . | Ethene, 1,1,2,2-tetrachloro- |
| U087 . . | O,O-Diethyl-S-methyl-dithiophosphate | U173 . . | Ethanol, 2,2'-(nitrosoimino)bis- |
| U088 . . | Diethyl phthalate | U004 . . | Ethanone, 1-phenyl- |
| U089 . . | Diethylstilbestrol | U006 . . | Ethanoyl chloride (C,R,T) |
| U148 . . | 1,2-Dihydro-3,6-pyridizinedione | U112 . . | Ethyl acetate (I) |
| U090 . . | Dihydrosafrole | U113 . . | Ethyl acrylate (I) |
| U091 . . | 3,3'-Dimethoxybenzidine | U238 . . | Ethyl carbamate (urethan) |
| U092 . . | Dimethylamine (I) | U038 . . | Ethyl 4,4'-dichlorobenzilate |
| U093 . . | Dimethylaminoazobenzene | U114 . . | Ethylenebis(dithiocarbamic acid) salts and esters |
| U094 . . | 7,12-Dimethylbenz(a)anthracene | | |
| U095 . . | 3,3'-Dimethylbenzidine | U067 . . | Ethylene dibromide |
| U096 . . | alpha,alpha-Dimethylbenzylhydro-peroxide (R) | U077 . . | Ethylene dichloride |
| | | U115 . . | Ethylene oxide (I,T) |
| U097 . . | Dimethylcarbamoyl chloride | U116 . . | Ethylene thiourea |
| U098 . . | 1,1-Dimethylhydrazine | U117 . . | Ethyl ether (I) |
| U099 . . | 1,2-Dimethylhydrazine | U076 . . | Ethylidene dichloride |
| U101 . . | 2,4-Dimethylphenol | U118 . . | Ethyl methacrylate |
| U102 . . | Dimethyl phthalate | U119 . . | Ethyl methanesulfonate |
| U103 . . | Dimethyl sulfate | U139 . . | Ferric dextran |
| U105 . . | 2,4-Dinitrotoluene | U120 . . | Fluoranthene |
| U106 . . | 2,6-Dinitrotoluene | U122 . . | Formaldehyde |
| U107 . . | Di-n-octyl phthalate | U123 . . | Formic acid (C,T) |
| U108 . . | 1,4-Dioxane | U124 . . | Furan (I) |
| U109 . . | 1,2-Diphenylhydrazine | U125 . . | 2-Furancarboxaldehyde (I) |
| U110 . . | Dipropylamine (I) | U147 . . | 2,5-Furandione |
| U111 . . | Di-n-propylnitrosamine | U213 . . | Furan, tetrahydro- (I) |
| U001 . . | Ethanal (I) | U125 . . | Furfural (I) |
| U174 . . | Ethanamine, N-ethyl-N-nitroso- | U124 . . | Furfuran (I) |
| U067 . . | Ethane, 1,2-dibromo- | U206 . . | D-Glucopyranose, 2-deoxy-2(3-methyl-3-nitrosoureido)- |
| U076 . . | Ethane, 1,1-dichloro | | |
| U077 . . | Ethane, 1,2-dichloro- | U126 . . | Glycidylaldehyde |
| U114 . . | 1,2-Ethanediylbiscarbamodithioic acid | U163 . . | Gluanidine, N-nitroso-N-methyl-N'nitro- |
| U131 . . | Ethane, 1,1,1,2,2,2-hexachloro- | | |
| U024 . . | Ethane, 1,1'-(methylenebis(oxy)bis(2-chloro- | U127 . . | Hexachlorobenzene |
| | | U128 . . | Hexachlorobutadiene |
| U247 . . | Ethane, 1,1,1-trichloro-2,2-bis-(p-methoxy phenyl) | U129 . . | Hexachlorocyclohexane (gamma isomer) |
| | | U130 . . | Hexachlorocyclopentadiene |
| U003 . . | Ethanenitrile (I,T) | U131 . . | Hexachloroethane |
| U117 . . | Ethane, 1,1'-oxybis-(I) | U132 . . | Hexachlorophene |
| U025 . . | Ethane, 1,1'-oxybis(2-chloro- | U243 . . | Hexachloropropene |
| U184 . . | Ethane, pentachloro- | U133 . . | Hydrazine (R,T) |
| U208 . . | Ethane, 1,1,1,2-tetrachloro- | U086 . . | Hydrazine, 1,2-diethyl |
| U209 . . | Ethane, 1,1,2,2-tetrachloro- | U098 . . | Hydrazine, 1,1-dimethyl- |
| U218 . . | Ethanethioamide | U099 . . | Hydrazine, 1,2-dimethyl- |
| U227 . . | Ethane, 1,1,2-trichloro | U109 . . | Hydrazine, 1,2-diphenyl- |

(continued)

## Table 4.5: (continued)

| Hazardous waste No. | Substance | Hazardous waste No. | Substance |
|---|---|---|---|
| U134 . . | Hydrofluoric acid (C,T) | U226 . . | Methylchloroform |
| U134 . . | Hydrogen fluoride (C,T) | U157 . . | 3-Methylcholanthrene |
| U135 . . | Hydrogen sulfide | U158 . . | 4,4'-Methylenebis(2-chloroaniline) |
| U096 . . | Hydroperoxide, 1-methyl-1-phenylethyl-(R) | U132 . . | 2,2'-Methylenebis(3,4,6-trichlorophenol) |
| U136 . . | Hydroxydimethylarsine oxide | U068 . . | Methylene bromide |
| U118 . . | 2-Imidazolidinethione | U080 . . | Methylene chloride |
| U137 . . | Indeno(1,2,3-cd)pyrene | U122 . . | Methylene oxide |
| U139 . . | Iron dextran | U159 . . | Methyl ethyl ketone (I,T) |
| U140 . . | Isobutyl alcohol (I,T) | U160 . . | Methyl ethyl ketone peroxide (R,T) |
| U141 . . | Isosafrole | U138 . . | Methyl iodide |
| U142 . . | Kepone | U161 . . | Methyl isobutyl ketone (I) |
| U143 . . | Lasiocarpine | U162 . . | Methyl methacrylate (I,T) |
| U144 . . | Lead acetate | U163 . . | N-Methyl-N'-nitro-N-nitrosoguanidine |
| U145 . . | Lead phosphate | U161 . . | 4-Methyl-2-pentanone (I) |
| U146 . . | Lead subacetate | U164 . . | Methylthiouracil |
| U129 . . | Lindane | U010 . . | Mitomycin C |
| U147 . . | Maleic anhydride | U059 . . | 5,12-Naphthacenedione, (8S-cis)- |
| U148 . . | Maleic hydrazide | | 8-acetyl-10-((3-amino-2,3,6-tri- |
| U149 . . | Malononitrile | | deoxy-alpha-L-lyxo-hexopyranosyl) |
| U150 . . | Melphalan | | oxy))-7,8,9,10-tetrahydro-6,8,11- |
| U151 . . | Mercury | | trihydroxy-1-methoxy- |
| U152 . . | Methacrylonitrile (I,T) | U165 . . | Naphthalene |
| U092 . . | Methanamine, N-methyl-(I) | U047 . . | Naphthalene, 2-chloro- |
| U029 . . | Methane, bromo- | U047 . . | Naphthalene, 2-chloro- |
| U045 . . | Methane, chloro- (I,T) | U166 . . | 1,4-Naphthalenedione |
| U046 . . | Methane, chloromethoxy- | U236 . . | 2,7-Naphthalenedisulfonic acid, 3,3'- |
| U068 . . | Methane, dibromo- | | ((3,3'-dimethyl-(1,1'-biphenyl)-4, |
| U080 . . | Methane, dichloro- | | 4' diyl))-bis(azo)bis(5-amino-4- |
| U075 . . | Methane, dichlorodifluoro- | | hydroxy)-, tetrasodium salt |
| U138 . . | Methane, iodo- | U166 . . | 1,4-Naphthoquinone |
| U119 . . | Methanesulfonic acid, ethyl ester | U167 . . | 1-Naphthylamine |
| U211 . . | Methane, tetrachloro- | U168 . . | 2-Naphthylamine |
| U153 . . | Methanethiol (I,T) | U167 . . | alpha-Naphthylamine |
| U225 . . | Methane, tribromo- | U168 . . | beta-Naphthylamine |
| U044 . . | Methane, trichloro- | U026 . . | 2-Naphthylamine, N,N'-bis(2-chloro- |
| U121 . . | Methane, trichlorofluoro- | | ethyl)- |
| U123 . . | Methanoic acid (C,T) | U169 . . | Nitrobenzene (I,T) |
| U036 . . | 4,7-Methanoindan, 1,2,4,5,6,7,8,8- | U170 . . | p-Nitrophenol |
| | octachloro-3a,4,7,7a-tetrahydro- | U171 . . | 2-Nitropropane (I) |
| U154 . . | Methanol (I) | U172 . . | N-Nitrosodi-n-butylamine |
| U155 . . | Methapyrilene | U173 . . | N-Nitrosodiethanolamine |
| U247 . . | Methoxychlor | U174 . . | N-Nitrosodiethylamine |
| U154 . . | Methyl alcohol (I) | U111 . . | N-Nitrosodi-N-propylamine |
| U029 . . | Methyl bromide | U176 . . | N-Nitroso-N-ethylurea |
| U186 . . | 1-Methylbutadiene (I) | U177 . . | N-Nitroso-N-methylurea |
| U045 . . | Methyl chloride (I,T) | U178 . . | N-Nitroso-N-methylurethane |
| U156 . . | Methyl chlorocarbonate (I,T) | U179 . . | N-Nitrosopiperidine |
| | | U180 . . | N-Nitrosopyrrolidine |
| | | U181 . . | 5-Nitro-o-toluidine |

(continued)

## Table 4.5: (continued)

| Hazardous waste No. | Substance | Hazardous waste No. | Substance |
|---|---|---|---|
| U193 . . | 1,2-Oxathiolane, 2,2-dioxide | U152 . . | 2-Propenenitrile, 2-methyl, (I,T) |
| U058 . . | 2H-1,3,2-Oxazaphosphorine, 2-(bis(2-chloroethyl)amino)tetrahydro-, oxide 2- | U008 . . | 2-Propenoic acid (I) |
|  |  | U113 . . | 2-Propenoic acid, ethyl ester (I) |
| U115 . . | Oxirane (I,T) | U118 . . | 2-Propenoic acid, 2-methyl-, ethyl ester |
| U041 . . | Oxirane, 2-(chloromethyl)- | U162 . . | 2-Propenoic acid, 2-methyl-, methyl ester (I,T) |
| U182 . . | Paraldehyde |  |  |
| U183 . . | Pentachlorobenzene | U233 . . | Propionic acid, 2-(2,4,5-trichlorophenoxy)- |
| U184 . . | Pentachloroethane |  |  |
| U185 . . | Pentachloronitrobenzene | U194 . . | n-Propylamine (I,T) |
| U242 . . | Pentachlorophenol | U083 . . | Propylene dichloride |
| U186 . . | 1,3-Pentadiene (I) | U196 . . | Pyridine |
| U187 . . | Phenacetin | U155 . . | Pyridine, 2-((2-(dimethylamino)ethyl)-2-thenylamino)- |
| U188 . . | Phenol |  |  |
| U048 . . | Phenol, 2-chloro | U179 . . | Pyridine, hexahydro-N-nitroso- |
| U039 . . | Phenol, 4-chloro-3-methyl- | U191 . . | Pyridine, 2-methyl- |
| U081 . . | Phenol, 2,4-dichloro- | U164 . . | 4(1H)-Pyrimidinone, 2,3-dihydro-6-methyl-2-thioxo- |
| U082 . . | Phenol, 2,6-dichloro- |  |  |
| U101 . . | Phenol, 2,4-dimethyl- | U180 . . | Pyrrole, tetrahydro-N-nitroso- |
| U170 . . | Phenol, 4-nitro- | U200 . . | Reserpine |
| U242 . . | Phenol, pentachloro- | U201 . . | Resorcinol |
| U212 . . | Phenol, 2,3,4,6-tetrachloro- | U202 . . | Saccharin and salts |
| U230 . . | Phenol, 2,4,5-trichloro- | U203 . . | Safrole |
| U231 . . | Phenol, 2,4,6-trichloro- | U204 . . | Selenious acid |
| U137 . . | 1,10-(1,2-Phenylene)pyrene | U204 . . | Selenium dioxide |
| U145 . . | Phosphoric acid, Lead salt | U205 . . | Selenium disulfide (R,T) |
| U087 . . | Phosphorodithioic acid, 0,0-diethyl-, S-methyl ester | U015 . . | L-Serine, diazoacetate (ester) |
|  |  | U233 . . | Silvex |
| U189 . . | Phosphorus sulfide (R) | U089 . . | 4,4'-Stilbenediol, alpha, alpha'-diethyl- |
| U190 . . | Phthalic anhydride |  |  |
| U191 . . | 2-Picoline | U206 . . | Streptozotocin |
| U192 . . | Pronamide | U135 . . | Sulfur hydride |
| U194 . . | 1-Propanamine (I,T) | U103 . . | Sulfuric acid, dimethyl ester |
| U110 . . | 1-Propanamine, N-propyl- (I) | U189 . . | Sulfur phosphide (R) |
| U066 . . | Propane, 1,2-dibromo-3-chloro | U205 . . | Sulfur selenide (R,T) |
| U149 . . | Propanedinitrile | U232 . . | 2,4,5-T |
| U171 . . | Propane, 2-nitro- (I) | U207 . . | 1,2,4,5-Tetrachlorobenzene |
| U027 . . | Propane, 2,2'oxybis(2-chloro) | U208 . . | 1,1,1,2-Tetrachloroethane |
| U193 . . | 1,3-Propane sultone | U209 . . | 1,1,2,2-Tetrachloroethane |
| U235 . . | 1-Propanol, 2,3-dibromo-,phosphate (3:1) | U210 . . | Tetrachloroethylene |
|  |  | U212 . . | 2,3,4,6-Tetrachlorophenol |
| U126 . . | 1-Propanol, 2,3-epoxy- | U213 . . | Tetrahydrofuran (I) |
| U140 . . | 1-Propanol, 2-methyl- (I,T) | U214 . . | Thallium (I) acetate |
| U002 . . | 2-Propanone (I) | U215 . . | Thallium (I) carbonate |
| U007 . . | 2-Propenamide | U216 . . | Thallium (I) chloride |
| U084 . . | Propene, 1,3-dichloro- | U217 . . | Thallium (I) nitrate |
| U243 . . | 1-Propene, 1,1,2,3,3,3-hexachloro- | U218 . . | Thioacetamide |
| U009 . . | 2-Propenenitrile | U153 . . | Thiomethanol (I,T) |

(continued)

Table 4.5: (continued)

| Hazardous waste No. | Substance | Hazardous waste No. | Substance |
|---|---|---|---|
| U219 . . | Thiourea | U231 . . | 2,4,6-Trichlorophenol |
| U244 . . | Thiram | U232 . . | 2,4,5-Trichlorophenoxyacetic acid |
| U220 . . | Toluene | U234 . . | sym-Trinitrobenzene (R,T) |
| U221 . . | Toluenediamine | U182 . . | 1,3,5-Trioxane, 2,4,6-trimethyl- |
| U223 . . | Toluene diisocyanate (R,T) | U235 . . | Tris(2,3-dibromopropyl) phosphate |
| U222 . . | o-Toluidine hydrochloride | U236 . . | Trypan blue |
| U011 . . | 1H-1,2,4-Triazol-3-amine | U237 . . | Uracil, 5(bis(2-chloroethyl)amino)- |
| U226 . . | 1,1,1-Trichloroethane | U237 . . | Uracil mustard |
| U227 . . | 1,1,2-Trichloroethane | U043 . . | Vinyl chloride |
| U228 . . | Trichloroethene | U239 . . | Xylene (I) |
| U228 . . | Trichloroethylene | U200 . . | Yohimban-16-carboxylic acid, 11,17-di- |
| U121 . . | Trichloromonofluoromethane | | methoxy-18-((3,4,5-trimethoxy- |
| U230 . . | 2,4,5-Trichlorophenol | | benzoyl)oxy)-methyl ester, |

**Characteristic of Reactivity:** A solid waste exhibits the characteristic of reactivity if a representative sample of the waste has any of the following properties:

(1) It is normally unstable and readily undergoes violent change without detonating.

(2) It reacts violently with water.

(3) It forms potentially explosive mixtures with water.

(4) When mixed with water, it generates toxic gases, vapors or fumes in a quantity sufficient to present a danger to human health or the environment.

(5) It is a cyanide or sulfide bearing waste which, when exposed to pH conditions between 2 and 12.5, can generate toxic gases, vapors or fumes in a quantity sufficient to present a danger to human health or the environment.

(6) It is capable of detonation or explosive reaction if it is subjected to a strong initiating source or if heated under confinement.

(7) It is readily capable of detonation or explosive decomposition or reaction at standard temperature and pressure.

(8) It is a forbidden explosive as defined by DOT, or a Class A or Class B explosive as defined by DOT (see *Explosives and Blasting Agents* earlier in this chapter).

**Characteristic of EP Toxicity:** A solid waste exhibits the characteristic of EP toxicity if, using the test methods described in Footnote 11 the extract from a representative sample of the waste contains any of the contaminants listed in Table 4.6 at a concentration equal to or greater than the respective value given in that table. Where the waste contains less than 0.5 percent filterable solids, the waste itself, after filtering, is considered to be the extract.

Table 4.6: Maximum Concentration of Contaminants
for Characteristic of EP Toxicity

| EPA Hazardous Waste Number | Contaminant | Maximum Concentration (mg/ℓ) |
|---|---|---|
| D004 | Arsenic | 5.0 |
| D005 | Barium | 100.0 |
| D006 | Cadmium | 1.0 |
| D007 | Chromium | 5.0 |
| D008 | Lead | 5.0 |
| D009 | Mercury | 0.2 |
| D010 | Selenium | 1.0 |
| D011 | Silver | 5.0 |
| D012 | Endrin (1,2,3,4,10,10-hexachloro-1,7-epoxy-1,4,4a,5,6,7,8,8a-octa-hydro-1,4-endo, endo-5,8-dimethano naphthalene) | 0.02 |
| D013 | Lindane (1,2,3,4,5,6-hexachloro-cyclohexane, gamma isomer) | 0.4 |
| D014 | Methoxychlor [1,1,1-trichloro-2,2-bis(p-methoxyphenyl)ethane] | 10.0 |
| D015 | Toxaphene ($C_{10}H_{10}Cl_8$ technical chlorinated camphene, 67 to 69% chlorine) | 0.5 |
| D016 | 2,4-D, (2,4-dichlorophenoxyacetic acid) | 10.0 |
| D017 | 2,4,5-TP Silvex (2,4,5-trichloro-phenoxypropionic acid) | 1.0 |

## Empty Containers

For RCRA, a container or inner liner removed from a container that previously contained a hazardous waste is *empty* if it contains *less than one inch* of residue at the bottom of the container or inner liner. This "one inch rule" does not apply to containers that previously held *acutely hazardous materials*. These are those commercial chemicals listed in Section 261.33(e) (Table 4.4). The rule also does not apply to containers which held hazardous wastes which are compressed gases; these are considered empty for RCRA, and hence not regulated, if they have been vented to atmospheric pressure.

A container that previously held an acutely hazardous chemical (see Table 4.4) is a hazardous waste *when discarded* or *intended to be discarded* unless it has been triply rinsed. Thus, a container holding *any* residue of an acutely hazardous material is subject to all the regulatory requirements of a hazardous waste.

## Some RCRA Exemptions

A waste which is hazardous solely because it exhibits one or more of the hazard characteristics is exempted from regulation when it is being beneficially used, reused or legitimately recycled or reclaimed. On the other hand, a hazardous waste which is a sludge or is a listed waste is subject to some regulatory requirements when it is being beneficially used, reused or legitimately recycled or reclaimed.[13]

Certain generators of hazardous wastes are exempted from the regulations because the quantity of hazardous waste generated in one calendar month is below a set quantity limit. This limit has been set at 1000 kilograms per month for hazardous waste in general and at one kilogram per month for a chemical listed in section 261.33(e) (Table 4.4). During the month in which less than these quantities are generated the "small quantity generator" is exempted from RCRA requirements. However, the wastes must be disposed of in an approved facility. Furthermore, the generator is not exempted from determining if the waste is hazardous regardless of the amount produced.

Furthermore, Section 261.4(b) of the regulation lists solid wastes which are not to be considered hazardous wastes. The entries in this list have increased considerably since the regulation became effective and the reader will do well in checking the latest version of this list before classifying a solid waste as hazardous.

## CLASSIFICATION UNDER FIFRA REGULATIONS

Under the Federal Insecticide, Fungicide and Rodenticide Act (FIFRA), the Environmental Protection Agency controls the use of pesticides to safeguard the health of the public and to prevent adverse effects on the environment. This is accomplished through the process of pesticide registration or premarket clearance. The regulations applicable to the registration of pesticides are found in the *Code of Federal Regulations* Title 40, Sections 162-180.

In its strictest sense, classification of a pesticide refers to the classification of its use. It is the responsibility of the EPA's Administrator to classify a pesticide for general or restricted use. A pesticide is classified for restricted use if it is determined that when applied in accordance with its directions for use, warnings and cautions, it may generally cause unreasonable adverse effects on the environment including injury to the applicator. In such cases, additional restrictions may include a requirement that the pesticide be applied only by or under the direct supervision of a certified applicator.

As far as labeling is concerned, classification of a pesticide refers to its categorization in toxicity groups, depending on the toxicological properties of the material. These toxicity categories, I through IV, dictate the text required on the front panel of labels and in many cases, the placement of a pesticide into a restricted use classification. Table 4.7 indicates the relationship of hazard indicator to Toxicity Category with the category being assigned on the basis of the highest hazard shown by any of the indicators.

It should be noted that the end point for oral and dermal $LD_{50}$ as well as for inhalation $LC_{50}$ is 14 days.[12]

A complete discussion of labeling under FIFRA is presented in Chapter 10.

### Table 4.7: FIFRA Toxicity Categories

| Hazard Indicators | Categories | | | |
|---|---|---|---|---|
| | I | II | III | IV |
| Oral $LD_{50}$ (mg/kg) | Up to and including 50 | From 50 through 500 | From 500 through 5,000 | Greater than 5,000 |
| Inhalation $LC_{50}$ (mg/$\ell$) | Up to and including 0.2 | From 0.2 through 2 | From 2.0 through 20 | Greater than 20 |
| Dermal $LD_{50}$ (mg/kg) | Up to and including 200 | From 200 through 2,000 | From 2,000 through 20,000 | Greater than 20,000 |
| Eye effects | Corrosive, corneal opacity not reversible within 7 days | Corneal opacity reversible within 7 days; irritation persisting for 7 days | No corneal opacity, irritation reversible within 7 days | No irritation |
| Skin effects | Corrosive | Severe irritation at 72 hours | Moderate irritation at 72 hours | Mild or slight irritation at 72 hours |

## SUMMARY

The regulations pertaining to the classification of industrial products for transportation and as waste have been reviewed (DOT and RCRA regulations respectively). In addition, the classification of pesticides has been briefly presented. These classifications' schemes differ, greatly in some cases, depending on the type and degree of protection the regulatory framework was designed to provide.

Under the DOT and the RCRA schemes, it is the responsibility of the shipper or generator to appropriately perform the classification of the material in question, after reviewing its physical, chemical and toxicological properties. On the other hand, under the FIFRA regulatory approach, the manufacturer of a pesticide, after reviewing the required data, proposes to the Agency the classification, label text, etc., the material should have in his/her judgement, with the Agency making the final determination.

The following list of tests was developed to serve as a guide to the basic data needed to classify most products. It was not designed to be an exhaustive list. However, evaluation of the data listed suffices, in most cases, for the development of an appropriate classification.

### Basic Information Needed for Product Classification

*GENERAL*

> Chemical Name
>
> Structure
>
> Description of Manufacturing Process
>
> Identification and Concentration of Impurities

## PHYSICAL/CHEMICAL PROPERTIES

Physical State

Boiling Point/Melting Point

pH

Viscosity

Density

Flash Point

Vapor Pressure

Metal Corrosivity

Water Reactivity/solubility

Thermal Stability Assessment

## TOXICOLOGY

Oral $LD_{50}$

Dermal $LD_{50}$

Inhalation $LC_{50}$

## FOOTNOTES

1. ASTM: American Society for Testing and Materials
   1916 Race Street
   Philadelphia, PA 19103
2. Organic Peroxide Producer's Safety Division
   Society of the Plastic Industries, Inc.
   355 Lexington Avenue
   New York, NY 10017
3. Flash point means the minimum temperature at which a liquid gives off vapor within a test vessel in sufficient concentration to form an ignitable mixture with air near the surface of the liquid.

   A number of closed-cup, flash test procedures can be used, depending on the viscosity of the liquid under investigation.

   For homogeneous liquids, having a viscosity less than 45 SUS (Saybolt Universal Seconds) at 100°F (37.8°C) and not forming a surface film while under test, one of the following methods are to be used:
   a) Standard Method of Test for Flash Point by Tag Closed Tester (ASTM D56-70),
   b) Standard Method of Test for Flash Point of Aviation Turbine Fuels by Setaflash Closed Tester (ASTM D3243-73), or
   c) Standard Methods of Test for Flash Point of Liquids by Setaflash Closed Tester (ASTM D3278-73).

   For all other liquids, one of the following methods must be used:
   a) Standard Method of Test for Flash Point by Pensky-Martens Closed Tester (ASTM D93-71), and methods b and c above.

For a liquid that is a mixture of compounds that have different volatility and flash points, its flash point shall be detemined as specified above on the material in the form in which it is to be shipped. If it is determined by this test that the flash point is higher than 20°F (−6.67°C) a second test shall be made on a sample of the liquid evaporated from an open beaker (or similar container), under ambient pressure and temperature (20° to 25°C) conditions, to 90 percent of its original volume or for a period of 4 hours, whichever comes first. The lower flash point of the two tests shall be the flash point of the material.

4. See 46FR 25492, May 7, 1981 and 46FR 31294, June 15, 1981, respectively.
5. Method for testing corrosion to skin.

Corrosion to the skin is measured by patch-test technique on the intact skin of the albino rabbit, clipped free of hair. A minimum of six subjects are to be used in this test.

1. Introduce under a square cloth patch, such as surgical gauze measuring not less than 1 inch by 1 inch and two single layers thick, 0.5 milliliter (in the case of liquids) or 0.5 gram (in the case of solids and semi-solids) of the substance to be tested.
2. Immobilize the animals with patches secured in place by adhesive tape.
3. Wrap the entire trunk of each animal with an impervious material, such as rubberized cloth, for the 4 hour period of exposure. This material is to aid in maintaining the test patches in position and retards the evaporation of volatile substances. It is not applied for the purpose of occlusion.
4. After 4 hours of exposure, the patches are to be removed and the resulting reactions are to be evaluated for corrosion.
5. Following this initial reading, all test sites are washed with an appropriate solvent to prevent further exposure.
6. Readings are again to be made at least at the end of a total of 48 hours (44 hours after the first reading).

Corrosion will be considered to have resulted if the substances in contact with the rabbit skin have caused destruction or irreversible alteration of the tissue on at least two out of each six rabbits tested. Tissue destruction is considered to have occurred if, at any of the readings, there is ulceration or necrosis. Tissue destruction does not include merely sloughing of the epidermis, or erythema, edema, or fissuring. (see Chapter 6 for a further discussion).

6. NACE: National Association of Corrosion Engineers
   2400 West Loop South
   Houston, TX 77027

7. Document AD/A-059 077
   National Technical Information Service
   U.S. Department of Commerce
   Springfield, VA 22161

8. Bureau of Explosives,
   Association of American Railroads
   American Railroads Building
   1920 L Street, N.W.
   Washington, DC 20036

9. This list may be revised from time to time by Notice published in the *Federal Register* to identify additional agents.

10. *Test Methods for the Evaluation of Solid Waste, Physical/Chemical Methods,* (1980), EPA publication number SW-846.
    Environmental Protection Agency
    Solid Waste Information
    26 W. St. Clair Street
    Cincinnati, Ohio 45268

11. EP Toxicity Test Procedure

    A. *Extraction Procedure (EP)*

    1. A representative sample of the waste to be tested (minimum size 100 grams) shall be obtained. For detailed guidance on conduction the various aspects of the EP see *Test Methods for the Evaluation of Solid Waste, Physical/Chemical Methods.*

    2. The sample shall be separated into its component liquid and solid phases using the method described in "Separation Procedure" below. If the solid residue* obtained using this method totals less than 0.5% of the original weight of the waste, the residue can be discarded and the operator shall treat the liquid phase as the extract and proceed immediately to Step 8.

        *The percent solids is determined by drying the filter pad at 80°C until it reaches constant weight and then calculating the percent solids using the following equation:

        $$100\% \text{ solids} = \frac{(\text{weight of pad} + \text{solid}) - (\text{tare weight of pad})}{\text{Initial weight of sample}}$$

    3. The solid material obtained from the Separation Procedure shall be evaluated for its particle size. If the solid material has a surface area per gram of material equal to, or greater than, 3.1 cm$^2$ or passes through a 9.5 mm (0.375 inch) standard sieve, the operator shall proceed to Step 4. If the surface area is smaller or the particle size larger than specified above, the solid material shall be prepared for extraction by crushing, cutting or grinding the material so that it passes through a 9.5 mm (0.375 inch) sieve or, if the material is in a single piece, by subjecting the material to the "Structural Integrity Procedure" described below.

    4. The solid material obtained in Step 3 shall be weighed and placed in an extractor with 16 times its weight of deionized water. Do not allow the material to dry prior to weighing. For purposes of this test, an acceptable extractor is one which will impart sufficient agitation to the mixture to not only prevent stratification of the sample and extraction fluid but also insure that all sample surfaces are continuously brought into contact with well mixed extraction fluid.

    5. After the solid material and deionized water are placed in the extractor, the operator shall begin agitation and measure the pH of the solution in the extractor. If the pH is greater than 5.0, the pH of the solution shall be decreased to 5.0±0.2 by adding 0.5 N acetic acid. If the pH is equal to or less than 5.0, no acetic acid should be added. The pH of the solution shall be monitored, as described below, during the course of the extraction and if the pH rises above 5.2, 0.5N acetic acid shall be added to bring the pH down to 5.0±0.2. However, in no event shall the aggregrate amount of acid added to the solution exceed 4 ml of acid per gram of solid. The mixture shall be agitated for 24 hours and maintained at 20° − 40°C (68° − 104°F) during this time. It is recommended that the operator monitor and adjust the pH during the course of the extraction with a device such as the Type 45-A pH Controller manufactured by Chemtrix, Inc., Hillsboro, Oregon 97123 or its equivalent, in conjunction with

a metering pump and reservoir of 0.5N acetic acid. If such a system is not available, the following manual procedure shall be employed:

(a) A pH meter shall be calibrated in accordance with the manufacturer's specifications.

(b) The pH of the solution shall be checked and, if necessary, 0.5N acetic acid shall be manually added to the extractor until the pH reaches 5.0±0.2. The pH of the solution shall be adjusted at 15, 30 and 60 minute intervals, mvoing to the next longer interval if the pH does not have to be adjusted more than 0.5 pH units.

(c) The adjustment procedure shall be continued for at least 4 hours.

(d) If at the end of the 24-hour extraction period, the pH of the solution is not below 5.2 and the maximum amount of acid (4 ml per gram of solids) has not been added, the pH shall be adjusted to 5.0±0.2 and the extraction continued for an additional four hours, during which the pH shall be adjusted at one hour intervals.

6. At the end of the 24 hours extraction period, deionized water shall be added to the extractor in an amount determined by the following equation:

$$V = (20)(W) - 16(W) - A$$

V = ml deionized water to be added
W = weight in grams of solid charged to extractor
A = ml of 0.5N acetic acid added during extraction

7. The material in the extractor shall be separated into its component liquid and solid phases as described under the Separation Procedure described below.

8. The liquids resulting from Steps 2 and 7 shall be combined. This combined liquid (or the waste itself if it has less than ½ percent solids, as noted in Step 2) is the extract and shall be analyzed for the presence of any of the contaminants specified in Table 4.6 using the Analytical Procedure described below.

### Separation Procedure

Equipment: A filter holder, designed for filtration media having a nominal pore size of 0.45 micrometers and capable of applying a 5.3 kg/cm$^2$ (75 psi) hydrostatic pressure to the solution being filtered, shall be used. For mixtures containing nonabsorptive solids, where separation can be effected without imposing a 5.3 kg/cm$^2$ pressure differential, vacuum filters employing a 0.45 micrometers filter media can be used. (For further guidance on filtration equipment or procedures see *Test Methods for Evaluating Solid Waste, Physical/Chemical Methods.*)

Procedure*

(i) Following manufacturer's directions, the filter unit shall be assembled with a filter bed consisting of a 0.45 micrometer filter membrane. For difficult or slow to filter mixtures a prefilter bed consisting of the following prefilters in increasing pore size (0.65 micrometer membrane, fine glass fiber prefilter, and coarse glass filter prefilter) can be used.

(ii) The waste shall be poured into the filtration unit.

(iii) The reservoir shall be slowly pressurized until liquid begins to flow from the filtrate outlet at which point the pressure in the filter shall be immediately lowered to 10-15 psig. Filtration shall be continued until liquid flow ceases.

(iv) The pressure shall be increased stepwise in 10 psi increments to 75 psig and filtration continued until flow ceases or the pressurizing gas begins to exit from the filtrate outlet.

(v) The filter unit shall be depressurized, the solid material removed and weighed and then transferred to the extraction apparatus, or, in the case of final filtration prior to analysis, discarded. Do not allow the material retained on the filter pad to dry prior to weighing.

(vi) The liquid phase shall be stored at 4°C for subsequent use in Step 8.

*This procedure is intended to result in separation of the "free" liquid portion of the waste from any solid matter having a particle size > 0.45μm. If the sample will not filter, various other separation techniques can be used to aid in the filtration. As described above, pressure filtration is employed to speed up the filtration process. This does not alter the nature of the separation. If liquid does not separate during filtration, the waste can be centrifuged. If separation occurs during centrifugation, the liquid portion (centrifugate) is filtered through the 0.45 μm filter prior to becoming mixed with the liquid portion of the waste obtained from the initial filtration. Any material that will not pass through the filter after centrifugation is considered a solid and is extracted.

### Structural Integrity Procedure

Equipment: A Structural Integrity Tester having a 3.18 cm (1.25 in.) diameter hammer weighing 0.33 kg (0.73 lbs.) and having a free fall of 15.24 cm (6 in.) shall be used. This device is available from Associated Design and Manufacturing Company, Alexandria, VA 22314, as Part No. 125, or it may be fabricated to meet specifications.

Procedure

1. The sample holder shall be filled with the material to be tested. If the sample of waste is a large monolithic block, a portion shall be cut from the block having the dimensions of a 3.3 cm (1.3 in.) diameter × 7.1 cm (2.8 in.) cylinder. For a fixated waste, samples may be cast in the form of a 3.3 cm (1.3 in.) diameter × 7.1 cm (2.8 in.) cylinder for purposes of conducting this test. In such cases, the waste may be allowed to cure for 30 days prior to further testing.

2. The sample holder shall be placed into the Structural Integrity Tester, then the hammer shall be raised to its maximum height and dropped. This shall be repeated fifteen times.

3. The material shall be removed from the sample holder, weighed, and transferred to the extraction apparatus for extraction.

### Analytical Procedures for Analyzing Extract Contaminants

The test methods for analyzing the extract are as follows:

For arsenic, barium, cadmium, chromium, lead, mercury, selenium, silver, endrin, lindane, methoxychlor, toxaphene, 2,4-D (2,4-dichlorophenoxyacetic acid) or 2,4,5-TP (2,4,5-trichlorophenoxypropionic acid): "Test Methods for the Evaluation of Solid Waste, Physical/Chemical Methods".

For all analyses, the methods of standard addition shall be used for quantification of species concentration.

12. See 43FR 37355, August 22, 1978

13. As this Chapter goes to press a new definition of solid waste and new standards for recycled hazardous wastes have been proposed. See 48FR 144T2, April 4, 1983.

# 5

# Labels and Medicine

**Richard Moriarty**
*National Poison Center*
*Children's Hospital*
*Pittsburgh, PA*

## INTRODUCTION

One can probably say, with a reasonable degree of certainty, that the totally "safe" product has yet to be invented. This statement is particularly applicable to those products formulated from chemicals. All products have the potential for misuse and/or abuse. Those individuals who are given the responsibility of creating labels for products must make every attempt to lessen the chance of such occurrences. This can be accomplished by providing easily understood and precise direction and warning statements on the product's label. Despite such efforts, product misadventures will occur. While most will result in little harm or damage, a few will produce tragic consequences.

Recognizing that such exposures will and do occur, most manufacturers acknowledge their responsibility to provide advice within the body of a product's label that is designed to lessen the effects of such occurrences. First aid is the term usually applied to such advice. Indeed, the term frequently identifies that particular section of the product's label. Simply defined, first aid is that action or series of actions that should be instituted immediately following an exposure to a product that will result in a cessation or a reduction of the harmful acute effects of that exposure.

Providing proper first aid advice is a task more easily stated than accomplished. Too often, the task is relegated to a person with little medical or toxicological background. Indeed, the person's total knowledge of the product in question may be sparse. Faced with this dilemma, it is not surprising that the label writer often turns to governmental agencies, trade associations or industrial committees for aid. While such groups may be able to provide some guidance it must be recognized that the advice given is basically "generic" in nature. The suggested first aid statements that are provided are designed to be applied to groups of products that share similar characteristics. Usually such characteristics are based on chemical composition. In some cases the "fit" is good. In others, it is not. Furthermore, there appears to be a growing movement on the part of such groups to place the responsibility for the development of such advice where many feel it rightfully belongs, with the manufacturer of the product. It is difficult to counter the argument that the maker of the product knows or should know that product best. Further, it is held by some that a manufacturer should be ultimately responsible for any harm caused by his product. If these assertions are sound, then one must conclude that the development of first aid statements should and must be the responsibility of the manufacturer. In approaching the development of such statements, it is imperative that the label writer first address the following questions:

1. What are the problems that are most likely to occur if the product is misused or abused?

2. Who is most likely to be affected by these problems?

In order to provide the answers to these questions, a longer list of queries must be made. Areas that require examination include:

1. *Is the precise formulation of the product known?* Often it is not. Relatively few manufacturers formulate their products from raw materials. The usual situation is for the manufacturer to obtain many, if not all, of the required ingredients from other suppliers. Frequently, a particular ingredient is selected on the basis of its performance and not its chemical composition. In other words, the manufacturer may be interested in adding a material to the product that will allow it to spread more easily, increase shelf life, or improve product stability. The manufacturer has little interest in the formulation of that particular material as long as it performs to his specifications. Furthermore, the supplier may be reluctant to share the formulation of that particular ingredient with anyone.

2. *Can the acute toxic effects of a product be determined on the basis of the product's formulation?* Obviously, if the precise formulation of the product is not known, the answer to this question is no. Even with complete knowledge of the formulation, predicting the acute toxic effects of a product may be difficult. Combinations of ingredients coupled with processing techniques may lead to the formation of entirely new chemical combinations that may alter the expected acute toxicity of the product. For exam-

ple, the addition of an improved surfactant to a toxic pesticide formulation will likely increase the toxic effect by enhancing its absorption from the gut, if the material is accidentally ingested.

3. *What are the acute effects of the product as determined from animal studies?* Even with the knowledge of a product's precise formulation and an unquestionable prediction of the acute toxic effects of each ingredient of the product, it would be the foolhardy manufacturer who would either not conduct acute animal toxicity studies on the product itself or review pertinent studies on similar products. Indeed, it is such studies that frequently form the basis on which one determines the acute toxicity effects that may be associated with misuse or abuse of the product. The approach to such studies and their interpretations are detailed in Chapter 6. Even a casual inspection of that chapter would indicate that such testing can be time-consuming, expensive, and the results, at times, misleadng. Species variability may mask or falsely accentuate an acute toxic effect. First aid statements based upon such findings could result in erroneous advice being given.

4. *What are the known acute toxic effects of the product based on human exposures?* Initially, one might say that such information usually does not exist since our society does not condone human experimentation involving products other than drugs. Conversely, it could be argued that such information could be obtained from unplanned human exposures but that these would only occur after the product had been placed on the market. Thus, the information would not be available to the label writer at the time that it was needed. While the former statement is usually true the latter is usually not. It is the unusual product that does not have several close "relatives" if not a "twin" already in the market place. Thus, such human exposure information probably exists. One source of such information is the National Poison Center Network whose resources will be described later in this chapter. Obviously, such exposure information is undeniably superior to all others since humans form the principal audience that is to be addressed by the first aid statements. Manufacturers should make every effort to obtain such information. As previously mentioned, the sole dependency on animal studies may lead to erroneous conclusions.

5. *Who is the most likely to be exposed to the product?* The answer to this question frequently is determined by the

manner in which the product is to be used. It is often reasoned that products primarily intended for industrial use rarely find their way into the home. Products designed for the laboratory probaby will not be used on a farm. However, first aid advice that is based on these assumptions may miss its mark. Commercial products are taken home. At times people with relatively little technical knowledge may be exposed to products intended to be used by highly trained individuals. The best approach to utilize in answering this question is that of the "lowest common denominator". One should assume that a child may be exposed to the product and determine if such an exposure would cause a significantly different or more severe problem than with an adult exposure. If this is the case then the first aid advice should reflect these concerns. Assume that the person reading the first aid advice has a fourth grade reading level and write the instructions accordingly.

6. *What form is the exposure likely to take and what degree of severity can be expected?* A number of factors must be taken into consideration in order to answer these questions. The form of the product is of critical importance. Liquids are more likely to be ingested or to cause skin or eye exposures than are solids. The converse is usually true of sprays or aerosol products. Inhalation problems may occur regardless of the form of the product. The amount of volume of the product is an equally important consideration. Manufacturers tend to utilize the same first aid statements across a product line. This is an acceptable practice only if, indeed, the exposure problems and their severity do not change as the amount of available product changes. Products that are to be diluted or mixed with other components deserve special attention. The label writer must clearly indicate what conditions are being addressed by the first aid statements. Finally, packaging is a critically important area that is too often overlooked. Package designs that enhance the likelihood of a particular type of exposure should be avoided. If this is not possible then the first aid advice given must adequately reflect and address those types of exposures that are likely to be caused by the packaging.

## DEVELOPMENT OF FIRST AID STATEMENTS

Once the nature and severity of the problems that can occur with an acute exposure to the product have been determined, the label writer is now in a position to develop appropriate first aid statements.

Fortunately, relatively few products produce life-threatening situations following acute exposures. Such situations include:

1. Cardiac arrest.

2. Respiratory arrest.

3. Severe hypotension.

Cardiac arrest can be defined generally as a cessation of the contractions of the heart. Such a situation is obviously life-threatening, and if it persists for more than several minutes it can cause moderate to severe brain damage or death.

Respiratory arrest is the cessation of breathing. Again, if this condition persists for more than a few minutes it can produce dire consequenses. While not as immediately life-threatening, exposures that produce a slowing of the respiratory rate may eventually produce similar results.

A precipitous and usually sudden drop in the blood pressure is the condition defined as hypotension. With such a situation the blood flow to the brain, heart, and kidneys can be drastically reduced. This limited blood flow causes these organs to be deprived of oxygen. Again, such a situation, if prolonged, can result in moderate to severe damage to these organs.

While an acute exposure to a product may produce only one of these conditions it must be understood that the body's homeostatic mechanisms are such that the occurrence of one of these conditions may precipitate the onset of the others.

If toxicity studies in animals or previous human exposure experience indicates that a product is or may be capable of producing any of these situations, that fact must be clearly stated on the label. At times such information may appear in the "Warning Section" of the label. Since that section may be located on another part of the label some distance from the first aid statement or that section may be obscured for a variety of reasons, it is prudent to repeat such information at the beginning of the first aid statements.

In providing first aid advice for such situations, one must first consider if the exposure itself might still be in progress and thus would need to be interrupted. An example of such a situation might be a massive exposure to a pesticide that was absorbed through the skin resulting in respiratory arrest. Obviously, such an exposure must be terminated in order to prevent prolongation of symptoms or the development of new ones. The termination of the exposure must be accomplished either rapidly so as not to delay the institution of the management of the life-threatening situation or in concert with such management. Furthermore, one must consider that those rendering the first aid might themselves be exposed and thus become victims. The label writer must determine if any or all of these situations apply to the product in question and provide suitable first aid advice.

After giving instructions addressing the protection of the rescuer and, if necessary, the means of interrupting the exposure, the label writer must then provide advice regarding the management of the life-threatening situation. In most cases this means the institution of cardio-pulmonary resuscitation or CPR. Fortunately this skill is being acquired by more and more people. Thus providing the advice "start CPR" in the first aid instructions should have meaning to many people. However, in this writer's opinion it should be mandatory to indicate at the same time why the advice is being given. The statement "start CPR if breathing has stopped" is more meaningful than simply stating "start CPR". Furthermore, it provides guidance which, in a panicky situation, could prevent overtreatment.

As will be discussed later, few antidotes exist for the management of acute exposures. However, when such specific therapy is available this fact must be noted. Frequently this information is relegated to the "physician's note" section of the first aid statements. While this approach is reasonable such positioning of the information may cause it to be difficult to find in an emergency situation. It is prudent that such information should also be placed prominently within the first aid statements. Utilizing this approach not only insures that those providing the initial care of the patient will have knowledge of the existence of specific therapy, it may also encourage those who deal with highly toxic products to maintain a readily available supply of these specific antidotes.

Acute exposures generally occur through one or a combination of the following routes:

1. Ingestion

2. Skin and eye contamination

3. Inhalation

Ingestions constitute the most common route of exposure in acute situations. In developing first aid statements to provide management information for such exposures, the label writer must first determine if the ingestion of a small amount of the product (more than 5 to 10 cc or more than 5 to 10 g) of the product will produce potentially harmful results.

If it is determined, as based on animal studies and/or human experience, that a given product is capable of producing problems if ingested in small amounts or more, then appropriate first aid must be provided. Generally there is agreement among most medical authorities that the first action can consist of giving the patient fluids. The usual recommendation is to provide one to two glasses of fluid. However, while in agreement with this general advice, this writer feels that such instructions need to be more specific. Glasses vary in the volume of the fluid that they can hold. Furthermore, without giving specific advice as to the fluid that should be given, the patient may be administered a fluid that may later cause problems. It is far better and certainly more precise to advise that the patient should receive eight to sixteen ounces of water. If water is not available then other fluids such as soda, fruit juice, etc. may be used. As will be discussed shortly, milk is to be avoided. Nothing should be given by mouth to a patient exhibiting signs of central nervous system depression such as drowsiness, sleep or coma.

Administration of fluids not only dilutes the potential toxic material, which may help in reducing the total amount of material absorbed, but it also sets the stage for the next treatment recommendation namely, emptying of the stomach. In the vast majority of ingestions, the standard approach is to remove the offending material from the stomach as quickly as possible. This usually is accomplished by inducing vomiting. All poison center medical authorities would agree that the most efficient way to accomplish this is to use Syrup of Ipecac. Syrup of Ipecac is a prescription drug. Howeveral, all States, through their respective Pharmacy Boards, permit one ounce of Syrup of Ipecac to be dispensed without a prescription. However, the law does require that the Syrup of Ipecac be contained in a fully labeled bottle which provides instructions for administration and contraindications.

Syrup of Ipecac produces vomiting in the patient through at least two mechanisms namely, stimulation of the vomiting center in the brain and direct irritation of the lining of the stomach. Because of this latter effect, milk should not be administered with Syrup of Ipecac. Milk is a demulcent that reduces the irritating effect of the Syrup of Ipecac and, as has been shown in several studies, interferes with this action. Ipecac works best if the stomach is somewhat full. Since ingestions rarely occur immediately following a meal, there is need to add additional volume to the contents of the stomach. This is the rationale for recommending that eight to sixteen ounces of water or fruit juice be administered with the Syrup of Ipecac. Studies have shown that administration of such fluids before or after the administration of the Syrup of Ipecac causes no change in the efficacy of the Ipecac.

If the first aid instructions indicate that vomiting should be induced by using Syrup of Ipecac, then the dose to be administered should be specifically stated. The usual dose is 15 cc. Some authorities have recommended using larger doses of Ipecac, however, there is no evidence that such an approach enhances the action of the drug. Vomiting occurs in approximately 85-90% of the patients within twenty minutes following the administration of the initial dose of Ipecac. If vomiting has not occurred within that time period, then a second dose of Ipecac followed by 8-16 ounces of water should be administered. The second dose of Ipecac increases the overall success rate of induction of emesis to approximately 95%. It must be remembered that there will be a small percentage of patients who will not vomit following the administration of two adequate doses of Syrup of Ipecac. It is not reasonable to provide advice regarding the management of such patients (within the first-aid statements).

There is toxicity associated with Syrup of Ipecac, namely the occurrence of abnormal heart rhythms. However, such problems are only observed with massive ingestions of Ipecac, usually 100 to 150 cc or more and thus are rarely seen.

Syrup of Ipecac has an extraordinarily long shelf life, usually three to five years. As previously stated, it is the emetic of choice. To suggest the use of other "emetics" such as egg whites, soapy water, or salt water in the first aid advice is counterproductive, and at times, dangerous. These materials possess little emetic properties and in the case of salt water, if vomiting is not achieved, the salt can be absorbed by the body and in of itself, produce salt poisoning which at times can cause death.

If, based upon animal studies or human experience, the judgement is made that the best way to manage an acute ingestion of a given product is to induce vomiting then in this writer's opinion the first aid statements should clearly state this and should also provide the specific recommendation that Syrup of Ipecac be used along with specific instructions for its use. The argument that such specific recommendations should not be given since Syrup of Ipecac may not be readily available at the scene is somewhat analogous to saying that advice to wear goggles while handling a product should not be given since the goggles may not be readily available. It is generally believed that if a message is consistently and persistently given, that message will eventually be understood. If those individuals who are administering first aid recognize that their task would have been made easier if Syrup of Ipecac had been available, then there is the strong likelihood that Ipecac will be available for the next exposure. However, it is reasonable for the label writer to assume that Syrup of Ipecac may not indeed be available. To address this possible situation, the label writer should provide an alternative method of inducing vomiting. The alternative method that is generally recommended is to give the patient 8-16 ounces of water and then to induce vomiting by gagging the patient by stimulating the back of the patient's throat with a finger. The recommendation that a "blunt object" should be used is to be avoided. In a panicky situation such advice may not be clearly understood and the use of an object which in reality has a sharp point or edge may be employed thus causing potential injury to the patient.

Vomiting can usually be induced up to four hours following an ingestion. However, it is usually not necessary to state a time limit for the induction of vomiting in the first aid instructions unless it is known that the product in question is absorbed so rapidly that the induction of emesis past a certain time would probably not be helpful. There appears to be no age limit regarding the induction of vomiting.

As with the administration of fluids, vomiting should not be induced in patients exhibiting signs of central nervous system depression. In such situations the patient's gag reflex may be depressed, and there is the possibility of aspiration of the material into the lungs should vomiting occur. Similarly, vomiting should not be induced in patients who have ingested a product that is acidic or alkaline in nature. In these cases vomiting can lead to tissue being re-exposed to the product resulting in additional damage. Furthermore, in the act of vomiting, tremendous pressures are generated within the stomach and the esophagus (the tube that connects the mouth to the stomach), such pressures may cause tissue which has been "thinned" because of an exposure to an acid or alkaline material to rupture. The induction of vomiting following the ingestion of a petroleum distillate continues to be a controversial area. In this writer's opinion the removal of such material from the stomach is rarely necessary. The only possible exception to this would be a situation in which the petroleum distillate is acting as a "carrying agent" for a more toxic material. In such a circumstance it appears to be reasonable to remove the material through the induction of vomiting using Syrup of Ipecac. The label writer should seek specific medical advice if confronted with such a situation.

Other means of emptying the stomach exists, such as gastric lavage which consists of using fluids to wash out the stomach and, apomorphine which is an injectable emetic. Since both procedures can only be utilized in the hospital setting, there is no need to incorporate either into first aid instructions.

It is a common misconception that vomiting "empties" the stomach. Numerous studies have shown that approximately 40-50% of the ingested material is

Table 5.1:  Adsorption Efficacy of Activated Charcoal

| Excellent Adsorption | No Adsorption | Possible Adsorption |
|---|---|---|
| Most organic substances | Boric acid | DDT |
| Acetylsalicyclic acid | Ferrous sulfate | Tolbutamide |
| Chlorpheniramine maleate | Sodium hydroxide | Malathion |
| Colchicine | Potassium hydroxide | |
| Dextroamphetamine sulfate | Mineral acids | |
| Diphenylhydantoin | Sodium metasilicate | |
| Iodine | Cyanide | |
| Phenol | Small doses of alcohol | |
| Alcohol (with massive doses of ethyl or methyl) | | |
| Primaquine | | |
| All liquids which are insoluble in aqueous solution, i.e., kerosene | | |

removed with emesis. Gastric lavage removes even less. Because of these findings, it has become increasingly common for poison center personnel to recommend the oral administration of activated charcoal following cessation of vomiting. Activated charcoal adsorbs a number of chemicals and drugs (See Table 5.1) and thus it can be administered to bind remaining toxic materials in the stomach. The dose of activated charcoal is 25 grams to 50 grams. It must be administered *after* Syrup of Ipecac is given and vomiting has occurred since prior administration would cause the Syrup of Ipecac to be adsorbed to the charcoal and rendered ineffective. As with Ipecac, it is this writer's opinion that activated charcoal can play a major role in the early management of the acute ingestion, thus instructions regarding its use should be included in the first aid advice.

Numerous studies have shown that there is little to be gained from the use of demulcents (such as egg white, milk, milk of magnesia), neutralizers (such as juice, vinegar or bicarbonate solutions), or oils in the management of an acute ingestion. Indeed, utilization of such "therapeutic" measures as these may cause more harm. These materials may induce vomiting in patients where vomiting is contraindicated. Neutralizing agents may liberate large amounts of gas that can rupture the stomach or the esophagus or generate heat resulting in additional tissue damage. Oils may be aspirated into the lungs and cause pneumonia which may be more life-threatening than the ingestion itself.

Skin and eye contamination usually need only be of concern to the label writer if the product is capable of producing tissue irritation and/or burns. However, one must consider the possibility of the development of a toxic reaction due to the absorption of the product through the skin. Organophosphate pesticides are an example of a group of products causing such problems. Skin and eye exposure to some products may cause allergic manifestations. Again, the writer must depend upon animal studies or human experience to provide such information. For those products capable of causing toxic reactions due to skin absorption or allergic reactions, the product's label should indicate that such hazards exist. The label should also provide instructions that either eliminate or minimize the user's exposure to the product. If a toxic reaction occurs either due to skin absorption or an allergic response, then first aid directions should provide advice that will terminate the exposure and then direct the user to seek medical attention. Similarly, for products judged to produce tissue irritation and/or burns, first aid should first address methods of terminating the exposure. This entails either moving the patient from the source of the exposure or removing the irritating material from the patient. The former is usually easily accomplished. However, the latter can at times be more difficult to achieve since it requires complete decontamination of the site(s) of exposure. Clothing, watches, jewelry and contact lenses must be removed if contaminated with the toxic material, since all may retain the material and thus be a source of continuing exposure. Care must be taken so that the person removing the contaminated items does not himself, become exposed. Once exposure has been terminated, the affected area should be thoroughly washed with a stream of lukewarm water for at least fifteen minutes. Consideration of such advice would indicate that relatively large volumes of water must be utilized. If such volumes are not readily available then any other fluid with a relatively neutral pH can be used.

Eye exposures, usually involving acidic or alkaline material, can be particularly difficult to manage. Such materials frequently are trapped under the eyelids. The patient rarely is capable of keeping his eye open sufficiently to allow the stream of fluid to adequately circulate under the eyelids and remove the offending material. In such circumstances, it is better for someone else to hold the patient's eyelid open and to direct the stream of fluid across the surface of the eye.

For those products which have been determined to produce problems via inhalation, there is relatively little first aid advice that can be given. Again, one must be sure to provide advice that will result in the elimination of the exposure while minimizing the risk of additional exposures to those providing help. Frequently, terminating the exposure and removing the patient to a source of fresh air is all that is necessary. Occasionally, such exposures may produce sufficient upper airway or lower airway irritation to cause the patient to have difficulty breathing. Obviously, if the patient ceases to breathe because of the exposure, then CPR should be instituted. For those patients experiencing mild to moderate respiratory problems, the administration of oxygen, if available, is helpful.

Rarely is a product introduced into the body by accidental injection. When this does occur the product is usually being used in a system employing high pressure equipment, such as airless paint sprayers. In the experience of this writer, the material injected rarely is sufficiently toxic to warrant concern. However, such exposures can result in massive tissue destruction since the injected material can compress blood vessels in the area of the injection. Such compression can interrupt the blood supply to the affected area and with such interruption, tissue can die quickly. First aid advice must provide such information and direct the patient to obtain professional medical care promptly.

There appears to be consensus regarding the order in which first aid statements are presented on the product's label. As previously mentioned, those products that are capable of causing a life-threatening situation following an acute exposure should have that information and any pertinent therapeutic advice prominently displayed at the beginning of the first aid instructions. Advice regarding the management of ingestions or skin absorption usually comes next, followed by first aid advice pertaining to eye and/or skin exposures, followed by recommendations for the management of inhalation exposures. It is this writer's opinion that this standard convention of presenting the order of advice is probably best. It is at least consistent and this in of itself, allows those who are providing first aid management of the patient to find the information they need quickly.

There are relatively few products for which the first aid advice provides sole and complete treatment of an exposure. For most products first aid is just that, namely that aid which is administered first. Frequently, the exposure is such that the patient should or must receive additional medical care or evaluation. To accomplish this, the advice that is usually given is to have the patient call or see a physician. While such advice is useful, it may not always be the best that can be offered. Physicians who are knowledgeable of the medical management of industrial exposures may not readily be available. It may be more reasonable to instruct the patient or those rendering care to contact the Poison Center immediately. Over the past decade many poison centers have made tremendous strides in improving the quality of their service. The centers have moved from their traditional position of providing management advice for acute exposures that occur in children to providing comprehensive advice for a wide variety of circumstances. The modern poison center is capable of managing acute and chronic exposures in children and adults, drug abuse problems, industrial exposures and environmental spills. Because of the increased competency of these centers it is reasonable that a poison center should be contacted immediately following the exposure. The center's personnel can determine if the first aid that has already been given is sufficient to manage the exposure. If it is not, then they are in a position to recommend additional treatment advice and to arrange for the patient to move to a nearby medical facility for further evaluation and care. Most major poison centers have developed an on-going liaison with the emergency facilities of hospitals in their communities. Utilizing such arrangements, the poison center can and does provide comprehensive care for such exposures.

While there has been a similar improvement in the capability of emergency rooms in providing better care for the poisoned patient, there continues to exist a fair amount of variation among emergency rooms. Therefore, based upon the present state of the art, it is reasonable to advise a patient to seek additional care first from a poison center, then an emergency room and finally a physician, after the initial first aid management has been instituted.

As previously mentioned, for those products that are highly toxic or for which specific medical therapy exists, it is reasonable to include with the first aid instructions a "note" to physicians. Such a note should include the following items:

1.  *Recommendations For Specific Medical Treatment*-Usually this means the inclusion of drugs that can specifically aid in the management of the exposure. As previously

noted, there are relatively few antidotes available. However, when they do exist they should be noted. Thus, this section would include information regarding the use of a cyanide first aid kit for exposure to cyanide and cyanide-like substances. It would also include specific instructions for the use of atropine in the management of highly toxic cholinesterase-inhibiting compound exposures.

2. *Signs and Symptoms*-It is reasonable to provide the physician with specific information regarding the signs and symptoms that the patient may exhibit following an exposure. It is also helpful to quantitate these statements so that the physician may be able to determine the extent of the exposure. It is also useful to provide a time frame for the appearance and/or disappearance of these signs and symptoms.

3. *Specific Dosage Information*-When specific antidotes exist or when drugs can be utilized in the management of the exposure, the specific dosage of these drugs should be given in the physician's note. Since such notes usually appear on the most toxic products, time is of the essence in providing adequate treatment for the exposure. It is helpful to provide the physicians with all the information that is needed in order to treat the patient adequately.

4. *Contraindications*-If it is known that there are certain drugs that are capable of producing complications if used in the management of the exposure to a given product, then this information must be noted in the physician's note.

At times those label writers that are developing the "physician's note" section of the first aid advice literally believe such a note will only be read by a physician. In reality, the note is frequently read by people who have some degree of medical training but who are not physicians. Therefore the label writer should exercise caution in writing such a note so that the technical language contained in it is not such as to make it incomprehensible to a paramedic, who may be attempting to treat the patient.

Determining the appropriateness of the first aid advice for a given product can be a formidable task. It requires the compilation and interpretation of technical information coupled with a thorough understanding of the packaging, marketing, and use of the product. At times one has the impression that the development of first aid advice is a "cut and paste" operation. Too often, the label writer assumes that if a first aid statement is adequate for one product, it ought to be adequate for all similar products. Nothing could be further from the truth. A product, even though it has the exact same formulation as another product, may have its own set of problems because it is being used by a different audience. In order to recognize these potential problems a company should make use of the expertise of people who are knowledgeable in the area of clinical toxicology. All too often, a company relies on the medical judgment of people who have rarely, if at all, managed cases of exposures to such products. In order for the first aid advice to be as useful as possible, the label writer must have the input of the real world.

One of the major functions of first aid advice is to communicate information to people who need it. Unfortunately, this is rarely totally accomplished. The National Poison Center Network has, for a number of years, reviewed first aid and warning statements of a large number of products. These studies have shown that many labels contain outdated, and at times, dangerous medical advice. When the advice is sound it is often incomplete. Providing the advice to "wash the eye thoroughly" if contamination occurs does not provide complete information. It would be better to advise "wash the eye thoroughly with lukewarm water for fifteen minutes". Such a statement provides specific advice that removes any margin for error. Other studies conducted by the National Poison Center Network have shown that most first aid and warning statements require at least an eleventh grade reading level in order to comprehend the statement. Many people who have need of this information do not possess that degree of reading capability. It has been shown repeatedly that such statements can be rewritten at a fourth or fifth grade reading level while maintaining the accuracy and completeness of the advice given. Furthermore, such statements rarely require more space than the original advice.

# 6

# Acute Toxicity

Donald G. MacKellar
*Toxigenics, Inc.*
*Decatur, IL*

## INTRODUCTION

It is particularly appropriate for acute toxicity to be discussed in a book on chemical labeling, as historically, the results of acute toxicity testing have been used to classify chemicals and to allow selection of appropriate statements of hazard to be used on labels. In the early days of toxicology, acute testing was often "quick and dirty", consisting of dosing groups of animals, observing the number of deaths, and estimating an $LD_{50}$. Reports of these tests comfortably fit on one page. The present day acute toxicity test is a much more sophisticated process; it is done under precisely controlled conditions, group sizes and dose ranges are adjusted to give statistically interpretable results, observation periods are extended, and detailed observations made so that secondary or delayed effects are identified. Necropsies are performed on animals dying during the tests and on animals sacrificed at the end of the observation period. In some cases, this allows an estimation of the target organ or system and the identification of more subtle effects. Reports of these new tests commonly run from twenty to fifty pages. The new tests in addition to providing much more information, make it much easier to compare tests done by different laboratories.

Acute studies frequently include a package planned to evaluate the hazard of a material due to exposure by differing routes of administration and, occasionally, several species will be tested by one or more routes.

Several terms that will be used throughout this chapter are defined below.

1. *Acute Exposure:* Acute exposure is defined as a single exposure or smaller multiple exposures occurring over 24 hours or less.

2. $ED_{50}$: (Effective dose 50), the dose at which 50% of the test population shows the observed effect. The dose is usually reported as milligrams per kilogram of body weight and species specified.

3. $LD_{50}$: The same as the $ED_{50}$ when the observed effect is lethality.

4. *LD Lo:* This is defined as the lowest dose at which death has been observed.

5. $LC_{50}$: In inhalation studies this is the concentration of test article which causes death in 50% of the test population exposed for 1 or 4 hours. This is reported either in milligrams per liter, milligrams per cubic meter, or in parts per million by volume.

6. *Dose Response:* This is one of the most general principles of both acute and chronic toxicology; it implies that the effect observed is caused by the material being tested and that the degree of response observed is related to the amount of test material administered. This assumes that there is a site or sites with which the chemical reacts to produce the response and that the degree of response is related to the concentration of the chemical at the reactive site and, therefore, that the concentration at the site is related to the administered dose.

7. *Limit Tests:* These are regulatory limits where a specified level of test material is administered and causes no adverse effect and the material is judged safe for the regulated use and no further testing is required.

## ROUTES OF ADMINISTRATION

Materials to be tested for acute toxicity can be administered to the animals by one or more of a number of routes. These routes correspond to the various ways exposures can occur in humans. The regulatory agencies frequently require standard packages of acute studies which include the routes listed below. In any case, the routes selected should be related to a possible human exposure. It would be useless, for example, to test a viscous grease with a low vapor pressure for inhalation toxicity, as this would not be a possible route of exposure.

### Oral Administration

This is usually done by gavage, (direct intubation into the stomach) or by administration of encapsulated material. This allows accurate measurement of the amount administered and an accurate time of administraton. It is not usually done by mixing with the feed or drinking water, as the material would frequently not be acceptable to the animal and the amount consumed would not be known. This method of administration omits one area of exposure which is important in human accidental exposure, and that is absorption in the mouth or buccal area; however, there is no reasonable way to overcome this.

The animals are fasted prior to the administration of test article so that there is a relatively consistent absorption rate and there is reduced opportunity for the test material to react with the stomach contents. Absorption of the material·can occur in the stomach or in the intestine, depending upon the molecular weight of the material, on the lipophilicity of the material, the irritation caused to mucosal membranes and its degree of ionization at the pH of the stomach or of the intestine. It is usual in these tests to use an animal such as a rodent, which does not have an emesis response; otherwise, it would be difficult to measure an amount of material retained. The oral $LD_{50}$ has been determined for more compounds than any other value; and, because of this large data base, this figure has proven very valuable in comparing and ranking new chemicals for their degree of hazard.

### Dermal Toxicity

Dermal toxicity, not related to irritation or surface effect, is a toxicity caused by material which is absorbed through the skin, transported within the body and causes a systemic effect in another location. It is a common route for human accidental exposure. Toxicity by this route often involves a dose related response as described above, and requires penetration through the very complex skin. The skin contains barrier layers designed to protect the body from both hydrophilic and lipophilic materials, as well as the mechanical barrier of the stratum corneum, the hard outer layer. For this reason, the material is left in contact with the skin for a significant period of time, usually 4 to 24 hours depending on the agency guidelines. Some of the guidelines provide for a variation where the skin is slightly abraded, not deeply enough to draw blood, but to penetrate the stratum corneum. This is usually done as part of a limit test. Liquids are usually administered on an "as is" basis, without dilution. Powders are generally finely ground and moistened with water or normal saline. On some occasions, powders are diluted with a vehicle such as corn oil to improve skin contact. For some special purposes, the powder or liquid may be diluted with a material such as dimethyl sulfoxide, which promotes more rapid absorption through the skin. If a vehicle is used in this type of test, a vehicle control should be run unless the vehicle is water. The results of acute dermal studies are usually reported as an $LD_{50}$, except for irritation or corrosion which are discussed later.

### Inhalation

Exposure by the inhalation route is the most complex and resource consuming variation of the acute toxicity test. An exposure chamber must be available which is capable of holding all the animals to be tested without undue crowding. The equipment must be available to supply clean air, conditioned for temperature and humidity at a rate adequate to maintain approximately 15 air changes per hour. Exposures will be in a range of 1 to 4 hours and it is necessary to have test article generation and monitoring equipment adequate to maintain the desired concentration over this length of time. Gases and vapors can be metered into the air stream at a controlled rate, because it is usually necessary to analytically determine the concentration at the breathing zone of the animals. Dust and liquid materials which do not vaporize must be supplied as aerosols, which must be distributed uniformly through the test chamber. With liquid and solid aerosols it is necessary to measure the particle size distribution in the aerosol so that the number of particles respirable into the different areas of the lung can be calculated. Gas mixture concentrations are determined analytically and reported in parts per million, volume to volume, and are confirmed by the nominal gas concentrations obtained from the ratio of gas used to the air passed through the chamber over a measured period of time. For liquid and solid aerosols the nominal concentration is obtained the same way as above and reported as milligrams per liter or milligrams per cubic meter. Because of the impaction and adherence of the test article to the sidewalls of the chamber, condensation of test article particles, and other possible contaminants as well as gravitational settling of the larger particles, the nominal concentration is seldom even close to the analytical concentration. The analytical concentration may be determined by inserting a probe at the level of the breathing zone of the animals and depositing the aerosol sample on a weighed filter for gravimetric estimation or for extraction and direct analytical determination. Alternatively, the sample can be drawn through a liquid absorber and the absorber submitted for analysis.

Absorption of test materials into the body through the lung system is very complex; in the case of gases it is assumed that the material enters the lung thoroughly mixed with air and is in contact with all sections of the lung interior. Depending on the nature of the gas, it may pass through the lung wall into the blood stream, or a portion of it may dissolve into the lung. A portion may be cleared from the lung without being absorbed and a portion of this may be transferred to the gastrointestinal tract through the lung-clearing mechanism. Aerosol particles, either solid or liquid, penetrate into the lung through a distance dependent on their size. Most particles less than 5 microns in diameter are considered respirable and will penetrate into the lung. Larger particles are filtered or impacted higher in the respiratory system. Larger particles and some of the smaller particles will be cleared rapidly from the lung by normal ciliary action unless they dissolve and penetrate the lower wall promptly. Clearing rates of 3 to 5 centimeters per minute have been measured for some insoluble substances. Unless there is a strong cough reflex, the cleared material from the lung frequently enters the gastrointestinal tract where it contributes to toxicity through that route.

As previously mentioned, inhalation toxicity is reported as an $LC_{50}$, the lethal concentration. Obviously the toxic effects of inhaled materials are related to the weight of the material inhaled in relation to the body weight of the animal and, on occasion, it may be desirable to estimate this number. This can be done using the formula below:

$$\text{Retained Dose} = \frac{C\ t\ V_m\ f}{w} \quad \text{(in mg/kg)}$$

$f$ = fraction of dose retained
$C$ = concentration (mg/$\ell$)
$t$ = exposure time (minutes)
$V_m$ = minute volume (ml/min)
$w$ = animal weight (g)

## Ocular Toxicity Tests

Acute eye studies are almost always intended to evaluate the irritation or injury that the chemical causes to the eye itself or surrounding tissues. There are a few compounds, however, which can cause systemic injury or even death by absorption into the eye or its surrounding tissues. Most of these compounds will also exhibit acute toxicity through other routes. The tests for eye irritation are described thoroughly in various guidelines which are discussed later in this chapter. The species of choice to be used is usually the rabbit. One eye is used as a control, the other eye receives a specified amount of the test material. Periodic observation is made and the results are evaluated according to a scoring system devised by Draize, et al[1]. The damage to various areas of the eye, the cornea, iris, and conjunctivae are rated and entered into the scoring table and a composite score calculated.

This test is under severe criticism by animal rights groups because of injury and pain to the animals. Most laboratories will not test known corrosive materials and will abort the test as soon as injury develops.

## Dermal Irritation

Skin irritation testing is described in several guidelines which are discussed later. While it is not a dose related toxicity test, it is usually part of the acute toxicology package. Specification for these tests differs from those in the dermal toxicity study, in that there are smaller patches with less occlusion and for some guidelines a different time of exposure. Skin irritation is scored on a table[2] giving weighted responses for varying degrees of erythema, edema, necrosis, and dehydration. The score obtained by summarizing this table is applicable to classification under regulatory guidelines. Skin irritation is also observed and recorded during dermal toxicity tests, but results obtained in this test may not compare with the test done by the specific irritation protocol. Likewise, there are occasional deaths due to systemic toxicity during the irritation tests, however, they only serve as a suggestion to proceed further with dermal toxicity studies.

## Intravenous or Intramuscular Injection

These methods are seldom used for industrial or household chemicals; they do not approximate a normal exposure. They are used, however, for evaluation of drugs on occasions when the drug may be used by injection or when a quick estimate of the toxicity of the drug is required to compare with its therapeutic effects. $LD_{50}$ or $ED_{50}$ can be calculated from these injections, graded doses are used, and a dose response curve is often obtained.

## SPECIES SELECTION

Theoretically, any species could be used for acute toxicity tests, however, it is important that there be a sufficient body of background information on the species to be able to estimate the hazard of a chemical which is being tested relative to other materials. This would be true unless the species selected was a potential target for harm in the use of the chemical where a direct correlation would then exist. Other items to consider in selecting a test species are the availability of animals of standard strain and quality, the availability of facilities to properly house and test the animal selected, and last but not least, *cost*—cost of the animal, cost of the facilities to house it, and cost of the test article to be administered. The cost should be such that it is feasible to use large enough groups of animals for each test to get good statistical results. The cost of a dog study can be 50 to 100 times the cost of a test utilizing rodents using the same number of animals. Attention should also be paid to physical variables which affect the type of test being done—for example, administration of a test chemical to a dog by capsule or by gavage for an oral study may result in prompt emesis with attendant loss of the compound. Rodents do not have this reflex, and once the dose is administered it is retained.

A toxicologist doing acute testing for labeling purposes is ordinarily working under one or more of the regulatory guidelines. These guidelines either prescribe the test species or limit the choice to one of two or three species. These guidelines are discussed later in the chapter.

Within each species there are possible variations in strain, sex, age, and size. It is important to select animals of a strain whose response to the test in question is well known. For acute toxicity testing, the animals selected should be young adults where organ maturity is relatively complete and aging processes have not begun. Dose responses will differ according to the age and weight of the test animal. Young rodents do not have their full complement of mixed function, hepatic-oxidative enzymes, but they do have most of the conjugating enzymes. Older animals tend to obesity, which can modify the distribution and storage of chemicals in the body. The degeneration of liver and kidneys and regression of other organs and tissues may have begun in older animals resulting in a different pattern and rate of metabolism. Animals from both sexes are included in acute toxicity tests and are usually tested in equal proportions at each test level. Female animals used should be nonpregnant and nulliparous. There are a number of occasions when the different sexes react differently to a test compound. Frequently the females are more sensitive. In these cases it may be necessary to calculate a separate $LD_{50}$ for each sex so that the dose response and slope of the $LD_{50}$ curve can be calculated with statistical significance.

## ANIMAL HUSBANDRY

This refers to the care of the animals prior to and during the tests. This is described in detail in the *Guide for the Care and Use of Laboratory Animals,* (NIH) Publication 78.23. The ethical and regulatory requirements to provide good care for the animals are important; however, good science also dictates that unusually good care be taken of test animals because varying degrees of stress on the animals result in variations in response to a test compound. Cages should be of sufficient size to prevent crowding. Most acute test protocols will permit group caging of animals as long as the group sizes are uniform and there is adequate room in the cages. Overcrowding can result in social stresses on the animals, fighting, and possible cannibalism. Nonuniform feed distribution can result and observations of weight gain are biased. Further, in group caging, it is more difficult to maintain a unique identity for each animal. Single caging avoids many of these problems and facilitates observation of the animal during routine checks. There is some stress on rodents in single caging because they are normally a gregarious species. Single caging is somewhat more expensive in equipment and floor space, but it is generally thought that this is worthwhile.

Animal cages and rooms should be kept clean and sanitary to minimize the spread of disease and to prevent the development of odors. Cage liners or bedding should be changed sufficiently often to prevent build up of waste, and periodically the animals should be transferred to clean sanitized cages and the dirty cages washed in a high temperature cage washer. Animal rooms should be designed and maintained to prevent the access of insects and other vermin and particularly, to prevent intrusion of wild rodents.

The carefully bred strains of animals used for toxicology testing may be much more uniform in their responses, but they are more sensitive to environmental stresses than their wild cousins. Temperature, humidity, and airflow in the animal quarters must be tightly controlled and adjusted to the optimum value for the individual species. Ventilation in the animal rooms should be equivalent to 12 to 15 changes per hour without drafts on the animals.

Animals received from a supplier or from the breeding area of an integrated laboratory are quarantined for a specific period of time to assure that they are in good health and that they are not affected by a communicable disease. This quarantine period can vary from one to two weeks for rodents to as long as several months for monkeys.

Waste material from the animal rooms should be carefully packaged so that other areas of the laboratory are not contaminated when hauling it to the waste disposal area.

## CONDUCT OF THE TEST

Some of the guidelines which are presented later in this chapter are quite detailed as to the way the test should be conducted. However, there are some general considerations which will apply to most of these guidelines. In the present day, most testing in acute studies is done under the applicable portions of the FDA Good Laboratory Practices Regulations. A full discussion of these is beyond the scope of present work. However, in brief, they provide the minimum requirements for managing a toxicology laboratory and for conducting and documenting the course of a toxicology study so that its quality can be determined. A detailed protocol describing the work to be done is required, as are standard operating procedures specifying the way various operations are carried out in the laboratory. These studies should be adequately planned in advance so that the number of animals per group and the number of test levels will be adequate for the later statistical evaluation of the responses.

Healthy animals obtained after quarantine are randomized by sex into the desired test groups so that bias due to minor weight variations and individual idiosyncrasies is not allowed to affect the final result. Doses are carefully calculated and where necessary, prepared with diluent or vehicle and administered to the animals. Animals are then carefully observed for their response to the test article. This observation is far more detailed than the simple observation of mortality. Animal weights are obtained periodically and any other physically observable effect is noted. One large testing laboratory which uses computer recordings of observations has a standardized observation glossary which lists 84 different observations which can be made. Each of these observations can be further qualified by its severity and, where appropriate, by its location on the animal. Most guidelines require that the time of onset and the duration of each of these effects be recorded. Any animals that die during the course of the test are submitted for gross autopsy and any unusual finding is noted. Animals surviving the observation period are sacrificed and necropsied in a similar manner. At the completion of the test, the results are calculated as shown in the next section.

## CALCULATION OF RESULTS

In order to estimate the toxicity of the compound one must have data obtained at several dose levels and must see that there is a dose-related response. Further, there must be sufficient animals at each dose level so that the idiosyncracies of a few individual animals will not confuse the results. Excluded from this section are the consideration of eye and skin irritation tests, which have their own specialized scoring system previously discussed.

The calculations in this section apply to many toxic effects: emesis, convulsions, lethargy, death or other effects. These calculations will be discussed on the basis of lethality, even though the points discussed will apply just as well to the other effects. Lethality is frequently used as an index of toxicity in acute studies. Lethality is easily measured and quantified and there is no question of degree. The lethal effects of chemicals have been measured for a long time so there is a large body of data to use in comparing the effects one measures with others to aid in estimating their effect on man.

The relationship proposed by Trevan[3] appears so frequently that it must be considered fundamental. This is a sigmoid dose-response curve shown in Figure 6.1. The ordinate on this curve is mortality percent; the abscisa is the dose of test article in milligrams per kilogram on a logarithmic scale. This type of response is observed in tests done with any of the routes of administration shown above. It is important that the dose levels be selected so that at least three of the doses result in partial kills. The type of curve shown in Figure 6.1 usually has a relatively straight portion between 16 and 84%. This portion of the line is used to draw the connecting lines that show the $LD_{50}$. It also provides for the measurement of the slope. If one were to consider a statistically normal population, these values, 16 and 84, represent the limit of one standard deviation from the mean.

Plotting curves of the type shown in Figure 6.1 from the data usually obtained in acute toxicity tests is often quite difficult because these responses seldom come at convenient points on the curve. Accordingly, over the years there have been numerous attempts to provide simplified plotting and

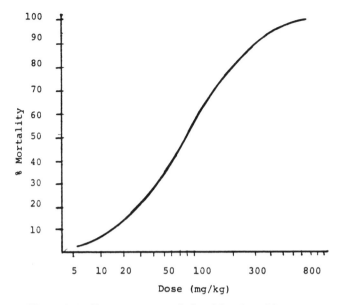

**Figure 6.1:** Dose response relationship, sigmoid curve.

measuring techniques which are easier used than the sigmoid curve. Two of these are in more general use than others, and they will be discussed here. Either of these can be done by hand, or by more or less sophisticated computer techniques.

If one takes the same data shown in Figure 6.1 and presents the frequency of mortality plotted against the log of the dose, one has (as shown in Figure 6.2) the bell-shaped curve, typical of a normal distribution. This shows that a large percentage of the animals died at levels near the median dose and that there was a decreasing number which were either more or less resistant to the compound being tested.

In a normally distributed population, the mean ±1 standard deviation represents 63% of the population; the mean ±2 S.D. covers 95.5% of the population; and the mean ±3 S.D. includes 99.7% of the population. The dose responses being considered are usually normally distributed, therefore, it's possible to convert the percent response to the deviations from the mean. These are often called normal equivalent deviations (NED). The NED for a 50% response would be 0, that for an 84% response would be +1, and 16% response would be −1. To avoid use of negative numbers, Bliss[4] suggested that the NED be changed by the addition of 5 to the value. He called these units "Probit" units. Thus, 50% response becomes a probit of 5. The interesting result of this is shown in Figure 6.3, where the data used in drawing the curve shown in Figure 6.1 are converted to probit units and plotted against the log of the dose. The data now appears as a straight line in which the $LD_{50}$ can easily be graphically determined. The slope of the dose response line is also easily obtained. Graph paper with the ordinates spaced in probability units and the

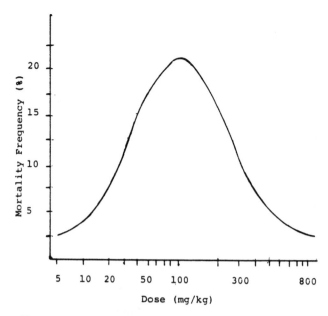

**Figure 6.2:** Dose response relationship, normal frequency distribution.

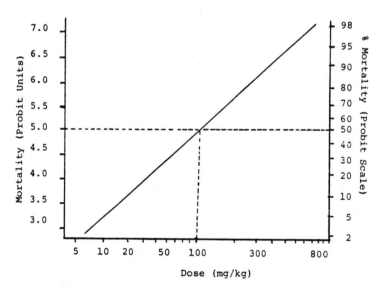

**Figure 6.3:** Dose response relationship, mortality in probit units.

absisca in log units is readily available for a manual plotting of acute toxicity data. Current mathematical methods adaptable to computer use are available for fitting data to these curves. The method of Litchfield and Wilcoxon[5] is quite satisfactory but requires a fairly large number of animals, about 50, in uniformly sized groups. Other methods such as the "Moving Average" Method of Thompson and Weil[6], work reasonably well with smaller numbers of animals and nonuniform group sizes. The computer programs will report the $LD_{50}$, the slope of the dose-response curve, and compute the statistical significance of the $LD_{50}$. A typical computer printout of a $LC_{50}$ calculation by the method of Litchfield and Wilcoxon[5] is shown in Figure 6.4. The slope of the dose response curve is useful in assessing the hazard of a compound and also in some cases, can be used to infer the type of mechanism taking place. If the slope of the dose response curve is flat, one can expect occasional deaths at those levels well below the $LD_{50}$. If the slope of the dose response curve is steep, the effect noted at those levels is slightly below the $LD_{50}$.

STUDY NO.  420-0000

ACUTE INHALATION TOXICITY STUDY - RAT

LITCHFIELD - WILCOXON LC-50

| DOSE (MG/L) | OBSERVED DEATHS (%) | EXPECTED DEATHS (%) | RESIDUAL | CONTRIBUTIONS TO CHI SQUARE |
|---|---|---|---|---|
| 17.50 | 20.0 | 19.9 | 0.1 | 0.0000 |
| 23.50 | 40.0 | 40.3 | -0.3 | 0.0000 |
| 30.00 | 60.0 | 60.0 | 0.0 | 0.0000 |
| 40.00 | 80.0 | 79.9 | 0.1 | 0.0000 |

TOTAL NUMBER OF ANIMALS  40          TOTAL     0.0001

NUMBER OF GROUPS  4

TOTAL CONTRIBUTIONS TO CHI SQUARE  0.001

CHI SQUARE (P=0.05) FOR 2 DEGREES OF FREEDOM IS 5.991

THE DATA ARE NOT SIGNIFICANTLY HETEROGENOUS.

LC-16 = 16.23 MG/L                     SLOPE FUNCTION  = 1.6332

LC-50 = 26.50 MG/L                              N       = 40

LC-84 = 43.29 MG/L                         F(LC-50)     = 1.2397

CONFIDENCE LIMITS OF THE SLOPE FUNCTION = 1.1172 - 2.3875

        LC-50 = 26.50 MG/L     (21.38 - 32.86)

        LC-1  =  8.45 MG/L     ( 3.38 - 21.16)

        LC-99 = 83.12 MG/L     (33.20 - 208.11)

**Figure 6.4:** Report of computer calculation of $LC_{50}$ by the method of Litchfield and Wilcoxon.[5]

## TEST PROTOCOLS AND GUIDELINES

In a volume oriented toward chemical labeling, it is important to know the requirements of various agencies concerned with labeling. These guidelines or regulations are all generally based on the principles previously discussed. However, each of them have their idiosyncracies and attempted standardization has not yet been wholly successful. A number of these are included as an appendix. In this appendix they are organized by *route* of administration and by *agencies*. It is important to note that even when using the harmonized guidelines, IRLG and OECD, each agency may have separate requirements as to what form of the material may be tested. FDA and TSCA usually prefer the product in its commercially pure form. EPA Pesticide Division may, in addition, request that the material be tested in its final usage form.

The two EPA guidelines presented should be considered with care. They are proposed guidelines and have not been approved as the final rule. Their proposal in 1979 resulted in a large number of comments. Most of these comments were directed to portions of the guidelines other than the acute sections so it is possible that when the revised guidelines are posted again, the acute section will not be very different. The Interagency Regulatory Liaison Group, IRLG, was made up of representatives of most of the participating agencies except DOT, and it is the intent that tests done under these guidelines will be acceptable to any of the participating agencies and that any specific guidelines each agency issues would be consonant with the IRLG guidelines. These IRLG guidelines are more general in nature and allow considerable latitude for individual interpretation. The guidelines presented by the Organization for Economic Cooperation and Development (OECD) are similarly general in nature. The OECD guidelines bind the member countries to accept toxicological data from another member country if it is done according to these guidelines. This implies that the U.S. would accept foreign work done according to these guidelines, but not necessarily work done in this country using the same guidelines.

Problems in interpreting guidelines can usually be solved by telephone or mail contact with the agency in question.

### Guidelines

The following pages contain the testing guidelines specified by the various agencies. They are presented in order of route of administration by agency. New guidelines from EPA office of Pesticide program are scheduled for release in 1983.

### Acute Oral Toxicity

A-1.  Department of Transportation—DOT 49 CFR 173.343a1

   (1) *Oral toxicity.* Those which produce death within 48 hours in half or more than half of a group of 10 or more white laboratory rats weighing 200 to 300 grams at a single dose of 50 milligrams or less per kilogram of body weight, when administered orally.

A-2. Environmental Protection Agency, Toxic Substances Control Act (EPA-TSCA) 44 FR 44066-7 (7/26/79)

(a) *Study Design.*

(1) *Species.* Testing must be performed with the laboratory rat.

(2) *Sex and age.* Young adult male and female animals must be used.

(3) Number of animals and selection of dose levels.

(i) A trial test is recommended for the purpose of establishing a dosing regimen which must include one dose level higher than the expected $LD_{50}$. If data based on testing with at least 5 animals per sex are submitted showing that no toxicity is evident at 5g/kg, no further testing at other dose levels is necessary. If mortality is produced, the requirements of paragraph (a)(3)(ii) of this section must apply.

(ii) Enough animals per dose level and sufficient dose levels spaced appropriately must be used to produce test groups with mortality rates between 10 percent and 90 percent and to permit the calculation of the $LD_{50}$ for males and females with a 95 percent confidence interval of 20 percent or less. At least 3 dose levels of the test substance in addition to controls (if any), must be tested. Though the group sizes may vary for each dose level, each group must contain equal numbers of male and female animals.

(4) *Control animals*

(i) A concurrent vehicle control group is recommended if the vehicle or diluent used in administering the test substance would be expected to elicit any important acute toxicologic response, or if there are insufficient data on the acute effects of the vehicle.

(ii) A concurrent untreated control group is not required.

(5) *Dosing.* All animals must be dosed by gavage. All animals must receive the same concentration of dosing solution. They should also receive about the same volume of dosing solution, which should not exceed 4-5 ml. per animal.

(6) *Duration of test.* The animals must be observed for at least 14 days after dosing, or until all signs of reversable toxicity subside, whichever occurs later.

(b) *Study Conduct*

(1) *Fasting.* Food shall be withheld from the animals the night prior to dosing.

(2) *Observation.* The animals must be observed frequently during the day of dosing and checked at least every 12 hours throughout the test period. The following must be recorded. Nature, onset,

severity, and duration of all gross or visible toxic or pharmacological effects, e.g., abnormal or unusual cardiovascular, respiratory, excretory, behavioral, or other activity, as well as signs indicating an adverse effect on the central nervous system (paralysis, lack of coordination, staggering), pupillary reaction; and time of death. The weight of each animal must be determined at least semi-weekly (3-4 day intervals) throughout the test period, and at death.

(3) *Sacrifice and necropsy.* All test animals living at the termination of the observation period must be sacrificed. All test animals, whether dying by sacrifice or during the test must be subjected to a complete gross necropsy following their death, in accordance with § 772.100-2(b) (7), Subpart A. All abnormalities must be recorded.

(c) *Data reporting and evaluation.* In addition to the information required by § 772.100-2 (b) (8), Subpart A, the test report must include the following information:

(1) Tabulation of response data by sex and dose level (i.e., number of animals dying per number of animals showing signs of toxicity per number of animals exposed)

(2) Time of death after dosing

(3) $LD_{50}$ for each sex for each test substance calculated at the end of the observation period (with method of calculation specified)

(4) 95 percent confidence interval for the $LD_{50}$ and

(5) Dose-response curve and slope.

## A-4. Consumer Product Safety Commission (CPSC)—16 CFR part 1500

CPSC does not provide a detailed guideline for oral studies. Their definitions regarding toxicity are as follows:

"Highly toxic" means any substance which produces death within 14 days in half or more than half of a group of 10 or more laboratory white rats each weighing between 200 and 300 grams, at a single dose of 50 milligrams or less per kilogram of body weight, when orally administered.

"Toxic" means any substance that produces death within 14 days in half or more than half of a group of white rates (200 - 300 grams) when a single dose of from 50 milligrams to 5 grams per kilogram is administered orally.

## A-5. Interagency Regulatory Liaison Group (IRLG)—Testing Standard 1981

### I. General Considerations

#### A. Good Laboratory Practices

Studies should be conducted according to good laboratory practice regulations (e.g., "Nonclinical Laboratory Studies, Good Laboratory Practice Regulations," *43 FR 59986, 22 December 1978).*

B. Test Substance

    1.  The specific substance or mixture of substances to be tested should be determined in consultation with the responsible agency. As far as is practical, composition of the test substance should be known. Information should include the name and quantities of all major components, known contaminants and impurities, and the percentage of unidentifiable materials to account for 100% of the test substance.

    2.  Ideally, the lot of the substance tested should be the same throughout the study. The test sample should be stored under conditions that maintain its stability, strength, quality, and purity from the date of its production until the tests are complete.

C. Animals

    1.  Recommendations contained in DHEW pub. no. (NIH) 74-23, entitled *Guide for the Care and Use of Laboratory Animals*, should be followed for the care, maintenance, and housing of animals.

    2.  Healthy animals, not subjected to any previous experimental procedures, must be used.

    3.  The test animal shall be characterized as to species, strain, sex, weight and/or age. Each animal must be assigned an appropriate identification number.

    4.  Animals may be group-caged for this test unless the pharmacological action of the test substance dictates otherwise, but the number of animals per cage should not prevent continued and clear observation of each animal. When signs of morbidity or excitability are observed in group-caged animals during the test, such animals should be moved to separate cages.

    5.  Animals should be assigned to groups in a manner that minimizes bias and assures comparability of pertinent variables (for example, weight variation in animals used in a test should not exceed ± 20% of the mean weight).

D. Dead Animals, Necropsy, and Histopathology

Animals should be observed as necessary to insure that there is minimal loss due to cannibalism or similar management problems. Where possible, necropsy must be performed soon after an animal is sacrificed or found dead to minimize loss of tissues due to autolysis. When necropsy cannot be performed immediately, the animal must be refrigerated at temperatures low enough to minimize autolysis and not cause freezer burn. If histopathological examination is to be conducted, tissue specimens should be placed in appropriate fixative when they are taken from the animal.

## II. Specific Considerations

A. Test Preparation

    1.  Animals: Althouth several mammalian test species may be used for the test, the rat is preferred. When attempting to estimate hazards to young humans, additional studies designed to consider the developmental stage of the test animal in relation to anticipated human exposure should be performed.

    2.  Fasting: Prior to administration of test substance, food should be withheld from rats overnight. For other rodents with higher metabolic rates a shorter period of fasting is appropriate.

    3.  Limit test: A trial test is recommended to establish the need for further testing. If a test at a dose of at least 5 g/kg body weight using the procedures described for

this study produces no compound-related mortality, then a full study using 3 dose levels may not be necessary.

4. Number and sex: At least 10 animals, 5 per sex, should be used at each dose level. Nonpregnant, nulliparous females should be used.

5. Dose levels: At least three and preferably four dose levels should be used to produce a range of toxic effects and mortality rates (*e.g.*, a range from 10 to 90% and bracketing the expected $LD_{50}$). The data should be sufficient to produce a dose response curve and permit an acceptable determination of the $LD_{50}$.

6. Controls: Controls are generally not required, since dose response during an $LD_{50}$ may serve as an internal control. If a vehicle or solvent of uncharacterized toxic potential is used, an acute oral toxicity test should be done using the solvent.

B. Test Procedure

1. Route of administration: Ideally, the dose should be administered in a single dose by gavage or capsule. Because of the physical/chemical nature of the test substance, doses may be administered in a suspension or capsules in divided doses over a period of 24 hours.

2. Dosage: The dose is administered via soft rubber or polyethylene tubing or a ball-tip needle. The maximum volume of aqueous solution that can be given in one dose depends on the rodent's size and should not exceed 2 ml/100g body weight. For nonaqueous liquids and suspensions, the volume should not exceed 1 ml/100g body weight. When possible, variability in test volume should be minimized, with concentrations being adjusted accordingly.

3. Observation period: The observation period should be at least 14 days. Although a 14-day observation period is sufficient for most compounds, animals demonstrating visible signs of toxicity after 14 days may be held longer.

4. Clinical observations: A careful clinical observation of each animal should be made at least once a day. Additional observations should be made daily with appropriate actions taken to minimize loss of animals to the study, *e.g.*, necropsy or refrigeration of those animals found dead and isolation or sacrifice of weak or moribund animals. All toxicological and pharmacological signs should be recorded including time of onset, intensity, and duration. The time of death should also be noted. Individual records should be maintained for each animal.

5. Weight change: Animals must be weighed individually on the day the test substance is administered, weekly thereafter, and prior to sacrifice.

6. Necropsy: A complete gross necropsy should be performed on all animals that die during the course of the test. Consideration should be given to gross necropsy of the animals sacrificed at termination of the test. Microscopic examination of gross lesions should also be considered. If the substance will not be subjected to additional acute or multiple dose testing that includes gross necropsy, or if the results of this test are to be used for labeling purposes, complete gross necropsy should be performed on the remaining animals at termination of the test.

## Acute Dermal Toxicity

B-1. DOT 49 CFR 173.43a3

*Toxicity by skin absorption.* Those which produce death within 48 hours in half or more than half of a group of 10 or more rabbits tested at a dosage of 200 milligrams or less per kilogram body weight, when administered by continuous contact with the bare skin for 24 hours or less.

B-2. EPA—TSCA 44 CFR 44067 7/26/79

§ 172.112-22—Acute dermal toxicity study.

(a) *Study design.*

(1) *Condition of test substance.* If the test substance is a liquid, it must be applied as a liquid. If the test substance is a solid, it must be slightly moistened (made pasty) with physiological saline before application.

(2) *Species.* Testing must be performed with at least one mammalian species, preferably albino rabbits. An alternative species may be used if the sponsor can provide sufficient data and/or rationale to demonstrate that it is a more appropriate species for a specific test substance.

(3) *Age.* Young adult male and female animals must be used.

(4) *Number of animals and selection of dose levels.*

(i) A trial test is recommended for the purpose of establishing a dosing regimen which must include one dose level higher than the expected $LD_{50}$ and at least one dose level below the expected $LD_{50}$. If data based on testing with at least 5 animals per sex with abraded skin are submitted showing that no toxicity is evident at 2 g/kg, no further testing at other dose levels is necessary. If mortality is produced the requirements of paragraph (a)(4)(ii) of this section apply.

(ii) The number of animals per dose level and the number and spacing of the levels must be chosen to product test groups with mortality rates between 10 percent and 90 percent, and permit appliciation of the $LD_{50}$ (abraded skin and intact skin) of males and females with a 95 percent confidence interval of 20 percent or less. At least 3 dose levels of the test substance, in addition to controls, must be tested. Though the group sizes may vary for each dose level, the groups must contain equal numbers of male and female animals.

(5) *Control animals* A concurrent untreated control group of animals is required. A concurrent vehicle or diluent used in administering the test substance would be expected to elicit any important acute toxicologic response, or if there are insufficient data on the acute effects of the vehicle.

(b) *Study conduct.*

(1) *Application.* In all animals, the application site must be free of hair. In addition, the application sites of abraded-skin groups must be abraded in such a way as to penetrate the stratum corneum but not the dermis. The test substance must be kept in contact with the skin of at least 10 percent of the body surface (for rabbits) for at least 24 hours. (See Draize (1944) for equivalent sq. cm. of body surface.) The preferred appliciation site is a band around the trunk of the test animal. A wrapping material

such as gauze covered by impervious, nonreactive rubberized or plastic material should be used to retard evaporation and keep the test substance in contact with the skin. At the end of the exposure period, the wrapping should be removed and the skin wiped (but not washed) to remove any test substance still remaining.

(2) *Duration of observation*. Animals must be observed for at least 14 days after dosing or until all sign of reversible toxicity in survivors subside, whichever occurs later.

(3) *Observation*. Animals must be observed frequently during the day of administration of the test and checked at least every 12 hours throughout the test period. The following must be recorded: Nature, onset, severity, and duration of each toxic and pharmacologic such as abnormal unusual cardiovascular, respiratory, excretory, behavioral, or other activity, as well as signs indicating an adverse effect on the central nervous system (paralysis, lack of coordination staggering); pupillary reaction; and time of death. The weight of each animal must be determined at least semi-weekly (3-4 day intervals) throughout the test period, and at death.

(4) *Sacrifice and necropsy*. All animals living at the termination of the observation period must be sacrificed. All test animals, whether dying by sacrifice or during the test, must be subjected to a complete gross necropsy following their death, in accordance with § 772.100-2(b)(7), Subpart A. All abnormalities must be recorded.

(5) *Histopathology*. Examination of skin must include histological examination of treated tissue in accordance with § 772.100-2(b)(7), Subpart A.

(c) *Data reporting and evaluation*. In addition to the information required by §772.100-2(b)(8) and paragraphs (b)(3), (4), and (5) of this section, the test report must include the following information.

(1) Tabulation of response data by sex and dose level (i.e., number of animals dying per number of animals showing signs of toxicity per number of animals exposed)

(2) Time of death after treatment;

(3) Time of recovery for fully recovered animals;

(4) $LD_{50}$ for each sex for each test substance for animals with abraded skin and for animals with intact skin, calculated at the end of the observation (with method of calculation specified);

(5) 95 percent confidence interval for each $LD_{50}$ and

(6) Dose-response curve, and slope (with confidence limits).

B-4. CPSC 16 CFR 1500.40

(a) *Acute dermal toxicity (single exposure).* In the acute exposures, the agent is held in contact with the skin by means of a sleeve for periods varying up to 24 hours. The sleeve, made of rubber dam or other impervious material, is so constructed that the ends are reinforced with additional strips and should fit snugly around the trunk of the animal. The ends of the sleeve are tucked, permitting the central portion to "balloon" and furnish a reservoir for the dose. The reservoir must have sufficient capacity to contain the dose without pressure. In the following table are given the dimensions of sleeves and the approximate body surface exposed to the test substance. The sleeves may vary in size to accomodate smaller or larger subjects. In the testing of unctuous materials that adhere readily to the skin, mesh wire screen may be employed instead of the sleeve. The screen is padded and raised approximately 2 centimeters from the exposed skin. In the case of dry powder preparations the skin and substance are moistened with physiological saline prior to exposure. The sleeve or screen is then slipped over the gauze that holds the dose applied to the skin. In the case of finely divided powders, the measured dose is evenly distributed on cotton gauze which is then secured to the area of exposure.

### Dimensions of Sleeves for Acute Dermal Toxicity Test
### (Test Animal-Rabbits)

| Diameter at Ends, cm | Overall Length, cm | Range of Weight of Animals, g | Average Area of Exposure, cm$^2$ | Average of Total Body Surface, % |
|---|---|---|---|---|
| 7.0 | 12.5 | 2,500-3,500 | 240 | 10.7 |

(b) *Preparation of test animal.* The animals are prepared by clipping the skin of the trunk free of hair. Approximately one-half of the animals are further prepared by making epidermal abrasions every 2 or 3 centimeters longitudinally over the area of exposure. The abrasions are sufficiently deep to penetrate the stratum corneum (horny layer of the epidermis) but not to disturb the derma; that is, not to obtain bleeding.

(c) *Procedures for testing.* The sleeve is slipped onto the animal which is then placed in a comfortable but immobilized position in a multiple animal holder. Selected doses of liquids and solutions are introduced under the sleeve. If there is slight leakage from the sleeve, which may occur during the first few hours of exposure, it is collected and reapplied. Dosage levels are adjusted in subsequent exposures (if necessary) to enable a calculation of a dose that would be fatal to 50 percent of the animals. This can be determined from mortality ratios obtained at various doses employed. At the end of 24 hours the sleeves or screens are removed, the volume of unabsorbed material (if any) is measured, and the skin reactions are noted. The subjects are cleaned by thorough wiping, observed for gross symptoms of poisoning, and then observed for 2 weeks.

## B-5. IRLG—Acute Dermal Toxicity Test 1981

### I. General Considerations

#### A. Good Laboratory Practices

Studies should be conducted according to good laboratory practice regulations (*e.g.*, "Nonclinical Laboratory Studies, Good Laboratory Practice Regulations," 43 FR 59986, 22 December 1978).

#### B. Test Substance

1. The specific substance or mixture of substances to be tested should be determined in consultation with the responsible agency. As far as is practical, composition of the test substance should be known. Information should include the name and quantities of all major components, known contaminants and impurities, and the percentage of unidentifiable materials, if any, to account for 100% of the test sample.

2. Ideally, the lot of the substance tested should be the same throughout the study. The test sample should be stored under conditions that maintain its stability, strength, quality, and purity from the date of its production until the tests are complete.

#### C. Animals

1. Healthy animals, not subjected to any previous experimental procedures, must be used.

2. The test animal shall be characterized as to species, strain, sex, weight and/or age. Each animal must be assigned an appropriate identification number.

3. Recommendations contained in DHEW pub. no. (NIH) 74-23, entitled *'Guide for the Care and Use of Laboratory Animals,'* should be followed for the care, maintenance, and housing of animals.

4. Animals should be caged individually for this test.

5. Animals must be assigned to groups in such a manner as to minimize bias and assure comparability of pertinent variables (for example, weight variation in animals used in a test should not exceed ± 20% of the mean weight).

#### D. Dead Animals, Necropsy, and Histopathology

Animals should be observed as necessary to insure that there is minimal loss due to cannibalism or management problems. Where possible, necropsy must be performed soon after an animal is sacrificed or found dead to minimize loss of tissues due to autolysis. When necropsy cannot be performed immediately, the animal must be refrigerated at temperatures low enough to minimize autolysis. If histopathological examination is to be conducted, tissue specimens should be placed in appropriate fixative when they are taken from the animal.

## II. Specific Considerations

A. Test Preparation

1. Animals: The rat, rabbit or guinea pig may be used; however, the young, adult, albino rabbit weighing 2.0 to 3.0 kg is the preferred species because of its size, ease of handling and restraint, and skin permeability. If rats or guinea pigs are used, rats weighing 200 to 300 g or guinea pigs weighing 350 to 450 g are suggested. Selection of other species may be acceptable but must be justified.

2. Preparation of skin: Approximately twenty-four hours before testing, fur from the back of animals should be clipped so that no less than 10% (about 240 cm$^2$) of the total body surface area is available for application of material.

3. Limit test: A trial test on abraded skin is recommended. The abraded area may be prepared by making four epidermal incisions with a clean needle through the stratum corneum (not deep enough to disturb the derma or produce bleeding), but other acceptable methods may be used. If a test at a dose of 2 g/kg body weight (or 2 ml/kg) or more, using the procedures described for this study, produces no compound related mortality, then a full study using 3 dose levels may not be necessary.

4. Number and sex: Equal numbers of animals of each sex with intact skin are required for each dose level. The number of animals per dose depends on the level of statistical confidence desired. Five animals per sex per dose are recommended in most cases. Females should be nulliparous and nonpregnant.

5. Dose levels: If mortality occurs in the limit test, an additional test should be conducted using rabbits with intact skin. At least three dose levels spaced appropriately to produce test groups with a range of toxic effects and mortality rates should be used.

B. Test Procedure

1. Test substance: Liquids should be tested directly, but solid test substances should be moistened sufficiently with normal saline or tap water to make a paste that will insure good contact with the skin. For some applications, it may be appropriate or necessary to use other vehicles. If a carrier or diluent is used, it should be non-irritating and of known low toxicity. When such vehicles are used, consideration should be given to the effects of those vehicles on absorption of the test substance.

2. Dosage: When technically feasible, the maximum quantity of substance plus vehicle to be applied is 2 g/kg body weight. The test substance should be applied uniformly over the dorsal surface area.

3. Administration (application): The test substance must remain in contact with the skin throughout the exposure period of 24 hours. Liquid or solid substances should be held in contact with the skin with a porous gauze dressing and non-irritating tape. The test site should be covered with an impermeable material such as plastic film or rubberized cloth in a semi-occlusive fashion, allowing air to pass between the skin and the covering. Occlusive skin dressings may enhance penetration of the test substance and should be used only when testing for effects that may occur under similar conditions in humans.

   During exposure, animals should be prevented from ingesting or inhaling the test substance. Restrainers, such as Elizabethan collars, that permit animals to move about their cages should be used for this purpose. Immobilization is not recommended.

At the end of the exposure period, all residual material should be removed by washing, using an appropriate solvent. About one half hour later, and once again at 72 hours, the exposed area should be examined, and all lesions noted and graded (Table I).

4. Observation period: The observation period must be at least 14 days. Although a 14-day observation period is sufficient for most compounds, animals demonstrating visible signs of toxicity at 14 days may be held longer.

5. Clinical observations: A careful clinical observation of each animal should be made at least once a day. Additional observations should be made daily with appropriate actions taken to minimize loss of animals to the study, *e.g.*, necropsy or refrigeration of those animals found dead and isolation or sacrifice of weak or moribund animals. All toxicological and pharmacological signs should be recorded including time of onset, intensity, and duration. The time of death should also be noted. Individual records should be maintained for each animal.

6. Weight change: Animals must be weighed individually on the day the test substance is administered, weekly thereafter, and at death or sacrifice.

7. Necropsy: A complete gross necropsy should be performed on all animals that die during the course of the test. Consideration should be given to gross necropsy of the animals sacrificed at termination of the test. Microscopic examination of gross lesions should also be considered. If the substance will not be subjected to additional acute or multiple dose testing that includes gross necropsy, or if the results of this test are to be used for labeling purposes, complete gross necropsy should be performed on all remaining animals at termination of the test.

## Acute Inhalation

### C-1. DOT 49 CFR 173.343a2

(2) *Toxicity on inhalation.* Those which produce death within 48 hours in half or more than half of a group of 10 or more white laboratory rats weighing 200 to 300 grams, when inhaled continuously for a period of one hour or less at a concentration of 2 milligrams or less per liter of vapor, mist, or dust, provided such concentration is likely to be encountered by man when the chemical product is used in any reasonable foreseeable manner.

### C-2. EPA—TSCA 44 FR 44067-9 (7/26/79)

(a) *Study design.*

(1) *Secies, sex, and age.* Testing must be performed with the laboratory rat. Young adult male and female animals must be used.

(2) Number of animals and selection of dose levels.

(i) A trial test is recommended for the purpose of establishing a dosing regimen which include one dose level higher than the expected $LC_{50}$ and at least one dose level below the expected $LC_{50}$. If data based on testing with at least 5 animals per sex are submitted showing that no toxicity is evident at 5 g/l, no further testing at other dose levels is necessary. If mortality is produced, the requirements of paragraph (a)(2)(ii) of this section apply.

(ii) The number of animals per dose level, and the number and and the spacing of dose levels must be chosen to produce test groups with mortality rates between 10 percent and 90 percent, and to permit calculation of the $LC_{50}$ with a 95 percent confidence limit of 20 percent or less. At least 4 dose levels of the test substance, in addition to controls, must be tested. Though the group sizes may vary for each dose level, the group must contain an equal number of male and female animals.

(3) *Duration of test.* In selecting the exposure period, allowance must be made for changed concentration equilibration time. Where there is no difficulty in maintaining a steady concentration of the test substance in the chamber(s), the exposure period must be at least 1 hour. Where there is some difficulty in maintaining a steady concentration the exposure period must last up to 4 hours. The animals must be observed for 14 days, or until all signs of reversible toxicity subside, whichever occurs later.

(4) *Use of solvent.* A solvent may be added to the substance, if necessary, to help generate an exposure atmosphere. If a product's labeling instructions specify the use of a particular solvent, that solvent is preferred. If no solvent is specified in the product's labeling instructions, the solvent, if any, which is used to formulate the product should be used.

(5) *Control groups.*

(i) A concurrent untreated control group is required.

(ii) If any solvent, other than water, is used in generating the exposure atmosphere, a vehicle control group must be tested. The vehicle control group must be exposed to an atmosphere containing the greatest concentration of solvent present in any test system.

(b) *Study conduct.*

(1) *Exposure Chamber design and operation.*

(i) Inhalation exposure techniques described in this section are based on the use of whole-body inhalation numbers which allow the experimental animals to receive whole-body dermal exposure, as well as the exposure by inhalation. In some cases, the investigators will want to use other inhalation exposure techniques involving face masks head-only exposure, intratracheal instillation, or other similar techniques which reduce or include added dermal and oral exposures. Some alternative techniques are described by Phalen, 1976. When alternative techniques are used, the procedures and results must be reported in a manner similar to that required with the use of whole body inhalation chambers.

(ii) Animals must be tested in a dynamic air flow exposure chamber. The chamber design must be chosen to enable production of an evenly distributed exposure atmosphere throughout the chamber. The chamber design also should minimize crowding of the test animals and maximize their exposure to the test substance.

(2) *Operation measurements.* The following measurements must be taken with care to avoid major fluctuations in the air concentrations or major discrepancies in the operation of the chambers.

  (i) Air flow. The rate of air flow through the chamber must be measured continuously.

  (ii) Chamber concentrations

    (A) Nominal concentrations must be calculated for each run by dividing the amount of the test substance used for the generating system by the air flowing through the chamber during the exposure.

    (B) Actual chamber concentrations must be determined by samples of chamber air taken near to the breathing zone of the animals as frequently as necessary to obtain an averaged integrated external exposure which is representative of the entire exposure period. The system used to generate the vapor, gas, or aerosol should be such that the chamber concentrations and particle size distributions are controlled under stable conditions, reflecting the current state-of-the-art, and should not vary in a range greater than 30 percent of the average (range/mean equal to or less than 30 percent).

  (iii) *Temperature and Humidity.* The temperature must be maintained at $24\pm2°C$, and the humidity within the chamber at 40-60 percent. Both must be monitored continuously.

  (iv) *Oxygen.* The rate of air flow through the chamber must be adjusted to insure that the oxygen content of exposure atmosphere is at least 19 percent.

  (v) *Particle Size Measurement.*

    (A) *General.* In the case of gases and vapors, particulate sampling should be carried out at intervals to insure the animals are not being exposed unknown and unexpected particulate materials. Aerosol particle size measurements should be made on samples taken at the breathing level of the animals. These analyses should be carried out using techniques and equipment reflective of the state-of-the-art. All of the suspended aerosol (on a gravimetric basis) should be accounted for, even when most of the aerosol is not respirable.

(B) *Sizing Analysis.* The sizing analysis should be in terms of equivalent aerodynamic diameters and should be represented as geometric mean (mean) diameters and their geometric standard deviations (see NIOSH syllabus in the Appendix to this section), as calculated from log probability graphs or computer programs. The size analyses should be carried out frequently during the development of the generating system to insure proper stability of aerosol particles, and only as often thereafter during the exposure as is necessary to determine adequately the consistency of particle distributions to which the animals are exposed, maintaining at least 20 percent of the particles at 10 microns or less. At a minimum, these analyses should be carried out once per hour for each level of exposure for gasous test substances, twice per hour for liquid test substances, and 4 times per hour for dusts and powders.

(3) *Observation.* The animals must be observed frequently during the day of dosing and checked at least every 12 hours throughout the test period, for at least 14 days after dosing or until all signs of reversible toxicity subside, whichever occurs later. The following must be recorded: Nature, onset, severity, and duration of all gross or visible toxic or pharmacologist effects, i.e., abnormal or unusual cardiovascular, respiratory, excretory, behavioral, or other activity, as well as signs indicating an adverse effect on the central nervous system (paralysis, lack of coordination, staggering; pupillary reactions; and time of death. The weight of each animal must be determined on the day of dosing, 2, 3, 4, 7, and 14 days after dosing, weekly thereafter, and at death.

(4) *Sacrifice and Necropsy.* All animals living at the termination of the observation period must be sacrificed. All test animals, whether dying by sacrifice or during the test, must be subjected to a complete gross necropsy following their death, in accordance with § 772.100-2(b)(7), Subpart A, Examination must include nasal passages, trachea, bronchi, and lungs, and any other tissues known to be affected by the test substance. All abnormalities must be recorded.

(5) Preservation of tissues and histopathology examination. The following are required:

(i) Those tissues designated in paragraph (b)(5)(ii) of this section must be placed in suitable fixative as soon as possible. Tissues and microscopic slides must be prepared according to the standards set forth in § 772.100-2(b)(7)(ii) and (iii), Subpart A. Tissue samples, tissue blocks, and microscopic slides must be preserved and held in accordance with § 772.100-1(i)

(ii) The following tissues must be examined microscopically:

    (A) Lungs, liver, and kidneys at all dose levels

    (B) Any tissue or organ that appears abnormal, at any dosage level, as determined in the necropsy examination.

(iii) The histopathology findings must be recorded and reported as required by paragraph (c)(10) of this section.

(c) *Data reporting evaluation.* In addition to information required by § 772.100-2(b)(8), Subpart A, and paragraphs (b)(3) and (b)(4) of this section, the test report must include the following:

(1) Vapor pressure and particulate size (median size with geometric standard deviation).

(2) Description of the chamber design and operation, including type of chamber, its dimensions, the source of makeup air and its conditioning (heating or cooling) for use in the chamber, the treatment of exhausted air, the housing and maintenance of the animals in the chambers, and similar related information. Equipment for measuring temperatures and humidity, the generating system, and the methods of analyzing airborne concentrations of particle sizing must be described;

(3) The following operation data must be tabulated both individually and in summary form, using means and standard deviations (with or without ranges) in tabular form. The data summaries must be grouped according to experimental groups, and nonexpected differences (such as in temperature and airflow) and must be tested for statistical significance. .

    (i) Airflow rates through the chamber;

    (ii) Chamber temperature and humidity;

    (iii) Nominal concentrations;

    (iv) Actual concentrations; and

    (v) Median particle sizes and their geometric standard deviations and percent of particles 10 microns or less.

(4) Tabulation of the response data (number of animals dying per number of animals exposed) at each exposure level by sex, and time of death after dosing;

(5) Tabulation of the body weights on the day of dosing, 2, 3, 4, 7, and 14 days after dosing, weekly thereafter, and at death.

(6) The $LC_{50}$ calculated on an exposure of one hour, for each sex for each test substance;

(7) Specification of the method used for $LC_{50}$ calculation;

(8) The 95 percent confidence interval for the $LC_{50}$.

(9) The dose-response curve and slope (with confidence limits); and

(10) The histopathology findings including a complete record of lesions and abnormalities observed, and the histological diagnosis and characterization of each kind of lesion or abnormality observed, naming those which apparently caused death or morbidity.

C-4. Consumer Product Safety Commission 16 CFR 1500.3

CPSC does not provide a detailed guideline for inhalation testing. Their classification definitions are as follows.

"Highly Toxic" means any substance which produces death in half or more than half of a group of white rats, each weighing between 200 and 300 grams when a concentration of 200 ppm by volume or less of a gas or vapor or 2 milligrams per liter or less of mist or dust is inhaled continuously for 1 hour or less.

"Toxic" as above except limits are 200 to 20,000 ppm (gas) and 2 to 200 milligrams per liter (mist or dust).

## Eye Irritation

D-2. EPA TSCA 44CFR 44070-1 (7/26/79)
§ 772.112-24 Primary eye irritation study

(a) *Study Design.*

(1) *Condition of test substance.*

(i) If the test substance is a liquid, it must be placed in the eye undiluted, in accordance with paragraph (b) of this section.

(ii) If the test substance is a solid or granular product, it must be ground into a fine dust or powder. The substance must not be moistened before it is placed in the eye in accordance with paragraph (b) of this section.

(2) *Species.* Testing must be performed with the albino rabbit.

(3) *Age and condition of animals.* Young adult animals should be used. The eyes must be examined using fluorescein dye procedures at least 24 hours before application of the test substance. Animals showing preexisting corneal injury are to be eliminated.

(4) *Number of animals.* At least nine animals must be used.

(5) *Number and selection of dose.* A dose of 0.1 ml of liquid or 100 mg of solid must normally be applied to each test eye. Smaller quantities may be used when the standard quantities would be lethal, or when 100 mg of the solid cannot feasibly be administered in the eye.

(6) *Caging.* Caging must be designed to minimize exposure to sawdust, wood chips, and other extraneous materials that might enter the eye.

(b) *Study Conduct.* The test substance must be placed on the everted lower lid of one eye; the upper and lower lids are then to be gently held together for 1 second before releasing to prevent loss of material. The other eye, remaining untreated, serves as a control. The treated eyes of six rabbits must remain unwashed. The remaining three rabbits receive test material, and then the treated eye is flushed for one minute with lukewarm water starting no sooner than 20-30 seconds after instillation. A local anesthetic to reduce pain in test animals may be used prior to administration of the test substance, provided that evidence can be presented indicating no significant difference in toxic reaction to the test substance will result from use of the anesthetic.

(c) *Observation and scoring.*

    (i) *Observation.* Readings of ocular lesions must be made at 24, 48, and 72 hours after treatment and at 4 and 7 days after treatment. Readings must be made every 3 days thereafter, if injury persists, for at least 13 days after treatment or until all signs of reversible toxicity subside. Grading and scoring of irritation are to be performed in accordance with the following tables [from Draize, J. H., et al. (1965)]. The most serious effects, such as pannus or blistering of the conjunctivae and other effects indicative of corrosive action must be reported separately.

    (ii) Table of scale of weight scores for grading the severity of occular lesions.

        I. Cornea

            (A) Opacity—Degree of Density (Area Taken for Reading) Scattered or diffuse area—details of iris clearly visible—1.

            Easily discernible translucent area, details of iris slightly obscured—2.

            Opalescent areas, no details of iris visible, size of pupil barely discernible—3.

            Opaque, iris invisible—4.

            (B) Area of Cornea Involved.

            One quarter (or less) but not zero—1.

            Greater than one quarter–less than one-half—2.

            Greater than one-half–less than three quarters—3.

            Greater than three quarters up to whole area—4.

            Score equals $A \times B \times 5$ Total maximum = 80.

        II. Iris

            (A) *Values*—Folds above normal, congestion, swelling, circum-corneal injection (any one or all of these or combination of any thereof), iris still reacting to light (sluggish reaction is positive)—1.

No reaction to light, hemorrhage; gross destruction (any one or all of these)—2.

Score A×5 total possible maximum = 10.

III. Conjunctivae

(A) Redness (Refers to Palpebral Conjunctivae Only). Vessels definitely injected above normal —1

More diffuse, deeper crimson red, individual vessels not easily discernible—2.

Diffuse beefy red—3.

(B) *Chemosis.*—Any swelling above normal (includes nictiating membrane—1.

Obvious swelling with partial eversion of the lids—2.

Swelling with lids about half closed—3.

Swelling with lids about half closed to completely closed—4.

(C) *Discharge.*—Any amount different from normal (does not include small amount observed in inner canthus of normal animals)—1.

Discharge with moistening of the lids and hair just adjacent to the lids—2.

Discharge with moistening of the lids and considerable area around the eye—3.

Score (A+B+C)×2 Total maximum = 20.

The maximum total score is the sum of all scores obtained for the cornea, iris, and conjunctivae.

(d) *Data reporting and evaluation.* In addition to the information required by § 772.100-2(b) (8), Subpart A, the test report must include the following information:

(1) pH value of each test substance.

(2) In tabular form, the following data for each individual animal and the averages and range for each test group (eyes washed and unwashed);

(i) The primary eye irritation score at 24, 48, and 72 hours and 4 and 7 days and any other readings; and

(ii) Description of any serious lesions.

D-4. CPSC 16 CFR 1500.42

(a) (1) Six albino rabbits are used for each test substance. Animal facilities for such procedures shall be so designed and maintained as to exclude sawdust, wood chips, or other extraneous materials that might produce eye irritation. Both eyes of each animal in the test group shall be examined before testing, and only those animals without eye defects or irritation shall be used. The animal is held firmly but gently until quiet. The test material is placed in one eye of each animal by gently pulling the lower lid away from the eyeball to form a cup into which the test substance is dropped. The lids are then gently held together for one second and the animals is released. The other eye, remaining untreated, serves as a control. For testing liquids, 0.1 milliliter is used. For solids or pastes, 100 milligrams of the test substance is used, except that for substances in flake, granule, powder, or other particulate form the amount that has a volume of 0.1 milliliter (after compacting as much as possible without crushing or altering the individual particles, such as by tapping the measuring container) shall be used whenever this volume weighs less than 100 milligrams. In such a case, the weight of the 0.1 milliliter test dose should be recorded. The eyes are not washed following instillation of test material except as noted below.

(2) The eyes are examined and the grade of ocular reaction is recorded at 24, 48, 72 hours. Reading of reactions is facilitated by use of a binocular loupe, hand slit-lamp, or other expert means. After the recording of observations at 24 hours, any or all eyes may be further examined after applying fluorescein. For this optional test, one drop of fluorescein sodium opthalmic solution U.S.P. OR equivalent is dropped directly on the cornea. After flushing out the excess fluorescein with sodium chloride solution U.S.P. or equivalent, injured areas of the cornea appear yellow; this is best visualized in a darkened room under ultraviolet illumination. The eyes may be washed with sodium chloride solution U.S.P. or equivalent after the 24-hour reading.

(b) (1) An animal should be considered as exhibiting a positive reaction if the test substance produces at any of the readings ulceration of the cornea (other than a fine strippling), or opacity of the cornea (other than a slight dulling of the normal luster), or inflammation of the iris (other than a slight deepening of the folds (or regae) or a slight circumcorneal injection of the blood vessels), or such substance produces in the conjunctivae (excluding the cornea and iris) and obvious swelling with partial eversion of the lids or a diffuse crimson-red with individual vessels not easily discernible.

(2) The test shall be considered positive if four or more of the animals in the test group exhibit a positive reaction. If only one animal exhibits a positive reaction, the test shall be regarded as negative. If two or three animals exhibit a positive reaction, the test is repeated using a different group of six animals. The sec-

ond test shall be considered positive if three or more of the animals exhibit a positive reaction. If only one or two animals in the second test exhibit a positive reaction, the test shall be repeated with a different group of six animals. Should a third test be needed, the substance will be regarded as an irritant if any animal exhibits a positive response.

(c) To assist testing laboratories and other interested persons in interpreting the results obtained when a substance is tested in accordance with the method described in paragraph (a) of this section, an "Illustrated Guide for Grading Eye Irritation by Hazardous Substances" will be sold by the Superintendent of Documents, U.S. Government Printing Office, Washington, D.C. 20402. The guide will contain color plates depicting responses of varying intensity to specific test solutions. The grade of response and the substance used to produce the response will be indicated.

## D-5. IRLG 1981

### I. General Considerations

A. Good Laboratory Practices

Studies should be conducted according to good laboratory practice regulations (*e.g.*, "Nonclinical Laboratory Studies, Good Laboratory Practice Regulations," 43 FR 59986, 22 December 1978).

B. Test Substance

1. The specific substance or mixture of substances to be tested should be determined in consultation with the responsible agency. As far as is practical, composition of the test substance should be known. Information should include the name and quantities of all major components, known contaminants and impurities, and the percentage of unknown materials to account for 100% of the test sample.

2. Ideally, the lot of the substance tested should be the same throughout the study. The test sample should be stored under conditions that maintain its stability, strength, quality, and purity from the date of its production until the tests are complete.

C. Animals

1. Healthy animals, without eye defects or irritation and not subjected to any previous experimental procedures, must be used.

2. The test animal shall be characterized as to species, strain, sex, weight and/or age. Each animal must be assigned an appropriate identification number.

3. Recommendations contained in DHEW pub. no (NIH) 74-23, entitled *Guide for the Care and Use of Laboratory Animals,* should be followed for the care, maintenance, and housing of animals.

4. Animals should be individually caged for this test.

5. Animals must be observed as necessary to insure that none is lost due to management problems.

D. Documentation

Color photographic documentation may be used to verify gross and microscopic findings. If photographs are taken, the equipment and film must be of sufficient quality to permit controlled, close-up color photography of the eye to yield clear, sharp-focus images that literally fill the camera field.

## II. Specific Considerations

A. Test Preparation

1. Testing should be performed on young, adult, albino rabbits (male or female) weighing about 2.0 - 3.0 kilograms. Other species may also be tested for comparative purposes.

2. For a valid eye irritation test, at least 6 rabbits must survive the test for each test substance.

3. Animal facilities should be designed and maintained to exclude intense and direct light, sawdust, wood chips, or other extraneous materials that might produce eye irritation.

4. A trial test on 3 rabbits is suggested. If the substance produces corrosion, severe irritation, or no irritation, then no further testing is necessary. However, if equivocal responses occur, testing in at least 3 additional animals should be performed. If the test substance is intended for use in or around the eye, then testing on at least 6 animals should be performed.

B. Test Procedure

1. Both eyes of each animal in the test groups must be examined using optical instruments, fluorescein, ultraviolet light, or other appropriate means within 24 hours before substance administration.

2. For most purposes, anesthetics should not be used; however, if the test substance is likely to cause extreme pain, local anesthetics may be used prior to instillation of the test substance for humane reasons. In such cases, anesthetics should be used only once, just prior to instillation of the test substance; the eye used as the control in each rabbit should also be anesthetized.

   For substance administration, the animal is held firmly but gently until quiet. The test substance is placed in one eye of each animal by gently pulling the lower lid away from the eyeball (conjunctival cul-de-sac) to form a cup into which the test substance is dropped. The lids are then gently held together for one second and the animal is released. The other eye, remaining untreated, serves as a control. Vehicle controls are not included. If a vehicle is suspected of causing irritation, additional studies should be conducted, using the vehicle as the test substance.

3. For testing liquids, 0.1 milliliter is used. For solid, paste, particulate substances (flake, granule, powder, or other particulate form), the amount used must have a volume of 0.1 milliliter weighing not more than 100 mg. The volume measurement should be taken after gently compacting the particulates by tapping the measuring container in a way that will *not* alter their individual form. The weight of the 0.1 milliliter test dose must be recorded.

4. For aerosol products, the eye should be held open and the substance administered in a single, short burst for about one second at a distance of about 4 inches directly in front of the eye. The velocity of the ejected material should not traumatize the eye. The dose should be approximated by weighing the aerosol can before and

after each treatment. For other liquids propelled under pressure, such as substances delivered by pump sprays, an aliquot of 0.1 ml should be collected and instilled in the eye as for liquids.

After the 24-hour examination, the eyes may be washed, if desired. Tap water or isotonic solution of sodium chloride (U.S.P. or equivalent) should be used for all washings.

C. Observations

1. The eyes should be examined at 24, 48, and 72 hours after treatment. At the option of the investigator, the eyes may also be examined at 1 hour and at 7, 14 and 21 days. In addition to the required observations of the cornea, iris, and conjunctivae, serious lesions such as pannus, phlyctena, and rupture of the globe should be reported. The grades of ocular reaction (Table I) must be recorded at each examination.

2. After the recording of observations at 24 hours, the eyes of any or all rabbits may be further examined after applying fluorescein stain. For this optional examination, one drop of fluorescein sodium ophthalmic solution (U.S.P. or equivalent) is dropped directly on the cornea. After flushing out the excess fluorescein with tap water or isotonic solution of sodium chloride (U.S.P. or equivalent), injured areas of the cornea appear yellow and are best seen under ultraviolet illumination.

3. A record of the discharge from treated eyes is not required; however, any exudate above normal may be recorded as additional information.

4. An animal has exhibited a positive reaction if the test substance has produced at any observation one or more of the following signs:

   (a) ulceration of the cornea (other than a fine stippling)

   (b) opacity of the cornea (other than a slight dulling of the normal luster),

   (c) inflammation of the iris (other than a slight deepening of the rugae or a light hyperemia of the circumcorneal blood vessels), or

   (d) an obvious swelling in the conjunctivae (excluding the cornea and iris) with partial eversion of the eyelids or a diffuse crimson color with individual vessels not easily discernible.

D. Evaluation

1. The test result is considered positive if four or more animals in either test group exhibit a positive reaction. If only one animal exhibits a positive reaction, the test result is regarded as negative. If two or three animals exhibit a positive reaction, the investigator may designate the substance to be an irritant. When two or three animals exhibit a positive reaction and the investigator does not designate the substance to be an irritant, the test shall be repeated using a different group of six animals. The second test result is considered positive if three or more of the animals exhibit a positive reaction. Opacity grades 2 to 4 and/or perforation of the cornea are considered to be corrosive effects when opacities persist to 21 days.

2. If only one or two animals in the second test exhibit a positive reaction, the test should be repeated with a different group of six animals. When a third test is needed, the substance will be regarded as an irritant if any animal exhibits a positive response.

**TABLE I**

**Grades for Ocular Lesions**

**CORNEA**

Opacity: degree of density (area most dense taken for reading)

No ulceration or opacity . . . . . . . . . . . . . . . . . . . . . . . . . . . . . . . . . . . . . . . . . . . 0
Scattered or diffuse areas of opacity (other than slight dulling of normal luster,
   details of iris clearly visible . . . . . . . . . . . . . . . . . . . . . . . . . . . . . . . . . . . . . . 1 *
Easily discernible translucent areas, details of iris slightly obscured . . . . . . . . . . . . . . . . 2
Nacreous areas, no details of iris visible, size of pupil barely discernible . . . . . . . . . . . . 3
Opaque cornea, iris not discernible through the opacity . . . . . . . . . . . . . . . . . . . . . . . 4

**IRIS**

Normal . . . . . . . . . . . . . . . . . . . . . . . . . . . . . . . . . . . . . . . . . . . . . . . . . . . . . . . 0
Markedly deepened rugae, congestion, swelling, moderate circumcorneal hyperemia,
   or injection, any of these or any combination thereof, iris still reacting to light
   (sluggish reaction is positive) . . . . . . . . . . . . . . . . . . . . . . . . . . . . . . . . . . . . . . 1 *
No reaction to light, hemorrhage, gross destruction (any or all of these) . . . . . . . . . . . . . 2

**CONJUNCTIVAE**

Redness (refers to palpebral and bulbar conjunctivae excluding cornea and iris)

Blood vessels normal . . . . . . . . . . . . . . . . . . . . . . . . . . . . . . . . . . . . . . . . . . . . . . 0
Some blood vessels definitely hyperemic (injected) . . . . . . . . . . . . . . . . . . . . . . . . . . 1
Diffuse, crimson color, individual vessels not easily discernible . . . . . . . . . . . . . . . . . . 2 *
Diffuse beefy red . . . . . . . . . . . . . . . . . . . . . . . . . . . . . . . . . . . . . . . . . . . . . . . . . 3

Chemosis: lids and/or nictitating membranes

No swelling . . . . . . . . . . . . . . . . . . . . . . . . . . . . . . . . . . . . . . . . . . . . . . . . . . . . 0
Any swelling above normal (includes nictitating membranes) . . . . . . . . . . . . . . . . . . . . 1
Obvious swelling with partial eversion of lids . . . . . . . . . . . . . . . . . . . . . . . . . . . . . . 2 *
Swelling with lids about half closed . . . . . . . . . . . . . . . . . . . . . . . . . . . . . . . . . . . . 3
Swelling with lids more than half closed . . . . . . . . . . . . . . . . . . . . . . . . . . . . . . . . . 4

* Readings at these numerical values or greater indicate positive responses.

## Skin Irritation or Corrosion

E-1. DOT-49 CFR

Skin Corrosion

(1) A material is considered to be destructive or to cause irreversible alteration in human skin tissue if when tested on the intact skin of albino rabbit by the technique described in Appendix A to this part, the structure of the tissue at the site of contact is destroyed or changed irreversible after an exposure period of 4 hours or less.

    1. Corrosion to the skin is measured by patch-test technique on the intact skin of the albino rabbit, clipped free of hair. A minimum of six subjects are to be used in this test.

2. Introduce under a square cloth patch, such as surgical gauze measuring not less than 1 inch by 1 inch and two single layers thick, 0.5 milliliter (in the case of liquids) or 0.5 gram (in the case of solids and semisolids) of the substance to be tested.

3. Immobilize the animals with patches secured in place by adhesive tape.

4. Wrap the entire trunk of each animal with an impervious material, such as rubberized cloth, for the 4 hour period of exposure. This material is to aid in maintaining the test patches in position and retard the evaporation of volatile substances. It is not applied for the purpose of occlusion.

5. After 4 hours of exposure, the patches are to be removed and the resulting reactions are to be evaluated for corrosion.

6. Following this initial reading, all test sites are washed with an appropriate solvent to prevent further exposure.

7. Readings are again to be made at least at the end of a total of 48 hours (44 hours after the first reading).

8. Corrosion will be considered to have resulted if the substance in contact with the rabbit skin has caused destruction or irreversible alteration of the tissue. Tissue destruction is considered to have occurred if, at any of the readings, there is ulceration or necrosis. Tissue destruction does not include merely sloughing of the epidermis, or erythema, edema, or fissuring.

E-2. EPA-TSCA 44 FR 44071 § 772.112-25 Primary dermal irritation study

(a) *Study Design.*

(1) *Condition of test substance.*

(i) If the substance is a liquid, it must be applied undiluted.

(ii) If the test substance is a solid, it must be slightly moistened with physiological saline before application.

(2) *Species.* Testing must be performed in at least one mammalian species, preferably the albino rabbit. Selection of other species and strains may be acceptable, but must be justified.

(3) *Age.* Young adult animals must be used.

(4) *Number of Animals.* At least six animals must be used.

(5) *Number and selection of dose levels.* A dose of 0.5 ml of liquid or 0.5g of solid or semisolid is to be applied to each application site.

(6) *Control groups.*

(i) A vehicle control group is required if the vehicle is known to cause any toxic dermal reactions or if there is insufficient information about the dermal effects of the vehicle.

(ii) Separate animals are not required for an untreated control group. Each animals serves as its own control.

(b) *Study Conduct.* The test substance must be introduced under 1-inch square gauze patches. The patches must be applied to two intact and two abraded skin sites on each animal. In all animals, the application sites must be clipped free of hair. In addition, the abrasion must penetrate the stratum corneum, but not the dermis. A wrapping material such as gauze covered by an impervious, nonreactive rubberized or plastic material should be used to retard evaporation and keep the test substance in contact with the skin. The animals should be restrained. The test substance must be kept in contact with the skin for 24 hours. At the end of the exposure period, the wrapping should be removed and the skin wiped (but not washed) to remove any test substance still remaining. It may be necessary to rinse off the material if colored test substances are used.

(c) *Observation and scoring.* Animals must be observed and signs or erythema and edema must be scored at 24 hours and 72 hours after application of the test substance. The irritation is to be scored according to the technique of Draize, J. H. (1959). Observation for irritation and scoring of any irritation must continue daily until all irritation subsides or is obviously irreversible.

(d) *Data reporting and evaluation.* In addition to the information required by § 772.100-2(b) (8), Subpart A, the test report must include the following information:

(1) pH value of each test substance.

(2) In tabular form, the following data for each individual animal and averages and ranges for each test group:

(i) Scores for erythema and edema at 24 hours, at 72 hours, and at any subsequent observations; and

(ii) Primary skin irritation scores according to the technique of Draize.

E-4. CPSC 16 CFR 1500.41

§ 1500.41 Method of testing primary irritant substances

Primary irritation to the skin is measured by a patch-test technique on the abraded and intact skin of the albino rabbit, clipped free of hair. A minimum of six subjects are used in abraded and intact skin tests. Introduce under a square patch, such as surgical gauze measuring 1 inch by 1 inch and two single layers thick, 0.5 milliliter (in the case of liquids) or 0.5 grams (in the case of solids and semisolids) of the test substance. Dissolve solids in an appropriate solvent and apply the solution as for liquids. The animals are immobilized with patches secured in place by adhesive tape. The entire trunk of the animal is then wrapped with an impervious material, such as rubberized cloth, for the 24-hour period of exposure. This material aids in maintaining the test patches in position and retards the evaporation of volatile substances. After 24 hours of exposure, the patches are removed and the resulting reactions are evaluated on the basis of the designated values in the following table:

| Skin Reaction | Value* |
|---|---|
| Erythema and eschar formation: | |
| No erythema | 0 |
| Very slight erythema (barely perceptible) | 1 |
| Well-defined erythema | 2 |
| Moderate to severe erythema | 3 |
| Severe (beet redness) to slight eschar formations (injuries in depth) | 4 |
| Edema formation: | |
| No edema | 0 |
| Very slight edema (barely perceptible) | 1 |
| Slight edema (edges of area well defined by definite raising) | 2 |
| Moderate edema (raised approximately 1 mm) | 3 |
| Severe edema (raised more than 1 mm and extending beyond the area of exposure) | 4 |

*The value recorded for each reading is the average value of the six or more animals subjected to the test.

Readings are again made at the end of a total of 72 hours (48 hours after the first reading). An equal number of exposures are made on areas of skin that have been previously abraded. The abrasions are minor incisions through the stratum corneum, but not sufficiently deep to disturb the derma or to produce bleeding. Evaluate the reactions of the abraded skin at 24 hours and 72 hours, as described in this paragraph. Add the values for erythema and eschar formation at 24 hours and 72 hours for intact skin to the values on abraded skin at 24 hours and 72 hours (4 values). The total of the eight values is divided by four to give the primary irritation score; for example:

| Skin Reaction | Exposure Time (hours) | Evaluation Value |
|---|---|---|
| Erythema and eschar formation: | | |
| Intact skin | 24 | 2 |
| Intact skin | 72 | 1 |
| Abraded skin | 24 | 3 |
| Abraded skin | 72 | 2 |
| Subtotal | | 8 |
| Edema formation: | | |
| Intact skin | 24 | 0 |
| Intact skin | 72 | 1 |
| Abraded skin | 24 | 1 |
| Abraded skin | 72 | 2 |
| Subtotal | | 4 |
| Total | | 12 |

Thus, the primary irratation score is $12 \div 4 = 3$

## REFERENCES

1. Draize, J. H., *Appraisal of Safety of Chemicals in Foods, Drugs & Cosmetics,* The Association of Food and Drug Officials of the United States, pages 49–51 (1959).
2. Draize, J. H., *Appraisal of Safety of Chemicals in Foods, Drugs & Cosmetics,* The Association of Food and Drug Officials of the United States, 46–48 (1959).
3. Trevan, J. W., *Proc. R. Soc Lond (Biol)* 101: 483–514 (1927).
4. Bliss, C. I., *Ann App. Biol* 22: 134–167; 306–333 (1935).
5. Litchfield, J. T. and Wilcoxon, F., *J Pharmacol Exp. Ther.* 96: 99–113 (1949).
6a. Weil, C. S., *Biometrics* 8: 249–263 (1952).
6b. Weil, C. S. and Thompson, W. R., *Biometrics* 8: 51–54 (1952).
6c. Thompson, W. R., *Bacteriol Rev* 11: 115–145 (1947).

## SUGGESTED READING

*Acute Toxicology in Theory and Practice,* Brown, V. K., New York: Wiley (1980).

*Appraisal of the Safety of Chemicals in Foods, Drugs and Cosmetics,* by The Association of Food and Drug Officials of The United States.

*Clinical Toxicology of Commercial Products,* Fourth Ed., Gosselen, Hodge, Smith and Gleason, Williams and Wilkins, Balto, (1981).

*Guide for the Care and Use of Laboratory Animals,* U.S. Dept. of Health and Human Services (NIH) 78.23.

*Guidelines for Testing of Chemicals,* OECD, Paris, 1981, 2, Rue Andre Paseol 75775, Paris Cedet 16 France.

*Guidelines for the Assessment of Drug and Medical Device Safety in Animals,* Pharmaceutical Manufacturers Assoc., (1977).

*New Concepts in Safety Evaluations, Vol. 1,* Mehlman, Shapiro and Blumenthal, Washington, DC: Hemisphere Press (1976).

*Principles and Procedures for Evaluating the Toxicity of Household Substances,* No. 1138, National Academy of Sciences, Washington, D.C.

*Toxicology, The Basic Science of Poisons, 2nd Ed.,* Cassarett and Doull, New York: Macmillan (1980).

# 7

# Chronic Toxicity

Anthony J. Garro
*The City College of New York*
*and*
*The Mount Sinai School of Medicine*
*New York, New York*

## PRINCIPLES OF CHRONIC TOXICITY

### Introduction

It is clear, both from human experience and from animal experimentation, that some chemical toxicities become evident only after repeated exposures over long periods of time or do not manifest themselves until months or years have elapsed from the time of exposure. Such toxic effects include: the induction of cancers (carcinogenesis), germ line and somatic cell mutations (mutagenesis), birth defects (teratogenesis), depressed germ cell production, and neurotoxic or behaviorial effects. The purpose of this chapter is to describe and discuss some of the more common methods used to identify substances which pose a chronic toxic hazard to man. The discussion, particularly with respect to experimental systems, will focus largely on methods used to identify potential carcinogens and mutagens. This emphasis is not meant to imply that these two problems are more important than other chronic effects but rather to reflect the intensity of research and development which has taken place in the last 10 years in the areas of mutagen and carcinogen detection.

It is important to note at the outset that several of the chronic toxic effects of concern, such as carcinogenesis, mutagenesis, and teratogenesis, differ in a fundamental way from acute toxic effects since their pathological endpoints are detectable only after the proliferation of a population of cells which are derived from the target cells.[1] These self replicating chronic effects may differ from acute toxicities not only with respect to their delayed appearance but also with respect to aspects of their dose response relationships. In the case of a carcinogen, for example, while the frequency of cancers induced in an exposed population generally will be dose dependent, the intensity of the carcinogenic effect (tumor growth and invasiveness) in individuals may be independent of the exposure dose and related to other factors which may either promote or inhibit tumor progression.

Even though much remains to be learned regarding the basic mechanisms of tumor progression, the general recognition that carcinogenic events result in a loss of normal cell regulation, either through modification of the genome and/or other molecular control mechanisms, has led to the development of a variety of tests designed to detect substances which pose a carcinogenic hazard. Inasmuch as many of the "short-term" or rapid screening tests for potential carcinogens detect changes induced at the genomic level, these same tests also identify potential mutagens. It is noteworthy, however, that these short-term tests generally are not useful in identifying teratogens, compounds which are relatively much more toxic to a developing fetus or embryo than to the adult organism.[2]

### General Design of Chronic Toxicity Studies: Carcinogenesis as a Model

The basic purpose of a chronic toxicity study is to assess the hazard of a potentially lifelong exposure of humans to a chemical agent. In designing such a study many variables must be taken into consideration, such as the choice of experimental animals, test dosages, route of administration, etc. These variables have been examined, with respect to the identification of chemical carcinogens, by an Interagency Regulatory Liaison Group (IRLG) which consisted of representatives from the Consumer Products Safety Commission (CPSC), Environmental Protection Agency (EPA), Food and Drug Administration (FDA), Occupation Safety and Health Administration (OSHA), National Cancer Institute (NCI), and National Institute of Environmental Health Sciences (NIEHS). This group has suggested that in order to thoroughly assess a chemical's carcinogenic potential in experimental animals the tests conducted should include: (a) two species of rodents; (b) both sexes of each; (c) adequate controls; (d) a number of animals sufficient to provide adequate resolving power to detect a carcinogenic effect; (e) treatment and observation times extending over most of the lifetime of the animals at a dose range including one level likely to yield maximum expression of carcinogenic potential; (f) detailed pathologic examination, and (g) statistical evaluation of results.[3]

Chemical carcinogens are identified in this type of bioassay as chemicals which cause either an increase in the incidents of tumors or a decrease in the latency period of tumor development. Compounds which decrease the latency period of tumor development without increasing tumor incidence may not be carcinogenic alone but act as cocarcinogens or promoters of tumor development.[4] Cocarcinogens and promoters are thought to be important factors in cancer development in humans,[5] and their identification may thus be important to cancer prevention. The basis for the IRLG recommendations are discussed below. For a detailed description of proposed federal regulations on carcinogenicity bioassays see EPA[6,7] CPSC[89] DOL.[10]

**Animal Model:** Rodents such as mice, rats, and hamsters are generally used for lifelong chronic toxicity studies for several reasons. First, their natural life spans are relatively short thus allowing a lifetime study to be completed in 2 to 3 years versus the 7 to 10 years it would take if larger animals such as dogs were used. Second, rodents are relatively inexpensive and easy to care for. Even as such, however, recent estimates of the cost of assaying a single chemical for carcinogenic activity, which requires on the order of 600 animals, is approximately $400,000.[7] Third, there are inbred strains readily available which are homogenous for such traits as background cancer rates, site specific carcinogenesis, lifespans, and response to husbandry systems. It should be kept in mind, however, that inbred strains are less representative of human populations than outbred strains which are genetically heterogeneous. This may be of some importance with respect to the shape of dose response curves, especially at low carcinogen concentrations.[11]

**Number of Species:** It is suggested that two species and both sexes of each species be included in a test. The use of at least two species is related to the fact that species-specific differences have been documented in experimental animals with respect to susceptibility and organ tropisms with known carcinogens. For example, 2-naphthylamine, which produces bladder tumors in man, produces hepatomas in mice, bladder tumors in hamsters, and no tumors in some strains of rats.[12] Similarly, 4-amino-biphenyl, another human bladder carcinogen, produces hepatomas in mice and mammary tumors in rats, while aflatoxin B1, a suspected human carcinogen, although carcinogenic in rats, is not in several mouse strains.[14] The purpose of using both sexes of the test species is to assess the possible effects of hormonal differences on carcinogenesis. Sex related differences in drug and carcinogen metabolism have been described,[15-17] and hormonal effects on tumor development are well known.[18-20]

**Numbers of Animals:** The IRLG report recommends that the number of animals be "sufficient to provide an adequate resolving power to detect a carcinogenic effect". What, however, is a sufficient number? In the case of a positive carcinogenic effect, large numbers of animals are not needed to generate statistically significant results. For example, a potent carcinogen such as dimethylnitrosamine can produce an 80% incidence of liver tumors[21] at relatively low concentrations (0.05 mg/kg, and this is readily detected using groups as small as 15 treated and control animals. A group of 15 animals, however, would not be adequate to detect a carcinogen which produced a cancer incidence of 10% in the tested animals. The number of animals needed to detect induced cancer incidences of 10%, 1%, and 0.1% at a 95% confidence level in animals with a control tumor incidence of 1% are given in Table 7.1.

As shown in this table, it would take a sample population, not including controls, of about 1,000 animals to reasonably exclude the possibility that a substance at a given test dose was not producing cancer in 1% of the exposed animals. This level of sensitivity in fact is rarely achieved since laboratory space limitations and economic factors impose practical ceilings on the numbers of animals which can be included in a test. For example, a test conducted in ac-

Table 7.1:  Approximate Numbers of Animals Required to Detect
Carcinogenic Effects at a 95% Confidence Interval Against a
Spontaneous Background Cancer Incidence of 1%*

| Induced Tumor Incidence | Proportion of Animals . . . with Tumors . . . | | Approximate Sample Size (each) |
|---|---|---|---|
| | Control | Treated | |
| 0.001 | 0.01 | 0.011 | 80,000 |
| 0.01 | 0.01 | 0.02 | 1,100 |
| 0.10 | 0.01 | 0.11 | 40 |

*Adapted from: T. Cairns[22]

cordance with the suggested IRLG guidelines, i.e., two species of animals exposed to two concentrations of the test substance, would involve 600 animals (100 on dose A, 100 on dose B, and 100 controls for each species), and would identify a substance as carcinogenic only if it induced cancer in 5-10% of the exposed animals. Considering that a substance would be considered a major health hazard if, at a given exposure level, it caused cancer in 0.1% of the people so exposed (approximately 100,000 extra cancers in the U.S.), it can be seen that typical rodent bioassays are relatively insensitive. This lack of sensitivity is a major factor in choosing the dose levels at which substances are tested for carcinogenic activity.

**Dose Levels and Prerequisite Subchronic Testing:** Carcinogenic chemicals generally exhibit a positive dose response relationship. One means, therefore, of circumventing the relative insensitivity of rodent bioassays that is due to the limited numbers of animals on test is to increase the dose levels of the test compound. The rationale is that a compound, which at low concentrations might induce cancer in 0.1% of an exposed human population, at 10 times or 100 times that concentration may produce a detectable increase in tumor incidence (10%) in a test animal population.

In order to test substances under "conditions likely to yield maximum tumor incidence",[23] it is recommended that two dose levels be used, one being the estimated maximum tolerated dose (EMTD) and the other ½ to ⅓ the EMTD. The EMTD is defined as the highest dose which can be administered to the test animals for their lifetime and that is estimated not to produce: (a) clinical signs of toxicity or pathologic lesions other than those related to a neoplastic (carcinogenic) response; (b) alterations of the normal longevity of the animals from toxic effects other than carcinogenesis; and (c) more than a relatively small percent inhibition of normal weight gain (not to exceed 10%).[24]

A dose which is some fraction of the EMTD is used in addition to the EMTD in case the EMTD produces unacceptable toxic effects in the chronic study. For example, a true carcinogen which seriously shortened life-span at the EMTD might be missed as a carcinogen if there was insufficient time for tumors to develop. Thus negative results obtained at test doses above the EMTD are generally considered inadequate. In contrast, positive results, i.e., carcinogenesis, observed above the EMTD are considered acceptable as evidence of carcinogenic activity unless there is some basis for questioning this conclusion. Also positive results obtained at or below the EMTD provide evidence for carcinogenic activity.[3] The use of two dose levels also is important in assessing, in at least a limited way, the dose response effect of the test substance.

The EMTD is determined on the basis of a subchronic toxicity study in which small groups of test animals are exposed to a fairly broad dose range of the test substance for a period of approximately 10% of the life span of the animals (90 days for rodents). In turn, the dose levels which are tested in the 90 day subchronic study are themselves based on information generated in acute toxicity studies of the type described in the preceeding chapter. In addition to providing the EMTD the acute and subchronic tests also are utilized to obtain information on the metabolism of the test substance, the extent to which the test substance and its metabolites are distributed and bioaccumulated in the host, and which host organ systems may be affected by exposure. This information may be of importance in choosing the species or strains to be used in the chronic study as well as the route of administration.

**Route of Administration:** The route of administration is an important consideration in carcinogenicity tests since it may influence the tumorgenicity of the test substance. For example, in rats, dibutylnitrosamine administered subcutaneously primarily induces urinary bladder tumors but when given orally induces hepatic and pulmonary tumors in addition to bladder tumors.[25,26] This type of effect most likely is to related to changes in absorption, circulation, tissue concentration, and metabolism which result from the different routes of administration. Another dramatic demonstration of this effect is seen with cycasin, a naturally occurring carcinogen found in the cycad nut. Cycasin is the β-glucoside of methylazoxymethanol and is noncarcinogenic in adult rats when administered by intraperitoneal injection but is carcinogenic when administered orally.[27] The carcinogenicity of cycasin is dependent upon its hydrolysis by a β-glucosidase, supplied in this case by gut microorganisms, to yield free methylazoxymethanol, which is itself a known carcinogen.

The most important consideration in choosing a route of administration is to select one which leads to absorption and distribution of adequate doses of the test substance. For highly volatile compounds inhalation is appropriate while for most other compounds oral administration provides a relatively easy means of monitoring the administered doses. It is not considered necessary that the route of administration mimic that of the potential human exposure, especially if there is evidence for absorption and distribution of the test substances in humans. Perhaps a good example of this is the various hair dye components which had passed rodent skin painting carcinogenicity tests,[28,29] but were shown to be mutagenic by bacterial assay and were known to be absorbed through the skin. More recent feeding studies conducted by the NCI Bioassay program indicate that at least one of these hair dye components, namely 2,4-diamino-anisole, is carcinogenic in rats and mice. In essence, positive results, regardless of route of administration, may be considered as evidence for carcinogenesis with potential relevance for human exposure provided that the route of administration does not produce metabolites or degradation products which never occur with human exposure.

There is a specific situation, however where carcinogenesis in animals does not appear to portend a potential human hazard. This is the induction of sarcomas at sites of subcutaneous implantation or injection of certain foreign bodies.[30,31] Humans are not realistically exposed to these materials by this route. There also are, however, chemical carcinogens which do not require metabolic activation, such as the nitrosamides[32] which tend to produce tumors primarily at the sites in which they are innoculated. Thus caution should be exercised in interpreting studies in which tumors appear only at the innoculation site.

In cases such as this, demonstration of genetic activity or the lack of it in a short-term test system capable of detecting direct acting carcinogens would provide useful information for the interpretation of the rodent carcinogenesis data.

## DETECTION OF MUTAGENIC SUBSTANCES AND APPLICATION OF MUTAGEN SCREENING TESTS FOR THE DETECTION OF CHEMICAL CARCINOGENS

### Introduction

Exposure to DNA-reactive substances may affect not only the health of the exposed individuals but in addition may also affect the health of future generations. This problem was reviewed by a Department of Health, Education, and Welfare (DHEW) Subcommittee on Environmental Mutagenesis and by the National Academy of Sciences whose reports[33,34] have served as a basis for proposed regulatory guidelines in this area.[35,36] The following discussion is based in large part on the Flamm DHEW Subcommittee Report.[33]

The primary objective of mutagen testing is to identify substances which have the potential to produce heritable genetic changes in man. In addition, since a good correlation has been observed between results obtained in some mutagenesis assays and carcinogenic activity,[37-41] the more rapid mutagenesis assays are proving to be useful adjuncts to the life-long rodent assays in identifying potential carcinogens.

Because of the different types of genetic lesions which may occur in humans it is recommended that a battery of tests with different genetic endpoints be used in mutagen screening. In order to understand the basis on which different tests are chosen, it is necessary to have some understanding of the different types of human genetic disorders which occur. These disorders can be classified into 2 major categories, gene mutations and chromosomal aberrations.

At the DNA level gene mutations can be subdivided into point mutations and small interstitial deletions. Point mutations involve changes in one or a small number of the $10^3$ to $10^5$ purine and pyrimidine base-pairs which make up a gene. These changes can result from base-pair substitutions or base-pair additions or deletions. Small interstitial deletions involve larger numbers of base pairs but are still too small to be detected by light microscopy. Gene mutations can manifest themselves as dominant or recessive gene disorders. In the case of dominant mutations the trait associated with the mutation will be expressed in all individuals carrying the mutant gene and an average of 50% of the children of such an individual will inherent and express the mutant gene. In contrast, traits associated with recessive mutations are not expressed in individuals who are heterozygous for the mutant gene, i.e., carry the mutant gene on one chromosome and a normal gene on the homologous chromosome. Recessive mutations are expressed only in individuals who are homozygous for the mutation, i.e., those individuals who inherited a copy of the mutant gene from each parent both of whom were heterozygotes. Sex-linked or X-chromosome-associated traits are an exception to this rule in that they are generally expressed in all males who carry the mutant gene on their X chromosome.

Recessive mutations, because they generally are unexpressed, may go undetected for many generations until, by chance, a diseased child is born to parents who are silent carriers of the mutations. At the present time there are approximately 1,000 human dominant gene disorders and a similar number of recessive gene disorders which have been identified. Bilateral retina blastoma, an inherited cancer, and Huntington chorea, the disease which killed the famous folksinger Woody Guthrie, are two examples of dominant gene disorders; examples of disease due to recessive gene mutations include: sickle cell anemia, cystic fibrosis, and Tay-Sacks disease.

Chromosomal aberrations, which make up the second broad category of human genetic disorders, include changes in chromosome number and major changes in chromosome structure such as large deletions, duplications, inversions, and translocations. These changes are of such a magnitude as to be detectable by cytological techniques using light microscopy. Changes in chromosome number result from a failure of paired chromosomes to separate properly (nondisjunction) during germ cell formation whereas changes in chromosome structure are generally thought to result from chromosomal breaks. Down's syndrome, which is one of the more familiar diseases in this category of genetic disorders, is caused by the presence of three copies (trisomy) of the 21st chromosome in cells of diseased individuals.

Basically all of the tests which are used for mutagen screening detect different consequences of damage to the same target molecule, DNA. These tests can be divided into categories based on the level of DNA or chromosome organization at which the effects of the damage are manifested. There are: (a) genetic tests which detect point mutations; (b) cytologic and genetic tests which detect chromosome aberrations, and (c) biochemical and physical tests which detect primary DNA damage. Within these three categories there are a variety of submammalian and mammalian test systems capable of detecting one or more aspects of the consequences of DNA damage. Examples of these test systems are listed in Table 7.2 which is perforce incomplete with respect to many of the newer, less-well validated systems. A recent review[41] has identified over 100 assays which have been examined as means of identifying potential mutagens and carcinogens.

It is beyond the scope of this chapter to detail the methodologies of even the sampling of tests listed in Table 7.2. For methodological details the reader is referred to the excellent texts edited by Hollaender[42] and Kilbey et al.[43] The purpose of the review presented here is to note some of the basic biological differences which may influence the extrapolation of different test results to humans and to provide a range of time and cost estimates for the various tests.

Table 7.2:  Examples of Commonly Used Test Systems for Detecting Mutagens

   (A)  Tests for Point Mutations
       1.  Microbial
          a. Bacteria:
             a. *Salmonella typhimurium* reversion assay
             b. *Escherichia coli* reversion and forward mutation assays
          b. Yeast:
             *Saccharomyces cerevisiae* reversion and forward mutation assays
       2.  Insect
          Drosophila, sex-linked recessive lethal test
       3.  Tissue Culture
          a. Mouse lymphoma forward mutations
          b. Chinese hamster forward mutations
          c. Diploid human fibroblast forward mutations
       4.  In vivo Mammalian
          Mouse specific locus test
   (B)  Tests for Chromosomal Mutations
       1.  Microbial
          *S. cerevisiae* mitotic recombination assay
       2.  Insect
          Drosophila dominant lethal test
       3.  Tissue Culture
          a. Cytogenetic analysis of cultured mammalian cells
          b. Sister chromatid exchange analysis
       4.  In vivo Mamalian
          a. Direct cytologic examination of somatic tissues (micronucleus test)
          b. Mouse heritable translocation test
          c. Mouse dominant lethal test
   (C)  Primary DNA Damage
       1.  Microbial
          *E. coli pol A* test
       2.  Tissue Culture
          Unscheduled DNA synthesis

## Microbial Tests

There are several well characterized bacterial and yeast assays which have been used for mutagen identification. The microbial assays are relatively sensitive, rapid, and inexpensive. With any one of these tests it is feasible for a single laboratory with 2 technicians to screen 70-75 chemicals per year at the cost of $300 to $500 per test compound. These assays do not provide any information regarding a substance's capacity to reach and damage germinal cells in mammals. This is an important consideration in evaluating a substance's capacity to produce heritable mutations in humans.˙This same shortcoming, however, applies to all of the assays described below with the exception of the *in vivo* mammalian genetic assays and some of the *in vivo* mammalian cytogenetic assays.

The general protocol of the assays is to expose a large number ($10^8$ to $10^9$) of indicator cells to a range of concentrations of the substance. Four to five concentrations can easily be tested in a single experiment to obtain a dose response curve. The treated cells are plated on various media which either select for the growth of mutant cells or impart a distinctive phenotype to the mutant colonies. The bacterial assays require a 2 day incubation period for the mutant colonies to reach a countable size whereas yeast, because of their slower growth rate, require a 6 to 7 day incubation. Exposure to the test substance can be performed either in liquid suspensions or by incorporating both the cells and test substance directly into the culture medium. The power of these assays is related to two factors, namely, the ability to conveniently handle large populations (imagine trying to deal with $10^8$ mice or rats), and the ability to apply selective growth conditions which allow the detection of rarely occurring mutants within these large populations.

This point is illustrated in the following example. The *Salmonella* reversion assay developed by Ames and his coworkers[44] is the most widely used of the microbial mutagenesis test. The bacterial strains employed in this assay require histidine for growth, as a result of different types of point mutations in various genes of the histidine biosynthetic pathway. When plated on a histidine deficient medium, the His⁻ cells are unable to replicate and only those cells which have reverted to a His⁺ phenotype, either spontaneously or through exposure to a mutagen, produce colonies. In the absence of any mutagen, the approximate number of His⁺ colonies which develop on a plate containing about $10^9$ His⁻ cells ranges from about 10 to 200 depending on the indicator strain. This is a background mutation frequency of $7 \times 10^{-8}$ to $2 \times 10^{-7}$. In actual tests the lower limit of mutagen activity detectable is one which produces a doubling in the number of His⁺ colonies produced on the selective medium.

Several modifications have been introduced into the *Salmonella* strains in order to increase their sensitivity towards mutagens. These modifications include: (a) mutations which increase the permeability of the bacterial cell wall thus allowing bulky compounds such as the polycyclicaromatic hydrocarbons to enter the cells; (b) mutations which inactivate excision repair of DNA damage thus forcing larger amounts of mutagen damaged DNA to be repaired by more error-prone pathways yielding a higher number of mutants per DNA hit; (c) incorporation of a particular plasmid (a small circular DNA molecule) which produces a product that further increases the frequency of error-prone DNA repair.

The *E. coli* test system works in much the same way as the *Salmonella* assay with the following difference: In addition to having genetic markers which can be utilized in a reversion assay,[45,46] there are also strains which can detect the induction of forward mutations in several different genes.[47] The theoretical advantage of a forward mutation assay over a reversion assay is that a forward mutation assay can detect both base-substitution and frameshift mutagens as long as they lead to the inactivation of the gene activity being assayed.

In a reversion assay, restoration of a mutated gene activity only occurs if the mutagen produces mutations of the same type, (base-substitution or frameshift) as the revertible mutation carried by the indicator strain. In the Ames assay this problem is circumvented by using different indicator strains, some of which are reverted by base-substitution mutagens and others which are preferentially reverted by different types of frameshift mutagens.[44]

*Salmonella* strains suitable for use in a forward mutation assay have been developed,[48,49] but such strains do not appear to be any more useful than the standard Ames strains.[50]

The yeast assays[51-53] differ in an important fundamental respect from the bacterial assays in that yeasts are eucaryotic organisms. Yeasts thus possess a membrane bound nucleus, which is not found in bacteria, as well as multiple chromosomes and a chromosome organization which is similar to that of higher eucaryotic organisms such as mammals. As eucaryotic organisms, yeasts can be used both to assay substances for their ability to induce point mutations and to detect the effects of chemicals on higher levels of chromosome organization. Yeasts, however, resemble bacteria and differ from mammals in that they possess a rigid cell wall which acts as a diffusion barrier and at present there are no cell wall deficient mutants available which might increase the permeability of the cells. The relative impermeability of yeast apparently makes them less sensitive than the Ames *Salmonella* strains for detecting the mutagenic activity associated with a number of carcinogens.[40]

## Mammalian Cells in Culture

There are a number of mammalian cell culture systems which have been developed to detect the induction of gene mutations. The best characterized of these systems involve established rodent cell lines with stable karyotypes. These include mouse lymphoma L5178Y cells;[54] Chinese hamster V79 cells;[55] and Chinese hamster ovary cells.[56] Attempts also have been made to use normal diploid human fibroblast cells,[57] but these attempts have not met with the same degree of success as that obtained with the established rodent lines.[41] All of the above assays detect the induction of forward mutations in several genetic loci. Cells which have sustained a mutation in one of these loci can be selected on the basis of their resistance to killing or growth inhibition by either various toxic purine or pyrimidine analogs or the cell poison ouabain. The test protocols are similar to those of the microbial assays. Large numbers of cells are exposed to the test substance and, after culturing for 2 to 4 days on a nonselective medium to allow for the expression of mutant phenotypes, the cells are plated on media which select for the outgrowth of mutant clones. It takes approximately 2 weeks from the time the cells are exposed to the test material to the time the selective plates can be scored for mutant induction. The assays, while not taking much longer than the microbial assays, are technically more difficult to perform. Cell cultures, however, also can be used in conjunction with cytological techniques to identify compounds which induce chromosome abberations.[58]

Validation studies of this type of analysis, using known mutagens and nonmutagens, have demonstrated that *in vitro* cytologic procedures can successfully identify many classes of chemical mutagens.[41] One such study, which utilized cultured Chinese hamster cells, detected chromosome breaking activity with the food additives, sodium saccharin, and sodium nitrate, the pesticide Sevin (naphthyl-N-methylcarbamate) and the teratogen urethane (ethyl carbamate) all of which were negative in bacterial tests.[59]

Another cytological technique which has been used successfully to monitor the effect of DNA damaging agents involves the detection of sister chromatid exchanges (SCEs). This approach has been facilitated by the development of new chromosome staining techniques which permit visualization of regions on sister chromatids which have undergone reciprocal exchanges.[61-63] Many studies have shown that the frequency of SCEs in cultured cells is increased by exposure of the cells to radiation, mutagens, and carcinogens.[41]

## DNA Repair-Dependent Assays of DNA Damage

All cells possess DNA repair systems capable of acting on various types of DNA damage. A number of assay systems have been developed which take advantage of different aspects of DNA repair processes to identify DNA damaging agents (For reviews see References 41–43). For example, DNA repair-deficient mutants of bacterial and eucaryotic cells are more sensitive than repair-proficient strains, to the lethal and growth inhibiting effects of radiation and DNA-reactive chemicals. Several bacterial assays have been developed around this basic principle and of these, the *E. coli pol A* assay developed by Rosenkranz and his coworkers has been the most widely used.[64,65] DNA damaging agents also can be identified by their ability to activate DNA repair systems and this can be detected both in tissue culture cells and *in vivo*. The principally used method for monitoring the activation of DNA repair is the detection of unscheduled DNA synthesis. This type of repair-dependent synthesis can be observed by the incorporation of radioactively labeled DNA precursors in mutagen treated cells in which normal replicative DNA synthesis has been blocked.

## Metabolic Activation

The active forms of most chemical mutagens and carcinogens are not the parent compounds themselves, but rather metabolites of these compounds.[18] The enzymes responsible for most of this metabolism are part of the microsomal cytochrome P-450 dependent mixed function oxidase system. These enzymes are found in many mammalian tissues but are present in highest concentrations in the liver. Since these enzymes are not found in the microorganisms used in the mutagen assays and are not significantly active in the mouse lymphoma, Chinese hamster and human fibroblast tissue culture cells, an exogenous source of these enzymes, must be added to the above mutagen assays. The most commonly used source of activating enzymes is a supernatant of rat liver homogenates which has been centrifuged at 9,000 $g$ to remove mitochondria and nuclei.[44] This so-called S-9 fraction is generally prepared from rats which had been pretreated with Aroclor 1254, a polychlorinated biphenyl-inducer of microsomal mixed function oxidase enzyme activities.[66,67] It recently has been suggested that mixtures of rat and hamster liver homogenates be used as the S-9 component of the assays because of the differential capacity of these enzyme preparations to activate various promutagens.[68-70]

It also is possible to use intact cells as a source of xenobiotic metabolizing enzymes. Activation for tissue culture systems can be provided by primary rat liver cells[71] and irradiated hamster embryo cells which are plated as a feeder layer for the indicator cells.[72,73] In the case of compounds which yield a spectrum of metabolites, some of which are active and others not, the use of intact cells rather than tissue homogenates may mimic more closely the type of metabolism which occurs *in vivo*. Whole animals have in fact been used for metabolic activation in host mediated assays.[74] In this procedure microorganisms are introduced into the peritoneal cavity, blood, or liver of mice or rats which are also treated with a test mutagen. After a period of time, ranging from minutes to hours, the microorganisms are recovered and plated on selective media to score the numbers of mutants generated. It is noteworthy, however, that validation studies have not shown any advantage in the use of host mediated assays over the use of S-9 fractions for mutagen activation.[40]

**Insect Assays**

The fruit fly *Drosophila melanogaster* is one of the best characterized of all the eucaryotic genetic systems. It has been readily adapted for the screening of potential chemical mutagens and depending on the type of genetic markers assayed, can be used to detect the induction of point mutations, interstitial deletions, and chromosomal aberrations.[75] Point mutations are assayed using the sex-linked recessive lethal test, whereas chromosome aberrations and unrepaired breaks can be assayed either by dominant lethal tests or by tests for specific translocations, or test for loss of the X or Y chromosomes. *Drosophila* has a generation time of 10 to 12 days and thus a sex-linked recessive lethal assay which requires two generations to complete would require 20 to 24 days; a dominant lethal assay, on the other hand, could be completed in a week since the end point of the assay is the failure of fertilized eggs to hatch. *Drosophila* is capable of metabolically activating promutagens and procarcinogens but it is not known to what extent the spectrum of metabolites generated from these compounds resembles those generated by mammalian metabolism. It is estimated that a single laboratory could complete sex-linked recessive lethal assays on 40-48 test compounds a year and at the same time conduct approximately 50 dominant lethal assays.[76]

**In Vivo Mammalian Cytogenetic Tests**

It is generally recognized that *in vivo* mammalian cytogenetic tests should be included as an important component of test batteries for the detection of human mutagens.[33,34] Chromosomal abnormalities are a frequently encountered genetic defect, occurring in one out of every 200 live human births and one out of every three spontaneous abortions[33,41] and cytogenetic procedures provide a rapid means of evaluating a substance's capacity to produce chromosomal abberations under *in vivo* activating conditions. The tests can easily be conducted as a component of standard subacute and chronic toxicity tests and when peripheral blood lymphocytes are used as the cell source, the methodology can be applied to detect possible toxic effects of chemical exposures in humans.[77] Cytologic analysis of peripheral blood lymphocytes also has been used to monitor the effects of cancer chemotherapies which employ DNA damaging drugs.[78]

A variety of tissues can be utilized for cytologic analysis of mutagen treated animals with bone marrow being one of the more commonly used cell sources.

The micronucleus test, which analyzes the effects of mutagen exposure on developing bone marrow red blood cells,[79] is probably the most rapid of the cytogenetic assays and compares well in sensitivity with cytogenetic analyses of more complex cell types.[41]

In essence the test screens mammalian red blood cells, which in their mature form lack a nucleus, for remnants of chromosomal fragments generated by mutagen exposure. These fragments are left behind when the nuclei are extruded from the maturing cells and they form the micronuclei for which the test is named. Experimentally, the test is normally conducted by administering the test substance by a route which enables it to enter the blood stream. In a typical assay four different dosages with five mice or rats per dose may be used. The bone marrow cells are isolated about 30 hours later and stained slides prepared. Approximately 1,000 red blood cells per animal are scored and the total test takes about 1 week to complete.

## In Vivo Mammalian Genetic Assays

The only way to determine whether an activated chemical mutagen actually reaches and produces damage in germinal tissue is to do experiments with living animals. There are several cytologic and genetic tests, which primarily rely on mice as the experimental organism, that can be used for this type of analysis. Most of the tests detect chromosomal aberrations rather than point mutations which are far more difficult to detect and require test populations of up to 50,000 animals.

The production of heritable chromosomal aberrations can be assayed by the dominant lethal test[80] or by the heritable translocation test.[81,82] The endpoint measured in the dominant lethal test is male sterility or a reduction in litter sizes in matings of treated males with nontreated females. In this assay mid-term pregnant females are dissected to determine the fate of fertilized ova. The failure of fertilized ova to develop after implantation is thought to be due to gross chromosomal damage. This cannot be confirmed, however, since the embryos scored are nonviable. The assay generally is conducted with 10 test males which are matched with 10 negative and 5 positive (known-mutagen-treated) controls. The males are mated over an 8 week period with 3 females per week generating a total of about 600 pregnant females to be examined by dissection. The 8 week period corresponds to the time it takes for sperm cells, which were at different stages of development at the time of treatment, to reach maturity. It is estimated to take approximately 90 man hours to complete the dissection analysis. The dominant lethal test is one of the more frequently performed mammalian assays and has been used as the safety evaluation of many pesticides. Hollstein et al.,[41] comment, however, that this method in general has not been effective in detecting known mutagens and carcinogens other than some potent direct-acting alkylating agents.

The heritable translocation test, although similar in some respects to the dominant lethal test, differs in that the offspring of the treated males are allowed to reach maturity and are thus available for cytologic determination of the type of chromosomal aberration inherited. In this assay treated males are mated weekly to different females over an 8 week period as is done in the dominant lethal test. The first generation (F1) male offspring from these matings are then used for new matings to identify those with a sterile or semisterile phenotypes. The spermatocytes from the affected males are then examined cytologically to determine the genetic basis for the phenotype. The test takes approximately 6 months to conduct and involves scoring litter sizes for several hundred matings of F1 males.

It should be noted that it is also possible to cytologically examine the dividing spermatocytes of the treated males themselves for evidence of chromosomal damage. The spermatocytes are examined 50 to 100 days following treatment of the animals to allow cells which were spermatozoa at the time of treatment to mature into spermatocytes. Although this approach provides information regarding a substance's capacity to damage germinal tissue, it does not provide information with respect to the heritability of the lesions observed.

Another method of monitoring heritable chromosomal abnormalities is the mouse sex chromosome-loss test.[83] This assay differs from the dominant lethal test and heritable translocation test in that it is thought primarily to measure nondisjunctions. In this assay progeny of treated male and female mice are examined for sex-linked phenotypic differences.

The only *in vivo* mammalian assay for the detection of induced point mutations is the mouse specific locus test.[84-86] The test involves monitoring first generation offspring of matings between mutagen-treated males (and matched nonexposed controls) with females that are homozygous recessives at 7 loci which affect such traits as coat or eye color and ear length. The biggest disadvantage of the test is that it requires analyzing on the order of 50,000 progeny and requires approximately 18 months to complete. It may be possible to significantly decrease the numbers of animals which need to be examined by increasing the number of loci at which mutations could be detected. The feasibility of utilizing biochemical markers for this purpose is under investigation.[87,88]

Given the biological diversity and number of test systems available for detecting mutagenically active chemicals, it is important to define the criteria by which a substance would be classified as a human mutagen. In their discussion of this problem the Flamm DHEW Subcommittee on Chemical Mutagenesis[33] noted that epidemiological studies are of almost no value in assessing a chemical's potential to cause heritable genetic alterations in humans. The Subcommittee therefore defined what it considered to be a set of "current optimal criteria" for identifying human mutagens on the basis of currently available methodologies. They suggested that a chemical be treated as a human mutagen if it produced germ cell damage detectable in mice by any of the following: heritable translocation, sex chromosome loss, or specific locus test. As an alternative to the specific locus test they suggested that positive results in any two tests, which would detect either point mutations in eukaryotic microorganisms or point mutations in mammalian cells in culture or recessive lethal mutations in insects, together with evidence for the presence of the test substance or its metabolites in mammalian germinal tissue, would be sufficient for declaring a compound a human mutagen.

The Flamm Subcommittee also noted that it would be impractical if not impossible to test all chemicals, to which humans are exposed, by the *in vivo* mammalian assays noted above. A tier approach was suggested in which chemicals would first be analyzed by several of the more rapid and less expensive screening assays and only those compounds giving positive results in the initial test battery would be tested further. They concluded that "Although definitive proof of non-mutagenicity is not possible, failure of a chemical to give positive results in several test systems capable of indicating a range of genetic effects will, as a practical matter and for regulatory purposes, result in treating a chemical as a nonmutagen."

## Application of Mutagen Screening Tests for the Detection of Chemical Carcinogens

Interest in the development of short-term tests for the identification of chemical carcinogens has been stimulated by a variety of factors, particularly the accumulating epidemiological evidence that environmental determinants, including such elements of lifestyle as smoking, drinking, and dietary habits are a major cause of cancer.[89,90] Also, the high cost and 3 year period necessary to complete a rodent carcinogenesis test make it impractical to use this type of assay as a first-order screening test for large numbers of chemicals. As noted earlier, a recent review[41] has identified over 100 short-term test systems which have been reported as capable of detecting chemicals with carcinogenic potential. It should not be surprising, considering the electrophilic, DNA-reactive nature of most chemical carcinogens, that many of these short-term tests are based on the ability of most carcinogens to induce mutations and other secondary phenomenon, such as repair synthesis, associated with DNA damage. Validation studies in which hundreds of chemicals, including known carcinogens and noncarcinogens, were tested in various short-term assays have confirmed the usefulness of a number of these assays but also have shown that there is no single short-term assay that can reliably detect all classes of chemical carcinogens.[40,41] A battery of short-term assays is therefore needed to decrease the possibility that screening tests will miss potential carcinogens or erroneously classify noncarcinogens. Which of the many short-term tests should be included in a test battery is a question which is still under investigation. The results of one such study which was a joint effort, begun in 1972 by the U.S. National Cancer Institute, the Japanese National Institute of Hygenic Sciences, and the British Imperial Chemical Industry, has recently been published and reviewed by Poirier and De Serras.[40]

In this study 102 compounds, including known carcinogens and noncarcinogens selected by the NCI, were tested in the following five assays: the Ames *Salmonella* reversion assay; the *E. coli polA* growth inhibition test; mitotic recombination in the yeast *S. cerevisiae*; DNA repair in cultured human fibroblasts; and mutagenesis in mouse lymphoma L5178Y cells. (All of these tests were reviewed in the preceeding discussion of mutagen assays.) The results of the study indicated that for the chemicals surveyed the most effective test combination consisted of the *Salmonella* and *E. coli polA* assays which together detected 82% of the organic carcinogens tested. The range of carcinogens detected with *Salmonella* was not improved by host mediated activation.

The two mammalian cell systems failed to generate sufficiently reproducible results and the *S. cerevisiae* assay was not sufficiently sensitive presumably because of the relative impermeability of the yeast cell wall. At the present time similar validation studies are being conducted using other *in vitro* mammalian cell assays including cell transformation tests which use as an endpoint the transformation of cultured cells into cells which exhibit the growth characteristics of tumor cells.[91]

Of all the short-term assays, the Ames *Salmonella* test is clearly the best validated and most widely used and will continue to be so for a number of years to come. The assay has an excellent record for identifying organic carcinogens, particularly carcinogens which are in the aromatic amine and polycyclic hydrocarbon classes.[37-41] In addition, positive results in the Ames assay and several other short-term mutagenesis assays, have been predictive of carcinogenic activity in rodent assays for a number of compounds with widespread human exposure. Examples of these include: the food preservative furylfuramide AF-2 which was used extensively in Japan from 1965 to about 1977; the "flame-retardant" tris-B-P (tris[2,3-dibromopropyl]phosphate) which was used to treat children's sleepwear from 1972 to 1977; and aromatic amine components of hair dye preparations.

There are, however, classes of chemical carcinogens that are not readily detected in the Ames assay and it is instructive to look at what these carcinogens are. First of all there is the class of chemical carcinogens referred to as promoters or "late acting" carcinogens which do not exhibit the electrophilic characteristics of DNA-reactive carcinogens and do not act as "initiators" of the carcinogenic process. These promoters act, through as yet unknown mechanisms, as enhancers of neoplastic progression in tissues which have been exposed to initiating-type carcinogens.[4] Examples of tumor promoters include the phorbol esters which promote carcinoma development in polycylic hydrocarbon-treated skin[92] and phenobarbitol which acts as a promoter of hepatic tumor development.[93] It also has recently been suggested that the weak carcinogenic activity of saccharin, which was detected in rodent studies[94] is due to its capacity to act as a promoter.[95,96]

Since they are not DNA-reactive, it is not surprising that promoters are not detected in mutagenesis-based carcinogen assays such as the Ames test. There are, however, several reports which indicate that chemicals with promoter activity may be detectable in cell transformation-based assays.[96,97]

Heavily chlorinated hydrocarbon carcinogens such as dieldrin, DDT, and carbon tetrachloride are not positive in either the Ames test, the *E. coli polA* test, or yeast mitotic recombination assay.[37,39,40] The reason for the lack of positive results for these compounds in all of the microbial assays is not known but may be related to their highly lipophilic nature which may result in their entrapment in microbial cell wall membranes.

In contrast to the above results, DDE has been detected as a mutagen in cultured Chinese hamster ovary cells.[98] The carcinogenic estrogen, diethylstilbestrol (DES), also is missed by all of the mutagenesis-based assays which may reflect the operation of another mechanism of cancer induction in the case of hormone induced cancers.[99] It should be noted, however, that a weak DNA binding activity for DES has been detected in an *in vivo* rodent assay.[100]

Finally, the Ames assay is not particularly useful for detecting carcinogenic metals and minerals. With the exception of chromate, the assay failed to give a positive reaction with either cadmium, nickel, or beryllium salts or asbestos.[37,41] Salts of cadmium, nickel chromium, and beryllium were positive, however, in an *in vitro* cell transformation assay.[101]

This brief review of what may be classified as Ames test "false negatives", i.e., compounds with documented carcinogenic activity in a rodent bioassay but giving a negative result in the Ames test, can be used to illustrate two important points regarding the short-term tests. The first of these is the previously mentioned need to use a battery of tests which measures different endpoints. For example, many of the compounds which were missed in the Ames test were picked up by cell transformation assay. The second point is that when chemical structural information is available on a test substance, this information should be correlated with test performance data in constructing an appropriate test battery. When they are available, it is also advisable to include known carcinogenic and noncarcinogenic structural analogs of the test substance, to serve as positive and negative controls in the analysis.

The validation studies also have indicated a 10-15% incidence of "false positives", i.e., compounds giving positive mutagenesis results with previous histories of being negative in carcinogenesis bioassays. Most of these false positives gave quantitatively weak mutagenesis reactions, and many are structural analogs of known carcinogens.[102] It has been suggested that most of these false positives may be weak carcinogens that are being missed by the animal assays. A false positive incidence of 10% in a single assay is not of overwhelming concern, however, as the false positives should be sorted out by other tests in the battery.

## Quantitative Comparisons of Mutagenic and Carcinogenic Activities

A greater than million-fold range in carcinogenic potency has been observed in rodent bioassays. For example, the daily doses required to produce a 50% tumor incidence in lifetime feeding studies for aflatoxin B1, 2-acetyl aminofluorene and trichlorethylene are approximately 1 μg, 1 mg, and 1 g respectively. Since a similar range in mutagenic potencies also has been observed in the *Salmonella* assay, the question has arisen whether a correlation exists between mutagenic potency in short-term tests and carcinogenic potency. Such a correlation could prove useful in making human risk assessments.

Initial work by Meselson and Russell[103] has suggested that a fairly good correlation exists between the two potencies across a broad range of chemical classes. This initial study included only 14 chemicals, however, and it has been pointed out that the correlation between mutagenic and carcinogenic activities does not always hold within chemical classes such as the nitrosamines.[104] It is clear that further work will be necessary to determine the extent of this type of correlation. Unfortunately, one of the principal difficulties in conducting this comparison appears to be the limited number of animal studies for which adequate quantitative information is available.[103]

## NONCARCINOGENIC CHRONIC EFFECTS: TERATOGENIC/ REPRODUCTIVE EFFECTS AND NEUROTOXIC EFFECTS

### Introduction

In addition to malignancies there is a wide spectrum of chronic diseases which may be induced or aggravated by chemical exposures. Some of the diseases cited in the proposed EPA guidelines on chronic toxicity testing include lung conditions such as chronic bronchitis, pneumoconiosis, and emphysema; chronic liver disease, such as cirrhosis; chronic kidney disease, such as nephrosis and chronic interstitial nephritis; chronic nervous conditions, such as impaired mental or motor activity and neuropathies; chronic skin effects; bone demineralization; cataracts; hematopoietic and cardiovascular conditions; and generalized emaciation and debilitation.[7] The developing embryo also can be highly sensitive to the toxic effects of some compounds in the absence of adult toxicities. This was dramatically illustrated by the thalidomide tragedy of 20 years ago which resulted in approximately 10,000 malformed children.[105] Similarly, the sterility produced in workers occupationally exposed to the nematocide and fumigant, 1,2-dibromo-3-chloropropane[106] illustrates the potentially unique sensitivity of reproductive organs to certain toxic substances.

The following sections review some of the proposed testing guidelines relevant to these types of toxicities.

### Nononcogenic Chronic Effects

The proposed EPA guidelines[7] for this type of testing differs in several respects from carcinogenesis tests. First of all, with respect to the test species, it is suggested that a nonrodent, usually the dog, should be included in addition to the rat. The principal reasons given for this are that nononcogenic chronic effects are sometimes missed in rodents and having nonrodent data may assist in establishing human exposure limits. The dose levels to be tested also cover a broader range than those suggested for carcinogenesis tests. The guidelines propose that three dose levels be used with the highest producing demonstrable toxic effects and the lowest producing nonobservable toxicity with the possible exception of tumorgenesis. Each test group should include at least 50 animals (25 of each sex) and the animals should be dosed for at least 30 months for the rats and at least 24 months for the dogs.

The proposed guidelines indicate that, in addition to routine examinations and observations for obvious toxicity and behavioral changes, clinical laboratory tests should be performed at intervals of 3, 6, 12, 18, and 24 months. These tests include hematological studies, blood chemistries, urinanalyses, and residue analyses of tissues. At the completion of the study complete necroscopies are to be performed on all animals.

The EPA guidelines also provide a suggested protocol for a combined carcinogenic/noncarcinogenic chronic effects study which would include three species, two rodent (rat and mouse), and a nonrodent. Needless to say, these types of studies are expensive. The estimated cost of a nononcogenic chronic effects study is $550,000 and that of the combined effects study is $800,000. These costs do not include the prechronic-dose range-finding studies which are estimated at $100,000 to $130,000.[7]

**Teratogenic Effects**

The proposed EPA guidelines for the teratogenesis studies[35] stipulate that two mammalian species (rat, mouse, hamster, or rabbit) should be included in this type of test. All the test animals must be young, mature, pregnant females; 20 or more animals should be used when the test species are either rats, mice, or hamsters, and at least 12 animals in the case of the rabbits. Again three dose levels are suggested with the highest dose exhibiting some maternal toxicity and the lowest inducing no observable maternal adverse effects. The test substance should be administered daily beginning at or before the time of implantation and continuing through the period of major organogenesis. The fetuses must be delivered approximately 1 day prior to term by caesarean section and both the parent and fetus examined by complete necroscopies.

At the present time there are no rapid "lower tier", i.e., nonmammalian screening systems which have been validated as capable of detecting teratogenic substances. The basic problem is that an applicable screening test "must differentially detect substances to which the conceptus is uniquely susceptible".[2] Virtually any toxic substance is capable of affecting the fetus at doses that are close to those which are toxic to the adult, but such "coeffective" teratogens would not necessarily pose a developmental hazard.

A potentially useful teratogen screening assay which utilizes the subvertebrate *Hydra attenuata* has been described.[2] Regeneration of adult organisms from disrupted cell suspensions of *H. attenuata* involves many cellular processes analogous to those which take place during embryogenesis and which may be disrupted by teratogenic compounds. The usefulness of this assay system remains to be validated by testing against a broad range of known human teratogens.

**Reproductive Effects**

Reproductive effects testing need not be done in more than one species and among rodents the rat is preferred.[35] It is suggested that a minimum of 10 males be exposed per test dose and each unit should contain enough females to produce approximately 20 litters. Three dose levels are to be read as in the case of the teratologic studies. The EPA standards propose that the test substance be administered to two generations of animals. The F0 generation is to be dosed starting at 40 days after birth and continuing for 100 days after which time they are bred. The F1 generation is similarly dosed and bred giving rise to an F2 generation. In addition to monitoring litter sizes and the health of the young, all the animals (F0, F1, and F2 generations) are to be necropsied and examined histopathologically.

## Neurotoxic Effects

The EPA has not as yet formulated guidelines for behavioral/neurotoxicity health effects. Both types of effects can be measured in rodent assays and the National Institute of Environmental Health Sciences is reported to be evaluating the relative sensitivities of different types of assays to detect neurotoxic effects.[107] The EPA is currently distributing for comment a series of draft guidelines for neurotoxic health effects testing proposed by the Organization for Economic Cooperation and Development (OECD).[108] These guidelines include testing protocols for both acute delayed neurotoxicities and subchronic neurotoxicities. In acute delayed tests the substance to be examined is administered a single acute dose and the test animals are then observed for a 21 day period. The procedure is then repeated one time. In the subchronic test the animals are dosed daily for a period of 90 days.

The experimental species recommended by OECD guidelines is the adult domestic hen. During the tests the animals are to be observed for behavioral abnormalities, locomotor ataxia, and paralysis. Twice a week the animals are subjected to a period of forced motor activity (ladder climbing). Animals that show signs of neurotoxic effects are sacrificed and examined by gross necropsy and histopathologically.

## EPIDEMIOLOGICAL CONSIDERATIONS AND APPLICATION OF CHRONIC TOXICITY TEST DATA TO HUMANS

### Epidemiology as a Toxicological Tool

If asked what system to use (microbial, insect, or mammalian, *in vivo* or *in vitro*) to assess the hazard to humans associated with chronic low dose exposure to a substance already in circulation, an epidemiologist's answer very likely would be *in vivo* and in humans. To accomplish this the epidemiologist would set up what is referred to as a retrospective cohort study. In such a study attempts would be made to identify separate cohorts of exposed and nonexposed individuals (or cohorts with distinctly different levels of exposure), who were matached for as many other variables as possible such as age, smoking and drinking habits, socio-economic status, etc., and then analyze the cohorts for differences in disease patterns.

The case-control study is another important approach which has been used successfully by epidemiologists to identify factors associated with particular disease processes. Case-control studies differ from cohort studies in that they start with people who are already ill. By questioning people who have a particular disease, and appropriate controls who do not, about their personal life styles, work habits, and other factors which the investigator has postulated to be relevant to the disease process, it may be possible to identify antecedent factors which are associated with the disease.

Case-control and retrospective cohort studies have been very successful in identifying large scale determinants of cancer such as smoking which may be responsible for as much as 30% of all cancers in developed countries.[90] These types of studies also have been successful in identifying several occupational carcinogens such as 2-napthylamine, asbestos, and vinyl chloride. Although only small numbers of individuals were exposed in these occupational cases it was possible to detect an increased cancer incidence because of the rare nature of the tumors induced. Generally, epidemiological studies are relatively insensitive in situations where the substance under investigation does not cause an unusual disease which is infrequently seen in the control population. For example, in the United States where bowel and breast cancer are relatively common, it would be difficult to epidemiologically identify a substance which increased the incidence of one of these diseases by 10 or 25%. It would take approximately a 100% increased incidence over background frequencies (a doubling of risk) to identify a carcinogen affecting either of these sites.[109]

Furthermore, because of the long latent period of 20-30 years between initial exposure and the appearance of chemically induced cancers, and the even greater period of time it may take before the accumulation of recessive germline mutations become evident, an epidemiologic approach is impractical for assessing carcinogenic or mutagenic hazards associated with a newly synthesized chemical.

In addition, although providing data which allow correlations to be made between particular exposures and disease states, epidemiological studies do not provide information regarding the mechanism underlying the association. Furthermore, if the association involves a substance which consists of a complex mixture of chemicals, epidemiologic techniques cannot identify the toxic chemical(s) within the mixture. As a case in point, numerous epidemiological studies have indicated that chronic excessive consumption of alcoholic beverages is associated with an increased risk of cancers of the upper alimentary and respiratory tracts and that this risk is synergistically increased by tobacco usage.[110] It is not known, however, whether it is the alcohol itself, or some other components of the drinks which contain low levels of carcinogens, that is responsible for the association.

Furthermore, if it is the alcohol, how is it working? Alcohol is known to be noncarcinogenic in animal studies but it may act as a cocarcinogen via its effects on microsome mediated carcinogen activation.[111-113] Consumption of alcoholic beverages during pregnancy also has been associated epidemiologically with a number of teratological effects which have been classified together as the fetal alcohol syndrome.[114] The role of alcohol in this syndrome, which may be the most common cause of learning difficulties in the United States, also is not known at present.

From the above considerations it should be clear that there are situations in which one must rely on data generated in nonhuman systems to assess the hazard to humans of certain chronic chemical exposures. In the preceeding sections a sampling of the more commonly used assays which have been developed for this purpose have been described. What remains to be discussed are the factors which influence the interpretation and applicability of the rodent and short-term biohazard test data humans.

## Qualitative Validation of Rodent Bioassays for the Identification of Chemical Carcinogens

In 1971 the International Agency for Cancer Research (IACR) initiated a program whose purpose was to evaluate the evidence for carcinogenic risk of chemicals to humans.[115] Based on an evaluation of epidemiologic studies, some 25 chemicals and 5 industrial processes, out of a total of 368 suspect

Table 7.3:  Chemicals and Industrial Processes that have been
Epidemiologically Associated with Human Cancers[5,115-118]

|  |  |
|---|---|
| (1) | Aflatoxins |
| (2) | 4-Aminobiphenyl |
| (3) | Arsenic compounds |
| (4) | Asbestos |
| (5) | Auramine (manufacture of) |
| (6) | Benzene |
| (7) | Benzidine |
| (8) | Beryllium (certain compounds) |
| (9) | Bis(chloromethyl) ether |
| (10) | Cadmium-using industries |
| (11) | Carbon tetrachloride |
| (12) | Chloromethyl ether |
| (13) | Chromium (chromate-producing industries) |
| (14) | Cyclophosphamide |
| (15) | Diethylstilbestrol |
| (16) | Dimethylcarbamoyl chloride |
| (17) | Dimethylsulfate |
| (18) | Ethylene oxide |
| (19) | Hematite mining (radon?) |
| (20) | Isopropyl oils |
| (21) | Melphalan |
| (22) | Mustard gas |
| (23) | 2-Naphthylamine |
| (24) | Nickel (refining) |
| (25) | N,N-bis(2-chloroethyl)-2-naphthylamine |
| (26) | Oxymetholone |
| (27) | Phenacetin |
| (28) | Phenytoin |
| (29) | Soot, tars (polycyclic aromatic hydrocarbons) |
| (30) | Vinyl chloride |

chemicals reviewed, were noted as being associated with the occurrence of cancer (Table 7.3). Of particular relevance to the present discussion is the fact that all of the chemicals and processes listed in Table 7.3, with the exception of arsenic, are carcinogenic in rodent bioassays.[115,119] In the case of arsenic, which is associated with skin cancer when ingested in drinking water and lung cancer after occupational exposure, there is evidence that it inhibits DNA repair.[120] This effect on DNA repair would be consistent with arsenic's being a cocarcinogen which could explain the epidemiological observations and the failure to detect its activity in rodent bioassays.[121] It is also of interest to note that animal carcinogenicity data was available for 4-aminobiphenyl, diethylstibestrol, mustard gas, vinyl chloride, and aflatoxins prior to the epidemiologic findings of an association of the compounds with cancer in humans.

The value of observations such as those cited above clearly would be diminished if the testing procedures used in the rodent bioassays generated many "false positive" results. The concept that most chemicals would be carcinogenic when tested at the high concentrations used in the rodent assays is, however, not supported by experimental evidence. For example, of some 140 pesticides and industrial compounds which were tested because of suspicions regarding their possible carcinogenicity, only 10% were found to be positive in rodent assays.[118] Similarly, only about 20% of the approximately 6,000 compounds listed in the NCI's Survey of Compounds Which Have Been Tested for Carcinogenic Activity were reported to be carcinogenic. Furthermore, since the chemicals on the NCI survey list were selected on the basis of their similarity to known carcinogens, the frequency of 10 to 20% positives is likely to be higher than what would be found in a random sampling of chemicals.

### Validation of Animal Assays for Identifying Other Types of Chronic Chemical Exposure-Related Health Effects

To the best of this author's knowledge there are no surveys, corresponding to those conducted by the IARC and the NCI for the evaluation of chemical carcinogenesis data, that have been conducted to evaluate the predicative value of chronic-exposure animal bioassays for the identification of teratogens, neurotoxins, or compounds which may affect fertility. Relatively few human teratogens have been identified epidemiologically, the known ones being thalidomide, vitamin D, some androgens and estrogens, and several cancer chemotherapeutic agents such as the antimetabolites methyl-folic acid and aminopterin and alkylating agents such as nitrogen mustards, and bisulfran.[105] Most of these compounds produce teratogenic effects in animals with the possible exception of aminopterin, which in a rat study failed to produce malformations but did increase the frequency of fetal deaths and resorptions.[122] Part of the difficulty in evaluating the impact of chronic chemical exposures on teratogenesis in humans is due to the high background incidence of birth defects which has been estimated to range from 2 to 10 percent depending upon whether or not one includes mental deficiencies and learning defects in the calculation.

With respect to neurotoxins, there are a number of industrial chemicals (acrylamide, n-hexane, methyl n-butyl ketone, cresyl phosphate), pharmaceuticals (nitrofuradantoin, isoniazid), and pesticides (leptophos, Kepone®) which have been associated with neuropathic effects in humans (for reviews, see References 107, 123, 124). Subchronic exposure studies in rodents and other animals such as cats have been used to identify and study the mechanism of action of neurotoxic chemicals which produce paralysis and behavioral changes in exposed animals. Studies are currently underway to evaluate the relative sensitivities of behavioral tests and morphological assays of peripheral and central nervous system axon morphology for detecting the earliest signs of chemically induced neuropathies.[107]

### Quantitative Risk Estimates

Although it seems clear that animal tests can be used with a reasonable degree of confidence to identify chemicals with carcinogenic activity, the application of such test data in the quantitative estimation of human risk is a more difficult process. The formulation of risk estimates requires extrapolations from effects observed in different species which are generally exposed to much higher dosage levels than those to which the human population would be exposed. There are uncertainties associated with these extrapolations, primarily because of the limited amount of data available in two areas of concern, namely, (1) the comparative sensitivities of humans and test animals to all classes of chemical carcinogens, and (2) the question of whether there are thresholds for chemical carcinogens and therefore "safe" doses.

**Comparative Dose Response Data:** There are only a limited number of cases in which it has been possible, even on a crude level, to compare animal and human dose responses to the same carcinogens. The paucity of data in this area is related to the fact that in most epidemiological studies the actual doses of the toxic substance is not known with certainty and that while complete dose response curves can be obtained in animal studies, such studies are very expensive. A systematic attempt to estimate human exposures and compare human and animal dose response relationships for several known carcinogens has been made and is contained in a report by the National Academy of Sciences and National Research Council on Environmental Studies Board on Pest Control.[125] The carcinogens for which responses were compared in the study were: benzidine; chlornaphazine; diethylstilbestrol; aflatoxin B1; vinyl chloride, and cigarette smoke.

If the animal test data were used to predict human cancer incidence on a dose per body weight basis, the animal results would have corresponded to the epidemiological findings in the cases of benzidine, chlornaphazine, and cigarette smoke. For aflatoxin, diethylstilbestrol, and vinyl chloride the animal data would have predicted higher human cancer incidences than those observed by factors of 10, 50, and 500 respectively. In these cases it was noted, however, that the epidemiologically observed incidences of aflatoxin, vinyl chloride, and diethylstibestrol-related human cancers may be serious underestimates for two reasons. First of all, these compounds were recognized as human carcinogens because of their association with relatively rare tumors. If they also increased the incidence of more commonly occurring tumors, this increased incidence would not be detected. Secondly, in the cases of vinyl chloride and diethylstilbestrol, the existing epidemiological data are for less than full lifetimes and therefore also may be underestimated.

On the basis of this albeit limited data, the Meselson Committee, which conducted the above study, has suggested as a working hypothesis that ". . . .in the absence of countervailing evidence for the specific agent in question, it appears reasonable to assume that the lifetime cancer incidence induced by chronic exposure in man can be approximated by the lifetime incidence induced by similar exposure in laboratory animals at the same total dose per body weight".[125]

**Thresholds in Chemical Carcinogenesis and Mutagenesis:** The question of whether there are thresholds or no response levels of exposure to chemical carcinogens and mutagens is one of the most vigorously debated issues in modern toxicology. At the root of this question is the uncertainty regarding the true shape of dose response curves to these classes of toxic agents at low exposure levels. As was previously noted, because of the practical limits on the numbers of animals generally used for chronic toxicity studies, test chemicals are administered at relatively high dose levels in order to detect significant responses in the small populations involved. The question raised by this type of testing protocol is whether a response, such as cancer induction, which was observed at a high dose level, would also occur at doses which may be a 100 to a 1,000 times lower than the lowest test dose. It is frequently argued, for example, that cancers seen in high dosage exposures would not occur at low dosages because of the bodies capacity either to detoxify the potentially carcinogenic chemical or repair the damage it produces. According to this argument, cancers seen at high dosage levels occur because the detoxificaion or repair processes have been saturated.

There have been several mathematical models developed for estimating the effects of exposures at dosage levels below those for which test data are available. It is beyond the scope of this chapter to review these models in any detail and for a more detailed discussion the reader is referred to reviews such as those by Schneiderman et al.[126] Cranmer,[127] and Hoel et al.[128] What will be discussed here are the biological bases and implications of some of these models, and a recently conducted, large-scale experiment which attempted to determine the shape of the dose response curve for mice fed low levels of the carcinogen 2-acetylaminofluorene (2-AAF).

There are basically two types of mathematical models commonly used to represent dose response relationships. One type which is referred to as a dichotomous or "yes-no" model[126] is concerned with whether or not a specific response, such as cancer, has occurred with increased incidence in the treated population. The second type, which is a time-to-response model, attempts to relate dose levels to the time of appearance of the measured effect.

There are a number of dichotomous models which have been proposed that are based on different biological phenomenon. The model proposed by Armitage and Doll[129], for example, emphasizes the multistage nature of chemical carcinogenesis. The convex dose response curve (Curve 2 in Fig. 7.1) would be consistent with this model. Convex dose response curves also are expected in genetically heterogeneous populations which would exhibit variations in susceptibility to carcinogens.[130] The multistage model also predicts that two independent carcinogens leading to the same cancer should multiply each others' effects when acting together, a situation that is seen in a number of human cancers.[126]

Another model, developed by Cornfield[131] attempts to take into account the pharmacologic aspects of carcinogen activating and inactivating reactions. This model could be used to explain the existence of low dose thresholds on the basis of a capacity to detoxify essentially all of the carcinogenic metabolites which may be generated at low dose exposures. Gehring et al.[132] have further emphasized the need for pharmacokinetic information in making high to low dose extrapolations and risk assessments. They have suggested that dose response data is meaningful only when expressed in terms of the amount of carcinogen metabolically activated in the body rather than the exposure dose.

Curve 3 in Figure 7.1 is an example of a dose response curve with a threshold level of exposure. Such a curve can be generated if linearity is assumed throughout the entire dose response range and the intercept on the dose axis is found by simple linear extrapolation of the observed responses (doses A-B). If the true shape of the dose response curve is convex (Curve 2) this type of extrapolation would lead to an underestimation of risk. If on the other hand the

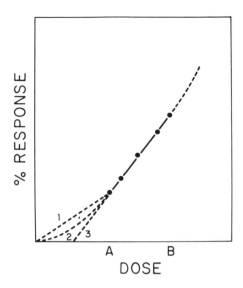

**Figure 7.1:** An observed dose response (A-B) with three hypothetical low dose extrapolations. Curve 1 assumes that the response to low doses approaches zero linearly, whereas curve 2 represents an asymptotic extrapolation to zero response. Curve 3 represents a linear extrapolation with the same slope as that of the observed dose response.

dose response is extrapolated to zero response in a linear fashion, as in Curve 1, it would lead to an overestimation of risk if the true dose response curve were convex in shape.

Time-to-response models also have important implications in a consideration of possible threshold mechanisms since they encompass the concept of "practical thresholds". Druckrey[133] has pointed out that the latent period or time from exposure to the observancy of tumors is inversely related to dose of carcinogen. Thus at very low doses, the median time to tumor appearance could extend beyond an individual's expected lifetime resulting in a practical threshold. It also has been noted, however, that the effect of dose on latent period may only occur at relatively high carcinogen concentrations and thus may not be appropriate for human risk estimates.[127]

**Results of the Low Dose 2-AAF Study:** In order to experimentally assess the shape of the dose response curve at low carcinogen doses, the National Center for Toxicological Research undertook a study involving some 24,000 mice which were fed 2-AAF over a dose range of 5 to 150 parts per million.[134] Two independent endpoints were monitored in the study, namely, the induction of bladder neoplasms and the induction of liver neoplasms. Animals were sacrificed at 18, 24, and 33 months into the study and the incidence of liver and bladder cancers was determined. Interestingly, two different types of dose response curves were observed (Figs. 7.2 and 7.3).

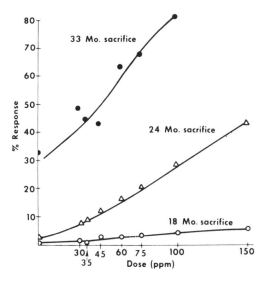

Figure 7.2: The prevalence of liver neoplasms in sacri-
ficed mice in respect to dose of 2-AAF. (Source: N.A.
Littlefield et al.,[135] reprinted with permission from the
*Journal of Environmental Pathology and Toxicology*.)

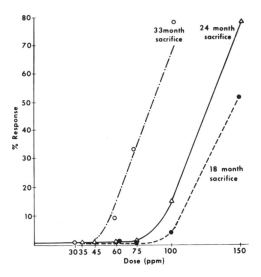

Figure 7.3: The prevalence of bladder neoplasms in
sacrificed mice in respect to dose of 2-AAF. (Source:
N.A. Littlefield et al.,[135] reprinted with permission
from the *Journal of Environmental Pathology and
Toxicology*.)

The liver tumors exhibited a nearly linear dose response consistent with the lack of a threshold dose whereas the incidence of bladder tumors dropped off sharply at the lower 2-AAF levels consistent with a threshold effect. The study also included a serial treatment schedule in which animals were fed 2-AAF for 9, 12, and 15 months after which time the carcinogen was removed from the diet and the animals sacrificed several months later. This treatment schedule revealed a further difference between the induction of the two types of tumors. The liver tumors continued to appear even after the 2-AAF was removed from the diet, whereas the bladder tumor incidence was greatly reduced following the removal of 2-AAF. Thus while liver tumors appeared very late in life (Fig. 7.2), the induction period was completed early. The induction of bladder tumors, on the other hand, apparently required a continued promoting effect of the carcinogen, a process which is similar to that described for smokers.[136,137]

The results of this study, which must be the largest of its kind ever conducted, suggest that even though two types of dose responses were observed, the nearly linear dose response for liver tumors supports the use of linear extrapolation rates at extremely low doses to assure conservative risk estimates.[138]

In addition, considerations of possible carcinogen thresholds should also take into account that the background incidence of cancer in the human population is close to 25%. And, as it is likely that many carcinogens produce the same kinds of cancers, the introduction of new carcinogens would simply act in an additive manner to this already high background.[126]

## REFERENCES

1. Saffiotti, U., Scientific bases of environmental carcinogenesis and cancer prevention: Developing an interdisciplinary science and facing its ethical implications. *J. Toxicol. Environmental Health* 2:14235–1447 (1977).

2. Johnson, E. M., Screening for teratogenic hazards: Nature of the problems. *Ann. Rev. Pharmacol. Toxicol.* 21:417–429 (1981).

3. Interagency Regulatory Liaison Group, Work Group on Risk Assessment Report, Scientific basis for identification of potential carcinogens and estimation of risks. *J. Natl. Cancer Inst.* 63:241–268 (1979).

4. Slaga, T. J., Sivak, A., Boutwell, R. K. (eds.), Mechanisms of Tumor Promotion and Cocarcinogenesis. *Carcinogenesis, a Comprehensive Survey,* Vol. 2. New York: Raven Press (1978).

5. Weinstein, I. B., The scientific basis for carcinogen detection and primary cancer prevention. *Cancer* 47:1133–1141 (1981).

6. Environmental Protection Agency, Proposed guidelines for registering pesticides in the U.S. Hazard evolution; humans and domestic animal. *Fed. Register* 43:37335–37403 (Aug. 22, 1978).

7. Environmental Protection Agency, Proposed health effects test standards for Toxic Substances Control Act test rules; Chronic health effects. *Fed. Register* 44:27356–27375 (May 9, 1979).

8. Consumer Product Safety Commission, Interim policy and procedure for classifying, evaluating, and regulating carcinogens in consumer products. Summary of policy and procedure. *Fed. Register* 43:25658–25665 (June 13, 1978).

9. Consumer Product Safety Commission, Interim statement of policy and procedure for classifying, evaluating, and regulating carcinogens in consumer products.

Clarifications of Commission intent and reopening of commencement period. *Fed. Register.* 43:60436–60438 (Dec. 28, 1978).

10. Department of Labor: Occupational Safety and Health Administration, Identification, classification, and regulation of toxic substances posing a potential occupational carcinogenic risk. *Fed. Register* 42:54147–54247 (Oct. 4, 1977).

11. Schneiderman, M. A., Decouflé, P., and Brown, C. C., Thresholds for environmental cancer: Biologic and statistical considerations. *Ann. N.Y. Acad. Sci.* 329:92–130 (1979).

12. International Agency for Research on Cancer, 2-Naphthylamine. *IARC Monogr. Eval. Carcinog. Risk Chem. Man* 10:51–72 (1974).

13. Clayson, D. B., Principles underlying testing for carcinogenicity. *The Cancer Bull.* 29:161–166 (1977).

14. International Agency for Research on Cancer, Aflatoxins. *IARC Monogr. Eval. Carcinog. Risk Chem. Man.* 10:51–72 (1976).

15. Gutmann, H. R., Malajka-Giganti, D., Barry, E. J., and Rydell, R. E., On the correlation between the hepatocarcinogenicity of the carcinogen N-2-fluorenylacetamide and its metabolic activation by the rat. *Cancer Res.* 32:1554–1561 (1972).

16. Song, C. S. and Kappas, A., Hormones and hepatic function. In: *Diseases of the Liver,* Fourth Ed. (L. Schiff, ed.), pp. 163–183, Philadelphia: Lippincott Co. (1975).

17. Seitz, H., Garro, A. J., and Lieber, C. S., Sex dependent effect of chronic ethanol consumption in rats on hepatic microsome mediated mutagenicity of benzo(a)pyrene. *Cancer Letters* 13:97–102 (1981).

18. Miller, J. A., Carcinogenesis by chemicals: An overview - G.H.A. Clowes Memorial Lecture. *Cancer Res.* 30:559–579 (1970).

19. Nobel, R. L., Tumors and hormones. In: *The Hormones,* (G. Pincus, K. V. Thurmann and E. B. Astwood, eds.) pp. 559–695, New York: Academic Press, Inc. (1964).

20. Weisburger, J. H., Chemical carcinogenesis. In: *Cancer Medicine,* (J. Holland and E. Frei, eds.) Philadelphia: Lea and Febiger (1973).

21. Magee, P. N. and Barnes, J. M., The production of malignant primary hepatic tumors in the rat by feeding dimethylnitrosamine. *Br. J. Cancer* 10:114–122 (1956).

22. Cairns, T., The $ED_{01}$ Study: Introduction, objectives, and experimental design. In: *Innovation In Cancer Risk Assessment ($ED_{01}$ Study),* (J. A. Staffa and M. A. Mehlman, eds.) pp. 1–8 Illinois:Pathotox Publishers Inc., (1979).

23. Food and Drug Administration Advisory Committee on Protocols for Safety Evaluation, Panel on carcinogenesis report on cancer testing in the safety evaluation of food additives and pesticides. *Toxicol. Appl. Pharmacol.* 20:419–438 (1971).

24. Sontag, J. M., Page, N. P., Saffiotti, U., Guidelines for carcinogen bioassays in small rodents. Natl. Cancer. Inst. Carcinogenesis Tech. Rep. Ser. No. 1. *Natl. Inst. Health* DHEW Publ. No. (NIH) 76-801. Washington, D.C.: U.S. Govt. Print. Off. (1976).

25. Druckrey, H., Preussman, R., Ivankovic, S., and Schmäl, D., Organotrope carcinogene wirkungen bei 65 verschieden N-Nitroso-verbindungen on BD-ratten. *Z. Krebsforsch* 69:103–201 (1967).

26. Ito, N., Hiosa, Y., Tojoshima, K., Okajima, E., Kamamoto, Y., Makiura, S., Yokota, Y., Sugihara, S., and Matayoshi, K., Rat bladder tumors induced by N-butyl-N(4-hydroxybutyl) nitrosamine. In: *Topics in Chemical Carcinogenesis,* (W. Nakaharu, S. Takayama, T. Sugimura, and S. Odashima, eds.), pp. 175–192, Tokyo Univ. Press, (1972).

27. Miller, J. A. and Miller, E. C. Carcinogens occurring naturally in foods. *Fed. Proc.* 35:1316–1321 (1976).

28. Ames, B. N., Kanimen, H. O., and Yamasaki, E., Hair dyes are mutagenic. Identification of a variety of mutagenic ingredients. *Proc. Natl. Acad. Sci. U.S.A.* 72:2423–2427 (1975).

29. Ames, B. N., Identifying environmental chemicals causing mutations and cancer. *Science* 204:587–593 (1979).

30. Bischoff, F. and Bryson, G., Carcinogenesis through solid state surfaces. *Prog. Exp. Tumor Res.* 5:85–97 (1964).

31. Brand, K. G., Foreign body induced sarcomas. In: *Cancer, A Comprehensive Treatise* (F. F. Becker, ed.), Vol. 1, pp. 485–511, New York and London: Plenum Press (1975).

32. Magee, P. N. and Barnes, J. M., Carcinogenic nitroso compounds. *Adv. Cancer Res.* 10:163–246 (1967).

33. Flamm, E. G., Chairman DHEW Report on Approaches to Determining the Mutagenic Properties of Chemicals: Risk to future generations. Prepared for DHEW by the DHEW Committee to Coordinate Toxicology and Related Programs (Working Group of the Subcommittee on Environmental Mutagenesis). *J. Environ. Path. and Tox.* 1:301–381 (1977). (Special issue: M. A. Mehlman, M. F. Cramner, and R. E. Shapiro, eds.).

34. National Academy of Sciences, Committee on Tox. Nat. Res. Council. Principles and procedures for evaluating the toxicity of household substances. Prepared for the Consumer Prod. Safety Commission. *Committee for the Revision of NAS.* Publication 1138, pp. 1–9; 86–98 (1977).

35. Environmental Protection Agency, Proposed health effects test standards for Toxic Substances Control Act test rules; General requirements; Acute and subchronic health effects; Mutagenic effects; Teratogenic/reproductive effects; other health effects-metabolism. *Fed. Register* 44:44054–44093 (July 26, 1979).

36. Prival, M. J., Genetic toxicology regulatory aspects. *J. Environ. Path. Toxicol.* 3:99–111 (1980).

37. McCann, J., Choi, E., Yamasaki, E., and Ames, B. N., Detection of carcinogens as mutagens in the *Salmonella*/microsome test: assay of 300 chemicals. *Proc. Natl. Acad. Sci. U.S.A.* 72:5135–5139 (1975).

38. Sugimura, T., Sato, S., Nagao, M., Yahagi, T., Matsushima, T., Seino, Y., Takeuchi, M., and Kawachi, T., Overlapping of carcinogens and mutagens. In: *Fundamentals in Cancer Prevention,* (P. N. Magee et al. eds.), pp. 191–215, Tokyo: Univ. of Tokyo Press; Baltimore, Md.: Univ. Park Press (1976).

39. Purchase, I.F.H., Longstaff, E., Ashby, J., Styles, J. A., Anderson, D., Lefervre, P. A., and Westwood, F. R., Evaluation of six short-term tests for detecting organic chemical carcinogens. *Br. J. Cancer* 37:873–879 (1978).

40. Poirier, L. A. and DeSerres, F. J., Initial National Cancer Institute studies on mutagenesis as a prescreen for chemical carcinogens: An appraisal. *J. Natl. Cancer Inst.* 62:919–926 (1979).

41. Hollstein, M., MaCann, J., Angelosanto, F. A., and Nichols, W. W., Short-term tests for carcinogens and mutagens. *Mutation Res.* 65: 133–226 (1979).

42. Hollaender, A. (ed.), *Chemical Mutagens, Principles and Methods for Their Detection,* Vols. 1–5, New York: Plenum (1971–1978).

43. Kilbey, B. J., Legator, M., Nichols, W. W., and Ramel, C. (eds.), *Handbook of Mutagenicity Test Procedures,* Amsterdam: Elsevier (1977).

44. Ames, B. N., McCann, J., and Yamasaaki, E., Methods for detecting carcinogens and mutagens with the *Salmonella*/mammalian-microsome mutagenicity test. *Mutation Res.* 31:347–364 (1975).

45. Clarke, C. H. and Wade, M. J., Evidence that caffeine, 8-methoxypsoralen and steroidal diamines are frameshift mutagens for *E. coli* K-12. *Mutation Res.* 28:123–125 (1975).

46. Green, M. H. L. and Muriel, W. J., Mutagen testing using TRP$^+$ reversions in *Escherichia coli. Mutation Res.* 38:3–32 (1976).

47. Mohn, G. R. and Ellenberger, J., The use of *Escherichia coli* K12/343(λ) as a multipurpose indicator strain in various mutagenicity testing procedures. In: *Handbook of Mutagenicity Test Procedures,* (B. J. Kilbey, M. Legator, W. Nichols, and C. Ramel, eds.), pp. 95–118 Amsterdam: Elsevier (1977).

48. Ruiz-Vásquez, R., Pueyo, C., and Cerdá-Olmedo, E., A mutagen assay detecting forward mutations in an arabinose-sensitive strain of *Salmonella typhimurium*. *Mutation Res.* 54:121–129 (1978).

49. Skopek, T. R., Liber, H. C., Krolewski, J. J., and Thilly, W. G., Quantitative forward mutation assay in *Salmonella typhimurium* using 8-azaguanine resistance as a genetic marker. *Proc. Natl. Acad. Sci.* U.S.A. 75:410–414 (1978).

50. Skopek, T. R., Liber, H. L., Kaden, D. A., and Thilly, W. G., Relative sensitivities of forward and reverse mutation assays in *Salmonella typhimurium*. *Proc. Natl. Acad. Sci.* U.S.A. 75:4465–4469 (1978).

51. Mortimer, R. K. and Manney, T. R., Mutation induction in yeast. In: *Principles and Methods for Their Detection* (A. Hollander, ed.), Vol. 1, pp. 289–310, New York: Plenum (1971).

52. Brusick, D. J. and Mayer, V. W., New development in mutagenicity screening techniques with yeast. *Environm. Hlth. Perspect.* 6:83–96 (1973).

53. Zimmerman, F. K., Procedures used in the induction of mitotic recombination and mutation in the yeast *Saccharomyces cerevisiae*. *Mutation Res.* 31:71–86 (1975).

54. Clive, D., Flamm, W. G., Machesko, M. R., and Bernheim, N. J., A mutational assay system using the thymidine kinase locus in mouse lymphoma cells. *Mutation Res.* 16:77–87 (1972).

55. Arlett, C. F. and Harcourt, S. A., Expression time and spontaneous mutability in the estimation of induced mutation frequency following treatment of Chinese hamster cells by ultraviolet light. *Mutation Res.* 16:301–306 (1972).

56. O'Neill, J. P., Brimer, P. A., Machanoff, R., Hirsch, G. P., and Hsie, A. W., A quantitative assay of mutation induction at the hypoxanthine-quanine phosphoriboxyltransferase locus in Chinese hamster ovary cells (CHO/HGPRT system): development and definition of the system. *Mutation Res.* 45:91–101 (1977).

57. Demais, R., Mutation studies with human fibroblasts. *Environm. Health Persp.: Experimental Issue* 6:127–136 (1973).

58. Nichols, W. W., Miller, R. C., and Bradt, C., *In vitro* anaphase and metaphase preparations in mutation testing. In: *Handbook of Mutagenicity Test Procedures* (B. J. Kilbey, M. Legator, W. Nichols, and C. Ramel, eds.), pp. 225–233 Amsterdam: Elsevier (1977).

59. Ishidate Jr., M. and Odashima, S., Chromosome tests with 134 compounds on Chinese hamster cells *in vitro* - a screening method for chemical carcinogens. *Mutation Res.* 48:337–354 (1977).

60. Zakharov, A. F. and Egolina, N. A., Differential spiralization along mammalian mitotic chromosomes I BudR-revealed differentiation in Chinese hamster chromsome. *Chromosoma* 38:341–365 (1972).

61. Latt, S. A., Sister chromatid exchanges, indices of human chromosome damage and repair: detection by fluorescence and induction by mitomycin C. *Proc. Natl. Acad. Sci.* U.S.A. 71:3136-3166 (1974).

62. Perry, P. and Wolff, S., New Giemsa method for differential staining of sister chromatids. *Nature* 261:156–158 (1974).

63. Korenberg, J. and Friedlander, E., Giemsa technique for the detection of sister chromatid exchanges. *Chromosoma* 48:355–360 (1974).

64. Slater, E. E., Anderson, M. D., and Rosenkranz, H. S., Rapid detection of mutagens and carcinogens. *Cancer Res.* 31:970–973 (1971).

65. Rosenkranz, H., and Poirier, L., Evaluation of the mutagenicity and DNA-

modifying activity of carcinogens and non-carcinogens in microbial systems. *J. Natl. Cancer Inst.* 62:873–892 (1979).

66. Alvares, A. P., Bickers, D. R., and Kappas, A., Polychlorinated biphenyls: A new type of inducer of cytochrome P448 in the liver *Proc. Natl. Acad. Sci. U.S.A.* 70:1321–1325 (1973).

67. Popper, H., Czygan, P., Greim, H., Schaffner, F., and Garro, A. J., Mutagenicity of primary and secondary carcinogens altered by normal and induced hepatic microsomes. *Proc. Soc. Exp. Biol. Med.* 142:727–729 (1973).

68. Nagao, M., Sugimura, T., Matsushima, T., Environmental mutagens and carcinogens. *Ann. Rev. Genet.* 12:117–159 (1978).

69. Raineri, R., Poiley, J. A., Pienta, R. J., and Andrews, A. W., Metabolic activation of carcinogens in the *Salmonella* mutagenicity assay by hamster and rat liver S9 preparations. *Environ. Mutag.* 3:71–84 (1981).

70. Weinstein, D., Katz, M., and Kazmer, S., Use of a rat/hamster S-9 mixture in the Ames mutagenicity assay. *Environ. Mutag.* 3:1–9 (1981).

71. San, R. H. C. and Williams, G. M., Rat hepatocyte primary cell culture-mediated mutagenesis of adult rat liver epithelial cells by procarcinogens. *Proc. Soc. Exp. Biol. Med.* 156:534–538 (1977).

72. Huberman, E. and Sachs, L., Mutability of different genetic loci by metabolically activated carcinogenic polycyclic hydrocarbons. *Proc. Natl. Acad. Sci. U.S.A.* 73:188–192 (1976).

73. Krahn, D. F. and Heidelberger, C., Liver homogenate-mediated mutagenesis in Chinese hamster V79 cells by polycyclic aromatic hydrocarbons and aflatoxins. *Mutation Res.* 46:27–44 (1977).

74. Malling, H. V., Host mediated assay, pro and con. In: *Molecular and Environmental Aspects of Mutagenesis* (L. Prakash, ed.), pp. 159–175, Springfield: Charles C. Thomas (1974).

75. Vogel, E. and Sobels, F. H., The function of *Drosophila* in genetic toxicology testing. In: *Chemical Mutagens: Principles and Methods for Their Detection* (A. Hollaender, ed.), Vol. 4, pp. 93–142, New York: Plenum (1976).

76. Würgler, F. E., Sobels, F. H., and Vogel, E., *Drosphila* as an assay system for detecting genetic changes. In: *Handbook of Mutagenicity Test Procedures* (B. J. Kilbey, M. Legator, W. W. Nichols, and C. Ramel, eds.), pp. 335–374, Amsterdam: Elsevier (1977).

77. O'Riordan, M. L. and Evans, H. J., Absence of significant chromosome damage in males occupationally exposed to lead. *Nature* 247:50–53 (1974).

78. Schnizel, A. and Schmid, W., Lymphocyte chromosome studies in humans exposed to chemical mutagens. The validity of the method in 67 patients under cytostatic therapy. *Mutation Res.* 40:139–166 (1976).

79. Schmid, W., The micronucleus test. In: *Handbook of Mutagenicity Test Procedures* (B. J. Kilbey, M. Legator, W. Nichols, and C. Ramel, eds.), pp. 235–242, Amsterdam: Elsevier (1977).

80. Bateman, A. J., The dominant lethal assay in the male mouse. In: *Handbook of Mutagenicity Test Procedures* (B. J. Kilbey, M. Legator, W. Nichols, and C. Ramel, eds.), pp. 325–334 Amsterdam: Elsevier (1977).

81. Léonard, A., Heritable chromosome aberrations in mammals after exposure to chemicals. *Radiat. Environm. Biophys.* 13:1–8 (1976).

82. Generoso, W. M., Cain, K. T., Huff, S. W., and Grossler, D. G., Heritable-translocation test in mice. In: *Chemical Mutagens: Principles and Methods for Their Detection* (A. Hollaender and F. J. DeSerres, eds.), Vol. 5, pp. 55–77, New York: Plenum (1978).

83. Russell, L. B., Numerical sex-chromosome anomalies in mammals: Their sponta-

neous occurrence and use in mutagenesis studies. In: *Chemical Mutagens: Principles and Methods for Their Detection* (A. Hollaender, ed.), Vol. 4, pp. 55–91, New York: Plenum (1976).

84. Russell, L. B., X-ray induced mutations in mice. *Cold Spr. Harbor Symp. Quant. Biol.* 16:327–336 (1951).

85. Carter, T. C., Lyon, M. F., and Philips, R. J. S., Induction of mutations in mice by chronic gamma irradiation; interim report. *Brit. J. Radiol.* 29:106–108 (1956).

86. Searle, A. G., The specific locus test. In: *Handbook of Mutagenicity Test Procedures* (B. J. Kilbey, M. S. Legator, W. Nichols, and C. Ramel, eds.), pp. 311–324, Amsterdam: Elsevier (1977).

87. Malling, H. V. and Valcovic, L. R., A biochemical specific locus mutation system in mice. *Arch. Toxicol.* 38:45–51 (1977).

88. Narayonan, K. R., Detection of biochemical mutants in mice. *Arch. Toxicol.* 38:61–73 (1977).

89. Epstein, S. S., Environmental determinants of human cancer. *Cancer Res.* 34:2425–2435 (1974).

90. Doll, R., Strategy for detection of cancer hazards to man. *Nature* 265:589–596 (1977).

91. Dunkel, V. C., *In vitro* carcinogenesis: A National Cancer Institute coordinated programme. In: Screening Tests for Chemical Carcinogens, Scientific Publication No. 1. *Int. Agency for Res. on Cancer.* Lyon, pp. 25–28 (1976).

92. Berenblum, I., Sequestial aspects of chemical carcinogenesis: Skin. In: *Cancer: A Comprehensive Treatise* (F. Becker, ed.), Vol. 1, pp. 323–344, New York: Plenum (1975).

93. Kitagawa, T., Pitot, H., Miller, E. C., and Miller, J. A., Promotion by dietary phenobarbitol of hepatocarcinogenesis by 2-methyl-N,N-dimethyl-4-aminobenzene in the rat. *Cancer Res.* 39:112–115 (1979).

94. Arnold, D. L., Charbonneau, S. M., Moodie, C. A., and Munroe, I. C., Long-term toxicity study with ortho-toluenesulfonamide and saccharin. *Toxicol. Appl. Pharmacol.* 41:164 (1977).

95. Murasaki, G. and Cohen, S. M., Effect of dose of sodium saccharin on the induction of rat urinary bladder proliferation. *Cancer Res.* 41:942–944 (1981).

96. Mondal, S., Brankov, D. W., and Heidelberger, C., Enhancement of oncogenesis in C3H/10T 1/2 mouse embryo cell cultures by saccharin. *Science* 201:1141–1143 (1978).

97. Mondal, S., Branknow, D. W., and Heidelberger, C., Two-stage chemical carcinogenesis in C3H/10T 1/2 cells. *Cancer Res.* 36:2254–2260 (1976).

98. Kelly-Garvert, F. and Legator, M. S., Cytogenetic and mutagenic effects of DDT and DDE in Chinese hamster cell line. *Mutation Res.* 223–229 (1973).

99. Hertz, R., The estrogen-cancer hypothesis with special emphasis on DES. In: *Origins of Human Cancer* (H. H. Hiatt, J. D. Watson, and J. A. Winsten, eds.), pp. 1665–1674, New York: Cold Spring Harbor Laboratory (1977).

100. Lutz, W. K., *In vivo* covalent binding of organic chemicals to DNA as a quantitative indicator in the process of chemical carcinogenesis. *Mutation Res.* 65:289–356 (1979).

101. Casto, B. C., Pierzynaki, W. J., Nelson, R. L., and DiPaolo, J. A., *In vitro* transformation and enhancement of viral transformation with metals. *Proc. Amer. Ass. Cancer Res.* 17:12 (1976).

102. McCann, J. and Ames, B. N., The *Salmonella*/microsome mutagenicity test: predictive value for animal carcinogenicity. In: *Origins of Human Cancer* (H. H. Hiatt, J. D. Watson, J. A. Winsten, eds.), pp. 1431–1450, New York: Cold Spring Harbor Laboratory (1977).

103. Meselson, M. and Russell, K., Comparison of carcinogenic and mutagenic potency. In: *Origins of Human Cancer* (H. H. Hiatt, J. D. Watson, and J. A. Winsten, eds.), pp. 1473–1481, New York: Cold Spring Harbor Laboratory (1977).

104. Kameswar, T., Young, J. A., Lijinsky, W., and Eplu, J. L., Mutagenicity of aliphatic nitrosamines in *Salmonella typhimurium. Mutation Res.* 66:1–7 (1979).

105. Becker, B. A., Teratogens. In: *Toxicology, The Basic Science of Poisons.* (L. J. Casarett, and J. Doull, eds.), pp. 313–332, New York: Macmillan (1975).

106. Legator, M. S., Chronology of studies regarding toxicity of 1,2-dibromo-3-chloropropane. *Ann. N.Y. Acad. Sciences* 329:331–338 (1979).

107. Schaumburg, H. H. and Spencer, P. S., Clinical and experimental studies of distal axonopathy - a frequent form of brain and nerve damage produced by environmental chemical hazards. *Ann. N.Y. Acad. Sci.* 329:14–29 (1979).

108. Environmental Protection Agency: Organization for Economic Cooperation and Development (OECD), Chemical program: Final reports on testing guidelines; Notice of availability. *Fed. Register* 45:26129–26130 (April 17, 1980).

109. Peto, R., Detection of risk of cancer to man. *Proc. R. Soc. Lond. B.* 205:111–120 (1979).

110. Lieber, C. S., Seitz, H. K., Garro, A. J., Worner, T. M., Alcohol-related diseases and carcinogenesis. *Cancer Res.* 39:2863–8886 (1979).

111. McCoy, G. D. and Wynder, E. L., Etiological and preventative implications in alcohol carcinogenesis. *Cancer Res.* 39:2844–2850 (1979).

112. Garro, A. J., Seitz, H. K., and Lieber, C. S., Enhancement of dimethylnitrosamine metabolism and activation to a mutagen following chronic alcohol consumption. *Cancer Res.* 41:120–124 (1981).

113. Seitz, H. K., Garro, A. J., and Lieber, C. S., Enhanced pulmonary and intestinal activation of procarcinogens and mutagens after chronic ethanol consumption in the rat. *Eur. J. Clin. Invest.* 11:33–38 (1981).

114. Streissguth, A. P., Landesman-Dwyer, S., Martin, J., and Smith, D. W., Teratogenic effects of alcohol in humans and laboratory animals. *Science* 209:353–361 (1980).

115. Tomatis, L., Agthe, C., Bartsch, H., Huff, J., Montesano, R., Saracci, R., Walker, E., and Wilbourn, J., Evaluation of the carcinogenicity of chemicals: A review of the monograph program of the International Agency for Research on Cancer (1971–1977) *Cancer Res.* 38:877–885 (1978).

116. IARC Monographs on the Evaluation of Carcinogenic Risk of Chemicals to Man. Volumes 1–20, *IARC*, Lyon, France (1979).

117. IARC Monographs on the Evolution of the Carcinogenic Risk of Chemicals to Humans. A Supplement; Chemicals, Groups of Chemicals and Industrial Processes Associated with Cancer in Humans. *IARC*, Lyon, France (1979).

118. Tomatis, L., The predictive value of rodent carcinogenicity tests in the evaluation of human risks. *Ann. Rev. Pharmacol. Toxicol.* 19:511–530 (1979).

119. Snyder, C. A., Goldstein, B. D., Sellakur, A. R., Bromberg, I., Laskin, S., and Albert, R. E., The inhalation toxicology of benzene: Incidence of hematopoietic neoplasms and hematotoxicity in AICR/J and C57BL/6J mice. *Toxicol. Appl. Pharmacol* 54:323–331 (1980).

120. Rossman, J. G., Meyn, M. S., and Troll, W., Effects of arsenite on DNA repair in *Escherichia coli.* In: *Environ. Hlth. Persp.* 19:229–233 (1977).

121. Rall, D. P., Validity of extrapolation of results of animal studies to man. *Ann. Rev. N.Y. Acad. Sci.* 329:85–91 (1979).

122. Murphy, M. L., Teratogenic effects in rate of growth-inhibiting chemicals. *Clin. Proc. Child. Hosp.* 18:307 (1962).

123. Epstein, S. S., Kepone®: hazard evaluation. The Science of the Total Environment. 9:1 (1978).

124. Xintaras, C., Burg, J. R., Johnson, B. L., Tanaka, S., Lee, S. T., and Bender, J., Neurotoxic effects of exposed chemical workers. *Ann. N.Y. Acad. Sci.* 329:30–38 (1979).

125. National Academy of Sciences/National Research Council Environmental Studies Board. Contemporary Pest Control Practices and Prospects: The Role of the Executive Committee 1. Pest Control: An Assessment of Present and Alternative Technologies. 1–438, 66–83. *NAS.* Washington, D.C.

126. Schneiderman, M. A., Decoufle, P., and Brown, C. C., Thresholds for environmental cancer: Biologic and statistical considerations. *Ann. N.Y. Acad. Sci.* 329:92–130 (1979).

127. Cranmer, M. F., Estimation of risks due to environmental carcinogenesis. *Med. Ped. Oncol.* 3:169–198 (1977).

128. Hoel, D. G., Gaylor, D. W., Kirschstein, R. L., Saffiotti, U., and Schneiderman, M. A., Estimation of risks of irreversible delayed toxicity. *J. Toxicol. Environ. Health.* 1:133–151 (1975).

129. Armitage, P. and Doll, R., Stochastic models for carcinogenesis. In: *Fourth Berkley Symposium on Mathematics and Probability.* (J. Neyman, ed.), pp. 19–38, Berkeley, Calif: University of California Press (1961).

130. Mantel, N., Heston, W. E., and Gurian, J. M., Thresholds in linear dose-response models for carcinogenesis. *J. Natl. Cancer Inst.* 27:203–215 (1961).

131. Cornfield, J., Carcinogenic risk assessment. *Science* 198:693–699 (1977).

132. Gehring, P. J., The relevance of dose-dependent pharmacokinetics in the assessment of carcinogenic hazards of chemicals. In: *Origins of Human Cancer* (H. H. Hiatt, J. D. Watson, and J. A. Winsten, eds.), pp. 187–203, New York: Cold Spring Harbor Laboratory (1977).

133. Druckrey, H., Quantitative aspects of chemical carcinogenesis. *Union Int. Contre. Cancer Monogr. Ser.* 7, pp. 60–78, New York: Springer-Verlag (1967).

134. Staffa, J. A. and Mehlman, M. A. (eds.), *Innovations in Cancer Risk Assessment (ED$_{01}$ Study),* Illinois: Pathotox Publishers Inc. (1979).

135. Littlefield, N. A., Farmer, J. H., Gaylor, J. H., and Sheldon, W. G., Effects of dose and time in a long-term, low-dose carcinogenic study. In: *Innovations in Cancer Risk Assessment (ED$_{01}$ Study)* (J. A. Staffa and M. A. Mehlman, eds.), pp. 17–34 Illinois: Pathotox Publishers Inc. (1979).

136. Levin, D. L., Devesa, S. S., Goodwin, J. D., and Silverman, D. T., *Cancer Risks and Rates,* Second Ed., U.S. Dept. of Health Education and Welfare, Washington, D.C. (1974).

137. Marx, J. L., Tumor promotors: Carcinogenesis gets more complicated. *Science* 201:515–518 (1978).

138. Gaylor, D. W., The ED$_{01}$ study: Summary and conclusions. In: *Innovations in Cancer Risk Assessment (ED$_{01}$ Study)* (J. A. Staffa and M. A. Mehlman, eds.), pp. 179–183, Illinois: Photox Publishers Inc. (1979).

# Part III

# Product Liability, Regulations and Labels

Perhaps nowhere else are the lines as clearly drawn on differing philosophies of labeling as in this section.

Various government agencies have established label regulations with which companies must comply. Some feel that as long as their lables comply with these regulations, they have satisfied their responsibilities. Many others view such regulations as only a "floor" in label development. That is, regulatory compliance is only the beginning.

These labelers are impressed with the need to satisfy product liability and worker protection needs. This theme is ably dealt with in Chapters 8, 11, 12, 13, and 15. There apparently is a certain antagonism between satisfying product liability needs (which are best served by listng all precautions, hazard statements and dangers associated with a chemical on its label) and worker protection.

Worker safety, according to recent social science research, is apparently enhanced by labeling only for major hazards and by establishing a general danger symbol, perhaps patterned on NFPA's 704 System or the DOT/UN Hazard "Diamond" Labels. While detailed labels do increase the amount of information available, they also raise anxiety levels and probably reduce reader retention and understanding.

One solution that appears reasonable is to restrict container labels to only the most pertinent hazards, utilizing NFPA-704 "Labels" for tanks and areas, and providing expanded material safety data sheets in the workplace.

Trade secrets, trademarks and patents are the province of Chapter 9. How to protect trade secrets, whether to patent and when to patent, and how to protect trademarks are fully discussed. Aspirin, Cellophane and Thermos are only three of many trademarks lost by companies over the years. Eastman-Kodak is a good example of a company currently pursuing an effective campaign to protect, in their case, Trade-dress.

Transportation regulations provide a separate focus for Chapter 14. DOT's classification scheme and the labels and placards that are specified by their regulations have had a wide ranging effect upon labeling in general. The DOT classes, covered in detail in Chapter 4 of Section II, have become the standard categories for label development. Many labelers take advantage of the well recognized DOT "Diamonds" by incorporating these "labels" in their product labels, when appropriate and lawful.

Labels are the central mechanism by which EPA regulates pesticides. It is the label, as pointed out in Chapter 10, that is registered. Label development in this area is often carried out after consultation with EPA label reviewers. The influence of DOT can be clearly recognized as the basis for FIFRA categorization. Although not clearly enunciated, EPA often chooses to treat those pesticides which meet DOT Poison B criteria as restricted use pesticides. Implicitly, EPA is saying that in spite of the best efforts, labels are not wholly effective in controlling the harmful effects of these products. This additional use control derives from CPSC actions in banning certain consumer products.

In Chapter 13's review of CPSC regulations, the DOT influence is again clear, although specific cut off points are adjusted for the consumer market. Here we encounter a new concept, products for which no adequate label can be prepared. Such products have been banned from consumer use. Extremely flammable contact adhesives are an example. In CPSC's final report, they declare that no label will effectively prevent harm to average consumers, even if the label is read, understood and complied with. EPA's restricted use category is a variant of this philosophy.

Chapters 11 and 12 discuss labeling under TSCA and RCRA. Both of these relatively recently enacted laws are concerned with disseminating information on the potential hazards of the chemicals they regulate. This legislation extends government regulation of chemical labeling into waste disposal, distribution, marketing, and production. As pointed out in Chapter 12, RCRA is especially designed to safeguard our environment by controlling the means by which chemicals are disposed. The basic mechanism is again a classification label scheme that owes much to DOT. While the categories are not exactly duplicated, they are very similar. A new concept introduced is EP toxicity. This test is based upon an Extraction Procedure that identifies the presence of heavy metals and pesticides and measures their levels. As Chapter 11 indicates, labeling under TSCA has, to date, only dealt with PCBs. EPA did participate in OSHA's label standard and must, under Section 6, consider labeling before taking any other action which is directed towards protecting human health or the environment.

Lastly, Chapter 15 covers the new proposed OSHA Label Standard. The author brings a special understanding to this area having served OSHA for eight years, most recently as Director of Health Standards.

For the first time Labeling regulations similar in scope to those instituted under CPSC, FDA and EPA are being required for all hazardous chemical products. While the focus is clearly on workplace labeling, OSHA's requirement that labels applied in the workplace must remain with the container, will, in effect, create distribution labeling. While it is probably true that product liability considerations and general social demand have created effective labeling programs at most chemical companies, there are, in fact, many gaps in general industry.

# 8

# Product Liability and Labels

David F. Zoll
*Chemical Manufacturers Association*
*Washington, D.C.*

## OVERVIEW

A manufacturer must look *beyond* the product liability statutes and regulations to determine the full scope of his duty to warn others of the potential risks his products may pose. He must also look at one of the most dynamic areas of law–those decisions of judges and juries which comprise the manufacturer's common law duty to warn others of the risks involved in using his product.

A manufacturer faces a frustrating task in trying to precisely define his responsibilities. Over time, the thrust of decided cases has been to consistently increase the scope and nature of the dangers he must identify, the uses of his product which he must foresee and the warnings he must sound.

And that trend is certain to continue. First of all, the philosophy of this area of the common law is *specifically designed* to accommodate society's expanding view of corporate responsibility. In addition, the manufacturer's hazard warning responsibilities will be determined after-the-fact, perhaps by a judge or jury hostile to business in general, but most certainly one having full knowledge of the injuries sustained by the plaintiff.

And in recent years, especially where questions of the policy values underlying the common affairs of life have been involved, judges have been increasingly inclined to submit issues to juries rather than to resolve the issues themselves as questions of law.[1]

Furthermore, certain legal procedures raise additional problems for a manufacturer trying to reverse a jury's findings. A trial judge need not *agree* with a jury verdict; in general, he needs only to be satisfied that proper procedures were followed and that reasonable men could not differ with the jury's verdict. In fact, the greater the danger from a particular product's "defect" - such as the defect of inadequate warnings, the less the *power* of a court to preclude a plaintiff's recovery by foreclosing a jury verdict of culpability.[2]

Lastly, the courts of appeal typically give great deference to a trial court's findings because that court heard the witnesses and the presentations of other evidence in the case.[3]

The common law duty to warn is an area of the law where a court decision which was first thought to be an aberration is *later* seen as the start of a new trend. It can be a costly arena for the unwary manufacturer–in adverse damage awards, adverse publicity, and damaged reputation. Product liability in general is one of the fastest growing areas of the law, with an estimated 500,000 cases being filed each year.[4] Expanding theories of liability and sizable damage awards are strong litigation incentives. There have been $500,000 awards for eye injuries and brain disorders; one jury awarded a plaintiff $930,000 for blindness caused by a defectively packaged drain solvent.[5] In one recent, significant case, discussed infra at 239-241, two plaintiffs recovered a total of $2,000,000 from three firms for the improper labeling of empty drums containing explosive vapors. In another case, discussed infra at 223-225, a plaintiff sought to recover a total of $11,000,000 from five manufacturers of an allegedly carcinogenic miscarriage preventative.

The manufacturer searching for a definition of his particular responsibilities must carefully scrutinize his own factual situation. Consequently, neither this nor any own published treatment of the duty to warn should be presumed to provide definitive answers to his questions. As an aid to the manufacturer, this chapter:

    A. Explains the social philosphy and legal principles which comprise the common law–to help the manufacturer better understand the emotional climate in which his actions may be evaluated;

    B. Uses a number of decisions by judges and juries to illustrate some of the more important duty-to-warn "rules"–to help the manufacturer understand how different factual situations will be treated in the courtroom;

    C. Offers a duty to warn check list–to assist the manufacturer in identifying, implementing and maintaining his own duty to warn;

    D. Discusses recent congressional initiatives in the product liability arena; and

    E. Identifies other important but collateral problems for the manufacturer.

The text of this chapter does not dwell on the details of legal theories and procedures. Expanded discussions of legal issues are contained in the footnotes together with references to other published sources for the reader's further study. In a few instances legal discussions are included in the text where such details seem to be genuinely important to the manufacturer's understanding of some aspect of his duty to warn.

## THE SOCIAL PHILOSOPHY AND PRINCIPLES OF THE COMMON LAW DUTY TO WARN

Why do court decisions turn out the way they do? A brief review of the social philosophy and principles of the common law duty to warn may assist a manufacturer in answering that question.

### The Common Law of Torts

The common law is designed to embrace society's changing views about the conduct of persons in society. The term "common law" refers to a large body of principles and rules which have authority as law simply because courts have enforced them as accepted customs and practices. A violation of a common law principle is not a violation of a statute or a regulation but it is a violation of law - an "illegal act", just the same. Insofar as common law principles have not been expressly abrogated by legislative acts — and states sometimes do substitute new definitions of particular rights and duties for those in the common law - the common law is a part of the legal philosophy of most of the United States.[6]

The "common law" embraces many principles relating to government and the security of persons and property. Within this arena is a separate group of actions - essentially a disconnected group for which the legal community has not developed a completely satisfactory definition, called "torts". Tort actions are concerned with *compensating* private persons who have *suffered a loss* to some legally protected interest (other than violations of contracts) *as a result of conduct by others which is regarded as socially unreasonable.*[7] For instance, torts include such "socially unreasonable acts" as assault, defamation, nuisance, and arrest without a warrant, to name a few.[8] Not only do torts include socially unreasonable conduct which society has rejected *over time,* but they also include violations of rights which a court at a particular point in time finds *should* be protected. The fact that the particular claim involved is a new one has not prevented courts from fashioning remedies for a plaintiff. Examples of the types of actions which have been created by the courts are: intentional infliction of mental suffering, denial of the right to vote, and infliction of prenatal injuries, again to name a few.[9] In sum, courts and juries reaching decisions based upon common law principles are providing the latest-in-time view of the perceived reasonableness of particular conduct and how injured persons should be compensated for losses resulting from that conduct.

### Negligence

There are certain causes of action within the field of torts which are relevant to a manufacturer's duty to warn of the risks of his products.[10] One of these was developed shortly after the turn of the 19th century, as accidents increased with the advent of the industrial revolution, and particularly with the expansion of the railroads. The tort cause of action which arose came to be called "negligence".

To define "negligence" and other tort principles it is helpful to rely upon the latest authoritative effort of legal scholars to describe and explain the rules followed by the various courts—the so called "Restatement" (Second) of Torts.[11] As defined in the Restatement, "negligence" is conduct which falls below the standard established by law for the protection of others against *unreasonable* risk of harm.[12] The standard of conduct for *avoiding* negligence is that of a *reasonable person* under similar circumstances.[13] Accordingly, negligent conduct may be either an act which a reasonable person should recognize as involving an *unreasonable risk* of causing harm to another or a *failure* to do an act which is necessary for the protection of another, where there is such a duty to act.[14]

Under common law principles of negligence, a supplier of a product will be liable for physical harm caused by a product, if the supplier *knew or had reason to know* that the product was likely to be dangerous when used for its intended purpose, had no reason to believe that the user would realize the danger, and failed to exercise reasonable care to inform the user of the dangerous condition or of facts which made the product likely to be dangerous.[15] A manufacturer must provide products that are reasonably safe for their foreseeable use.[16]

### Strict Liability

The common law duty to warn involves yet another cause of action–strict liability, which was first used as grounds for holding a defendant liable in 1963[17], was then expanded upon in the so-called "consumer movement" of the 1960s and 1970s and is now accepted by many of the states. This concept focuses not so much on the supplier's conduct as on the status of the product itself. Here, liability is imposed if, without a warning, the product would be "in a defective condition unreasonably dangerous to the user or consumer".[18] Whether a product is in such a "defective condition"–i.e., unaccompanied by warnings or instructions, depends upon a consideration of the same factors analyzed in determining negligence (e.g., foreseeability and seriousness of the harm) but here they are applied in a different manner. The injured plaintiff does not have to identify and prove a *particular failure* to exercise care *on the part of the manufacturer* and the manufacturer cannot defend himself by asserting that he acted in a reasonable manner. The focus of attention is on the condition of the *product*–the fact that it was in commerce without a warning and thereby caused an injury.

Strict liability is a two-part social policy enforced by the courts. The first part of the policy is "risk allocation"–manufacturers who benefit from the sale of their products in the market place are to compensate consumers who are injured by those products because the manufacturer is in a better position to bear the cost of such injuries–directly or through insurance.[19] The second concept is an "incentive to deterrence"–the manufacturer is in the best position to discover defects or dangers in his product and to take steps to guard against them.[20]

A major problem for a manufacturer confronting a strict liability allegation is that the law does not give him many legal *defenses*. The manufacturer will not be liable if the plaintiff filed his action too late in time (beyond the "statute of limitations") or, in general if he can show there was no causal relationship between the inadequate warning and the plaintiff's injury. For instance, the manufacturer will not be liable if the plaintiff misused the product[21] or if he simply ignored a perfectly adequate warning.[22]

But what if the injured plaintiff simply was stupid as to how he handled a product? In the law, such action by a plaintiff is more elegantly referred to as "assumption of risk"–he knew what he was doing was dangerous but he did it anyway, or "contributory negligence"–the plaintiff's own action amounted to negligence. There appears to be a considerable overlap between these two legal principles which need not concern us here.[23] What is important is that in a strict liability situation it is much more difficult for the manufacturer to take advantage of the plaintiff's conduct. There must be a finding that the plaintiff in fact appreciated the danger and then voluntarily and unreasonably proceeded to encounter the known danger.[24] If the plaintiff's only negligence, in a strict liability case, was in failing to *discover* the inadequacy of the warning or in *guarding against* it, the manufacturer will not be able to use that conduct as a shield against liability.[25]

## Other Theories of Liability

One area of the law which has been subject to a great deal of controversy and expansion recently is the efforts of courts to deal with duty to warn situations in which a number of manufacturers might be involved. The courts have over time developed philosophies and procedures for dealing with such situations. For instance, where a plaintiff can not identify which of two or more defendants caused an injury, the "burden of proof" in some cases may be shifted to the defendants for them to prove they were *not* responsible.[26] The assumption here is that the defendants may be in a better position to provide the needed evidence than the plaintiff.[27]

Where there is an alleged concert of action among defendants the law, as defined by the Restatement, has adopted the following principles:

> For harm resulting to a third person from the tortious conduct of another, one is subject to liability if he (a) does a tortious act in concert with the other or pursuant to a common design with him, or (b) knows that the other's conduct constitutes a breach of duty and gives substantial assistance or encouragement to the other so to conduct himself, or (c) gives substantial assistance to the other in accomplishing a tortious result and his own conduct, separately considered, constitutes a breach of duty to the third person[28]

The philosophy of this theory is that:

> those who, in pursuance of a common plan or design to commit a tortious act, actively take part in it, or further it by cooperation or request, or who lend aid or encouragement to the wrongdoer, or ratify and adopt his acts done for their benefit, are equally liable with him. [ ] Express agreement is not necessary, and all that is required is that there be a tacit understanding. . .[29]

This theory might be applicable to a situation where a group of manufacturers of a particular product *assisted and encouraged* each other to provide *inadequate* warnings of risk.

The theory of "enterprise liability" has been developed to cover situations in which a group of manufacturers adhere to a practice, procedure or standard which itself leads to the manufacture of a particular, unidentifiable injury-producing product.

A fact situation in which this theory was discussed (but, because of a number of procedural reasons may not be a solid *precedent* for application of the theory)[30] was *Hall v. E. I. DuPont de Nemours & Co., Inc.*[31] The defendants in *Hall* were virtually the entire U.S. blasting cap industry and its trade association. The class action plaintiffs were children injured in a series of blasting cap explosions across the country over a four-year period. The plaintiffs alleged that the industry practice of omitting warnings on individual blasting caps and the lack of other safety measures created unreasonable risks.

There was evidence in the case that the defendants, acting independently, had adhered to an industry-wide practice regarding blasting cap safety features, that they had delegated some aspects of safety investigation and design, such as labeling, to their trade association, and that there was industry-wide cooperation in the manufacture of blasting caps.

The Court indicated a joint liability theory could be established one of three ways: (1) the existence of an explicit agreement or joint action among the defendants regarding warnings and safety features; (2) the existence of parallel behavior sufficient to support an inference of agreement or cooperation; (3) evidence that, acting independently, the defendants adhered to an industry-wide standard or custom with regard to the safety features of blasting caps.[32] The court commented upon the fact that the dynamics of market competition frequently result in explicit or implicit safety standards, codes and practices which are widely adhered to in an industry.[33] It then stated:

> Where such standards or practices exist, the industry operates as a collective unit in the double sense of stablizing the production costs of safety features and in establishing an industry-wide custom which influences, but does not conclusively determine, the applicable standard of care. See Prosser, Law of Torts § 33 at 166–68 (4th ed. 1971) (on relationship of industry custom to standard of care). As our decision in *Hall* below indicates, the existence of industry-wide standards or practices alone will not support, in all circumstances, the imposition of joint liability. But where . . . individual defendant-manufacturers cannot be identified, the existence of industry-wide standards or practices could support a finding of joint control of risk and a shift of the burden of proving causation to the defendants.[34]

It then discussed various situations in which the courts confronted damages caused by a group-such as in air or water pollution, where "the only feasible method of ascertaining risks, imposing safeguards and spreading costs is through joint liability or other methods of joint risk control"[35] Addressing the facts in the case before it, the court said:

> The allegations in this case suggest that the entire blasting cap industry and its trade association provide the logical locus at which

precautions should be taken and liability imposed. It is unlikely that individual manufacturers would collect information about the nation-wide incidence and circumstances of blasting-cap accidents involving children, and it is entirely reasonable that the manufacturers should delegate this function to a jointly-sponsored and jointly-financed association.

In the event that the evidence warrants it, the imposition of joint liability on the trade association and its members should in no way be interpreted as "punishment" for the establishment of industry-wide institutions. Such liability would represent rather the laws's traditional function of reviewing the risk and cost decisions inherent in industry-wide safety practices, whether organized or unorganized. See, e.g., The T. J. Hooper, 60 F.2d 737 (2d Circ. 1932).

To establish that the explosives industry should be held jointly liable on enterprise liability grounds, plaintiffs, pursuant to their pleading, will have to demonstrate defendants' joint awareness of the risks at issue in this case and their joint capacity to reduce or affect those risks. By noting these requirements we wish to emphasize their special applicability to industries composed of a small number of units. What would be fair and feasible with regard to an industry of five or ten producers might be manifestly unreasonable if applied to a decentralized industry composed of thousands of small producers.[36]

While these theories do not directly threaten industry-wide organizations and institutions, they clearly indicate that the manufacturer must help assure appropriate quality and direction to the activities of such organizations.

Finally, there is the so-called "market-share theory", a controversial and, according to the dissenting opinion in *Sindell v. Abbott Laboratories*,[37] an unwarranted extension of traditional tort doctrine, in part because the plaintiff was not required to prove that any defendant caused or even probably caused the plaintiff's injuries.[38] The case is discussed here at length because it vividly illustrates how a court will develop new legal principles in order to provide relief to a severely injured plaintiff who is unable to collect for injuries because traditional theories or legal procedures do not precisely fit the plaintiff's situation.

The plaintiff in *Sindell* had a malignant bladder tumor as a result of her mother having taken the drug DES, a miscarriage preventative, during her pregnancy. The plaintiff could not identify the specific manufacturer of the DES which her mother had taken. Nevertheless, the plaintiff was allowed to hold a group of five manufacturers, which had produced the drug from an identical formula, liable for a proportion of the total judgment based upon their respective market shares *unless* a manufacturer could demonstrate that it *could not* have produced the product which had caused the plaintiff's injuries. Trial evidence showed that the group of five manufacturers:

1. knew or should have known that the drug was carcinogenic, and that there was a grave danger, after varying periods of latency, that it would cause cancerous and precancerous growths in daughters of mothers who took the drug;

2.  advertised and marketed the drug as a miscarriage preventative when they knew or should have known that it was ineffective to prevent miscarriages;

3.  failed to test DES for efficacy or safety; the tests they did rely upon which were performed by others indicated DES was not safe or effective; and

4.  in violation of FDA authorization, marketed DES on an unlimited basis rather than as an experimental drug and failed to warn of its potential danger.

The court noted that it was dealing with a fast spreading and deadly disease requiring radical surgery to prevent it from spreading, and a medication, where the consumer is virtually helpless to protect himself from "serious, sometimes permanent, sometimes fatal, injuries caused by deleterious drugs"[39]

Faced with the plaintiff's inability to identify the manufacturer which caused her injury on the one hand, and the seriousness of the injury to the defenseless plaintiff on the other, the court laid a foundation for its extension of tort law by discussing how various existing theories of liability utilized by the plaintiff were inadequate.

It then stated:

> In our contemporary, complex industralized society, advances in science and technology create fungible goods which may harm consumers and which cannot be traced to any specific producer. The response of the courts can be either to adhere rigidly to prior doctrine, denying recovery to those injured by such products, or to fashion remedies to meet these changing needs. Just as Justice Traynor in his landmark concurring opinion in *Escola v. Coca Cola Bottling Company* (1944) 24 Cal.2.d 453, 467–468, 150 P.2d 436, recognized that in an era of mass production and complex marketing methods the traditional standard of negligence was insufficient to govern the obligations of manufacturer or consumer, so should we acknowledge that some adaptation of the rules of causation and liability may be appropriate in these recurring circumstances. . .[40]

> The most persuasive reason for finding plaintiff states a cause of action is that advanced in *Summers:* as between an innocent plaintiff and negligent defendants, the latter should bear the cost of the injury. Here, as in *Summers,* plaintiff is not at fault in failing to provide evidence of causation, and although the absence of such evidence is not attributable to the defendants either, their conduct in marketing a drug, the effects of which are delayed for many years, played a significant role in creating the unavailability of proof.

> From a broader policy standpoint, defendants are better able to bear the cost of injury resulting from the manufacture of a defective product. As was said by Justice Traynor in *Escola,* "[t]he cost of an injury and the loss of time or health may be an overwhelming misfortune to the person injured, and a needless one, for the risk of injury can be insured by the manufacturer and distributed among the public as a cost of doing business." (24 Cal.2d p. 462, 150 P.2d p. 441; see also

Rest.2d Torts, § 402A, com. c, pp. 349–350.) The manufacturer is in the best position to discover and guard against defects in its products and to warn of harmful effects; thus, holding it liable for defects and failure to warn of harmful effects will provide an incentive to product safety. (*Cronin v. J.B.E. Olson Corp.* (1972) 8 Cal.3d 121, 129, 104 Cal. Rptr. 483, 501 P.2d 1153; *Beech Aircraft Corp. v. Superior Court* (1976) 61 Cal. App. 3d 501, 522–523, 132 Cal. Rptr, 541.) These considerations are particularly significant where medication is involved, for the consumer is virtually helpless to protect himself from serious, sometimes permanent, sometimes fatal, injuries caused by deleterious drugs.[41]

This decision has prompted considerable discussion among legal commentators,[42] and litigation on behalf of other DES consumers as well as other product areas in which the theory seemed applicable.[43] Initial reports of the decision jolted representatives of the insurance industry who viewed additional application of the theory as posing staggering consequences to insurance interests.[44] As indicated, it may have been influential in getting the U.S. Congress to provide relief for private companies burdened by large insurance costs and may yet provide support for proposed legislation designed to provide greater certainty and uniformity in product liability law.

What is important to recognize for the purposes of this chapter is the flexibility of the judicial system in providing relief for injured plaintiffs as against corporations which are deemed to have fallen short of their duty to warn.

## Review

The next section examines the rules a manufacturer should follow and the court decisions which have analyzed those rules, but first the following list summarizes some of the important aspects of the philosophy and principles of the common law duty to warn.

1. The common law is a system of rules of conduct which have been enforced by courts as legal obligations for the very reason that society has adhered to them as customs and practices over time.

2. The duty to warn is a multifaceted obligation imposed on manufacturers who, because they obtain benefits from the products they create and place into commerce, must bear the responsibility of informing users of their products and the public at large about the dangers those products may pose. Society believes those manufacturers are in a better position than users and the public to identify those dangers, to minimize them, and to bear the associated costs.

3. Society's shaping of the nature and scope of a manufacturer's common law duty to warn has been strongly influenced by the consumer and environmental movement and initiatives of the 1960s and 1970s. The prudent manufacturer will take into consideration the fact that not only

will the court or jury be evaluating his duty to warn after the fact, with an injured plaintiff in front of them, but that the climate of current public opinion is such that benefits of doubts will probably be resolved in the plaintiff's favor.

4. Product liability factors are becoming increasingly larger cost items in a corporation's budget process. Insurance costs are increasing; civil penalty awards are increasing. The number of product liability suits is increasing. The prudent manufacturer will initiate an ongoing preventative program to assure that he adequately fulfills his duty to warn.

The litigated cases which are discussed in the next section have been selected to illustrate particular aspects of the manufacturer's duty to warn and how that particular aspect was treated in the reported case. The manufacturer should note, however, that nearly all product liability cases involve a basic pattern of analysis. In order for the manufacturer to have been found liable there must have been, *in fact,* a *duty* to warn, a *breach* of that duty, and (except in *Sindell*) a *causal relationship* established between the manufacturer's breach and the plaintiff's injury. In conducting this analysis, regardless of the "legal theory" being employed, the courts will consider:

1. The *seriousness* of the harm posed by the product.

2. The *foreseeability* of the harm.

3. The *nature* of the warning which should have been given - including such factors as the feasibility of giving a warning, and the product users' knowledge.

4. The *extent of the warning*–was it sent far enough into the chain of commerce?

## REVIEW OF COURT DECISIONS

### The Risk of Harm–What Must A Manufacturer Know About His Own Product?

#### General Rules

A Manufacturer will be held to the knowledge of an expert regarding his product. He may be responsible for keeping abreast of the latest discoveries, scientific advances and other information about the possible harmful effects of his products. His responsibility for conducting tests will be measured by the degree of that potential harm.[45]

\* \* \*

While certainly not the first case to define a manufacturer's obligation to know the risks his product poses, *Borel v. Fibreboard Paper Products Corporation*[46] provides a good illustration of how courts interpret that obligation, especially for chemicals involving serious, latent and chronic effects. This is also the first case in which a worker used the product liability laws to recover for an occupational disease. Another distinction for the case is that it is regarded as having been the trigger for an avalanche of asbestos related cases which, at least through 1979, totalled 1,370.[47]

Plaintiff Borel, an insulation worker, sued numerous manufacturers of insulating materials containing asbestos for asbestosis and mesothelioma injuries caused by the defendant's alleged failure to warn Borel of the dangers involved in handling asbestos. Borel's exposures to asbestos occurred over a thirty-three year period between 1939 and 1969 (the year in which suit was filed); he died before the trial stage of the District Court action. His complaint alleged the following *negligent* acts by the manufacturers:

1. Failure to inform him regarding safe wearing apparel and proper protective equipment and the methods of handling and using the different products containing asbestos;

2. Failure to take reasonable precautions or to exercise reasonable care to warn him of the danger to which he was exposed;

3. Failure to test the asbestos products in order to ascertain the dangers involved in their use;

4. Failure to remove their products from the market after ascertaining that the products would cause asbestosis.

Notice the diversity and age of the following sources of record evidence of information regarding the health effects of asbestos reviewed by the court.

1. The first reported case of asbestosis was among English textile workers in 1924 - (citing a 1924 edition of the *British Medical Journal*).

2. In the next decade, numerous similar cases were observed and discussed in the medical journals (citing a 1930s publication).

3. By the mid-1930s the hazard of asbestosis as a pneumoconiotic dust was universally accepted (citing a 1936 edition of the Journal of the American Medical Association)

4. Asbestosis incidence in U.S. insulation workers was reported as early as 1934 - in the *British Journal of Radiology*.

5. A 1938 Public Health Service Bulletin "fully documented" the significant risk involved in asbestosis textile factories.

6. A 1945 edition of the Journal of Industrial Hygiene reporting on the *first* large scale survey of asbestos insulation workers which concluded that Navy shipyard workers involved in the asbestos covering of vessels was a *"relatively safe operation"*. That study was criticized, however, because of the relatively short exposure time experienced by the survey workers.

7. In 1947 the American Conference of Governmental Industrial Hygienists, a quasi-official body responsible for making recommendations concerning industrial hygiene, issued guidelines suggesting threshold limit values for exposure to asbestos dust 5

million parts per cubic feet of air (in 1968 it reduced that level to 2 million).

8. In 1965, the Annals of the New York Academy of Science published a study of 1,522 New York insulation workers in which almost half of the men examined had evidence of pulmonary asbestosis, and where at least 90% of those with forty years of exposure had abnormalities.

The defendants argued that Borel's injuries were not foreseeable until 1968, and that, because of the latent period of the disease, he must have contracted it well before then.[48]

The Court noted that asbestosis is a cumultative disease, so that Borel's most recent exposures could have contributed to his overall condition; the defendant's failure to warn, therefore, could have produced an actionable injury.[49]

Furthermore, expert witnesses testified that in the *1930s* the danger of inhaling asbestos was widely recognized, that there were literally "dozens and dozens" of articles on asbestos and its effects on man, that its dangers could be controlled.[50]

After laying this historical foundation regarding the health effects of asbestos the court reached the following conclusions regarding the manfucturers' obligations concerning knowledge of their products:

> . . .in cases such as the instant case, the manufacturer is held to the knowledge and skill of an expert. This is relevant in determining (1) whether the manufacturer knew or should have known the danger, and (2) whether the manufacturer was negligent in failing to communicate this superior knowledge to the user or consumer of its product. *Wright v. Carter Products, Inc.,* 2 Cir. 1957, 244 F.2d 53. The manufacturer's status as expert means that at a minimum he must keep abreast of scientific knowledge, discoveries, and advances and is presumed to know what is imparted thereby [Citing Keeton, Product Liability—Problems Pertaining to Proof of Negligence, 19 S.W.L.J. 26, 30-33 (1965)] But even more importantly, a manufacturer has a duty to test and inspect his product [the Court's lengthy footnote is repeated, *infra*[51]]

> The extent of research and experiment must be commensurate with the dangers involved. A product must not be made available to the public without disclosure of those dangers that the application of reasonable foresight would reveal. Nor may a manufacturer rely unquestioningly on others to sound the hue and cry concerning a danger in its product. Rather, each manufacturer must bear the burden of showing that its own conduct was proportionate to the scope of its duty.[52]

Against those principles, the court noted the evidence tended:

> to establish that none of the defendants ever tested its product to determine its effect on industrial insulation workers. Nor did any defendant ever attempt to determine whether the exposure of insulation workers or others to asbestos dust exceeded the A.C.G.I.H's recommended threshold limit values, or indeed, whether those standards were accurate or reliable.[53]

In *Karjala v. Johns-Manville Products Corp.*,[54] a post-*Borel* asbestos case, there was evidence that the defendant itself knew by the 1930s that persons who worked in asbestos plants were exposed to a substantial health hazard. Furthermore, trial evidence showed the existence of pre-1950s articles in which "some suggestion" was made of a connection between asbestosis and those who worked with asbestos fibers. The evidence was sufficient to impose on the manufacturer the knowledge of an expert regarding his product, the duty to test his product, and the duty to warn.[55] In sum, the manufacturer may have to warn when there is "some suggestion" of harm; he clearly can not wait until the evidence of harm is compelling.[56]

It is reasonable to assume that when a product's risk of harm is not comparatively severe, and additional testing would be costly, the courts will impose "reasonable" testing obligations on manufacturers. A case involving products other than chemicals will illustrate the point.

The plaintiffs in *Barfield v. Atlantic Coastline R.R.*,[57] were workmen involved in transferring oil from a dock to a ship by means of hoses. A latent defect in one of the hoses caused a hose to split, spilling oil and starting a fire. The Court reviewed evidence that the hoses had already undergone extensive tests and inspections and declined to hold the manufacturer liable for additional inspections. The court concluded that additional tests were not practical or economically feasible; the possibility of harm had been foreseen and appropriate measures taken.

On the other hand, a manufacturer will clearly be held responsible for knowledge which reasonable tests *would* have revealed. In *Chapman Chemical Co. v. Taylor*,[58] the court held that the manufacturer of a chemical dust used for spraying on crops should have been aware of the peculiar carrying quality of the dust.

How fast a manufacturer must add recently discovered information to warnings on products already in commerce may well be questions for a jury and depend upon the seriousness of the harm involved and the feasibility of giving a warning.[59]

### Foreseeability—What Must A Manufacturer Foresee About How His Product Will Be Used?

### General Rules

> The Manufacturer must consider the normal *uses* of his product and the *foreseeable misuses* of his product within the *environment* in which the product will be used.
>
> * * *
>
> The injurious result has got to be one of those consequences which is not entirely outside the range of expectation or probability[60]
>
> An intervening act by a third person (even if negligent) relieves the original negligent actor from liability only if the subsequent actor's act could not have been anticipated by the first actor in the exercise of due care[61]

What is it that a manufacturer should foresee about the uses and misuses of his product? Suppose he sells perfume in bottles which atomize the perfume into a mist and he knows or should know that the mist is flammable. It may very well be foreseeable that someone might spray perfume on their neck or face while smoking a cigarette thus igniting the perfume.[62]

In *Pease v. Sinclair Refinery Co.*,[63] a manufacturer of chemistry teachers' demonstration kits offered one kit which contained sample tubes of different liquids, one of which was supposed to be kerosene. Unfortunately, since kerosene has the same physical appearance as water, the manufacturer substituted water for kerosene in the tube, perhaps in an effort to save costs. A severe injury occurred when the water was inadvertently mixed with a chemical which was explosive in the presence of water. The court held that the manufacturer should have foreseen that, in a chemistry classroom setting, a number of chemicals might come in contact with each other, most certainly with water. The court balanced the gravity of the possible harm - explosion, against what it viewed as the ease with which the manufacturer could have provided a warning. Compare the results in that case, however, with the *Croteau v. Borden Co.*,[64] where a chemical manufacturer was held not liable to a *laboratory technician* for a failure to warn that if one of its chemicals was mixed with a wide variety of other chemicals an explosion might be produced.

A decision involving an apparently harsh reading of a manufacturer's duty to "foresee" is *Gardner v. Q.H.S., Inc.*[65] The owner of an apartment building sued the manufacturer of hair rollers when a tenant boiled her hair rollers in water too long, the water evaporated, and the rollers, which were filled with paraffin, ignited and caught fire. Part of an apartment building burned down.

The instructions accompanying the rollers indicated that they should be boiled in water for fifteen minutes just before they were used. The paraffin was designed to retain heat so that the heated rollers would curl hair when it was rolled and set. With the instructions was the following precautionary note:

> Use plenty of water. Do not let water boil away. Cautionary note: rollers may be inflammable only if left over flame in pan without water. Otherwise Q.H.S. Setting/Rollers are perfectly safe.

The injured plaintiff in the case put the rollers in a pan of water on an electric stove, turned the burner to high, went to lie down and fell asleep. The fire soon developed.

The court sent the issue of the adequacy of the warning to the jury. The court explained that some users of rollers could have concluded that "flame" did not include electrical heat and that there was no warning that there was a *strong possibility* that the paraffin could *ignite* if the water boiled away.[66]

In this case, of course, the manufacturer *did* foresee both the use and misuse of its product; it did not, in the court's view, give sufficient warning of the seriousness of the consequences of misuse.

This apparently harsh result may be explained by the format of the label. Notice that right next to the precautionary language is the concluding statement that except for the instructions identified, the rollers "are perfectly safe". As a general rule, it is wise to avoid placing boasts about a product next to warnings of the products—and certainly the fire in the home is a serious risk. Avoid the commingling of the text of warnings with advertising or promotional language.

The manufacturer's obligation to foresee misuse does not extend to foreseeing *abuse*. In these situations it is the misuse and not the nature of the warning which causes the harm. For instance, in *Collins v Rowe*,[67] recovery for carbon monoxide injuries was denied where the refrigerator which emitted the damaging chemical would not have done so if the plaintiff had not tampered with it.

A drain solvent manufacturer was not liable to an employee who was burned when he knocked a container of the solvent containing sulfuric acid off of a shelf in an unlighted room. Although the label in this case failed to include the word "Danger" as required by federal law, the plaintiff's helper's actions in storing the solvent carelessly, when he knew the solvent was dangerous, was the proximate cause of the harm.[68]

The manufacturer will be required to make reasonable as opposed to extraordinary assumptions about the ultimate disposition of his product in predicting risks. When initially well-designed blasting caps, located in a mine 30 years after their manufacture, exploded through spontaneous combustion, the manufacturer was not held liable for having failed to warn of the risks of spontaneous combustion. *Barrett v. Pablis Powder Co.*[69]

A decision which illustates both the obligation to be aware of the effects of a product and the uses which might be made of it is *LaPlant v. E.I. DuPont de Nemours & Co., Inc.*[70] The manufacturer produced a weed killer which the plaintiff used to kill weeds in a ditch adjacent to grazing land for his prize cattle. The labels on the weed killer indicated that it was harmless to livestock—which was true, insofar as the effects on livestock from directly ingesting the compound. However, the action of the compound in killing weeds resulted in the creation of salty-tasting nitrites. The cattle were attracted to the dying weeds; to them, the salty residues were poisonous and the manufacturer was held liable for the death of the plaintiff's cattle.

*Cooley v. Quick Supply Co.*,[71] involved the sale of dynamite fuses. Safety fuses were sold in 50 foot rolls, groups of which were placed in large cartons. Each *carton* contained a written warning which, among other things, referred to instructions on the use of the dynamite set out in a separate pamphlet which was also furnished with each shipment of dynamite. In this case, no measures had been taken to pass warnings on to ultimate users of the safety fuse nor, the court noted, had there been any inquiry made by the defendant to determine whether or not it was the retailer's practice to furnish information to the purchasers. There was evidence that many sales of the fuses were in lengths of less than 50 feet and that it was sold to anyone who requested it. Since the only purpose for which the fuses were bought was to explode dynamite the court found that the defendant should have foreseen that some purchasers would be unfamiliar with the proper use of dynamite and ignorant of the proper manner in which to use the safety fuse. This should have also made it foreseeable that the probability of injury, if the fuses were not properly used, would be significant. In this case, the fuses were misused and a horrible explosion resulted. The defendant took the position that it would have been burdensome if not impossible to give notice to each ultimate purchaser of the product. But the court indicated that the difficulty of giving notice is only one of the factors to be considered. It saw no reason why the defendant could not have taken measures to get a warning to those who needed it; it emphasized that the law demands only that the method used give reasonable assurance that the information will reach those whose safety depends upon them having it. The defendant in this case made no attempt to satisfy that rule. Furthermore, the court noted, the defendant could not absolve itself of liability by *assuming* the purchaser's judgment and caution without such warnings.

## What Kind of A Warning Must Be Given?

### General Rules

> The warning must be appropriate; implicit in the duty to warn is the duty to warn with a degree of intensity that would cause a reasonable man to exercise for his own safety the caution commensurate with the potential danger. From this it follows that the likelihood of an accident taking place and the seriousness of the consequences are always pertinent matters to be considered with respect to the duty to provide a sufficient warning label, and that there is a particular need for a sufficient warning where there is a representation that the product in question is not dangerous.[74]
>
> * * *
>
> A seller may be required to give directions or warnings on a product container. The seller may reasonably assume that those with common allergies, as for example to eggs, or strawberries, will be aware of them and he is not required to warn against them. Where, however, the product contains an ingredient to which a substantial number of the population are allergic, and the ingredient is one whose danger is not generally known, or if known is one which the consumer would reasonably not expect to find in the product, the seller is required to give warning against it, if he has knowledge, or by the application of reasonable, developed human skill and foresight should have knowledge, of the presence of the ingredient and the danger. Likewise, in the case of poisonous drugs, or those unduly dangerous for other reasons, warnings as to use may be required.[75]

The case law in this area dramatically illustrates a manufacturer's vulnerability to after-the-fact assessments by courts and juries. Since plaintiffs do not intentionally harm themselves, it may be relatively easy for a court or jury to identify some aspect of a communication which, had it been present, could have alerted the plaintiff to danger.

A prudent manufacturer will look at all of the potential communication methods he has at his disposal in developing a hazard communication system for his product: the message content, label size, print size, format, coloring, and placement, and instructions for use of the product and for remedies from improper use. It is a sad irony when a company develops a great expertise and spends enormous sums of money on advertising communications for its product and then is held liable for inadequate communication of the risks involved in using the product.

The warning must precisely fit the risk. Suppose a company manufactures a solvent which is used to clean machinery. It sells it in fifty-five gallon drums with big red labels on them which read as follows: "Keep away from fire, heat and open flame and lights. Caution. Leaking packages must be removed to a safe place. Do not drop." Assume the labels did *not* indicate that the vapors of the product were themselves explosive.

This particular fact situation was presented in *Simonetti v. Rinshed-Mason Co.*[76] The defendant-manufacturer's *customer* stored the solvent in the 55 gallon drums in a building, separate from the work area, which was used for the storage of various types of flammable materials. When the solvent was to be used as a machinery wash it was removed from the drums in one gallon safety cans and used in a ventilated area where fire extinguishers were accessible and "no smoking" signs posted. Evidence was presented at the trial that the plaintiff had been present during a discussion among workers regarding the safety of conducting welding in the vicinity of the fumes which hung in the workplace some hours after the machinery had been washed with the solvent. The plaintiff himself had expressed concern about doing any welding in the presence of the vapor fumes. There was also evidence that the plaintiff took no precautions as a result of those discussions. However, the court ruled that the plaintiff's conduct was for the jury to evaluate.

An explosion occurred when a 2,000 degree welding arc was used in the environment of fumes. The defendant, of course, tried to argue that the plaintiff was careless and foolhardy to have proceeded with the welding, but the issue of the plaintiff's conduct was sent to the jury.

At trial a witness testified that while the label on the drum was required by the U.S. Interstate Commerce Commission, the custom in the industry regarding such products was to use an additional label containing instructions as to how the product should and should not be used. The witness also testified that in similar situations in the industry one would find such additional warnings as "extremely flammable," "vapors explosive," "use in well ventilated area," "keep away from fire" and "no smoking."

The court concluded that while it was possible to infer from the label used that the product might be flammable under certain conditions, there was no warning that the product's vapors or fumes could ignite and cause a tremendous explosion hours after the substance had been used.

Another case which involved a general warning which failed to take proper account of all aspects of a product's vapors was *Tucson Industries, Inc. v. Schwartz.*[77] In this case, the plaintiff's eyes were severely damaged and partially blinded by fumes from a cement being used to bond tops onto tables in a room *adjacent* to the plaintiff's office. The cement *had* been labeled to warn of its flammability and the toxicity of its vapors. The harmful fumes reached the plaintiff by traveling through air conditioning ducts.

The court said that if the vapors were so harmful as to have caused the plaintiff's injuries that an adequate warning would have indicated, first, that the fumes could in fact cause blindness and, secondly, that ventilation precautions were needed to prevent the fumes from traveling within buildings.

Use of just the right words alone clearly may not be enough to satisfy the duty to warn. In *Hubbard-Hall Chemical Co. v. Silverman*[78] the defendant-manufacturer had provided an extensive series of written warnings on its labels for an extremely poisonous insecticide. The plaintiff did not contend that the details and preventative measures on the labels failed to adequately convey what should or should not have been done with respect to the product. Unfortunately, however, the workers who were killed after using the product had little or no knowledge of English. At trial, the workers' foreman testified that he had told the workers that if they did not wear sufficient protective equipment they would die. But since the workers were deceased, the court ruled, the jury was entitled to believe or *not* believe the foreman. It was the defendant-manufacturer who had the burden of showing the jury that adequate instructions had reached the deceased workers and the jury could give weight to the absence of a skull and cross bones or similar pictorial representations of danger on the label.

A good illustration of the need to consider a range of communication factors is presented by the recent case of *Ziglar v. E.I. Dupont de Nemours Co.*[79] In this case, a worker, thinking that water was in a translucent jug which was shaped similar to the type used in the retail sales of milk, drank some of the water-like contents and then died. Unfortunately, the jug contained an experimental pesticide. The manufacturer had complied with all applicable *mandatory* labeling requirements. However, the court found that not every "reasonable" warning precaution had been taken, and pointed to the following factors:

1. The pesticide was in fact similar to water in appearance;

2. There was no evidence that the deceased worker could read or write;

3. The skull-and-crossbones on the container was small - 4/17 of an inch square;

4. Instructions on the jug, which stated:

> If swallowed, give a tablespoon of salt in a
> glass of warm water and repeat until vomit
> fluid is clear

> were deemed inadequate because they placed too much emphasis on the need to obtain and take salt and not enough on the need to induce vomiting.

*Edwards v. California Chemical Company*[80] involved an insecticide composed principally of lead arsenate which caused arsenic poisoning in a worker who used the product. The label involved warned that the product was not to be inhaled, touched or taken internally. But the court was impressed with what it viewed as two overriding factors. First, on one side of the container was an extremely detailed set of instructions as to how, where, when and for what purpose the product served as an insecticide. However, on the warning side there was no information as to how the product could be used safely—such as the need to use a respirator and protective clothing.

The court stated:

> As a matter of reason and everyday common sense, the jury would be
> authorized, in light of these two labels, to believe that it would be

difficult, if not impossible, for a user to use the product in the fashions detailed by defendants without in some measure inhaling, touching, or ingesting some of the product.

In light of the specific instructions in one area and the failure in the other and more important area to prescribe protective clothing and a respirator and to tell the user *how to safely use the product,* the jury would be warranted in believeing that the label warning was inadequate. It could be inferred from the omission that the product was not sufficiently dangerous and virulent to require protective devices or else the label would have so stated and warned. In other words, the jury could determine that the duty to warn under these circumstances could only be fulfilled by putting the user on specific notice as to how the product might be safely used, in addition to its advices as to how it might be efficiently used.[81]

When the possible harm involved is drastic, even the fact that there is a statistically remote possibility of injury may not relieve the manufacturer of his duty to warn. The classic case on this issue is *Davis v. Wyeth Laboratories, Inc.*[82] The case involved a 39 year old man who developed paralytic polio as a result of taking a polio vaccine at a mass immunization clinic. The court held that, although the manufacturer did not know it at the time it *first* put its vaccines on the market, the vaccine posed a small but definite risk of causing polio in adults. That information had been discovered by an advisory committee established by the Surgeon General. The court noted that, in the cause of promoting the effort to those involved with the mass immunization, there was evidence that there was a complete lack of warning as to the potential danger to adults but, in addition, that the manufacturer had given assurances that the vaccine was safe for everyone. The court found that the manufacturer had a duty to warn consumers of the risks and that the failure to do so rendered the drug unfit and unreasonably dangerous within the meaning of section 402A of the Restatement. It distinguished the case from a prescription drug situation where the manufacturer can often rely upon information given to physicians.[83] This drug was not *dispensed* as a prescription drug but to all comers at a clinic without a physician to intervene. The court said:

Appellant would approach the problem from a purely statistical point of view: less than one out of a million is just not unreasonable. This approach we reject. When, in a particular case, the risk qualitatively (e.g., of death or major disability) as well as quantitatively, on balance with the end sought to be achieved, is such as to call for a true choice judgment, medical or professional, the warning must be given.[84]

Instructions regarding the remedies to be employed if harm occurs and proper use of a product may be just as important as the warning of the risk itself. In *Charles Pfizer & Co., Inc. v. Brauch,*[85] a chemical product contained a warning that its use should be discontinued if cattle reacted to it adversely but it failed to indicate what antidote should be employed in case of severe reactions. The manufacturer was held liable for the death of several cattle.

Similarly, in *Gongalez v. Virginia Chemical Co.,*[86] a chemical manufacturer was held liable when the pilot of a crop dusting airplane was overcome from the toxic compound he was using as a defoliant after his plane crashed. Although the label on the product cautioned avoidance of inhalation and personal contact, it did not contain a skull and cross bones, the word poison or the listing of an antidote (omissions which violated state and federal laws). Omission of the antidote greatly exaggerated the plaintiff's injuries–facts which strongly influenced the Court.

Where there is good evidence that a plaintiff was well aware of the risks in using a product–as in the case of *Daniels v. The Atlantic Refining Co.*,[87] involving a herbicide, the manufacturer will not be held liable regarding his duty to warn. On the other hand, in some situations, *no* amount of warning will suffice, such as where a product is inherently unsafe for a particular use.

For instance, in *D'Hehouville v. Pioneer Hotel Co.*,[88] an acrylic manufacturer was held strictly liable for the wrongful death of an attorney in a hotel fire caused by a carpeting which ignited easily and did not self extinguish. The court found that the carpet's inherent characteristics and not the absence of a warning made it an unreasonably dangerous as a carpeting fabric. No warning could have altered the flammability of the fabric or made it safe for the ultimate consumer.

It may be of assistance to know that there have been cases in which verbal communications from a manufacturer's representative to a customer have been sufficient to overcome deficiencies which might have existed in written communications. In *Thomas v. Arvon Products Company, Inc.*[89] the plaintiff was permanently blinded from inhaling toxic fumes from a varnish. In addition to using a label cautioning against the breathing of the vapor and suggesting ventilation, the manufacturer (1) sent a representative on several visits to the customer explaining the use and application of the product, (2) repeatedly emphasized to a foreman the need for ventilation, and (3) criticized the employer's ventilation. How practical or reasonable such communications may be will depend upon the facts of each situation, of course.

Finally, compliance with federal or state labeling requirements will not alone be sufficient to fulfill the common law duty to warn.[90] Violation of such mandatory requirements may establish negligence *per se*.[91]

## How Far Into the Chain of Commerce Must the Warning Be Sent?

### General Rules

> Precise rules cannot be established in advance which will determine whether a manufacturer who supplies a product for the use of a third person satisfies his duty to warn by telling that third person of the dangers of the product or of precautions required. Among the factors which the courts will consider are: (1) the likelihood or unlikelihood that harm will occur if the vendor does not pass on the warning to the users, (2) the trivial or substantial nature of the probable harm, (3) the probability or improbability that the particular purchaser will pass on the warning, (4) and the ease or burden involved in having the manufacturer give the warning to the ultimate user.[92]

\* \* \*

How far into the chain of commerce does a manufacturer have to send his product warnings? What if the product is sold in different types of packages or containers; what if it changes its form at different points of time in commerce? Does a manufacturer have to supply labels for each new form the product takes? What if he has good reason to believe that since the customer was a knowledgeable industrial user of the product that the customer should provide warnings to his employees? This is a particularly difficult and expanding area of the law.

In *Walker v. Stauffer Chemical Corporation*,[93] the defendant manufactured bulk sulfuric acid. After that product left the defendant it was compounded into a product by a customer, packaged and subsequently sold, through distributors, to a manufacturer of drain cleaning compounds. The compounding process involved a change in the chemical composition of the bulk acid; the acid was substantially altered not only as to its chemical composition but also as to the containers in which it was distributed. In this case the ultimate product was in no way considered to be the one in the same sulfuric acid which was distributed by Stauffer. The case arose as a result of injuries sustained from an explosion of the drain cleaner. The question before the court was whether Stauffer, the manufacturer of bulk sulphuric acid, was required to transmit warnings of the explosion potential of the cleaning compounds. The court in this case said:

> We do not believe it realistically feasible or necessary to the protection of the public to require the manufacturer and supplier of a standard chemical ingredient such as bulk sulfuric acid, not having control over the subsequent compounding, packaging or marketing of an item eventually causing injury to the ultimate consumer, to bear the responsibility for that injury. The manufacturer or seller of the product causing the injury is so situated as to afford the necessary protection.[94]

Seldom are the facts of litigated cases quite as clear as they appear to have been in *Walker*. A number of cases turn on the issue of the degree of reliance which a manufacturer may place upon the knowledgable supervisory personnel of the plaintiff. Cases have allowed companies to rely upon the fact that the customer was an industrial user and would pass warnings on to employees;[95] but, more recently distinctions are being made in the type of situations in which those assumptions are entertained. Where there is good evidence that the use of an item is to be closely directed by specialized personnel, such as technicians and engineers, as well as evidence of the fact that such specialized personnel have adequate information regarding the danger involved, courts may not impose an obligation on the manufacturer to see that the warnings are passed on. Such was the case in *Jacobsen v. Colorado Fuel and Iron Corp.*[96] where wires being used to retain concrete molds broke loose as a result of stress and killed a workman. Supervising personnel were seen as having adequate knowledge of the hazards involved. A similar result was reached in *Hopkins v. E.I. Dupont de Nemours & Co., Inc.*,[97] which held that a dynamite manufacturer was not negligent in failing to warn its customers of dangers from the premature explosion of dynamite placed in recently drilled holes before allowing the heat caused by the drilling to escape; the employer was shown to have knowledge of the danger.

But a contrary result was reached in *Jackson v. Coast Paint and Lacquer Co.*[98] where the defendant claimed it had no duty to warn the ultimate user that its paint product's fumes were toxic because the plaintiff's employer already had that knowledge. In rejecting the defendant's argument, the court said:

> At least in the case of paint sold in labeled containers, the adequacy of warnings must be measured according to whatever knowledge and understanding may be common to painters who will actually open the containers and use the paint; the possibly superior knowledge and understanding of painting contractors is irrelevant.[99]

The court then discussed the *Jackson* and *Hopkins* cases and stated:

> There are important distinctions between products such as the dynamite in *Hopkins* or the steel strand in *Jacobson* and the paint involved here. Paint is not a product "the use of which is to be directed by technicians or engineers." Further, it is a product so dispensed that warnings to the ultimate consumer can readily be given.[100]

In the famous *Borel* decision, discussed above, in response to the argument that the insulation contractors should have provided warnings to the plaintiff-worker, the court said:

> We agree with the Restatement: A seller may be liable to the *ultimate* consumer or user for failure to give adequate warnings. The seller's warnings must be reasonably calculated to reach such persons and the presence of an intermediate party will not in itself relieve the seller of its duty. In general, of course, a manufacturer is not liable for miscarriages in the communication process that are not attributable to his failure to warn or the adequacy of the warning. This may occur for example, where some intermediate party is notified of the danger, or discovers it for himself, and proceeds deliberately to ignore it and to pass the product on without a warning. (citations omitted)[101]

Another case, a close fact situation in which a Court of Appeals reversed a lower court decision and in which the manufacturer *was* permitted to rely upon the specialized knowledge of an employer was *Martinez v. Dixie Carriers.*[102] In *Martinez,* because of a limited marketing of a petrochemical transported by barge, the manufacturers were permitted to rely upon the fact that only professionals who were aware of handling precautions required for benzene and other substances would be in contact with the cargo of chemicals. The manufacturer had provided benzene warning cards and product identification to appropriate crew members. The manufacturer was therefore not liable for injuries of a shore-based crew member who was overcome by fumes while stripping the inside of a tank on the cargo vessel.

Another case allowing reliance on supervisory personnel is *Bryant v. Hercules, Inc.*.[103] Summary judgment was entered for the dynamite manufacturer in a strict liability action to recover for eight miners who were killed when fragments from a blast detonated a pile of dynamite that had been left in the vicinity of the blast. There was no duty to warn because the hazard was already known to the miners as well as the supervisory employees in the mine, and the manufacturer had in fact warned that possible explosives should be kept out of the line of a blast.

Perhaps the most important recent case in this area of the law is *Shell Oil Co. v. Gutierrez*,[104] which illustrates several principles, including the principle that a manufacturer's lack of access to the final form in which the product reaches the user is simply one of the considerations bearing on the existence and extent of the manufacturer's duty. In this case, labels were held deficient in that they failed to provide information on how to dispose of empty containers of xylene–a flammable liquid with explosive vapors.

Shell sold xylene in tankcars and trucks with labels marking the contents as "flammable–keep away from heat, sparks, flame." The xylene was sold to a distributor who stored it in tanks and then packaged it in 55 gallon drums. The distributor's labels marked the drums as "flammable liquid." A retailer, a Shell jobber, purchased the drums from the distributor and sold them, in this case, to Westinghouse. Westinghouse used the solvent liquid to clean varnish from electrical motors. All the trouble started when an empty drum wound up in the vicinity of a Westinghouse welder. The vapors in the empty drum exploded causing serious injuries.

There apparently was a great deal of testimony on the disposition made of the drums in the Westinghouse workplace. In general, returnable drums were not placed in the area of the plant where the plaintiff did his welding but were stored in a segregated area until they were returned to the distributor or manufacturer. In some unexplained way an empty xylene drum was moved from the segregated area to within a few feet from where the welder was working. There was testimony to the effect that Westinghouse use of empty drums of dangerous liquid for various purposes such as props for scaffolding, trash barrels, and so forth was not unusual among industrial users. One of the the big debates at trial was what knowledge the various parties had of the dangerous propensities of xylene.[105]

One of the distributors had been marketing petroleum products for 27 years; another had been purchasing xylene for several years. Evidence showed that Shell had distributed a great amount of written information to the intermediaries about the chemical properties of its products, including xylene, but xylene was not singled out or distinguished from its other solvents. The intermediaries were not advised of safe procedures for handling the product. Evidence showed that they both knew xylene was flammable and that an empty drum could explode. However, they were not advised to have empty drums returned nor to flush out the empty drums, so no instructions of that type were passed on to Westinghouse. Shell did tell the intermediaries where to obtain labels but not what they should say. The court was impressed with the fact that Shell did not require as a condition of sale that any labels be placed on these drums, and did not follow up after sale to see what, if any, labels were actually used.[106]

To refute Shell's argument that a warning to the distributor is all that can be required, the court quoted a comment to Section 388 of the Restatement (Second):

> Giving to the third person through whom the chattel is supplied all the information necessary to its safe use is not in all cases sufficient to relieve the supplier from liability. It is merely a means by which this information is to be conveyed to those who are to use the chattel. The question remains whether this method gives a reasonable assurance that the information will reach those whose safety depends upon their having it. All sorts of chattels may be supplied for the use of others, through all sorts of third persons and under an infinite variety of circumstances. This being true, *it is obviously impossible to state in advance any set of rules which will automatically determine in all cases whether one supplying a chattel for the use of others through a third person has satisfied his duty to those who are to use the chattel by informing the third person of the dangerous character of the chattel, or of the precautions which must be exercised in using it in order to make its use safe . . . .(Emphasis supplied)[107]*

The plaintiff did not claim that Shell had an absolute obligation to pass information directly to the plaintiff. Among the factors which the court said would be involved were: (1) the likelihood or unlikelihood that harm will occur if the vendor does not pass on the warning to the user, (2) the trivial or substantial nature of the probable harm, (3) the probability or improbability that the particular purchaser will pass the warning, and (4) the ease or burden involved in the manufacturer giving a warning to the ultimate user.[108]

The court felt there was adequate evidence in the record to support a finding that Shell failed to adequately warn of the danger of explosion, the possible precautions, or the type of labeling that would be appropriate.[109]

Shell had objected to the trial court's exclusion of expert testimony as to whether the "flammable liquid" label was adequate, although it allowed testimony on the various propensities of xylene. It was proper, the court said, to exclude testimony by an expert on a matter where the jury is competent to determine the fact in issue, where the facts can be intelligently described to and understood by the jurors so that they can form reasonable opinions for themselves. The court quoted from *Walton v. Sherwin-Williams Co., 191 F.2d 277 at 285–86 (8th Cir. 1951)*

> . . . the adequacy of a set of warnings or directions is not a scientific matter; whether or not a given warning is adequate depends upon the impression that it is calculated to make upon the mind of an average user of the product.[110]

In sum, the jury was as competent as any expert to determine whether the "flammable liquid" label was adequate to convey the hazard of explosion of an empty barrel and to instruct on safe methods of handling the barrel.[111]

One of the intermediary-defendants asserted in its appeal that it had a right to rely upon Westinghouse, the plaintiff's employer, properly instructing and warning its employees. The court rejected the argument stating that "Westinghouse was not engaged in a sophisticated technical speciality where the supervising personnel and the workmen themselves were highly trained experienced professionals with a high degree of expertise involving xylene."[112]

Another grounds of appeal was the argument that the intermediaries were not in fact the cause of the injuries since the injured workman did not read the warning that was given. The court rejected that reasoning, stating

> That the party who is injured might not have read or heeded a warning is not always sufficient to disprove the existence of a causal relationship between the injury and the defect. Adequate warning could have actuated a policy in handling "empties" which would have prevented the accident [citation omitted]. Had the label stated that the barrel should be washed with an inert solvent, stored with bungholes covered, presumably the barrel would not have been where it was.[113]

The court also commented on the physical aspects of the label, such as the conspicuousness, prominence and relative size of the print

> Here, the only label attached to the barrel was small in size, approximately 4" x 4".
>
> The jury could have determined that the physical aspects of this label were not inadequate in light of the foreseeable risk of injury, and that if a larger, more conspicuous label was attached, it would have been seen, read and heeded.[114]

## SUGGESTED "DUTY TO WARN" CHECKLIST

### Overview—The Need For an Information System

Failure to fulfill a duty to warn can have significant consequences for the manufacturer in terms of civil penalty payments, damaged commercial reputation, and adverse publicity in general. The number of issues the manufacturer must consider dictates that he establish some sort of routine–a *hazard communications system*, which assures that he obtain appropriate information on a continuing basis.

Establishing such a routine is not a simple task. If ever there were a corporate activity which involved a multidisciplinary communications challenge it is hazard communications. It is not an exaggeration to suggest that each of the following areas of expertise are likely to be involved in a manufacturer's assessment of his common law responsibility to warn of the risks his products may pose. The potential reasons for their involvement, *suggested* below, are merely *illustrative*.

**1. Advertising:** Claims made in all media and by all methods will have to be designed and reviewed to be sure they are not inconsistent with product risks and to be sure they do not, implicitly or explicitly, give false assurances regarding a product. Representations by salesmen and other manufacturers' representatives should be *pre*viewed and directed to assure consistency with and appropriateness with warnings of risks.

**2. Business and management:** Executive decisions will be required to establish a corporate hazard communication system, to allocate the money and manpower resources necessary to maintain the system and to resolve conflicts between the various disciplines within the system. Executive actions help provide the sense of corporate commitment needed to make the system work.

**3. Customer relations:** Real world "feed back" on how a product does or does not function can be obtained from customer communications, as can new product uses which the manufacturer did not anticipate. Both types of information may be important in updating a hazard communication system.

**4. Distribution:** Factors bearing upon the handling and condition of the product in delivery or transportation may indicate special communications problems such as label durability, visibility, or appropriateness for different environments (heat, pressure, etc.).

**5. Environmental affairs:** Information on environmental as well as health effects of a product should be identified. Degradation properties, bioaccumulation properties, acidity—those and other characteristics may have relevance to a hazard communication.

**6. Epidemiology:** Epidemiological information on workers exposed to a product during its manufacture may provide insights for hazard communication purposes. Epidemiological studies by others on individual substances in a product may also be important.

**7. Governmental affairs:** While this category of activity may cover different functions in different companies it usually involves personnel who are in a position to obtain current information on government and private research studies and similar information regarding the possible effects of various chemical substances and products.

**8. Health and Safety affairs:** As in 7, above, this category of activity includes persons in a position to obtain health effects information from government and private sources, and to alert the manufacturer to substances of concern to government authorities.

**9. Industrial hygiene:** The manufacturer's own industrial hygiene practices may have to take his product risks into account on a current basis.

**10. Industry intelligence; industry practice:** Manufacturer's representatives who have access to general industry information–through the exchange of regular commercial information, or through industry associations, symposia, seminars and similar communications should be linked into a manufacturer's hazard communication system because they may provide important information on the industry's hazard warning practices.

**11. Insurance:** Insurance against product liability has dramatically increased as a cost factor for manufacturers in recent years and appropriate budgeting and planning will be needed and a review of policy coverage will have to be conducted periodically.

**12. Inventory control:** Costs of initial hazard warning efforts and the timing, costs, and nature of revisions in those warnings will be important factors for the manufacturer to consider. One of the things he may want to analyze is his in-house or contractual supply of such things as labels and safety data sheets, the product inventory which is already labeled and yet to be labeled, and the estimated times and costs for total or partial revision of the label or other materials.

**13. Labor relations:** The manufacturer will want to establish appropriate communications with and procedures for those of his workers who come into significant exposure with products or substances posing risks.

**14. Law:** Communication with legal counsel should be maintained on a continual basis. Rapidly developing case law which may be relevant to the manufacturer's duty to warn should be analyzed and existing and revised hazard communications and procedures should be reviewed on a routine basis. Counsel should also advise how a document retention program can be utilized to establish a defense of the manufacturer's actions if they are challenged.

**15. Library sciences:** The manufacturer's access to subscriptions, interlibrary loans, and various publications should be evaluated and organized to provide appropriate reference material for use in hazard communication assessments.

**16. Medicine:** The latest medical information on the effects of substances on human health will be vital to the manufacturer's assessment of his duty to warn.

**17. Packaging:** Continual review of packaging design will be necessary to assure that such characteristics as durability, visibility and shape do not detract from the effort to communicate hazard and instructional information.

**18. Printing:** Costs and availability of printing and graphics services, in-house or by contract, will have to be evaluated for efficiency, cost, timing, and dependability of service.

**19. Purchasing:** Information obtained on the health and environmental effects of substances purchased as inputs to manufacturing or processing operations should be continually reviewed in order to update hazard communications.

**20. Quality assurance:** Product changes necessitated by quality assurance constraints should be reviewed for their possible consequences for the health or environmental effects of the product.

**21. Records retention and paper flow:** Record retention policies will have to be established and, as the nature of hazard communications change, revised. The nature and the timing of the routing of internal corporate communications will have to be reviewed so that those persons with ultimate hazard communication responsibility receive appropriate information in a timely fashion.

**22. Research and Development:** Research and development, planning and operation, will obviously be a vital component in a manufacturer's assessment and monitoring of his hazard communication responsibilities.

**23. Sales:** While advertising communications, as noted above, are important aspects of a manufacturer's assessment of his duty to warn, it may also be important to instruct sales representatives regarding a proper concern for their descriptions of potential product uses. Sales representatives may also provide important feed back regarding new uses customers find for a product and the new risks which may be involved.

**24. Training:** Some product risks may be such that the manufacturers will want to give workers or customers special training in the safe handling of the product.

**25. Transportation:** Special conditions involved in the transportation of a product may require specialized hazard communications; changes in those conditions should be monitored.

**26. Toxicology:** Toxicological information regarding a product will obviously be of crucial importance to the manufacturer.

With so many corporate activities potentially involved in hazard communications, the prudent manufacturer will establish a corporate communication *system* which *routinizes* the *generation* and *retrieval* of information needed to develop appropriate hazard communications. Of course, the sytem should be tailored to each manufacturer but it should be designed to provide information for initial and revised communications.

## Compile Health And Environmental Effects Information On The Product

The first step in designing a hazard communication system is to identify the actual and potential effects your products may pose to health or the environment. There are a wide range of sources one should consult for this information, including:

(1) *Internal company information* such as that produced as a result of research and development or product tests;

(2) Reputable *scientific publications* of professional societies and acdemic institutions or other learned journals;

(3) *Intelligence* obtained from various industry communications, meetings and publications;

(4) Information provided by *suppliers,* voluntarily or on your request;

(5) Information on *product performance* or characteristics developed from customers, distributors, or transporters;

(6) Allegations regarding products similar to one's own or its constituents in products liability or related *litigation;*

(7) New information generated in *government publications;*

(8) Information shared at symposia, workshops and *seminars.*

## Identify Mandatory Requirements

Another early step in designing a hazard communication system is to identify all of the mandatory requirements related to your product by federal, state and local authorities (and, of course, applicable international requirements) since those requirements will establish your minimum obligations. At the federal level, the following sources should be reviewed to determine their applicability to your products.

• *The Federal Hazardous Substances Act.*
  15 U.S.C. §§1261–1274;

• *Federal Environmental Pesticide Control Act of 1972*
  (formerly the Federal Insecticide, Fungicide and Rodenticide Act). 7 U.S.C. §136;

• *Hazardous Materials Transportation Act.*
  49 U.S.C. §1801–1812;

- *Toxic Substances Control Act.*
  15 U.S.C. §§2601–2629:

- *Occupational Safety and Health Act of 1970.*
  29 U.S.C. §651 *et seq.*;

- *Consumer Product Safety Act.*
  15 U.S.C. §2051 *et seq.*;

- *Food, Drug and Cosmetic Act.*
  21 U.S.C. §§301 *et seq.*; and the

- *Resource Conservation Recovery Act.*
  49 U.S.C. §§6901–6987.

## Identify Industry Customs and Practices

For products similar to one's own, and for its constituents, one should be aware of the hazard warning practices which have in fact been employed by other members of industry. This would include not only voluntary standards adopted or recommended by private organizations but existing practices whether or not they have been identified as customary. These should be reviewed to determine whether they should be regarded as *de facto* minimum standards.

## Identify Foreseeable Uses and Misuses Of The Product

The manufacturer should consider the manner in which his product *might reasonably be used* in the workplace in distribution, storage, transportation, and use. He will have to imagine the various scenarios in which the product's potential risks might become realized and identify the exact nature of the precautions required to prevent the hazard and the remedial steps which should be taken if it occurs. As a reviewing exercise, he should consider what, in the use of his product can/can not happen, may/may not happen, should/should not happen, must/must not happen, will/will not happen, as the product moves in the workplace and in commerce.

## Identify The Primary and Supplementary Methods of Communication Available

The manufacturer should consider the various methods of communication available, e.g., the label itself, safety data sheets, so-called cargo cards or similar transportation-related documents, package inserts, and so forth.

## Consider the Efficiency of the Specific Messages of Warning

The manufacturer should evaluate all aspects of the warning in terms of its efficiency as a communications effort. He should prioritize the warnings, giving greatest emphasis to the greatest risks. He should consider the ability of the possible audiences to receive the communication–e.g., illiterate or non-English speaking; supervised or unsupervised; handicapped; colorblind. He should consider environmental constraints on the message, e.g., poor lighting; stacked packing or storing; an outdoor environment or special physical factors which may effect durability such as heat, cold, pressure, rain, sun, etc.; or visibility. He should consider whether the words, format, print and design of the communication are accurate, conspicuous, unambiguous, appropriately vivid, relevant and legible.

## Use Common Sense

The manufacturer should also apply common sense to his hazard communication process. He must avoid placing so much verbal clutter on a label that it becomes difficult to understand or causes a disincentive to even read it. On the other hand, the label must contain enough information to warn and instruct a user of the product in an intelligent fashion and address the range of audiences likely to be involved.

## CONGRESSIONAL INITIATIVES IN THE PRODUCT LIABILITY ARENA

Of considerable interest to manufacturers will be the Product Liability Risk Retention Act of 1981 which passed the U.S. House of Representatives (as H.R. 21201) on July 28, 1981 and the U.S. Senate, in identical form (as S.1096) on September 11, 1981. The legislation, perhaps inspired by cases such as *Sindell,* discussed previously, was introduced by Senator Robert W. Kasten (R.-Wis.). Among other things, it removes numerous state insurance regulation barriers to the formation of self-insurance groups, and enables product sellers to get together and buy commercial liability policies on a group basis.

Of interest over the next months, perhaps years, will be the proposed Product Liability Act prepared by the Senate Commerce Committee. The central thrust of the bill is to create a *federal* product liability law which would *preempt* state liability statutes but *retain* the jurisdiction of the state courts.

## COLLATERAL ISSUES

Two other related areas of law are also expanding and should be taken into account by the manufacturer.

The first relates to a corporation's liability for the product liabilities of a company it purchases.

The so-called "line of product" theory has been adopted by the California Supreme Court in *Ray v. Alad Corp.,*[115] which holds that where the successor corporation acquires all or substantially all of the assets of the predecessor corporation for cash and continues essentially the same manufacturing operation as the predecessor corporation, the successor remains liable for product liability claims of its predecessor corporation. The theory has been applied even though the successor corporation expressly disclaims liability during the purchase transaction.

The same theory was followed in the recent New Jersey Supreme Court decision in *Ramirez v. Amstead Industries, Inc.*[116] in the process of which the court overruled the principal New Jersey case on corporate successor liability,

*McKee v. Harris-Seybold Co.*[117] To hold otherwise, the court said, would be inconsistent with the developing principles of strict products liability and unresponsive to the interests of persons injured by defective products in the stream of commerce.

The court acknowledged the adverse effects its ruling would have on the ability of small firms to obtain fair price transfers of their enterprises. It decided, however, that risk spreading and cost avoidance over time would become the norm for corporate planning.

In *Ramirez,* the defective product had been made in the *1940s.* The Court indicated that inequities in the situation would have to be addressed by the legislature.

Another area of interest will be developments in the law of insurance. In *Insurance Co. of North America v. Forty-Eight Insulations*[118] the Sixth Circuit held that, under applicable Illinois and New Jersey laws, insurance companies whose policies covered "bodily injuries" would be obligated to pay for insurance for time periods measured from the time of tissue damage caused by the inhalation of asbestos, rather than from the date of a manifestation of injury. It also held that, where a manufacturer was uninsured during a portion of the time that plaintiffs, suffering from a progressive lung disease, were exposed to asbestos products, the manufacturer would be required to bear a pro-rata share of the costs of defending the litigation.

## FOOTNOTES

1. See *Hall v. E.I. DuPont de Nemours & Co., Inc.,* 345 F. Supp. 353, 369, (1972) citing *Passwaters v. General Motors Corp.,* 454 F.2d 1270, 1275, n.5 at 1276 (8th Cir. 1972) as well as Noel, "Defective Products: Abnormal Use, Contributory Negligence, and Assumption of Risk", 25 *Vand. L. Rev.* 93, 128 (1972).
2. *Shell Oil Co. v. Gutierrez,* 581 P.2d 271, 279 (Ariz. App.) (1978).
3. *Borel v. Fibreboard Paper Products Corp.,* 493 F.2d 1076, 1092 (5th Cir. 1973), *Cert. den.,* 419 U.S. 869 (1974).
4. *CCH Products Liability Reports,* Vol. 1, p. 751.
5. *Id.*
6. *Black's Law Dictionary.*
7. Prosser, *Law of Torts* (4th ed. 1971).
8. *Id.*
9. *Id.*
10. A third *non-tort* theory of liability could be included in this discussion–"breach of warranty," which has aspects of both "torts" and "contracts". However, in recent years, changes in the law of strict liability have made it much easier for a plantiff to plead negligence and strict liability than breach of warranty. Since that form of pleading is no longer favored and is primarily a question of pleading strategy than a matter of practical consequence for manufacturers, discussion of it has been omitted.
11. The ongoing effort to prepare "restatements" of law began as a result of a grant to the American Law Institute in 1952 from the A. W. Mellon Educational and Charitable Trust of Pittsburgh. Among the fields of law embraced in this effort are agency, trusts, conflict of laws, and contracts as well as torts. Extensive efforts in the expanding field of torts has produced *two* "restatements".
12. Restatement (Second) of Torts, Section 282.

13. *Id.*Section 283.
14. *Id.*Section 284.
15. *Id.* Section 388, which states:

> One who supplies directly or through a third person a chattel for another to use is subject to liability to those whom the supplier should expect to use the chattel with the consent of the other or to be endangered by its probable use, for physical harm caused by the use of the chattel in the manner for which and by a person for whose use it is supplied, if the supplier:
>
> (a) knows or has reason to know that the chattel is or is likely to be dangerous for the use for which it is supplied, and
> (b) has no reason to believe that those for whose use the chattel is supplied will realize its dangerous condition, and
> (c) fails to *exercise reasonable care to inform them of its dangerous condition or of the facts which make it likely to be dangerous.* (Emphasis added.)

16. *Id.* Section 395.
17. *Greenman v. Yuba Power Products, Inc.,* 377 P.2d 897 (1963).
18. Restatement (Second) of Torts, § 402 (A) and Comment (j).
19. *Hall v. E.I. DuPont de Nemours & Co., Inc. supra,* n.1 at 368–369.
20. *Id.* at 369; and see Restatement (Second) of Torts, Section 402A, comment c which states:

> (c) On whatever theory, the justification for strict liability has been said to be that the seller, by marketing his product for use and consumption, has undertaken and assumed a special responsibility toward any member of the consuming public who may be injured by it; that the public has the right to and does expect, in the case of products which it needs and for which it is forced to rely upon the seller, that the reputable sellers will stand behind their goods; that the public policy demands that the burden of accidental injuries caused by products intended for consumption be placed upon those who market them, and be treated as a cost of production against which liability insurance can be obtained; and that the consumer of such products is entitled to the maximum of protection at the hands of someone, and the proper person to afford it are those who market the products.

21. *Collins v. Rowe,* 428 S.W. 2d 194 (Ky Ct. App. 1968).
22. *Trimble vs. Irwin,* Product Liability Reports ¶6073, 441 S.W. 2d 818 (Tenn. Ct. App. 1968).
23. See Product Liability Reports ¶436a.
24. Restatement (Second) of Torts, Section 402A, note n.
25. *Id.*
26. Restatement (Second) of Torts, Section 433(B). subsection (3).

> This theory is illustrated by the famous case of *Summers v. Tice,* 33 Cal.2d 80, 199 P.2d 1 (1948).
> In *Summers,* two hunters fired in the plaintiff's direction; one caused the plaintiff's injuries but it was not possible to tell which one. Since both hunters were negligent, the court shifted the burden of proof to the defendants, telling each to try to absolve himself. That case was subsequently embodied in Restatement (Second), Section 433B, subsection (3), which provides:
>
>> Where the conduct of two or more actors is tortious, and it is proved that harm has been caused to the plaintiff by only one of them, but there is uncertainty as to which one caused it the burden is upon each actor to prove that he has not caused the harm.

27. *Id.* comment f.
28. *Id.* Section 876.
29. Prosser, *supra,* see 49, p. 292.
30. See *Sindell v. Abbott Laboratories,* 607 P.2d 924, 934 fn. 22 ($980).
31. *Hall v. E.I. DuPont de Nemours & Co., Inc., supra,* n.1 at 353.
32. *Id.* at 374.
33. *Id.*
34. *Id.*
35. *Id.* at 377.
36. *Id.* at 378.
37. *Sindell v. Abbott Laboratories, supra,* n.30 at 924.
38. *Id.* at 938.
39. *Id.* at 936.
40. *Id.*
41. *Id.*
42. See the extensive discussion of *Sindell* and the enterprise liability theory in 46 *Fordham Law Rev.* 963.
43. See *Payton v. Abbott Labs.,* 512 F.Supp. 1031 (1981), denying recovery under Massachusetts law based on a similar theory because plaintiffs had not established that an agreement had existed between the defendants not to properly test or warn about DES, and they had not assisted each other in such activities as marketing of the drug.

    In *Hardy v. Johns-Manville Sales Corp.,* 509 F.Supp 353(1981), the Eastern District of the District Court for the Eastern District of Texas, ruling on motions related to discovery and collateral estoppel, announces an intention to adopt some sort of *Sindell* liability. The court granted discovery motions targeting market share.

    The *Sindell* approach was rejected in *Ryan v. Eli Lilly & Co.* 514 F.Supp. 1004 (1981). An enterprise liability theory was adopted in *Ferringno v. Eli Lilly and Co.,* 175 N.J. Super. 551, 420A.2d 1305 (1980).
44. "Impact of California Court Ruling In DES Suit Seen Staggering", *Journal of Commerce,* April 18, 1980.
45. *Borel v. Fibreboard Paper Products Corporation, supra,* n.3, pp. 1076, 1089.
46. *Id.*
47. *Insurance Co. North America v. Forty-Eight Insulations,* 633 F.2d 1212, 1215 (1980).
48. *Borel v. Fibreboard Paper Products Corporation, supra,* n.3 p. 1092.
49. *Id.*
50. *Id.*
51. See 1 Frumer & Friedman, Products Liability, § 6.01 [1] and cases cited; Noel, "Manufacturer's Negligence of Design or Directions for Use of Product", 71 Yale L. J. 816, 853 (1962). *See also Roginsky v. Richardson-Merrill, Inc.,* 2 Cir. 1967, 378 F.2d 832; *Tinnerholm v. Parke-Davis & Co.,* S.D.N.Y. 1968, 285 F.Supp. 432, aff'd., 411 F.2d 48; *Schenebeck v. Sterling Drug, Inc.,* D.D.Ark. 1968, 291 F.Supp. 368.
52. *Id.* pp. 1089–1090.
53. *Id.* at 1093.
54. *Karjola v. Johns-Manville Products Corp.,* 523 F.2d 155 (8th Cir. 1975).
55. *Id.* at 158.
56. *Toole v. Richardson-Merrill, Inc.,* Products Liability Reporter 5814, (Cal. Ct. App. 1967), 60 Cal. Rptr. 398.
57. *Barfield v. Atlantic Coastline R.R.,* 197 So.2d 545 (1967).

58. *Chapman Chemical Co. v. Taylor*, 222 S.W. 2d 820 (1949).
59. *Sterling Drug. Inc. v. Cornish*, 370 F.Supp. 82 (1966).
60. *Moran v. Faberge, Inc.*, 332 A.2d 11 (Md. 1975) and *Hall v. E.I. DuPont de Nemours & Co., Inc.*, supra n.1, 353, 361.
61. *Hall v. E.I. DuPont de Nemours & Co., Inc.*, supra at 367.
62. *Moran v. Faberge Inc.* supra.
63. *Pease v. Sinclair Refinery Co.*, 104 F.2d 183 (2d. Cir. 1939).
64. *Croteau v. Borden Co.*, 395 F.2d 238 (2nd Cir. 1968)., only the District Court opinion provides helpful analysis; see 277 F.Supp. 945 (E.D.Pa. 1968).
65. *Garner v. Q.H.S., Inc.*, 448 F.2d 2343 (4th Cir. 1971).
66. *Id.* at 243.
67. *Collins v. Rowe*, 428 S.W.2d 194 (Ky. C.A. 1968).
68. *Steagall v. Dott Mfg. Corp.*, 446 S.W. 2d 515 (Tenn.S.Ct. 1969).
69. *Barrett v. Pablis Powder Co.*, 86 Cal. App. 3d 560 (1978).
70. *LaPlant v. E.I. DuPont de Nemours & Co., Inc.*, 346 S.W. 2d 231.
71. *Cooley v. Quick Supply Co.*, 221 N.W. 2d 763 (Iowa, 1974).
72. *Hall v. E.I. DuPont de Nemours & Co., Inc.*, supra n.1.
73. *Id.*
74. *Griffin v. Planters Chemical Corp.*, 302 F.Supp. 937, 944, (D.S.C. 1969).
75. Restatement (Second) of Torts, Section 402A, Comment J.
76. *Simonetti v. Rinshed-Mason Co.*,200 N.W. 2d 354 (1972).
77. *Tuscon Industries, Inc. v. Schwartz*, 501 P.2d 936 (1973).
78. *Hubbard-Hall Chemical Co. v. Silverman*, 340 F.2d 402 (1st Cir. 1965).
79. *Ziglar v. E.I. DuPont de Nemours & Co.*, No. 79 CVS 286, (N.C. Court of Appeals decided July 21, 1981).
80. *Edwards v. California Chemical Company*, 245 So.2d 259 (1971).
81. *Id.* at 265.
82. *Davis v. Wyeth Laboratories, Inc.*, 399 F.2d 121 (9th Cir. 1968).
83. *Id.* at 130; and see *Sterling Drug Inc. v. Cornish*, 370 F.2d 82 (8th Cir. 1967).
84. *Id.* at 129–130.
85. *Charles Pfizer & Co., Inc. v Bouch*, Products Liability Reporter ¶5170, 365 S.W. 2d 832. (Tex. Ct. Cir. App. 1964).
86. *Gonzalez v. Virginia Chemical Co.*, Products Liability Reports ¶5338, 39 F.Supp. 567. (D.C.S.C. 1965).
87. *Daniels v. The Atlantic Refining Co.*, 295 Fed. Supp. 125. (1968).
88. *D'Hehouville v. Pioneer Hotel Co.*, (9th Cir. 1977) Products Liability Reporter ¶7954.
89. *Thomas v. Arvon Products Company, Inc.*, 227 A.2d 897 (Pa.1967).
90. *Rumsey v. Freeway Manor Minimax*, 423 S.W.2d 387 (1968); *Griffin v. Planters Chemical Corporation*, 302 F.Supp. 937, (D.C.S.C. 1969).
91. *Gonzalez v. Virginia-Carolina Chemical Co.*, 239 F.Supp. 567 (E.D.S.C. 1965).
92. *Shell Oil Co. v. Gutierrez*, supra, n.2, 278; *Cooley v. Quick Supply Co.*, supra, n.71.
93. *Walker v. Stauffer Chemical Corporation*, 96 Cal. Rptr. 803, (Cal. Ct. App. 1971).
94. *Id.*
95. *Younger v. Dow Corning Corporation*, 451 P.2d 177, 184 (Kansas 1969).
96. *Jacobsen v. Colorado Fuel and Iron Corp.*, 409 F.2d 1263 (9th Cir. 1969).
97. *Hopkins v. E.I. DuPont de Nemours & Co., Inc.*, 212 F.2d 623 (3rd Cir. 1954).
98. *Jackson v. Coast Paint and Lacquer Co.*, 499 F.2d 809 (9th Cir. 1974).
99. *Id.* at 812, 813.
100. *Id.* at 813, 914.
101. *Borel v. Fibreboard Paper Products Corporation*, supra, n.3., pp. 1076, 1091, 1092.
102. *Martinez v. Dixie Carriers, Inc.*, 529 F.2d 457 (5th Cir. 1976).
103. *Bryant v. Hercules, Inc.*, 325 F.Supp. 241 (D.C. Ky-1970).

104. *Shell Oil Co. v. Gutierrez, supra,* n.2.
105. *Id.* at 276.
106. *Id.*
107. *Id.* at 278.
108. *Id.*
109. *Id.* at 279.
110. *Id.*
111. *Id.*
112. *Id.* at 280.
113. *Id.*
114. *Id.* at 281.
115. *Ray v. Alad Corp.,* 560 P.2d 3 (1977).
116. *Ramirez v. Amstead Industries, Inc.,* No. A-12. (N.J.S. Ct., decided June 18, 1981).
117. *McKee v. Harris-Seybold Co.,* 118 N.J. Super. 480 (1972).
118. *Insurance Co. of North America v. Forty-Eight Insulations,* 633 F.2d 1212 (6th Cir. 1980).

# 9

# Patents, Trade Secrets and Trademarks

**James Toupin**
*Covington and Burling*
*Washington, DC*

This chapter deals principally with the laws of patents, trade secrets and trademarks as they affect the marketing of new products. Since the issuance of a patent largely obviates the need for secrecy over proprietary information contained in a patent application, the interrelated laws of patents and trade secrets have a major impact on the information companies reveal in marketing products. This chapter outlines the basic principles of the laws of patents and trade secrets and their impacts on labelling. Greatly simplified, the law of trademarks protects the good will associated with product brand names. This chapter will discuss the legal considerations in the choice and use of marks. It will also in passing deal with the law of copyrights as it affects particular marketing problems raised in the discussion of the other subjects this chapter covers.

## PATENTS

This section first outlines the basic provisions of the patent law, summarizing the nature of the protection a patent provides and the statutory provisions governing what inventions can receive patent protection. This presentation ends with discussions of a few of the recently controversial legal issues about the patentability of certain subject matters that may be of interest to the chemical industry—chemicals with obvious structures but unexpected properties, experimental chemicals and processes and chemicals requiring approval by agencies other than the United States Patent and Trademark Office before distribution, microorganisms created through genetic engineering, and computer programs used in conjunction with industrial processes. Second, this section presents a basic discussion of the disclosures that patent applicants must make and their implications for the retention of trade secrets. The third subsection discusses the uses of patent searches for companies considering new products. Fourth, the relationship of U.S. patent protection to foreign patents will be considered. The fifth subsection concerns the impact of the patent laws on the labelling of nonpatented products. The sixth discusses the specific requirements of the Patent Act for the marking of patented products.

## Basic Coverage of the Patent Act

A patent issued by the United States Patent and Trademark Office can grant the patent owner substantial economic advantages. Section 154 of the Patent Act sets forth in brief the scope of the patent right:

> Every patent shall contain . . . a grant to the patentee, his heirs or assigns, for the term of seventeen years, . . . of the right to exclude others from making, using or selling the invention throughout the United States . . . [35 U.S.C. § 154.]

The Act provides extensive remedies against those who infringe this exclusive right. Under § 283, courts may issue injunctions against infringements of rights secured by patents. Section 284 provides for damages for infringement, in no event less than a reasonable royalty for use of the invention by the infringer, together with interest and costs. The court may increase the damages by up to three times the amount of damages the court finds for the patent owner.

These remedies are available not only against persons who make, use or sell the patented invention in the United States, but also against those who actively induce infringement. 35 U.S.C. § 271(b). Moreover, contributory infringers are also subject to the remedies of the Act. In essence, a contributory infringer is one who sells to another an item specially suited for use in an infringing process or product, with knowledge that the item is specially suited for use in the infringement and is not a "staple" item suitable for a substantial noninfringing use. See *Label Licenses and Patent Misuse* later in this chapter.

Broad as these remedies can be, they are subject to substantial uncertainties. First, the issuance of a patent does not establish once and for all the patent owner's exclusive rights to the invention. Under § 282, a patent is only presumed to be valid. An accused infringer may defend a patent infringement suit not only by claiming noninfringement, but also by establishing in court that, despite the U.S. Patent and Trademark Office's action, the patentee was not entitled to the patent in the first place. In 1979, the acting Commissioner of Patents and Trademarks reported that, of the patents as to which courts had reached decisions holding them valid or invalid (a very small percentage of the patents actually issued), 55 percent had been held to be invalid. 989 O.G. 2 (December 4, 1979). Moreover, owners of valid patents can find themselves barred from enforcing their patent rights when they have misused their patents by seeking to use the patent to gain monopolistic advantages over unpatented products or have used their patents in committing an antitrust violation. See *Morton Salt Co.* v. *Suppiger Co.,* 314 U.S. 488, 52 U.S.P.Q. 30 (1942).

**Basic Requirements for Patentable Inventions:** Section 101 of the Patent Act[1] gives the basic definition of patentable inventions as follows:

> Whoever invents or discovers any new and useful process, machine, manufacture, or composition of matter, or any new and useful improvement thereof, may obtain a patent therefor, subject to the conditions and requirements of this title. [35 U.S.C. § 101.]

Although, as discussed below, there have been disputes as to the patentability of some inventions, the basic definitions of patentable inventions in the chemical area are fairly clear. The phrase "composition of matter" in § 101 relates primarily to chemical compositions. Patentable compositions of matter include chemical compounds and mixtures of ingredients. New methods of producing known products may be patentable as "processes". A 'process' also "includes a new use of a known process, machine, manufacturing composition of matter, or material." Thus, new uses of known chemicals are patentable as processes. Both DDT and propanil are among the many known chemicals for which process patents have issued.

**Novelty and Nonobviousness:** To be patentable, an invention must be new, useful and not obvious. Section 102 of the Act defines novelty quite technically. An invention is not novel if, before the applicant's invention, the invention was (a) known or used by others in this country (§ 102(a)); (b) patented or described in a printed publication anywhere in the world before the applicant's invention (§ 102(b)); (c) described in a patent on a United States patent application filed before the applicant's own invention (§ 102(e)); or (d) made in this country by another who has not abandoned, suppressed or concealed it (§ 102(g)).

The Act also contains incentives for prompt filing of patent applications. An invention becomes unpatentable, even if the applicant was the first to invent it, if the invention was patented or described in a printed publication anywhere in the world, or was in public use or on sale in this country, more than one year prior to the date of the United States patent application. 35 U.S.C. § 102(c).

Section 103 of the Act requires that an invention described in a patent application must not "have been obvious at the time the invention was made to a person having ordinary skill in the art to which [the] subject matter pertains." The fact that a company's experts may think that a new substance or process is highly novel does not necessarily exclude the possibility that the substance or products may not be "obvious". The Patent and Trademark Office, and courts in potential infringement suits, assume an encyclopedic awareness of technical literature and product developments that no individual expert may have. The courts and the U.S. Patent and Trademark Office, in determining whether an invention is obvious, first establish the scope of the prior art, then the differences between the prior art and the patent claim, and finally the level of ordinary skill in the art. *Graham* v. *John Deere Co.*, 383 U.S. 1 (1966).

**Special Patentability Questions:** *New but Obvious Chemicals with Nonobvious Properties*—Many of the crucial problems of defining patentable subject matters affect the chemical industry especially. The patentability of chemicals with obvious structures but with unobvious properties is a troublesome question on which the courts differ. In the landmark case of *In re Papesch,* 315 F.2d 381, 137 U.S.P.Q. 43 (C.C.P.A. 1963), the Court of Customs and Patent Appeals held that a homologue of an already-known chemical could be patented if the homologue has useful and nonobvious properties. The court reasoned that although the new compound was structurally obvious, its properties were sufficiently novel to warrant patenting. Patentability, the court held, should depend not on the similarity or difference of a compound's formula to the formulas of previous compounds, but on the similarity or difference of the properties of the compounds. The court noted that, but for the discovery of new properties, no one might have made the structurally obvious chemical. The Court of Customs and Patent Appeals has since extended *Papesch,* and most other courts have followed the *Papesch* decision. See *Eli Lilly and Co.* v. *Premo Pharmaceutical Laboratories, Inc.,* 630 F.2d 120, 207 U.S.P.Q. 719 (3d Cir.), *cert. denied,* 101 S. Ct. 573 (1980); *Eli Lilly and Co.* v. *Generix Drug Sales, Inc.,* 460 F.2d 1096, 174 U.S.P.Q. 65 (5th Cir. 1972); *Commissioner* v. *Deutsche Gold-und-Silber-Scheideanstalt Vormals Roessler,* 397 F.2d 656, 157 U.S.P.Q. 549 (D.C. Cir. 1968); *In re Stemniski,* 444 F.2d 581, 170 U.S.P.Q. 343 (C.C.P.A. 1971); *In re Ackermann,* 444 F.3d 1172, 170 U.S.P.Q. 340 (C.C.P.A. 1971).

However, some courts in infringement cases have challenged the validity of the reasoning in *Papesch.* They urge that granting patents for homologues, analogues, isomers and other chemicals for formulas that could be anticipated on the basis of previous knowledge gives an inventor too many rights, since he not only gains the exclusive right over the unexpected uses of the compound, but also over uses that would have been predictable on the basis of the prior art. These courts hold that the inventor should only receive a process patent for the new use of the chemical. *Carter-Wallace, Inc.* v. *David-Edwards Pharmacal Corp.,* 341 F. Supp. 1303, 173 U.S.P.Q. 65 (E.D.N.Y.), *aff'd on other grounds sub nom. Carter-Wallace, Inc.* v. *Otte,* 474 F.2d 529, 176 U.S.P.Q. 452 (2d Cir. 1972), *cert. denied* 412 U.S. 929 (1973); *Monsanto Co.* v. *Rohm & Haas Co.,* 312 F. Supp. 778, 164 U.S.P.Q. 556 (E.D. Pa.), *aff'd on other grounds,* 456 F.2d 592, 172 U.S.P.Q. 323 (3d Cir.), *cert. denied,* 407 U.S. 934, *U.S. reh. denied,* 409 U.S. 899 (1972).

*Experimental Chemicals and Processes and Products Needing Approval by Other Agencies before Marketing*—To be patentable, inventions have to be useful. U.S. Const. Art. 1 § 8; 35 U.S.C. § 101; *see also* 35 U.S.C. § 112. Generally, this requirement does not pose a substantial obstacle to obtaining patents. However, the requirement can be a problem for chemical and drug patent applicants.

In the landmark case of *Brenner* v. *Manson,* 383 U.S. 519 (1966), the Supreme Court stated that a new process for making known steroids was unpatentable, when the patent application did not disclose any use for the steroids except in laboratories that might test them to find uses for them. The Court of Customs and Patent Appeals has applied this decision to the patentability of chemical intermediates, or starting materials, that produce intended products. *In re Joly,* 376 F.2d 906, 153 U.S.P.Q. 243 (C.C.P.A. 1967); *In re Kirk,* 376 F.2d 936, 153 U.S.P.Q. 266 (C.C.P.A. 1967). The application for patents for new, nonobvious intermediates or processes must show that the resulting chemical compound has particular uses that would be obvious to persons skilled in the art to which the intermediate or process pertains, or the application must reveal the use. It is not enough that inventors of new processes, compounds, or intermediates know what applications experimenters will investigate. Allegations, for instance, that a substance is "biologically active" or that adjacent homologues have certain uses do not sustain a patent claim.

The Court of Customs and Patent Appeals has enunciated this doctrine despite the protests of some of its judges that there is a commercial market for new chemical compounds, intermediates and processes to test whether they have commercial uses. See *In re Kirk,* 376 F.2d 936, 947-68 (Rich, Smith, J.J., dissenting). Inventors or companies that want to market such new products to others for further investigation while keeping the products proprietary must rely on the law of trade secrets. Splitting work on a new product this way can lead to joint ownership of a patent that may ultimately issue on it between the person who first invented the compound or process and the person who found a use for it.

On the other hand, a chemical product need not actually be marketable to be sufficiently useful to be the subject of a patent. While a patent grants the patent owner the right to exclude others from using the invention, a patent does not grant its owner the right to use the invention if it is subject to regulation by agencies other than the U.S. Patent and Trademark Office. For example, while Patent and Trademark Office examiners may inquire whether a product is safe in determining whether it is useful, their inquiry is not as strict as the safety determinations that such agencies as the Food and Drug Administration and the Environmental Protection Agency make when passing on the marketability of new products such as drugs and pesticides. *In re Watson,* 517 F.2d 465, 186 U.S.P.Q. 11 (C.C.P.A. 1975); *In re Langer,* 503 F.2d 1380, 183 U.S.P.Q. 288 (C.C.P.A. 1974); *In re Hartop,* 311 F.2d 249, 135 U.S.P.Q. 419 (C.C.P.A. 1962). As a result, the Patent and Trademark Office can and does issue patents for new chemical products that other federal agencies have determined cannot yet be marketed.

Satisfying the requirements of such statutes as the Toxic Substances Control Act, the Federal Insecticide, Fungicide and Rodenticide Act and the Food, Drug and Cosmetics Act can consume a substantial portion of the seventeen-year patent term. While a company may try to keep new products as trade secrets until they are ready to be marketed, the failure promptly to file patent applications may allow competitors to obtain patents over similar inventions. The company would thereafter have to challenge such patents in court in order to market its own products.

*Microorganisms*—New and useful processes using microorganisms have long been recognized as patentable, as have the new and useful chemical products of such processes. However, since materials "found in nature" have been considered not patentable apart from the novel and nonobvious uses to which inventors put them, microorganisms posed a novel legal problem when, with the use of recombinant DNA, scientists began inventing unobvious, new and useful microorganisms not found in nature. The Supreme Court in *Diamond* v. *Chakrabarty,* 447 U.S. 303 (1980) recently settled some of the confusion in this area by holding that a microorganism produced by genetic engineering (in that case a bacterium used for cleaning oil spills) with markedly different characteristics from any found in nature, and having the potential for significant utility, is patentable as the result of human ingenuity and research. The *Chakrabarty* decision leaves open important questions in this field, since it left unaddressed such questions as whether mutant strains created not by genetic engineering but by the subjection of strains found in nature to mutation-ordering treatment or biologically pure cultures of impure microorganisms found in nature are patentable. Nonetheless, *Chakrabarty* should provide an economic impetus to research in genetic engineering.

*Computer Programs*—The patentability of computer programs (particularly as they relate to industrial processes) has been the subject of several recent Supreme Court cases, which have clarified the law in some respects, but have left major uncertainties. In *Gottschalk* v. *Benson,* 409 U.S. 63 (1972), the Court held that an algorithm for converting numerical information from binary-coded numbers to pure binary numbers for use in programming digital computers was not patentable, on the basis that patenting the mathematical formula would be tantamount to patenting an idea. The mathematical formula had no substantial practical application except in connection with digital computers. In *Parker* v. *Flook,* 437 U.S. 584 (1978), the Court had before it a patent application for a method of updating numerical alarm limits to signal inefficient or dangerous conditions in catalytic chemical conversions of hydrocarbons. The process consisted of measurement of process variables, calculation with an algorithm of an updated alarm limit value, and adjustment of the actual alarm limit. The Court held that since the algorithm was the only novel part of this process, the process was unpatentable.

In an opinion issued March 2, 1981, *Diamond* v. *Diehr,* 450 U.S. 175 (1981), the Supreme Court considered a patent application for a process involving continuous measuring of the temperature inside a press for molding synthetic rubber, feeding the measurements to a computer, which repeatedly calculated the proper cure time, and automatic opening of the press. A divided court, by a 5 - 4 decision, held that this process as a whole was patentable, since, unlike the process in *Gottschalk,* the noncomputer steps of the process were significant. The Court noted that the combination of steps resulted in a product superior to the use of any of the steps apart from the whole sequence. In view of the Court's rejection of patent claims for computer programs alone apart from significantly innovative industrial processes, the developers of computer programs are relying increasingly on copyright and trade secret protection.

## Patent Disclosure Requirements and Trade Secret Protection

Issuance of a patent obviates the legal necessity for keeping the information contained in the patent application secret. The Patent Act requires the patentee to make the knowledge behind the invention publicly available. Under § 112 of the Act, a patent applicant must include in the patent application a written description of the patent that is clear, concise, full and exact enough to enable someone skilled in the most relevant art to make and use the invention without undue experimentation. The patent application must also set forth the "best mode" for carrying out the invention that the inventor contemplates at the time of filing the application. Finally, the patent application must distinctively claim the subject matter that the inventor regards as his invention. Patent applicants must, in essence, define what invention they are claiming in as precise terms as possible so that others may know whether or not what they are doing or plan to do would infringe the patent.

Pending issuance of the patent, an applicant may maintain the information in the application as a trade secret. Until the patent issues, the Patent and Trademark Office keeps all information about the patent application secret. 35 U.S.C. § 122. Applications are not available to the public or to other inventors.[2] However, once the patent issues, the application and supporting documents, as well as Patent and Trademark Office action on the application, are filed in the Patent and Trademark Office search room. Such disclosure ends trade secret protection over the disclosed information.

The disclosures in patents sometimes expand a patentee's commercial opportunities. A patent can expand the possibility for cross-licensing. Thus, for instance, an inventor may obtain a patent for a chemical compound adequately disclosing the invention, how to make and use the invention and the best mode contemplated for its use at the time of filing. Another inventor may, for example, discover an entirely new use or make an improvement in the process which is novel and not obvious, or make a similar advance in the process of creating the compound. These inventions would be patentable by the new inventor. However, that inventor could not make commercial use of his invention without obtaining permission from the holder of the patent for the compound itself. At the same time, by virtue of his patent, the new inventor could prevent the original patentee from engaging in the new mode of use that he has created. In such circumstances, it is common for the new inventor to license the original patentee to use the new process or use in exchange for a license to use the compound. The owner of the patent in the underlying compound, who might not have ever thought of these new uses, obtains the benefit of their discovery.

In some situations, companies may be reluctant to distribute products without a patent. For example, a company may own a new chemical compound that is useful in its own industrial processes, which it would sell for use by the general public if it could retain its proprietary right in the chemical. However, if the chemical is readily analyzable, distribution would destroy the company's trade secret rights. Lack of a patent therefore is impeding sale of the product. The Patent and Trademark Office rules provide that, where lack of a patent is preventing distribution of a product, a patent applicant may apply to have its application "made special" and expedited.

The disclosure requirements also, of course, present risks. For instance, both the patenting of new microorganisms produced by genetic engineering and of processes that use microorganisms pose a difficult trade secret problem for potential applicants. Applicants often cannot describe these microorganisms adequately to meet the Patent Act's disclosure requirements. The U.S. Patent and Trademark Office allows for deposit of the microorganisms in depositories rather than full description in patent applications. However, these depositories, being independent of the Patent and Trademark Office, are not bound by its rules on the confidentiality of pending applications. Thus, if a patent applicant wants to retain trade secret protection for the microorganism in the event a patent does not issue, he should enter into an agreement with a recognized depository to limit access to the microorganisms during the pendency of the application to himself, his designees, and representatives of the Patent and Trademark Office, and to make the microorganism publicly available only if and when a patent issues.

The "best mode" provision of Section 112 may require the patentee to reveal information which it otherwise would hold as a trade secret. Patent protection only extends to the invention set forth in the patent claims. 35 U.S.C. §§ 112, 154. The best mode of using the invention may itself not be patentable, but may be useful commercial information which would be protectable under state trade secret law if kept secret. However, courts often hold patents invalid for failure to disclose the best mode that was being considered for use whether or not the mode had been finally developed at the time the patent application was filed. Patentees may keep secret new modes of use for the invention that they first contemplate after the patent application is filed, even though those new modes may be superior to the best mode contemplated at the time of filing.

A patent application can also reveal sensitive information about which employees of a company have gained specialized information in particular areas. A patent application must generally be filed by the person or persons who invented the subject matter for which the patent is sought. 35 U.S.C. §§ 102(f), 111, 115, 116. Exceptions exist only for very limited circumstances. 35 U.S.C. §§ 116, 117, 118. Thus, by reading patent applications, competitors can often find out who in a given company has contributed to particular projects or is skilled in particular areas. The law of unfair competition provides some protection against competitors hiring each other's employees for the purpose, or with the foreseeable result, of obtaining secret information about the other company or its inventions.

### The Importance of Patent Searches

It is important for companies to keep the patent system in mind, whether or not they choose to seek patents for their own products. When technical personnel create useful products or processes that seem to them to be significant advances on the prior art, the assistance of patent counsel very often becomes useful. First, companies need to determine whether use or sale of these new processes or products might infringe someone else's patent. This may involve a search of the relevant records of the United States Patent and Trademark Office (and of the patent offices of any foreign countries where the products may be used or marketed). Second, if the product seems to technical personnel to be novel and significant of the art, companies may want to consider whether to apply for a patent. In such a situation, a search should be made at the Patent and Trademark Office to determine whether the invention seems likely to be patentable. How extensive a search should be before a patent application is filed, as well as the decision whether to seek a patent at all, may depend upon how valuable the invention seems. For a very important item, patent lawyers may also conduct searches of the relevant trade literature to see whether the invention has been anticipated but not patented. Some companies monitor the

Patent and Trademark Office records both to keep track of commercially useful patents whose exclusive terms have expired and to keep track of new inventions to which the Patent Office grants exclusive rights.

### Protection in Foreign Countries

United States patents afford protection against infringement only in the United States, and the patents issued by foreign countries are similarly limited in territorial effect. As a result, companies that own new inventions should consider patenting their inventions in foreign countries where they may want to produce or sell the patented items or license others to do so. There are some significant differences in the patent laws of foreign countries, of which only a sampling can be discussed here. Some countries, unlike the United States, have requirements that patents must be used. Failure to use patent rights can, under the laws of some countries, result in governments granting compulsory licenses to other parties to use the patented subject matter in those countries. Some countries also require the payment of yearly fees to maintain patents. Failure to pay such fees may cause forfeiture of the patent. In the United States, once the filing fee is paid, no further fees are required for patents, although income generated by use of the patented invention and capital gains from sale on the patent may be taxed.

Under the Patent Cooperation Treaty, the United States has cooperated in coordinating filing procedures and standardizing application formats for specially designated "international patent applications." See 35 U.S.C. §§ 361-68. Moreover, the United States, like other signatories to the Paris Convention for the Protection of Industrial Property, treats applications for patents that were filed in foreign patent offices no longer than one year previously as though they had been filed in the U.S. Patent and Trademark Office at the same time. 35 U.S.C. § 119.

However, an invention must be patented under the United States Patent Act to obtain patent protection in the United States. The United States Patent Act prohibits premature applications for foreign patents for inventions made in the United States. An invention made in the United States cannot be patented here if, before a United States patent is granted, the inventor received a foreign patent on the basis of a patent application filed more than twelve months before the filing of an application for a United States patent. 35 U.S.C. § 102(d). If an application for a foreign patent on an invention made in the United States is filed less than six months after an application for a United States patent is filed (except when a license from the Commissioner of Patent and Trademarks allows the foreign filing), any patent issued by the U.S. Patent and Trademark Office is invalid. 35 U.S.C. §§ 184, 185.

### Label Licenses and Patent Misuse

In addition to more formal licensing, patent holders can license the use of their patented processes or machines by the ways in which they distribute unpatented goods. Thus, courts have held that a patent holder's sale of unpatented goods that have only one use constitutes an implied license to use the products they have bought in that way. *Rohm and Haas Co.* v. *Dawson Chemical Co.,* 599 F.2d 685, 203 U.S.P.Q. 1 (5th Cir. 1979), *rev'd on other grounds,* 448 U.S. 176 (1980); *B.B. Chemical Co.* v. *Ellis,* 117 F.2d 829, 48 U.S.P.Q. 487, (1st Cir. 1941), *aff'd,* 314 U.S. 495 (1942). Likewise, placing on containers of unpatented products instructions telling buyers how to use those products in a patented process grants an implied license to use the products in the manner described on the label. See *United States* v. *Univis Lens Co.,* 316 U.S. 241, 249-51; *Ansul Co.* v. *Uniroyal, Inc.* 306 F. Supp. 541, 565, (S.D.N.Y. 1969), *modified on other grounds,* 448 F.2d 872 (2d Cir. 1971), *cert. denied,* 404 U.S. 1018 (1972). Finally, so-called "label licenses" can be more explicit, stating the rates charged for use of the patented process and conditions for that use.

The patent laws affect the labelling of unpatented goods used in conjunction with patented machinery or processes. This is because the courts have enunciated a doctrine of "patent misuse", which, broadly, prohibits attempts to use patent rights to gain economic advantages over the sale of unpatented goods. A patent holder cannot enforce its rights until the effects of patent misuse are cured.

Under the patent misuse doctrine, courts have put restrictions on the methods of issuing label licenses. The courts have generally held that a patent holder may not license others to use its unpatented products in the patented manner, while refusing to license its patent to parties that purchase the unpatented goods from its competitors. Moreover, patentees have been found to have misused patents by charging higher royalty rates to licensees who use the unpatented products of competitors than to purchasers of their own products. See, e.g., *B.B. Chemical Co.* v. *Ellis, supra; Ansul Co.* v. *Uniroyal Inc.,* 448 F.2d 872 (2nd Cir. 1971), *cert. denied,* 404 U.S. 1018 (1972).

However, the court in *Hall Laboratories, Inc.* v. *Springs Cotton Mills, Inc.,* 112 F. Supp. 29 (W.D.S.C.), *aff'd on other grounds,* 208 F.2d 500 (4th Cir. 1953), upheld a label license that operated under the following conditions:

1. Anyone desiring to use the patented process could do so on payment of a uniform rate per pound of the unpatented product used in the process;

2. The user could purchase the product wherever he chose;

3. If the user chose to purchase the product from the process patentee, the royalty was included in the price and that fact was stated on the label;

4. If the user chose to purchase the unpatented product from another source, he could enter into a license agreement with the patent owner and pay royalties under the agreement; and

5. If the user of the process wanted to purchase the unpatented product for unpatented uses, he paid only the price charged for the product, not the extra charge for the patent-license royalty.

Recently, the Supreme Court, in *Dawson Chemical Co.* v. *Rohm & Haas Co.,* 448 U.S. 176 (1980), limited the reach of the patent misuse doctrine in a manner crucial to the chemical industry. In that case, Rohm & Haas Co., which holds the patent for the use of the unpatented chemical propanil as a pesticide, sued Dawson Chemical Co. for selling propanil with directions to purchasers as to how to apply the chemical to rice crops in accordance with Rohm & Haas' patent. Dawson Chemical defended the suit on the ground that Rohm & Haas had misused its patent by refusing to grant licenses that would allow use of the patented pesticide process in connection with propanil produced by others. Propanil has only one commercial use, that is, as a pesticide.

The Supreme Court held that Rohm & Haas had not engaged in patent misuse. Congress in 1952 had amended the Patent Act to state on its face that patent holders did not engage in patent misuse simply because they sold "nonstaple" articles for use with their patent, authorized others to use nonstaple articles with the patent, and sued others who sold nonstaple articles for use with the patent. 35 U.S.C. § 271(d). A "nonstaple" article is one that has no substantial commercial use apart from its use in connection with a patented subject matter. The court held that these amendments were intended to allow patent owners to tie the sale of nonstaple articles, like propanil, together with authorization to practice the patented process, and at the same time to allow patent owners to prohibit other producers of nonstaple products from selling the articles for use in connection with the patented process.

This decision greatly changes the nature of the inquiry companies must undertake in deciding to market patented processes. The first inquiry in many cases will now be whether the articles sold for use with the patented process are staples. If so, all of the restrictions of the patent misuse doctrine apply. If not, patent holders may require users of the patented process to buy from them the articles, such as chemicals, needed to engage in the patented process. Briefs by the government and by private intervenors assumed that a decision in favor of the patent owner in *Dawson Chemical Co.* v. *Rohm & Haas Co.* would allow patent owners to charge a higher royalty to licensees who use the nonstaple products of others. However, the court did not reach this question.

### Patent Marking

Two statutory provisions cover the use of symbols indicating that a product is patented. If the patentee is to recover damages in any action for infringement, 35 U.S.C. § 287 requires the patentee and persons making or selling any patented article for or under it to place on the patented article, package or label: (1) the word "patent" or the abbreviation "pat.", and (2) the number of the patent. If the statutory notice is not placed on the patented item, the patentee can only recover damages from the date that the infringer was notified that he was infringing the patent, if he continued to infringe thereafter.

The second section provides basic protection against the passing off of goods as patented or as the product of a patent holder. 35 U.S.C. § 292 imposes a maximum $500 fine for every use of the name or imitation of the name of the patentee, use of the patent number, and use of the words "patent" or "patentee" or the like, if such act was intended to imitate the mark of a patentee or to deceive the public by inducing them to believe that the thing was made or sold by or with consent of the patentee, or if the act was committed with the intent of leading others to believe falsely that an item was patented. The statute allows private parties to sue for and recover one half of the penalty. Thus, applicants for patents should be careful not to use the patent markings prior to the issuance of a patent, and patentees should be careful to cease distributing products with patent markings still on them after the patent has expired.

In addition intentional patent mismarking can jeopardize other rights. In *Surgitube Products Corp.* v. *Scholl Manufacturing Co. Inc.*, 158 F. Supp. 540, 116 U.S.P.Q. 253 (S.D.N.Y. 1958), *aff'd* 262 F.2d 824, 120 U.S.P.Q. 241 (2d Cir. 1959), where a patent owner had applied a patent notice to unpatented goods, the court held that this action was patent misuse and therefore barred enforcement of the patent. In *Preservaline Manufacturing Co.* v. *Heller Chemical Co.*, 118 Fed. 103 (N.D. Ill. 1902), the court found that the plaintiff was guilty of fraud in suggesting in circulars distributed after the patent had expired that its product was still patented, and the court therefore denied an injunction for trademark infringement and unfair competition.

Many companies put a marking such as "U.S. Patent Pending" on their products when they have filed a patent application, but a patent has not yet been issued. Companies find such markings useful both as a matter of public relations, since they are informing the public that they regard their product as a distinct advance on the art, and for competitive purposes, since they are informing other companies that they expect to receive exclusive rights in the novel elements of their product. 35 U.S.C. § 292 provides for a fine for marking upon, or affixing to, or using in advertising in connection with any article, for the purpose of deceiving the public, language indicating that a patent application has been made, when none has been filed, or language indicating that an application is pending when it is not.

Labels often contain, in addition to the statutory language, the prefix "U.S." in front of the word "patent". Such an additional marking is very often useful for goods patented in the United States that will be sold in an international commerce. Many countries also have their own marking systems, and also penalties for false marking. An addition of "U.S." can prevent the false assumption that an American-patented good sold in another country is patented in that country rather than in the United States. If foreign patents are obtained, the specific marking provisions of foreign countries should be consulted. Statutes of foreign countries vary in the wording that they suggest, in the place where patent markings should be put, and in the consequences of failure to mark. Generally, most countries impose the same consequences as does the United States for failure to include patent notices on patented articles. Failure to mark will generally preclude a recovery of damages, except if actual notice is given. However, some countries have provisions for fines for failure to mark design patents as patented, and some countries do not allow recovery of damages if the patent marking does not appear on the items, even if the infringer was actually informed of the patent.

## TRADE SECRETS

As an alternative to applying for patents, companies often try to protect the proprietary nature of the information developed in their business by keeping it secret. They may take this course for any of a variety of reasons. They may have no choice: information they want to protect may not be patentable, or they may have significant doubts about its patentability. The information, even if technically patentable, may not be so valuable that they want to go to the cost of applying for patents. Companies may regard patent protection as not sufficiently reliable.

For whatever reason the course of action is undertaken, keeping business information secret is a frequently practiced method of operation. There are three principal pitfalls to this method. First, keeping information secret affords no protection against others who develop the same information independently. Second, the scope of protection against disclosure of the secret information is often uncertain. Third, maintaining the secret can be a costly proposition.

### Basic Requirements and Relationship to Patent Protection

Unlike patent protection, which depends upon a grant of rights by the federal government, the protection of trade secrets depends primarily on the laws of the several states. In *Kewanee Oil Co.* v. *Bicron Corp.,* 416 U.S. 470 (1974), the Supreme Court held that the Patent Act did not preempt state trade secret law. In general, when someone with a duty to maintain the secret seeks to disclose or use improperly, or in fact discloses or uses improperly, trade secret information in a way damaging to the owner of the trade secret, the owner may seek injunctive relief or damages under state law. See 4 *Restatement of the Law of Torts,* 1-29 (1937). The scope and extent of trade secrets can vary in significant respects from state to state. The most generally accepted definition is contained in Comment (b) to § 757 of the *Restatement of the Law of Torts:*

> A trade secret may consist of any formula, pattern, device or compilation of information which is used in one's business, and which gives him an opportunity to obtain an advantage over competitors who do not know or use it. It may be a formula for a chemical compound, a process of manufacturing, treating or preserving materials, a pattern for a machine or other device, or a list of customers. [4 *Restatement of the Law of Torts* 5 (1937)]

As the Supreme Court noted in *Kewanee Oil Co.* v. *Bicron Corp.,* 416 U.S. at 484-91 (1974), under this definition a trade secret may be clearly unpatentable, doubtfully patentable or clearly patentable.

**Secrecy:** All states agree that, to be protectable as trade secrets, commercially valuable information must be kept secret. Though some courts have made statements to the contrary, this does not mean that the developer of information cannot tell anyone about it. However, it does in general mean that trade secret information can be disclosed only to limited classes of people for limited purposes, and reasonably prudent steps must be taken to keep the information from being disclosed. Companies should allow only employees with a need to know to have access to the trade secret information. Segregating specialized processes where secret information is revealed, restricting visitors to such areas, posting signs indicating areas where manufacturing processes are secret, stamping sensitive documents as secret or proprietary are all common precautions possessors of trade secrets take.

One important consequence of the fact that trade secret law only protects secrets is that protection is lost when a company distributes goods that reveal the trade secret. See 1 R.M. Milgrim, *Trade Secrets* § 2.05[2] (1980), and cases there cited. For the chemical industry, the sale of chemical compounds that are analyzable is the critical situation in which trade secrets may be lost through distribution. Despite the general rule, courts have held that where it is highly expensive or very difficult technically to analyze substances, companies can obtain protection against disclosure of secret information that makes the discovery of a formula significantly easier. For instance, in *Riteoff, Inc.* v. *Contact Industries, Inc.,* 43 A.D. 731, 350 N.Y.S.2d 690, 181 U.S.P.Q. 330 (2d Dept. 1973), the court held under New York law that the formula for a publicly sold spray cleaner was a trade secret when ascertainment of the ingredients and their percentages "would require a very detailed and difficult analysis". Moreover, distribution of chemicals generally will not disclose such information as the processes by which the chemicals were manufactured or intermediate compounds created in the process but not analyzable in the final compound. In some instances there may be trace elements crucial to the commercial value of the compound that chemical analysis does not reveal.

**Use:** As is not the case with patented inventions, a company with trade secret information must actually make use of the information in order to gain protection for it. See 1 R.M. Milgrim, *Trade Secrets* § 2.02 (1980). At a minimum, courts are unwilling to afford companies protection for ideas or information that have no commercial value, which the companies do not have the capacity to exploit, or for which they are not willing to take steps to find someone who can exploit them. Beyond this, the courts have taken a wide variety of views as to how actively a company must be trying to exploit trade secrets before the courts will protect them. The content of chemical substances or mixtures that are being used experimentally can be protected as trade secrets, even when the substances are still too experimental to qualify for patent protection.

**Classes of Persons Covered:** Trade secret law only gives companies causes of action against people who have information that they know or ought to know is trade secret information and that they know in some fashion they have responsibility to protect. *Restatement of Torts* §§ 757-759. Even in the absence of written contracts, employees are under obligations by virtue of implied contract or their confidential relationships to their employers not to reveal trade secrets. In general, the law assumes that officers, directors of corporations and technical experts will have a relatively high level of knowledge of what information is trade secret. The American Institute of Chemical Engineers' *Code of Ethics* provides that a chemical engineer "will not disclose information concerning the business affairs or technical processes of any present or former employer or client without his consent." While this provision may go a little beyond the scope of protection which trade secret law would afford companies, it shows that chemical engineers have a professional responsibility to maintain secrets. Nontechnical production personnel generally are not believed to have quite the same knowledge of what is valuable secrt information and what is not. However, all employees have some responsibility, enforceable in the courts, for not disclosing information that they know or ought to know is commercially valuable and secret. Marking information and locations as containing secrets can help in putting employees on notice that they are guardians of a company's proprietary information.

Contracts for maintaining trade secrets are common and often advantageous between employers and employees, even despite the general duty against disclosure by employees of trade secrets. These contracts can and often should vary according to the categories of employees covered. The most general provision is a covenant not to disclose trade secret information. In addition, many contracts include covenants not to compete. Covenants not to compete must be reasonable in scope, geographical coverage and duration, but this requirement varies significantly state-to-state. In general, however, what is reasonable may vary according to the position, knowledge and responsibility of the employee. Many contracts also include provisions in which employees assign to the employer inventions that they make during their employment.

Even where a duty not to disclose trade secrets arises by virtue of the employer-employee relationship, a trade secret agreement can have significant advantages. For instance, the contract can be used to help establish that the information disclosed is trade secret. The contract, within limits, can specify which state's law will govern construction of the agreement. The contract may expand the remedies that would otherwise be available.

Contracts also often provide the basis for trade secret protection downstream in the distribution of products. Independent contractors, unlike employees, as a general rule do not have a fiduciary responsibility independent of contract to maintain trade secrets. When companies do not themselves have the resources or commercial interests in exploiting a new process or product, they may license others to do so, without having obtained a patent. In such a case, the contract will often provide that the information being divulged will be maintained secret. Similarly, they may have independent contractors formulate or otherwise supply chemicals used in products the contents of which are trade secrets. These companies can be bound by contract not to disclose to others the nature of the substances they formulate. Independent sales agents, especially when exclusive, are sometimes in unique circumstances to analyze or reverse engineer products, and courts have enforced agreements that barred sales agents from reverse engineering or from releasing information conveyed to them by the owner of the trade secret for the purpose of the sale.

It should be noted that, in the absence of a contract, nonemployee union representatives are not covered by the employee duty to maintain trade secrets. In three very recent companion cases, *Minnesota Mining and Manufacturing Co. and Oil, Chemical and Atomic Workers,* 1981-82 CCH NLRB ¶18,892 (1982), *Colgate-Palmolive Co. and Oil, Chemical and Atomic Workers International Union,* 1981-82 CCH NLRB ¶18,893 (1982), and *Borden Chemical and International Chemical Union,* 1981-82 CCH NLRB ¶18,894 (1982), the National Labor Relations Board for the first time addressed the question whether it is an unfair labor practice for an employer to withhold from the collective bargaining unit lists of chemicals found in the workplace. In these cases, unions had requested lists of the chemicals to which workers might be exposed, and the companies had declined to deliver such lists on the grounds that they included information regarded as trade secret. The NLRB held that the employers could not withhold disclosure of those chemicals that were not trade secret. However, the Board did not in these cases decide that the employers must disclose those substances that are trade secret, but rather instructed the employers and unions to engage in collective bargaining in order to establish conditions satisfactory to both under which the information might be disclosed. The Board stated that if collective bargaining failed it would balance the interest of the employer in its trade secrets against the interest of the union in disclosure to decide whether to require the information. It is noteworthy that in the *Minnesota Mining and Manufacturing* case, the Board stated as one of the reasons for requiring disclosure the unions' suspicions of discrepancies between material safety data sheets provided to the employer and the warnings that employees actually received in the workplace.

**Extent of Protection:** It should be remembered that the law of trade secrets generally only protects companies from the breach of secrecy. As is not the case with patented inventions, others may use the same information if they obtain it independently. Two companies may maintain the same information secret and therefore have protectable rights against the improper disclosure of that information by people who have obtained the secret from them. However, if information is known generally in the trade, the fact that a company attempts to keep its own knowledge of that information a secret will generally not avail it at all in suits against people who will disclose that information obtained from it. In essence, the secret possession of generally known information is not regarded as giving a company a substantial commercial advantage.

Maintaining new product information as trade secrets can pose commercial risks when others may seek patents if they develop similar products. Exactly the extent of this risk is uncertain. Section 102(g) of the Patent Act, 35 U.S.C. § 102(g), provides that a person may obtain a patent over an invention, even if he was not in fact the first to invent, if the first inventor has concealed or suppressed the invention or otherwise not acted diligently. Thus, a second inventor may be able to obtain patent protection over an invention when the first inventor has relied on the trade secret doctrine for his protection. The reported cases to date have not addressed the question as to whether the second inventor, having obtained a patent, can thereafter sue the first inventor for infringement.

No single standard governs the ability of the various agencies of the federal government to disclose trade secret information that companies have submitted to them. In *Chrysler Corp.* v. *Brown,* 441 U.S. 281 (1979), the Supreme Court held that Exemption 4 of the Freedom of Information Act, 5 U.S.C. § 552(b)(4)—which excludes trade secrets and privileged or confidential commercial or financial information from the general obligation that FOIA places on agencies to disclose information—does not foreclose agencies from revealing such information. However, the Court held that the Administrative Procedure Act provides access to the courts for suits to prevent government agencies from disclosing trade secret information when some law does make disclosure illegal. In that case, the Court sustained a suit brought to prevent disclosures prohibited by the most general statute that bars the release of trade secret information, the Trade Secrets Act. That Act provides that an officer or employee of the United States who discloses trade secret information coming to him in the course of his employment or official duties is subject to fine, imprisonment and removal, unless his disclosure is authorized by law.[3] An agency may, under the Trade Secrets Act, reveal trade secrets it obtains if authorized to do so by other statutes. The Trade Secrets Act does not deal with all situations in which chemical manufacturers may submit data to the government. Thus, 7 U.S.C. § 136(h), which contains prohibitions on the disclosure of trade secret information submitted in the process of registering pesticides, contains its own penalties for violations, specifically supercedes the EPA in that context from the coverage of the Trade Secrets Act and provides broad authority for the disclosure of enumerated trade secrets under specific circumstances. The full range of statutes and regulations governing the disclosure by the government of trade secrets is beyond the scope of this chapter, but specific provisions should be consulted when necessary. However, where statutes prohibit the disclosure by officials of the federal government of trade secret information, companies now have a well-established avenue of relief for preventing such disclosures under the Freedom of Information Act.

### Trade Secrets and Labelling Hazardous Substances

Probably the most important trade secret consideration in the labelling of chemicals concerns the listing of ingredients. Listings of ingredients may reveal identities of chemical substances or the identity of chemical constituents of mixtures, or assist in the discovery of the proportion of substance mixes, when otherwise such information would be very expensive or impossible for competitors to discover through analysis. The current voluntary standard of the American National Standards Institute, Inc. for the precautionary labelling of hazardous industrial chemicals (ANSI Z129.1-1976) recommends that companies consider including the identity of products or hazardous components on warning labels. While some states and localities have labelling requirements, there at present is no general federal requirement for the identification of the contents of industrial chemicals. A fairly recent EPA proposal would have changed this situation. On December 31, 1979, the Environmental Protection Agency proposed regulations that identified specific chemical substances and designated categories of chemicals that would be subject to health and safety study reporting requirements under § 8(d) of the Toxic Substances Control Act. 44 Fed. Reg. 77470. If promulgated, the regulation would have subjected this information to disclosure under the provision of § 14(b) of the Act forbidding EPA from withholding safety and health studies from disclosure on the grounds that they are confidential business information. Presently, the Occupational Safety and Health Administration is reviewing comments on a proposal, 47 Fed. Reg. 12092 (March 19, 1982), of a hazards identification rule that would require the labelling of hazardous chemicals, broadly defined, used in the workplace.

In the absence of regulation, companies need to consider the extent to which label identification of contents would reveal valuable commercial secrets. Protection of these secrets involves different considerations depending upon whether hazardous labelling will be used within the trade secret owners' workplace or outside of it. Since employees can be bound to keep such information secret, many companies provide such information much more readily within their own workplaces than they do when they distribute the product to others. Nonetheless, maintaining trade secrets can pose substantial administrative problems even within the workplace. Courts may regard a trade secret as having been abandoned if revealed to those who do not need to know the information. Thus, the labelling of hazardous substances in the workplace may require restricting access to areas of plants to which access could otherwise be relatively free. Alerting employees to the need to keep such information secret may require added warnings as to the secret nature of such information. Employers may in certain circumstances decide to expand the universe of employees with whom they enter into explicit secrecy agreements.

As the law currently stands, employers have developed a variety of alternatives to placard and label notification. These alternatives serve a variety of purposes other than the protection of trade secrets. For instance, some employers believe that excessive labelling in the workplace can confuse workers, but nonetheless want to make information on hazards available.

These alternatives nonetheless can also serve to simplify the protection of trade secret information. For example, lists of substances found in the workplace can be posted on bulletin boards with specific warnings as to the secret nature of the information. Some employers include this information in training booklets rather than posting them. In other cases, employers may label hazardous chemicals with coded information, the meaning of which can be made available to employees on request. These are only examples of alternatives that can simplify trade secret problems. No single approach is likely to be ideal for every workplace.

Once products leave the plant and are distributed to others, the trade secret difficulties multiply. Many companies that otherwise follow the current hazardous warning suggestions of the American National Standard Institute omit the identity of substances where inclusion would reveal confidential information. OSHA regulations require shipyards to obtain an extensive list of information about all chemical products they use, including the "trade name" of mixtures and formulas for single chemicals, the chemical names of ingredients, the percentage that each ingredient represents of the whole mixture, physical data about chemicals or mixtures of chemicals, fire and explosion hazard data and reactivity data. 29 C.F.R. § 1915.97. OSHA requires use of its Form 20 for recording this information or use of an approved substitute.

Providing material safety data sheets is becoming an increasingly common phenomenon in the chemical industry for several reasons, even where material safety data sheets are not required by regulation. Providing the warning and physical information in detail is useful in minimizing product liability exposure. Users of chemicals have in recent years become increasingly concerned about potential worker compensation claims from their use of hazardous chemicals.

Many companies have developed alternatives to OSHA Form 20 for general use. These alternative forms often provide for the exclusion of trade secret information, particularly of the chemical names and percentages of ingredients. Such forms generally provide the telephone number of a company official who can be called in the event of a medical emergency. Such information can be provided in medical emergencies without jeopardizing trade secret protection to individuals such as doctors who need to know and who can be bound not to disclose the information.

The physical data—such as boiling point, vapor pressure, solubility in water, specific gravity and flash point—called for by material safety data sheets can assist in the identification of the specific compositions of chemicals. Thus, trade secret considerations can enter into decisions whether to report these data. However, in most cases, obtaining these data by testing the substance is not nearly as expensive as analyzing the composition of a complicated substance, and there are not substantial technical difficulties to establishing the key physical data points. Thus, physical data about chemicals are often not protectable as trade secret information when ingredient information is. The physical data are probably more useful than the identity and specific percentages of all the ingredients in mixtures, and also reveal less about the products. As a result, experience with material safety data sheets indicates that companies substantially less often omit these data, where they are available, than they omit the compositions of mixtures from material safety data sheets.

If promulgated, proposed OSHA regulations on identification and communication of hazards in the workplace, 47 Fed. Reg, 12,092 (March 19, 1982), discussed at length elsewhere in the volume, will stabilize the somewhat *ad hoc* practices that have grown up in the absence of regulation. Under that proposed standard, chemical manufacturers, defined as establishments where chemicals are produced for use or distribution, would have to make hazard determinations about the chemicals they produce and provide material safety data sheets to certain downstream employers. All employers in enumerated industries, including the chemical industry, would have to evaluate health hazards, mark the identity of hazardous chemicals, place appropriate warnings on containers, obtain and develop material safety data sheets for each hazardous chemical that it produces or uses, and provide employers with information and training about the hazardous chemicals in the workplace. A hazardous chemical that is part of a mixture would have to be identified if it is 1% of the mixture, unless the mixture is determined to be nonhazardous.

The proposal would provide that an employer may withhold the precise chemical name if (1) the employer can substantiate that it is a trade secret, (2) the chemical is identified by a generic compound that would provide useful information to a health professional, (3) all other required information on the properties and effects is contained in the safety data sheet, and (4) the material safety data sheet, available to workers, indicates which category of information is being withheld on trade secret grounds. An employer may not, however, withhold the chemical names of carinogens, mutagens, teratogens, or chemicals that cause irreversible damage to human organs from persons who have a "need to know." The proposal would provide that employers giving the names of trade secret chemicals may condition access on reasonable confidentiality agreements. Finally, a general exception requirement that information withheld as a trade secret be provided on a confidential basis to a treating physician who states in writing (except in emergency situations) that a patient's health problems may be the result of occupational exposure. A statement to this effect with the name of the manufacturer and an emergency telephone number must be included in the material safety data sheet.

A comparison of this proposal to the original OSHA proposal on this subject issued during the Carter Administration, 46 Fed. Reg. 4412 (1981), shows several points at which the new proposal is more protective of trade secrets. The original Carter Administration proposal would have required identification of carcinogens at any level of concentration. The original proposal would have required that dangerous intermediates in conduits in the workplace be identified by placards or labels which would have allowed such warnings to be visible to visitors who would not otherwise have known about intermediates created in the industrial process. The current proposal does not require such labelling on conduits but requires that pertinent information be otherwise available to the employee. Creating a suitable balance between providing employees with information to protect them from hazards and safeguarding commercially valuable trade secrets is a sensitive regulatory issue that is likely to be hotly debated before any proposal is finally adopted.

## TRADEMARKS

Turning to the law of trademarks, one comes to a very different set of concerns. When a new product is put on the market, companies very often want to give the product a distinctive name that will serve to distinguish their product from those of others. Both state and federal trademark law provide protection against the subsequent adoption by others of marks that are so similar that they will be likely to lead to confusion as to the source of sponsorship of the goods or services.

### Types of Marks Protected

A *trademark* is any word, symbol or device that is used by a manufacturer or merchant to distinguish his goods from those of others, and is affixed in any fashion to the goods, is applied to the goods as a label, tag or stamp, is applied to containers for the goods, or is used on displays of the goods. See 15 U.S.C. § 1127; 37 C.F.R. § 2.56. Brand names for goods are typical kinds of trademarks, but not the only kind. A label may have multiple trademarks. See, e.g., *Carter Wallace, Inc.* v. *Procter & Gamble Co.*, 434 F.2d 794 (9th Cir. 1970). Thus, it may bear both what is commonly understood as a brand name and also a "house mark", so long as both are so displayed that they are distinct and each identifies the product. Where a label of a product clearly enough uses a word as a distinctive term for an ingredient of the product, the term can become a trademark for the ingredient. A label can also include a slogan that is protectable as a trademark if the public identifies it as distinguishing the goods as coming from a particular source.

In addition, a fanciful design can serve as a trademark. For the design to be a trademark, the decorative features of the design must be incidental to the design's principal impact of identifying and distinguishing the goods. A background design must be sufficiently distinctive that it makes an impression on potential buyers of the goods independent of its association with the words on the label. Like descriptive words, descriptive designs do not receive trademark protection unless they have gained secondary meaning. See 1 J. T. McCarthy, *Trademarks and Unfair Competition,* 169-70 (1973).

This chapter will chiefly discuss the problems of using words as trademarks for goods. Trademark law, however, also protects other kinds of marks that are not called trademarks. Thus, the United States Patent and Trademark Office registers distinctive words, symbols or designs that are used in the sale or advertising of services that a manufacturer or merchant offers to others. The Lanham Trademark Act defines these as *service marks*. 15 U.S.C. § 1053, 1127. While the Patent and Trademark Office does not accept use on advertisements or other promotional materials as a trademark use, the Office recognizes that the principal way in which service marks are used is on advertisements and other promotional materials, since there is no tangible good on which to place a service mark. U.S. Patent and Trademark Office, *Trademark Manual of Examining Procedures* § 1301.06 (1979). For instance, a distinctive word, used to identify the service that a laboratory provides to others in analyzing their chemicals, will be a service mark when used to advertise those services. However, the distinctive name of a laboratory that analyzes a company's own chemicals will not be a service mark because those services are not provided to others, even if the company informs the public that it has such a laboratory.

Trademarks and service marks are the two principal types of marks that companies use to identify their products. Other marks serve still other purposes. Thus, the U.S. Patent and Trademark Office also registers *certification marks*. Certification marks are defined as marks used on or in connection with the products or services of persons other than the owner of the mark, to certify regional or other origin, material, mode of manufacture, quality, accuracy or other characteristics of such goods or services, or that the work or labor on the goods or services was performed by members of a union or other organization. 35 U.S.C. § 1127. The owner of a certification mark authorizes others to use his certification mark on his goods or services if their products meet the criteria he has set for the use of the certification mark. He must not produce the goods or perform the services, and unlike the owner of a trademark or service mark, must not control the nature and quality of the goods on which they are used. The certification mark owner undertakes a responsibility to check that the goods on which certification marks are placed meet the certification standards. The GOOD HOUSEKEEPING SEAL OF APPROVAL is perhaps the best-known certification mark.

In some cases, trademark owners find it useful to adopt certification marks when their specialized product is incorporated in the final good which they themselves do not produce or control. Perhaps the most famous example of such a mark is DuPont's mark TEFLON. DuPont owns trademark registrations for TEFLON and TEFLON II for its nonstick coating. In addition, it owns certification mark registrations for TEFLON, in conjunction with accompanying language. DuPont allows producers of final goods who use the TEFLON coating in accordance with DuPont's guidelines to place the certification mark on their products. Though a trademark can be part of a certification mark, a trademark and a certification mark cannot be identical. Thus, DuPont includes in its certification mark language indicating that DuPont's role in the production of the final good is the approval of the use of the TEFLON finish. See *In re Monsanto Co.*, 201 U.S.P.Q. 864 (T.T.A.B. 1978); *E.I. DuPont de Nemours & Co.* v. *Yoshida International, Inc.*, 393 F. Supp. 502 (E.D.N.Y. 1975).

Federal trademark law also recognizes other kinds of marks that commercial companies generally do not use. Thus, the U.S. Patent and Trademark Office registers *collective marks,* which are trademarks or service marks used by members of a cooperative, association or other collective group or organization. Collective marks include marks that are used to indicate membership in a union, an association or other organization. 35 U.S.C. §§ 1054, 1127.

The U.S. Patent and Trademark office may not register *trade names.* See *In re Walker Process Equipment Inc.,* 102 U.S.P.Q. 443 (Comm'r Patents 1954), *aff'd* 110 U.S.P.Q. 41 (C.C.P.A. 1956). Trade names, as defined under the Lanham Trademark Act,[4] are generally firm names adopted to identify businesses, vocations or occupations. 15 U.S.C. § 1127. These names do not qualify for registration if they do not meet any of the definitions of marks given above. Trade names can be identical to trademarks or service marks, but very often a trade name is not used as an identifier of a product. Absent specified regulations to the contrary, companies often do not put their trade names on their products at all. Even where business names are put on products, they do not become registrable as trademarks unless they are intended to be used in the marketplace to identify or distinguish the products. Although trade names are not registrable in the U.S. Patent and Trademark Office, and law of unfair competition provides protection of trade names against adoption of other names or marks that are likely to cause confusion.

### The Choice of Trademarks

Choosing a trademark, particularly for consumer products, often leads to a struggle between lawyers and marketing executives. Marketing executives very often want to choose a name for the product that will describe or suggest the qualities of their product as much as possible in order to use the chosen name to advertise the value of their company's product. Lawyers, on the other hand, know it is very hard to protect words that are descriptive or suggest qualities of a product against the use by others of similar words.

Standard trademark law ranks marks according to the degree to which the law will afford legal protection. Fanciful or arbitrary words receive the greatest protection. The protection of such words is relatively broad both with respect to preventing use of variations of the word and with respect to how close a product with a similar trademark needs to be to the company's trademarked product before the trademarks will be regarded as likely to cause damaging confusion. An example of an arbitrary word is KODAK, which has no meaning apart from its association with products put out by the Eastman Kodak Corporation.

Words that suggest, but do not describe qualities of the products receive the second broadest scope of protection. It is often difficult to distinguish between words that are suggestive and words that are descriptive of qualities of the products.

Words that are merely descriptive do not receive any legal protection unless the company putting out the product has advertised so persuasively in the relevant market that potential customers know that the descriptive word really identifies a product that comes from a particular source (even though the consumer may not know the name of the company that is the source of the product). This additional connotation is called a word's "secondary meaning." To be protected or registered, geographically descriptive words and surnames must also gain secondary meaning. See 15 U.S.C. § 1052(f).

Finally, some words cannot be protected as trademarks at all. These are "generic" words that have become in the public mind the common names of goods or services, and cannot become the proprietary identification of a product coming from a particular source. Grade designations of products may in some circumstances be regarded as generic if used generally in the trade. Standard chemical names are generic terms. Coined words can also be generic names for chemicals, in addition to their technical names. For example, 'DDT' and 'propanil', which have already been mentioned above in other contexts, are generic terms for pesticides that also have chemical names. The Department of Agriculture and the International Organization for Standardization have from time to time sent to the Patent and Trademark Office lists of coined words that have become common names for pesticides. Such organizations as the United States Adopted Names Council have submitted lists of generic terms for pharmaceuticals and other products. These lists are not definitive as to which words have become generic. The Patent and Trademark Office does not check the validity of such lists. The test of which words are generic is how they are used by the public or the relevant market or scientific community.

The Lanham Trademark Act bars the registration of marks that are confusingly similar to marks or trade names previously used in the United States, although concurrent registration is allowed where two parties, independent of each other, have established marks in different regions. Merely descriptive marks, surnames and geographically descriptive or misdescriptive marks cannot be placed on the Principal Register, which affords marks the benefits of the Lanham Act, but can be noted on a Supplemental Register. 15 U.S.C. § § 1052(e), 1091. The U.S. Patent and Trademark Office may not register immoral, deceptive or scandalous marks; marks that disparage or falsely suggest connections with persons, institutions, beliefs or national symbols; marks that simulate the flag or coat of arms of municipalities, states or countries; or marks that consist of or comprise the name of a particular living individual, except with his written consent, or of a deceased President of the United States, except with his widow's permission. 15 U.S.C. § § 1052(a), (b), (c).

### The Usefulness of Searches in Choosing Trademarks

In adopting a mark for a product, companies need to be sure that they are not infringing the prior rights of other businesses. Having to abandon a mark once marketing has begun can be a very expensive matter. Trademark searches should be conducted to make sure that no one has prior rights to marks or trade names that are identical or so close that confusion would arise. First, the records of the United States Patent and Trademark Office should be examined. By virtue of the Lanham Trademark Act of 1946, parties using trademarks are deemed to know about any registered mark in the Patent and Trademark Office. Thus, the fact that a company may not actually know of a registered mark will be no defense in an infringement suit. Companies may also want to conduct broader searches. States operate their own trademark registers, which give local rights similar to the rights that registration in the Patent and Trademark Office grants on a national basis. Many trademark owners register marks in states rather than federally, usually because their operations are so local that they have not engaged in interstate commerce as is required for registration in the United States Patent and Trademark Office, or because states register marks much more quickly than the United States Patent and Trademark Office does. Search services also offer a variety of options for trying to find out about names that are not registered but nonetheless are in use. They may, for example, check trade name registries that are privately maintained, as well as telephone listings to see if companies have chosen similar names for company names.

## Guidelines for Preventing Trademarks from Becoming Unprotectable

Companies that adopt trademarks for their products must be careful how they use the marks so that the marks do not lose their distinctiveness and become generic. The common names of quite a number of common products started out as trademarks, even as trademarks that were arbitrary. However, the public in time came to believe that the trademarks were the names of the products in general rather than trademarks representing products coming from a particular company. Thus, Bayer lost the right to protect the word 'aspirin' and DuPont lost the right to protect the word 'cellophane,' even though these words at first had no meaning apart from their meanings as trademarks. *Bayer Co.* v. *United Drug Co.,* 272 F. 505 (S.D.N.Y. 1921); *DuPont Cellophane Co.* v. *Waxed Products Co.,* 85 F.2d 75 (2d Cir. 1936). In a more recent case, a company challenged DuPont's right to protect its exclusive use of TEFLON, claiming that the public associates TEFLON with any nonstick coating, not merely DuPont's product. DuPont successfully defended this suit by showing that a substantial portion of the consumer public knew that TEFLON described nonstick coating coming from a particular company. *E.I. DuPont de Nemours & Co.* v. *Yoshida International, Inc.,* 393 F. Supp. 502 (S.D.N.Y. 1975).

To maintain the distinctiveness of trademarks, companies must take care in using them. There is no rulebook on the use of trademarks, and sufficiently widespread public use of a mark in a generic sense can cause a company to lose its trademark rights in the term despite heroic efforts to maintain its distinctiveness. However, certain guidelines for the use of marks have become generally accepted and are helpful in keeping marks from becoming generic. In advertising or other narrative copy, the mark should be used in a distinctive script. In this narrative, trademarks have been written in capital letters, which is a common way of distinguishing them. Marks should at a minimum be spelled with initial capitalization. In narrative copy, the mark should be used at least once in connection with a generic name for the product, preferably as an adjective. Thus, Xerox, Corp. is careful not simply to call its product XEROX, but rather to call it the XEROX copier. Some trademark owners have become even more elaborate in making sure that they do not identify their trademarks as the name of a product. One example, which at this writing is commonly heard on television, is the reference to SANKA brand decaffeinated coffee. There the trademark is not even used as a direct modifier of the name of the product. Trademarks should generally not be used as verbs.

Recently, the Federal Trade Commission, exercising its power under the Lanham Act, petitioned in the U.S. Patent and Trademark Office for cancellation of the trademark registration for FORMICA. The Commission alleged that FORMICA had come to be understood as a generic word for decorative plastic laminates. This petition was controversial, especially since none of the competitors of the owner of the FORMICA trademark registration had brought legal action seeking the ability to use the term. In § 18 of the Federal Trade Commission Improvements Act of 1980, 94 Stat. 374, Congress rescinded through fiscal year 1982, the FTC's budget authorization to seek trademark registration cancellation. The definitions in the *Webster's New International Dictionary of the English Language* (2d Ed. 1934) for 'aspirin' and 'Formica' (sic) illustrate the problems of maintaining the distinctiveness of trademarks. The dictionary defines 'aspirin' as "a white crystalline compound, the acetyl derivative, or the acetate, of salicylic acid, $C_9H_8O_4$, used as an antipyretic and analgesic like the salicylates but producing fewer undesirable effects. *Aspirin* was originally a trademark." 'Formica' is defined as "a trade-mark applied to a laminated phenolic insulating material; hence [*sometimes not cap.*], the materials bearing this trade-mark."* This entry, since there was no judgment that FORMICA had become generic, thus referred to FORMICA as a trademark, but noted that it is sometimes used without initial capitalization. Dictionary references are, of course, not dispositive of whether or not a term is generic.

Owners of trademarks for patented goods should be particularly careful to use their marks for these goods in a fashion designed to maintain their distinctiveness. In the absence of licensing, there is often only one source for a particular patented good. The case law indicates that it is particularly easy for trademarks for patented goods to become understood by the public as the name of the goods themselves.

In some cases, companies may find it useful to license others to produce products and distribute them under the company's own trademark. Licenses should be carefully drawn to maintain the trademark owner's rights in the mark. At a minimum, the trademark owner must maintain some control over the quality of the products that a licensee distributes using the licensed mark. Otherwise, a trademark owner may lose his rights to the mark. Also, a licensor should be careful to require a licensee to use the mark in a fashion designed to keep the mark from becoming generic. Protection of a trademark can commit companies to litigation. If others, without licenses, use the same or confusingly similar marks on related goods, a company can lose its exclusive rights to its own mark, since such use can cause the mark to become generic.

### Registering Trademarks

The U.S. Patent and Trademark Office both grants patents and registers trademarks and other marks. However, there are significant differences be-

---

*By permission. From Webster's New International Dictionary, Second Edition © 1934 by G. & C. Merriam Co., publishers of the Merriam-Webster® Dictionaries.

tween the need for letters patent and trademark registrations. As has been discussed earlier, the exclusive right in a patented invention only arises by virtue of the grant of a patent by the Patent and Trademark Office. On the other hand, the U.S. Patent and Trademark Office, in registering trademarks, confirms the rights that a trademark owner already holds and extends the protection which the trademark is afforded. Trademark owners can obtain legal protection through the court system for distinctive marks even if they are not registered.

Registration of a trademark on the Principal Register of the U.S. Patent and Trademark Office, however, gives a trademark owner certain advantages. First, it puts all other people adopting trademarks on notice as to his use of the mark. In the absence of registration, a trademark owner can find himself without a remedy if a subsequent user adopted a confusingly similar mark without knowledge of his prior use. Second, registration operates as a presumption of the validity of his ownership of the trademark in subsequent litigation, and an affidavit showing five years of continuous use after the trademark registration has issued can make his rights to use the marks incontestable for many purposes. 15 U.S.C. § § 1065, 1115. Third, registration gives access to federal courts for infringement suits. 15 U.S.C. § 1121. Fourth, the Lanham Trademark Act adds to the panoply of remedies available for trademark infringement, allowing recovery, for instance, of treble damages in some cases, for the profits an infringing company derived from the infringement, and for seizure by customs of infringing goods being imported into the United States. 15 U.S.C. § § 1116-1124.

Rights in the mark arise not from registration but from use of it as a trademark. In the United States, a trademark can only be registered after it has been used in commerce. 15 U.S.C. § 1051. The United States registration practice is on this point contrary to that of many foreign countries, where registration of a mark is necessary before the mark is used.

A trademark owner can maintain his federal registration as long as he still uses the mark. A trademark registration has a twenty-year term. Between the fifth and sixth anniversaries of registration, the trademark owner must file an affidavit showing that the mark is in use at that time or not in use for excusable reasons. If the affidavit is not filed, the Patent and Trademark Office cancels the registration. 15 U.S.C. § 1058. An affidavit showing that the mark has been in continuous use for five years renders the registrant's ownership of the mark incontestable for most purposes. 15 U.S.C. § 1065. At the end of each twenty-year registration term, the registrant can renew the registration for another twenty-year term, if he is still using the mark. 15 U.S.C. § 1059.

One consequence of the rule that rights in a trademark arise from the use of a trademark is that, by ceasing use, a trademark owner loses his rights in the mark. A brief, temporary hiatus in the use of a mark may not cause abandonment of rights in it if trademark owner intends to continue use, and if the hiatus occurs for excusable commercial reasons, such as strikes or other unavoidable commercial problems. However, companies cannot merely through registration reserve marks for their future use. Until recently, some large companies had tried to maintain their rights to marks that they were not actively using on goods in the marketplace, by operating so-called trademark maintenance programs. These programs involved obtaining federal trademark registrations on the basis of uses of the marks in minor distributions of goods and thereafter selling a deliberately small supply of goods bearing the mark each year. The purpose of these sales was not to create a trade in the goods, but rather to maintain rights to the mark. Recently, in *Procter & Gamble Co.* v. *Johnson & Johnson, Inc.*, 205 U.S.P.Q. 697 (S.D.N.Y. 1979), a district court held that such minor use programs did not give the operator of the program any rights in the marks, because, in the absence of a present intent to market the trademarked product, the registrant had not established the good faith commercial use necessary for trademark protection.

While the U.S. Patent and Trademark Office does not approve trademarks for use, it does check as to whether that use has been lawful. In particular, when presented with labels for which statutes require prior approval by other agencies, the Patent and Trademark Office examiners inquire whether the trademark owner has complied with the requirements of these other agencies. Examiners will check, for instance, whether applicants to register trademarks for pesticides have met EPA's requirements for registration under the Federal Insecticide, Fungicide and Rodenticide Act. The details of that Act are discussed elsewhere in this volume. However, the Act affects the choice of trademarks, since it requires companies to submit the "names" of their products for prior approval. EPA's regulations prohibit misbranding and the use of names, brands or trademarks that are false or misleadng; and require that the name, brand or trademark be placed on the front panel of the label. The regulations give as an example of misbranding a name that suggests one or more, but not all, of the active ingredients, even though the other ingredients are stated elsewhere on the label. Even though trademark law generally does not require use of trade names on labels, the name and address of the distributor or producer must appear on a pesticide label, and if the distributor's name appears, the label must show through appropriate language that the distributor is not the producer.[5]

### Trademark Registration Marking

The issuance of a federal registration allows the owner of a mark to use a capital "R" enclosed in a circle, the words "Registered in the U.S. Patent and Trademark Office" or "U.S. Reg. Off." in displaying the mark, 35 U.S.C. § 1111. These trademark registration insignia may not be used before the registration issues, or after it has expired. A trademark owner can obtain monetary damages, as opposed to injunctive relief, under the Lanham Trademark Act for infringement of the registered mark only from the date on which the infringer learned that he was infringing a registered mark. While a trademark owner can impart such knowledge by sending protest letters which identify his mark as registered, the most effective way to insure the availability of the full scope of remedies under the act is to use the ® or other symbols, which serve as notice to potential infringers. Intentional use of these symbols for a mark that has not been registered should be avoided, since it can give rise to suits for deceptive advertising.

Before trademarks are registered, companies often use the symbol "TM" with the trademarks ("SM" with service marks). These symbols indicate that the user of the term intends the term to be a trademark. Such symbols are particularly useful when a company has adopted a descriptive term as a trademark and is trying to establish its secondary meaning in the mark.

## Copyright and Trademark Protection of Labels Compared

Finally, companies may consider using the copyright laws to protect the original elements of labels. Failure to place the copyright notice on the labels before the goods are sold will generally forfeit all rights to copyright protection. The statutory copyright notice consists of: (1) the word 'Copyright,' the abbreviation 'Copr.' or the letter 'C' enclosed in a circle; (2) the year of publication and (3) the name of the copyright owner, or an abbreviation by which the name can be recognized, or a generally known alternative designation of the owner. 17 U.S.C. § 401. Registering a copyright in the Copyright Office is not necessary for copyrighting a work. However, a copyright owner must register the copyright in the Copyright Office before bringing suit against an infringer and remedies for infringements prior to the date of registration are limited.

Copyright protection for labels is rather limited. A work must have a minimal degree of originality in order to obtain copyright protection. Such protection extends only to the original expression, not to the facts or ideas expressed. In keeping with these principles, the Copyright Office has stated that protection does not extend to "words and short phrases such as names, titles and slogans, familiar symbols or designs . . . lettering or coloring, mere listing of ingredients or contents." 37 C.F.R. § 202.1(a).

However, copyright can provide protection for original elements of pictorial material or text, whether or not strictly unique or novel. One famous example illustrating the difference between trademark and copyright protection is *Kitchens of Sara Lee Inc.* v. *Nifty Foods Corp.*, 266 F.2d 541 (2d Cir. 1959), in which Sara Lee sued for trademark infringement of the famous work mark SARA LEE by defendant's use of the trademark LADY EILENE for cake, and lost its trademark infringement suit. However, Sara Lee also sued for copyright infringement on the basis that the defendant had copied the predominant part of the pictures on its labels—namely, prominent pictures of pieces of cake. (As noted above, such descriptive designs are not protectable as trademarks in the absence of secondary meaning.) The court found that the defendant had obviously put copies of these pictures on its own labels and awarded damages for copyright infringement. However, the court refused to award damages for the copying of the circular, rectangular or octagonal shapes of the aluminum packages or for the serving directions or ingredients listings, which it indicated were not copyrightable.

## FOOTNOTES

1. This chapter concerns patents issued for inventions covered by § 101, since they are of most interest in the chemical industry. However, it should be noted that in addition, § 161 provides for the issuance of patents to persons who discover or invent and asexually reproduce distinct and new varieties of plants, and § 171 provides for patents for new, original and ornamental designs for articles of manufacture.

2. Under 35 U.S.C. § 181, inventions affecting national security may not be patented and must be kept secret. The Commissioner of Patents and Trademarks may therefore refer to other governmental agencies on a confidential basis applications that he deems may affect national security.

3. 18 U.S.C. § 1905, the Trade Secret Act provides: "Whoever, being an officer or employee of the United States or of any department or agency thereof, publishes, divulges, discloses, or makes known in any manner or to any extent not authorized by law any information coming to him in the course of his employment or official duties or by reason of any examination or investigation made by, or return, report or record made to or filed with, such department or agency or officer or employee thereof, which information concerns or relates to the trade secrets, processes, operations, style of work, or apparatus, or to the identity, confidential statistical data, amount or source of any income, profits, losses, or expenditures of any person, firm, partnership, corporation, or association; of permits any income return or copy thereof or any book containing any abstract or particulars thereof to be seen or examined by any person except as provided by law; shall be fined not more than $1,000, or imprisoned not more than one year, or both; and shall be removed from office or employment."

4. This nomenclature is specialized to trademark practitioners. Thus, for example, the American National Standard on labelling hazardous industrial chemicals states that identification of a hazardous product or component should not be limited to a nondescriptive "trade name". This recommendation obviously is designed to include trademarks.

5. The U.S. Patent and Trademark Office checks for compliance with labelling requirements of other statutes, most of which are of less direct concern to the chemical industry: The Meat Inspecting Act, 21 U.S.C. §§ 71-96; Poultry and Poultry Products Inspection Act, 21 U.S.C. §§ 27, 457; The Federal Alcohol Administration Act, 27 U.S.C. §§ 201-212; The Federal Seed Act, 7 U.S.C. § 1551-1610; The Food, Drug and Cosmetic Act, 21 U.S.C. §§ 301, 321-392; and 26 U.S.C. § 7805, relating to cigars and cigarettes. The Trademark Manual of Examining Procedure, which guides decisions by examiners at the Patent and Trademark Office, also notes that the following statutes contain labelling requirements, but that examiners generally do not check for compliance with them: The Wool Products Labelling Act, 15 U.S.C. § 68; The Fur Products Labelling Act, 15 U.S.C. § 69; The Textile Fiber Products Act §§70-70k (of interest to the chemical industry because it covers manufactured as well as natural fibers); 15 U.S.C. § 297, relating to the use of trademarks and trade names in the transportation of goods made of gold or silver or their alloys; 15 U.S.C. §§ 1261-1273, relating to the labelling of hazardous substances (discussed elsewhere in this volume); The Consumer Products Safety Act, 15 U.S.C. § 2051 (discussed elsewhere in this volume); The Fair Packaging and Labelling Act.

# 10

# Pesticide Labeling Under the Federal Insecticide, Fungicide and Rodenticide Act (FIFRA)

**Steven D. Jellinek**
*Jellinek Associates, Inc.*
*Washington, D.C.*

## INTRODUCTION

This book discusses in some detail the various purposes of requirements for labeling in the chemical industry: indentification, warning, instruction, and so forth. In most cases—drugs, consumer products, transportation of hazardous materials—labeling requirements are expected to serve one or more of these purposes. It is safe to say, however, that in no case is labeling as central to the function of a class of chemicals nor as intertwined with their regulation as it is with pesticides.

The pesticide label is crucial to implementation of FIFRA—the Federal Insecticide, Fungicide, and Rodenticide Act (7 U.S.C. 136 et seq.). Just how crucial is demonstrated by the fact that use of a pesticide in a manner inconsistent with its label is a violation of the law. Furthermore, the pesticide label is the regulatory document of record in the implementation of FIFRA and that label's instructions and prohibitions have the force of law.

In spite of the prominent position it plays in the registration of pesticides—chemical company officials typically talk about "getting a label" when they mean obtaining EPA registration for a pesticide—labeling does not command major portions of the law itself nor of the regulations implementing the law. In fact, the most important provision of FIFRA effecting labeling—Section 2(q)—is a definition which, in curious negative legal draftmanship, goes on at length to describe the labeling requirements which, if not followed, result in a pesticide being "misbranded". Likewise, tucked away in Section 3 of the Act—which deals with procedures and requirements for registering pesticides—is Section 3(c)(5)(B) which authorizes the Administrator of EPA to register a pesticide if, among other things, "its labeling and other material required to be submitted comply with the requirements of this Act." EPA has brought these two provisions together by holding that the labeling provisions itemized in the statutory definition of "misbranding" are requirements of the Act that must be complied with. EPA's regulations implementing these requirements for registered pesticides are found at 40 CFR 162.10 in less than ten of the two-hundred sixty-two pages of regulations covering the pesticide program. Labeling requirements for pesticides used under an experimental use permit are found at 40 CFR 172.6. EPA's regulations are silent, for the most part, on labeling as it applies to nonregistered products (those shipped between producer's establishments, for disposal, or intended solely for export) and pesticide devices. However, the agency is in the process of preparing proposed registration guidelines covering pesticide labeling which will up-date and expand upon the current regulations and include coverage of pesticide products and devices. Draft versions of these guidelines were informally circulated for comment throughout 1981. They will be published sometime in 1982, either as formal proposal regulations, or as nonbinding guidance.

This chapter will be primarily devoted to a discussion of current regulatory requirements for labeling pesticides, with reference to EPA's soon-to-be proposed modifications, based on the August 1981 version of the draft proposed labeling guidelines (hereinafter referred to as "draft guidelines"). Before getting into detail on labeling, however, the chapter will review the background and context of the EPA pesticide program and the Agency's major responsibilities under FIFRA and under the pesticide amendments to the Federal Food, Drug, and Cosmetic Act.


## EPA's PESTICIDES PROGRAM

### Introduction

EPA's role in the regulation of pesticides can be roughly divided into two broad categories: first, the licensing of new pesticides and new uses of existing pesticides and, second, the review of existing pesticides. In both cases pesticides are evaluated against a statutory standard which calls for the protection of public health and the environment from the "unreasonable adverse effects" of pesticides. The term "unreasonable adverse effects" takes into account the economic, social, and environmental benefits of the pesticide, as well as its risks (Sec. 2(bb)). Thus, FIFRA's basic statutory decision-making scheme for the registration of pesticides, as well for their removal from the market place, involves the weighing and balancing of risks and benefits. Risk is usually expressed in terms of the potential, or probability, of health or environmental damage, while benefits are usually measured in the dollar values of factors such as crop yields, costs of alternative pest control measures, and consumer costs.

### Registration

Pesticides can not be sold in the United States unless they are registered with EPA. Under Section 3 of FIFRA, therefore, EPA operates a premarket licensing or clearance process for all pesticides sold in the U.S., no matter whether they are manufactured here or abroad. This process also applies to new uses of previous registered pesticides. Indeed, registration is generally granted for specific formulations (including active and inert ingredients) and use patterns of a given pesticide, not for blanket use. The term "use pattern" is defined by FIFRA as the manner in which the pesticide is applied and includes factors such as the target pest, the crop or animals treated, the application site, and the application technique and frequency. Applications for pesticide registration must be accompanied by supporting data covering the pesticide's physical and chemical properties, residue and environmental chemistry, and health and environmental effects, among other factors, and a proposed label which spells out in detail directions for use, warnings, and a variety of other information. Indeed, the draft label is the de facto substantive application form since the official application form (EPA No. 8570) is merely a cover sheet and checklist for transmitting the proposed label and its supporting data.

Any significant additions to or changes in the use (target pest, crop, application rate or technique, etc.) of a previously registered pesticide requires a new registration approving the additional or modified use. And, of course, such changes will be formally effected by appropriate changes on the label.

EPA reviews approximately 20,000 registration applications of various kinds each year. The data supporting these applications are developed by applicants for registration in accordance with testing guidelines prescribed by EPA. If the data are complete, and if EPA's review concludes that the product will perform its intended function without unreasonable adverse effects on health or the environment, when used in accordance with wide spread and commonly recognized practice, the Agency will register the product unconditionally under Section 3(c)(5) of FIFRA.

## Conditional Registration

Section 3(c)(7) of FIFRA, enacted in 1978, authorizes EPA to permit conditional registration of certain products, pending the development of a complete set of data. Products eligible for conditional registration include those that are the same as ones already registered, new uses of already registered chemicals, and, under limited circumstances, brand new chemicals. Under the conditional registration provisions, EPA can register a new product or use of an already registered pesticide if it finds that there will be no significant increase in the risk of unreasonable adverse effects to health or the environment. It can conditionally register a previously unregistered pesticide upon a finding that the public interest would be served and that risks would not be unreasonable during the period required to perform additional studies.

## Tolerances

Under Sections 408 and 409 of the Federal Food, Drug, and Cosmetic Act (FFDCA) EPA establishes tolerances, or maximum pesticide residue limits, for food and feed commodities that are marketed in the United States. Before a pesticide can be registered under FIFRA for use on a food or feed crop, EPA must either establish a tolerance for that use, or grant an exemption from the tolerance requirement. Determination of tolerances involves careful review and evaluation of residue chemistry and toxiology data to ensure that maximum residue levels established for food and feed are acceptable for human consumption. It includes consideration of the cumulative effect of the respective pesticide and related substances. Tolerances are set at a level no higher than necessary to permit the marketing of treated commodities.

## Special Registrations

The Special Registration Program responds to unexpected and temporary pest control problems and supports state governments in registering pesticides for intrastate use.

Under Section 18 of FIFRA, EPA may grant emergency exemptions to state or federal agencies authorizing the use of pesticides for purposes not included in their federal registrations. This authority provides EPA with a mechanism to resolve serious economic, health, or environmental emergencies caused by pest outbreaks for which registered pesticides are not available. Exemptions may include the use of previously cancelled pesticides or the use of pesticides on crops for which no tolerances have been established. In determining whether an emergency exists, EPA evaluates the risks and benefits of such pesticides.

Under Section 24 of FIFRA, EPA reviews state registrations of pesticides distributed and used only within the registering state to meet special local needs. EPA's review is intended to insure that state registrations meet FIFRA requirements and do not pose unreasonable adverse effects.

Under Section 5 of FIFRA, EPA may issue experimental use permits (EUPs) to pesticide registrants. These EUPs permit large scale experimentation to develop data for new pesticides or new uses of currently registered pesticides. In many cases the crop subjected to the experimental pesticide is destroyed. If the crop will be marketed after the experimental program is completed, a temporary tolerance for a safe residue level on the food or feed commodity must be established by the Agency before the EUP is issued. Both EUP and temporary tolerance decisions are based on an evaluation on human and environmental risks and benefits associated with a proposed use.

## Reregistration

Section 3 of FIFRA instructs EPA to reexamine, by current scientific standards, the health and environmental safety of the more than 41,000 currently registered pesticides and to reregister those which are found to meet the "no unreasonable adverse effects" standard. Instead of reviewing each of the 41,000 products on a case-by-case basis, however, the agency is developing comprehensive registration standards for each active ingredient common to numerous pesticide products. There are approximately 500 such active ingredients. Each standard sets out the conditions registrants must meet, including labeling, to reregister products containing that active ingredient. Once a standard is developed for a particular active ingredient, new pesticide products will also be registered in accordance with it.

## Rebuttable Presumption Against Registration (RPAR)

Under the authority of FIFRA Section 3, EPA has established certain risk criteria for identifying those pesticides which may be causing unreasonable adverse effects. The risk criteria include cancer, mutagenic affects, reproductive effects and hazard to nontarget species. Whenever valid scientific evidence demonstrate a pesticide meets or exceeds one of these risk criterion, EPA announces in the Federal Register that a rebuttable presumption against the registration of that pesticide has arisen. This notice includes a description of the scientific basis for EPA's findings and invites registrants, users, and the public to submit evidence in rebuttal or support of the presumption against registration. Unless evidence submitted rebuts the presumption, the agency then conducts an intensive risk/benefit review for each significant use of the RPAR chemical. On the basis of that review, the agency proposes a regulatory decision, and after further public comment, issues a final decision either to register, reregister, restrict, or cancel some or all of the uses of the pesticide. The proposed decision is reviewed by USDA and the EPA Scientific Advisory Panel. The final decision may be appealed to the Administrator.

### Imports and Exports

Under Section 17 of FIFRA, all pesticides being imported into the U.S., must comply with the provisions of the Act. The Department of the Treasury must notify EPA of the arrival of imported pesticides and may refuse entry to those which EPA determines are in violation of the Act.

Pesticides intended solely for export need not be registered. However, Section 17 does require that pesticides intended solely for export bear certain of the basic labeling statements required of pesticides sold in the U.S., and, if it is not registered under Section 3, the exporter must also have, prior to export, a statement signed by the foreign purchaser acknowledging that he understands it is not registered and cannot be sold in the U.S. Producers of export pesticides must also register establishments and keep required records and production data.

Whenever a registration, or cancellation or suspension of a product registration becomes effective or ceases to be effective, Section 17 requires EPA to so notify foreign governments through the State Department.

## INFORMATION THAT MUST BE INCLUDED IN PESTICIDE LABELS

Unless otherwise noted, all the requirements outlined in this and remaining sections are contained in 40 CFR 162.10. Appropriate subsection references will be noted in parenthesis throughout the text.

### General

In general labels on pesticides must include four different categories of information covering identification, ingredients, warnings, and use. The first category of information identifies the product and its manufacturer. Second, registrants must disclose the ingredients of the product on the label. Third, the label must include appropriate warnings to pesticide users in order to alert them to potential danger. And, finally the label must include directions for using the pesticide on the approved sites for the appropriate target pests, to insure that it is used in a way that prevents unreasonable adverse effects to publich health or the environment.

### Product Identification

EPA requires all pesticide products to bear a label which includes certain identifying elements (162.10(a)(1)).

**Product Name:** The name, brand, or trademark under which the pesticide is sold must appear on the front panel of the label (162.10(b)). False, misleading, and nonregistered names are prohibited. The draft guidelines elaborate with a series of restrictions designed to prevent the name or trademark from misleading consumers. In addition, the draft guidelines require that the label identify the broad pesticidal functions of the product, e.g., insecticide, herbicide, fungicide.

**Producers' Name:** Existing regulations require the label to include the name and address of the producer, registrant, or person for whom the product was produced. (162.10(e)). In its draft guidelines EPA is considering changing this provision to require that the name appearing on the label be that of the registrant or distributor. The agency's rationale for this change is that users will then be able to contact the party responsible for the product if questions or problems arise. The guidelines also encourage registrants to include telephone numbers or precise identification of company contact points on the label to facilitate communication by users.

**Product Registration Number:** Each label of a registered pesticide product is required to display the registration number assigned to the product at the time of registration (162.10(e)). The number must be preceded by the phrase "EPA Registration No." or "EPA Reg. No.". Registrants are admonished to take care that this phrase is not used in a manner which suggests or implies EPA endorsement of the product.

**Producing Establishments Registration Number:** Every pesticide product is required to bear the registration number of the final establishment at which the product was produced (162.10(f)). The number must be preceded by the phrase "EPA Est." or "EPA Establishment". This number may appear on the immediate container or the wrapper or outside container of the product. It must appear on the wrapper or outer container if the number on the immediate container cannot be read clearly through the outside wrapping. The draft guidelines provide additional recommendations on this provision but make no significant changes.

**Net Weight or Measure of Contents:** The net measure of contents of the product must appear on the label or container (162.10(d)). This measure must reflect the average content of the package, unless specifically stated to be the minimum content, and must be exclusive of wrappers, container, or other materials.

The current regulations require contents statements for pesticide liquids to be expressed in conventional American units of fluid ounces, pints, quarts, and gallons. Contents statements for solids, semisolids, or mixtures of liquids and solids must be in terms of weight expressed in avoirdupois pounds and ounces. Metric units may be added, but may not replace, the conventional units of measure described above.

The actual contents of the package is permitted to vary above a minimum content or around an average content, but only if it represents an unavoidable deviation consistent with good manufacturing practice. No deviation is permitted below a stated minimum content, nor may the average content of a shipment of packages fall below the stated average content.

The draft guidelines make only minor changes in these requirements, including modifications to make them consistent with the Fair Packaging and Labeling Act (16 CFR 500) and to permit the use of metric units in lieu of, or in precedence to, U.S. units.

**Ingredients Statement:** The label of every pesticide product must bear a statement of the ingredients of the products (162.10(g)). This statement must include the name and percentage by weight of each active ingredient and the total percentage of all inert ingredients. Inert ingredients are those which have no pesticidal activity and are used to dissolve and dilute active ingredients or to facilitate their application.

The ingredients statement normally must appear on the front panel of the label. The ingredients statement must also appear on any outside wrapping or container, unless it can be clearly read through such wrapping.

In some cases it may be impracticable to place the ingredients statement on the front panel of the label, particularly if the pesticide is packaged in extremely small or odd-shaped containers, or if the statement is unusually long. When faced with this situation, registrants may request permission from EPA to place the ingredients statement elsewhere on the package. In any case, the text of the statement must run parallel with the text on the panel on which it appears. Furthermore, it must be clearly distinguishable and separate from other text.

Each active ingredient must be listed by its accepted common name, if one exists, followed by its chemical name. Trademark or proprietary names may not be used unless EPA has accepted them as common names.

In addition to names, the percentages of each active ingredient must be stated and, along with the percentage of inert ingredients, must add-up to 100%. EPA does not permit labels to express percentages of active ingredients in ranges. Registrants are expected to list percentages with as much precision as possible, assuming the exercise of good manufacturing practice. If variation between manufacturing batches is unavoidable, the percentage value stated for each active ingredient must be the lowest which may occur. If a pesticide contains arsenic in any form, the ingredients statement must contain a statement of the percentages of the total and water soluble arsenic, both expressed as elemental equivalent.

Some pesticides are subject to rapid deterioration in chemical composition following manufacture. Such products are required to bear the statement "Not for sale or use after (date)" in a prominent position on the label. Furthermore, these products must meet all label claims up to the stated expiration date.

Although inert ingredients normally need not be separately identified, EPA may require registrants to list those inert ingredients that the agency determines may pose a hazard to man or the environment.

The draft guidelines generally expand upon the current requirements by providing examples and additional informative material.

## Warnings and Precautionary Statements

In addition to product identification, the second major category of information that must be included in pesticide labels encompasses a series of warnings and precautionary statements covering the human, environmental, and physical and chemical hazards posed by the product (162.10(h)). The intent of these statements is to alert the pesticide user to potential hazards from exposure, to inform him of precautionary measures which will help avoid or reduce the possibility of injury, and to instruct him on practical treatment or other remedial steps he can take if exposed.

For human hazard the statements required by EPA include the signal word, child hazard warning, statements of practical treatment, and any general precautionary statements which are applicable. For environmental hazard, the EPA requires statements on toxicity to nontarget fish, wildlife, birds, aquatic invertebrates, beneficial insects, and plants. Statements required for physical and chemical hazards focus primarily on flammability, but may include hazards related to explosion, oxidation, chemical reactions, or other characteristics of the product.

EPA has grouped warnings and precautionary statements into those that must be displayed on the front panel of the label and those which may appear elsewhere.

**Required Front Panel Warning Statements:** EPA requires that the human hazard signal word, the child hazard warning, and, for the most toxic pesticides, the statements of practical treatment, must appear on the front panel of the label. Furthermore, all of these required statements must be grouped together and must stand out from other text and graphics on the label so that ordinary purchasers and users will not easily miss them. Table 10.1 gives the minimum type size requirements for the front panel warning statements on labels of different sizes.

The basic framework for labeling the human hazard of a pesticide product rests on four toxicity categories established by testing the product in five different acute toxicity test systems: Oral (ingestion), dermal (absorption through the skin), inhalation, eye irritation, and skin irritation. The toxicity categories range from I, the most toxic, to IV, the least toxic.

*Human Hazard Signal Word (162.10(h)(1)(i))*—The signal word "DANGER" has been designated by EPA as the word which must appear on the labels of all pesticide products which meet the criteria of Toxicity Category I. Furthermore, if a product meets those criteria due to its oral, dermal, or inhalation

#### Table 10.1: Minimum Type Size for Front Panel Warnings

| Size of Label Front Panel (in$^2$) | Required Signal Word (all capitals) (Points) | "Keep Out of Reach of Children" (Points) |
|---|---|---|
| 5 and under | 6 | 6 |
| Above 5 to 10 | 10 | 6 |
| Above 10 to 15 | 12 | 8 |
| Above 15 to 30 | 14 | 10 |
| Over 30 | 18 | 12 |

Table 10.2: Toxicity Categories and Related Hazard Criteria

| Hazard Indicators | . . . . . . . . . . . . . . . . Toxicity Categories . . . . . . . . . . . . . . . . | | | |
|---|---|---|---|---|
| | I | II | III | IV |
| Oral LD$_{50}$ (mg/kg) | Up to and including 50 | From 50 through 500 | From 500 through 5,000 | Greater than 5,000 |
| Inhalation LC$_{50}$ (mg/ℓ) | Up to and including 0.2 | From 0.2 through 2 | From 2.0 through 20 | Greater than 200 |
| Dermal LD$_{50}$ (mg/kg) | Up to and including 200 | From 200 through 2,000 | From 2,000 through 20,000 | Greater than 20,000 |
| Eye effects | Corrosive; corneal opacity not reversible within 7 days | Corneal opacity reversible within 7 days; irritation persisting for 7 days | No corneal opacity; irritation reversible within 7 days | No irritation |
| Skin effects | Corrosive | Severe irritation at 72 hours | Moderate irritation at 72 hours | Mild or slight irritation at 72 hours |

toxicity, it must also bear the word "POISON," in red on a contrasting background, and the skull and crossbones must be placed in the immediate area of the word "poison". Products which fall into Toxicity Category I because of eye and skin irritation need not display the skull and crossbones, only the signal word "danger".

Products in Toxicity Category II must display the signal word, "WARNING". Those in Toxicity Categories III and IV are required to bear the word, "CAUTION". EPA regulations do not permit the use of more than one human hazard signal word on the front panel of a label, nor do they permit the label to bear the signal word of a higher or lower Toxicity Category without a specific determination by the agency. The draft guidelines make very few changes in the statements and format of this section of the regulations. The information has been expanded, organized for easier use, and supplemented with EPA recommended statements for practical treatment measures.

*Child Hazard Warning (162.10(h)(1)(ii))*—The statement "Keep out of reach of children" must be prominently displayed on the front panel of every pesticide product. Although EPA can waive the requirement, it will do so only when the registrant can demonstrate that the likelihood of contact with children is extremely remote at every stage of the product's life cycle or if, by virtue of its intrinsic characteristics, the pesticide is approved for use on infants or small children.

The draft guidelines make few changes from existing regulations, except to spell out normal EPA practice, such as the requirement that the child hazard warning appear on a separate line above the signal word.

*Statement of Practical Treatment (162.10(h)(1)(iii))*—EPA prefers the term "practical treatment" to "first aid" and generally does not permit the use of the word "antidote", since few pesticides have specific antidotes. The agency also prefers that all practical treatment statements appear on the front panel of the label, but only requires front panel placement for pesticides falling into Toxicity Category I on the basis of oral, dermal, or inhalation toxicity. With EPA's permission users may place this statement elsewhere on the label if a reference statement, such as "See statement of practical treatment on back panel", is placed on the front panel near the word "Poison" and the skull and crossbones. The statement of practical treatment for other Toxicity Categories need not be displayed on the front panel, but is required to appear on other parts of the label.

As noted above, the draft guidelines contain EPA's general recommendations for statements of practical treatment based on current medical advice. However, the agency encourages registrants to evaluate their own products and to choose the most appropriate method recommended by physicians and toxicologists.

**Other Required Warnings:** In addition to the required front panel warning statements, EPA requires a series of other warning statements and precautionary measures to be displayed on the label. These statements must be grouped together under the general heading "Precautionary Statements" and under the appropriate subheadings of "Hazards to Humans and Domestic Animals", "Environmental Hazard", and "Physical or Chemical Hazard."

*Human Hazard*—For hazards to humans and domestic animals, the required precautionary statement serves as an expansion of the front panel signal word. Precautionary statements must indicate the nature of the hazard involved, the route or routes of exposure, and the precautions that should be taken to avoid accident, injury, or damage (162.10(h)(2)(i)). The appropriate hazard signal word must immediately precede the precautionary statement paragraph. See Table 10.3 for examples of precautionary statements by toxicity category as they appear in EPA's current regulations.

*Environmental Hazard*—EPA requires registrants to include precautionary statements on pesticide labels when the pesticide poses a hazard to nontarget organisms, other than humans or domestic animals (162.10(h)(2)(ii)). As with human hazard statements, environmental statements must indicate the nature of the hazard and appropriate precautions to avoid potential accident, injury, or damage.

EPA's regulations outline a number of different circumstances which require environmental hazard statements. For example, pesticides which are designed to be used outdoors but are not registered for aquatic uses must include the following statement: "Keep out of lakes, ponds, or streams. Do not contaminate water by cleaning of equipment or disposal of wastes." The regulations also include a general provision requiring appropriate label cautions to protect pollinating insects from pesticides used in agriculture, forestry, and mosquito abatement programs. In addition to such broad requirements, EPA's regulations provide specific criteria for acute toxicity to wildlife, fish, and birds, for pesticides intended for outside use. If such a pesticide contains an active ingredient with a mammalian acute oral $LD_{50}$ of 100 or less, or with an avian acute oral $LD_{50}$ of 100 mg/kg or less, or a subacute dietary $LC_{50}$ of 500 ppm or less, it must bear the statement: "This Pesticide is Toxic to Wildlife." If it contains an active ingredient with a fish acute $LC_{50}$ of 1 ppm or less it must bear the statement: "This Pesticide is Toxic to Fish." Finally, if on the basis of field studies or accident history it can be demonstrated that use of the pesticide may be fatal to birds, fish, or mammals, EPA requires the term "toxic" to be replaced with "extremely toxic" in the label statements outlined above.

### Table 10.3: Precautionary Statements by Toxicity Category

| Toxicity Category | Oral, Inhalation or Dermal Toxicity | Skin and Eye Local Effects |
|---|---|---|
| I | Fatal (poisonous) if swallowed (inhaled or absorbed through skin). Do not breathe vapor (dust or spray mist). Do not get in eyes, on skin, or on clothing. (Front panel statement of practical treatment required.) | Corrosive, causes eye and skin damage (or skin irritation). Do not get in eyes, on skin, or on clothing. Wear goggles or face shield and rubber gloves when handling. Harmful or fatal if swallowed. (Appropriate first aid statement required.) |
| II | May be fatal if swallowed (inhaled or absorbed through the skin.) Do not breathe vapors (dust or spray mist.) Do not get in eyes, on skin or on clothing. (Appropriate first aid statements required.) | Causes eye (and skin) irritation. Do not get in eyes, on skin, or on clothing. Harmful if swallowed. (Appropriate first aid statement required.) |
| III | Harmful if swallowed (inhaled or absorbed through the skin.) Avoid contact with skin (eyes or clothing). (Appropriate first aid statement required.) | Avoid contact with skin, eyes or clothing. In case of contact, immediately flush eyes or skin with plenty of water. Get medical attention if irritation persists. |
| IV | (No precautionary statements required.) | (No precautionary statements required.) |

EPA's draft guidelines extend and expand significantly upon the environmental hazard criteria and warning statements of the current regulations. For example, they include new toxicity criteria for honey bees and aquatic invertebrates. They also put less emphasis on broad, general label statements, such as "Keep out of lakes, ponds, and streams," in favor of narrower and more precise statements tailored to specific pesticides and specific uses. Over the years EPA has concluded that vague and ambiguous broad statements often lead to confusion among users and are difficult to enforce.

*Physical and Chemical Hazards*—EPA regulations require that the term "flammable" appear on the label of any pesticide product whose flash point falls between 20°F and 80°F. See Table 10.4 for a description of flammability criteria and appropriate required warning and precautionary statements.

Although EPA does not spell out criteria or statements for physical and chemical hazards other than flammability, it may require such statements on the label if data submitted with the application for registration indicate a potential hazard. The draft guidelines raise the upper limit of the flashpoint criteria from 80°F to 100°F, consistent with the Hazardous Materials Transportation Act, provide specific criteria for labeling pesticles as "non-flammable", and expand upon the flammability statements in Table 10.4.

Table 10.4: Flammability/Explosive Criteria and Warning Statements

| Flash Point | Required Heat |
|---|---|
| . . . . . . . . . . . . . . . . . . (A) **Pressurized Containers** . . . . . . . . . . . . . . . . . . | |
| Flash point at or below 20°F; if there is a flashback at any valve opening | Extremely flammable. Contents under pressure. Keep away from fire, sparks, and heated surfaces. Do not puncture or incinerate container. Exposure to temperatures above 130°F may cause bursting. |
| Flash point above 20°F and not over 80°F or if the flame extension is more than 18 inches long at a distance of 6 inches from the flame | Flammable. Contents under pressure. Keep away from heat, sparks, and open flame. Do not puncture or incinerate container. Exposure to temperature above 130°F may cause bursting. |
| All other pressurized containers | Contents under pressure. Do not use or store near heat or open flame. Do not puncture of incinerate container. Exposure to temperatures above 130°F may cause bursting. |
| . . . . . . . . . . . . . . . . . (B) **Nonpressurized Containers** . . . . . . . . . . . . . . . . . | |
| At or below 20°F | Extremely flammable. Keep away from fire, sparks, and heated surfaces. |
| Above 20°F and not over 80°F | Flammable. Keep away from heat and open flame. |
| Above 80°F and not over 150°F | Do not use or store near heat or open flame. |

## Directions For Use

The third major cateory of information that must be included on pesticide labels encompasses a variety of topics that fall under the general heading of "Directions For Use." As noted very early in this chapter, pesticides are registered on the basis of their uses against specified pests and for specified crops. The use directions are, in fact, the practical embodiment of the EPA's registration decision.

**General Requirements:** EPA requires registrants state directions for use in terms that can be read and understood easily by those who will be using or supervising the use of the pesticide. The objective of this requirement is to provide assurance, that if followed, the directions protect users from personal injury, and prevent unreasonable adverse environmental effects.

The EPA does not specify any particular location on the label for use directions, as long as they are conspicuous enough to read easily. Furthermore, the agency may permit registrants to put directions for use on printed or graphic material accompanying the pesticide if it is not necessary for the directions to appear on the label and if the label directs the user to any accompanying instructions. In such cases the agency may require the collateral labeling to be securely attached to, or placed within the outside wrapper of, each package of the pesticide.

EPA's general requirement for specific use directions may be waived by the agency for certain categories of pesticides and users: pesticides which are intended for use only by manufacturers of products, other than pesticide products, in their regular manufacturing processes; pesticide products for which sale is limited to physicians, veterinarians, or druggists; pesticide products which are intended for use only by formulators in preparing pesticides for sale to the public. In each case, the general exception is qualified by a number of conditions designed to assure the safe and limited of the pesticide.

**Statement of Use Classification:** EPA's existing regulations, promulgated in 1975, followed through on Section 3(d) of FIFRA by establishing a system whereby each pesticide product would be labeled with a statement either of "Restricted Use" or "General Use". The classification of pesticides into these two categories was supposed to have taken place by October 22, 1976. In fact, that deadline was never met. EPA has instead classified 60 pesticide active ingredients, representing approximately 2,000 end-use products, as "Restricted Use" pesticides based primarily on their acute toxicity. Very few products have been classified for general use and the agency has been reluctant to do so due to concern about misleading users to believe that the product may be used for "general" purposes that do not appear on the label. This development has little practical effect on registrants. Labels for nonrestricted use pesticides simply do not bear a "General Use" classification.

Restricted use pesticides are pesticides which may only be applied by or under the direct supervision of certified applicators or are subject to other regulatory restrictions. Certified applicators must have successfully completed a special State or EPA training program on safe application of particularly hazardous pesticides.

If a pesticide use is classified as restricted, the statement "Restricted Use Pesticide" must appear on the front panel of the label, set in type of the same minimum size as that required for the human hazard signal words, see Table 4.1. It must be prominently displayed in relation to the other text and graphic material on the label. In some cases the label of a product bearing restricted uses may also bear nonrestricted uses. Such products must be labeled as if all uses were restricted.

EPA requires that a summary statement of the terms of the restriction appear immediately below the term "Restricted Use Pesticide." For use restricted to certified applicators the required summary statement is: "For retail sale to use only by Certified Applicators or persons under their direct supervision and only for those uses covered by the Certified Applicator's certification." The appropriate wording for other types of restrictions will be determined by EPA, by regulation, on a case-by-case basis.

**Contents of Directions for Use:** Current regulations list a number of items that must be included under the heading "Directions for Use" (162.10(i)(2)). One of the most important from a law enforcement point of view is the statement: "It is a violation of Federal law to use this product in a manner inconsistent with its labeling." This statement underscores the legal and regulatory force of the pesticide label itself and must be highlighted as a paragraph separate from the other text in the Directions for Use.

In addition to the "use inconsistent" statement, the Directions for Use must enumerate the site or sites of application, e.g., the crops, animals, areas, or objects to be treated by the pesticide, the target pest or pests associated with each site and pest combination. They must also include instructions on methods of applying the pesticide, including, when required, directions on dilution and or the use of special application equipment. In addition, the label must give users instructions on the timing of pesticide application, the number of applications necessary, and the frequency of application.

EPA's current regulations (162.10(i)(2)(viii)) require each pesticide product to bear a reentry interval statement that meets the requirements of 40 CFR 170, Worker Protection Standards for Agricultural Pesticides. Part 170 sets out general occupational safety and health standards for farm workers. It does not prescribe specific label statements. Its general standards prohibit farm owners from permitting workers to enter treated fields until pesticide sprays have dried or dusts have settled, set specific minimum reentry intervals for twelve active ingredients, and prohibit the application of pesticides while unprotected workers are in the treatment area. Part 170 also provides for more specific and stringent reentry conditions to be imposed on a case-by-case basis and by states. A number of pesticide uses—mosquito abatement, greenhouses, livestock, and golf courses—are exempted from Part 170, although appropriate label instructions and restrictions must be followed. EPA has generally not provided specific guidance on reentry interval statements to registrants beyond that contained in 162.10 or Part 170. Some states, California in particular, have developed extensive reentry restrictions tailored to local conditions and uses.

EPA's existing regulations (162.10(i)(2)(ix)) also require the label to include, in the Directions for Use section, specific instructions on the storage and disposal of pesticides and pesticide containers. Such instructions must meet the requirements of 40 CFR 165 and be grouped under the heading "Storage and Disposal." They must be of the same type as the child hazard warning (see Table 10.1). Unfortunately, both 162.10 and Part 165 were promulgated prior to the enactment of the Resource Conservation and Recovery Act (RCRA), and its subsequent regulations concerning the disposal of hazardous wastes (40 CFR 261 Subpart C). EPA has exempted most farmers and small generators of hazardous waste from the RCRA regulations, but has yet to resolve the overlaps and inconsistencies between RCRA and FIFRA with respect to pesticide waste storage and disposal.

The existing provisions of 40 CFR 165 Subpart C list a series of recommended practices that should be followed in the storage and disposal of pesticides. These recommendations are very general in nature and contain few specific suggestions for label statements.

Finally, the Directions for Use must contain, where appropriate, statements limiting or restricting uses in order to prevent unreasonable adverse effects or to ensure that tolerance residues will not be exceeded. Such limitations include preharvest or preslaughter intervals, e.g., "Do not apply within ____ days of harvest"; rotational crop restrictions, e.g., "Do not plant ____ within 12 months of application"; and geographic restrictions such as those limiting applications to certain states or excluding application from critical wildlife habitats.

In the area of Directions for Use EPA's draft guidelines contain extensive modifications and expansions of existing regulations and add a number of new provisions not found in current regulations. Major expansions are proposed covering reentry intervals and storage and disposal. Brand new provisions include separate sections for manufacturing use products, unit packages, pest control devices, and products shipped between registered establishments operated by the same producer, intended solely for export, shipped under an emergency exemption, and transported for disposal. Finally, the Agency also includes in its draft guidelines new proposals for labeling specific types of products. These provisions generally correspond to the types of labeling requirements that have been imposed on specific categories of products over the years but that have never been published formally by the Agency. The types of products for which the draft guidelines contain specific new provisions include: antimicrobial agents; organic pest control products; fungicides and nematicides; terrestrial herbicides, plant regulators, desiccants, and defoliants; invertebrate control products; and vertebrate control products.

## FORMAT OF PESTICIDE LABELS

### General

As noted earlier in this chapter, FIFRA establishes the pesticide label as the major instrument for providing pesticide users with information, warnings, and instructions for use, as well as the primary representation of EPA's regulatory decision on registration of the product. An effective format for and organization of the label's various elements is critical to the fulfillment of its various purposes. Given the amount, complexity, and variety of information required, however, that is not an easy task. In developing its regulations on labeling, EPA adopted a "format labeling concept" which was designed to make label information easy to identify and to encourage users to read and follow the various instructions and precautions on the label. An example of a label format for a typical "Restricted Use Pesticide" is displayed in Figure 10.1 and explained in accompanying notes. EPA's draft guidelines generally retain and follow the format requirements given below with the exception of a proposal to increase the minimum type size from six points to eight points, with corresponding increases in other label elements, such as the signal word.

### Placement of the Label

In general, the EPA requires that the label appear on or be securely attached to the immediate container of the pesticide product (162.10(a)(4)). The regulations define "securely attached" as meaning that a label can be reasonably expected to remain affixed to the container during the foreseeable conditions and period of use. If the product is packaged in an outside wrapper then either the label must be clearly readable through that wrapper, or if not, it must be securely attached to the wrapper or outside container.

Pesticides in transit in bulk are subject to the labeling, marking, and placarding provisions of 49 CFR 170-189, concerning transportation of hazardous materials. A copy of the accepted EPA label must be attached to the papers that accompany the shipment and must be left with the consignee at the time of delivery. Similarly, pesticides stored in bulk, in the custody of the user, must have a copy of the EPA's approved label attached securely to the container in the immediate vicinity of the discharge valve.

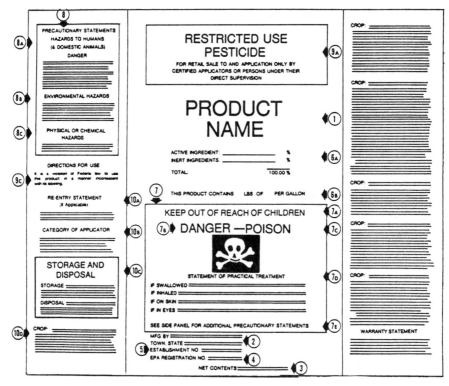

(Notes to Figure 10.1)

| Item | Label Element | Applicability of Requirement | Placement on Label Required | Preferred |
|---|---|---|---|---|
| 1 | Product Name | All Products | Front Panel | Center front panel |
| 2 | Company Name and Address | All Products | None | Bottom front panel or end of label text |
| 3 | Net Contents | All Products | None | Bottom front panel or end of label text |
| 4 | EPA Reg. No. | All Products | None | Front Panel |
| 5 | EPA Est. No. | All Products | None | Front panel, immediately before or following Reg. No. |
| 6a | Ingredients Statement | All Products | Front Panel | Immediately following product name |
| 6b | Pounds/Gallon Statement | Liquid products where dosage given as lbs ai/unit area | Front Panel | Directly below the main ingredients statement |
| 7 | FRONT PANEL PRECAUTIONARY STATEMENTS | All Products | Front Panel | |
| 7a | Keep Out of Reach of Children (Child Hazard Warning) | All Products | Front Panel | Above signal word |
| 7b | Signal Word | All Products | Front Panel | Immediately below Child Hazard Warning |
| 7c | Skull & Crossbones and word "POISON" (in red) | All products which are Category I based on oral, dermal or inhalation toxicity | Front Panel | Both in close proximity to signal word |

**Figure 10.1:** Sample label format.

| 7d | Statement of Practical Treatment | All products in Categories I, II and III | *Category I:* Front panel unless referral statements is used. *Others:* Grouped with side panel precautionary statements | Front panel for all |
|----|------|------|------|------|
| 7e | Referral Statement | All products where precautionary labeling appears on other than front panel | Front Panel | |
| 8 | SIDE/BACK PANEL PRECAUTIONARY STATEMENTS | All Products | None | Top or side of back panel preceding Directions for Use |
| 8a | Hazards to Humans and Domestic Animals | All Products in Categories I, II and III | None | Same as above |
| 8b | Environmental Hazards | All Products | None | Same as above |
| 8c | Physical or Chemical Hazards | All pressurized products; others with flash points under 150°F | None | Same as above |
| 9a | Restricted Block | All RESTRICTED products | Top center of front panel | Preferably blocked front panel |
| 9c | Misuse Statement | All products | Immediately following Statement of Classification or head of Directions for Use | |
| 10a | Re-entry Statement | All cholinesterase inhibitors | In the Directions for Use | Immediately after Misuse Statement |
| 10b | Category of Applicator | All RESTRICTED products | In the Directions for Use | Immediately after Re-entry Statement (when used) |
| 10c | Storage and Disposal Block | All products | In the Directions for Use | Immediately before specific directions for use or at the end of directions for use |
| 10d | Directions for Use | All products | None | None |

**Figure 10.1:** *(Continued)*

## Placement and Prominence of Label Statements

EPA requires that all information required to be on the label - words, statements, graphics - be clearly legible to a person with normal vision. Furthermore, they must be conspicuous in relation to other material on the label to the extent that the ordinary person would be likely to read the label under normal purchase and use circumstances.

All required label text must be on a clear contrasting background. EPA has encouraged the use of dark colored text against a lighter background rather than light against dark. In any event, the text must not be obscured or crowded. The minimum type size permitted by EPA is six-point, although the Agency encourages use of larger size type.

Finally, all required text must be in English, although EPA may require or the registrant may propose text in other languages. When another language is used on a label, it will be subject to the same requirements as the English language version.

## LABELING FOR EXPERIMENTAL USE PERMITS

In general, the labeling requirements for pesticides shipped and used under an experimental use permits (40 CFR 172.6) represent a skeleton version of the requirements for registered pesticides.

The label must prominently display the words: "For Experimental Use Only." In addition it must include: the Experimental Use Permit number; the establishment registration number; the statement "Not for sale to any person other than a participant or cooperator of the EPA-approved Experimental Use Program"; the same, brand, or trade-mark; and the name and address of the permittee, producer, or registrant. EPA also requires experimental use permitted pesticide labels to inform users of the net contents and ingredients, and to include appropriate warnings, precautionary statements, and limitations on reentry into treated areas. The label must also include directions for use, unless EPA has approved the experimental program itself as labeling, in which case the program documentation must be distributed with the product. Registered pesticides may also be approved for use under an experimental use permit. When this occurs the Agency may require that supplemental labeling accompany the product for its use in the experimental program.

## LABEL IMPROVEMENT PROGRAM

EPA's pesticide labeling regulations discussed in this chapter were originally promulgated in July of 1975. The Agency has gained a lot of experience since then in reviewing registration applications and in approving specific labeling language for individual pesticide products and for different crops, pests, and uses. Much of this experience has been captured and included in the draft pesticide labeling guidelines that have been referred to throughout the text of this chapter.

The draft guidelines however, if and when they are promulgated, are designed to govern labeling decisions made as of and after their promulgation. They are not effective retroactively to pesticide labels approved in the past under older and different standards. Theoretically, these older labels will be revised as part of the development of registration standards for the universe of pesticides in commerce (see discussion earlier in this chapter). Unfortunately the registration standards program can only review a small proportion of pesticides on the market in any given year and some pesticide labels will not be comprehensively reviewed for many years, perhaps as long as 15 years.

In view of this, and in view of the fact that many pesticide labels have not been reviewed at all since 1972, or earlier, EPA has established the Label Improvement Program as a means to bringing old or obsolete labels up-to-date, to delete incorrect recommendations, and to provide expanded information, in certain areas that are especially important to the protection of public health and the environment.

The Label Improvement Program was initiated by a notice in the Federal Register, dated June 5, 1980 (45 FR 37884). Under this program registants will be required to amend their registrations to modify product labels to conform with specifications received from EPA. As of the writing of this chapter, EPA has published three Label Improvement Program Notices covering the deletion of salt water emesis statements, fumigants, and termiticides. The first notice required registrants to discontinue recommending salt water as an emetic agent for inducing vomiting. Medical opinion no longer regards salt solutions as effective for that purpose and, indeed, believes that some children risk salt poisoning from this method of emesis. The second notice covered fumigation of vessels, box cars, trucks, warehouses and similar enclosed containers and structures. It required registrants of certain fumigants to amend their product labels in ways which would minimize the exposure of persons working in and around fumigated areas to dangerous levels of pesticides. The notice on termiticides is aimed at reduced homeowner exposure and minimizing the risk of groundwater contamination.

In addition to the three notices already promulgated, the Agency is planning future Label Improvement Notices on ingredients statements, analytical methods, eye irritation standards, antifouling paints, and farm worker safety.

## CONCLUSION

Pesticide labeling is at an unusually and, to registrants, confusing juncture in its development. EPA is formally operating on the basis of regulations that grew out of a period when the potential dangers of exposure were not as well recognized as they are today and when the methods and technology for using pesticides safely were not as advanced as they are today. At the same time EPA has launched several different efforts which could result in changing and up-dating labeling requirements for many pesticide products: the pesticide labeling guidelines, the Label Improvement Program, and the registration standards program. Members of the public who produce and use pesticides will have to remain alert to the various and changing developments in this area since it does not appear likely that any single pattern will emerge in the near future.

# 11

# Labeling Under the Toxic Substances Control Act (TSCA)

**Robert M. Sussman**
*Covington and Burling*
*Washington, D.C.*

The Toxic Substances Control Act (TSCA), Pub. L. 94-469, was signed into law on October 11, 1976 and became effective on January 1, 1977. One of the most complex and detailed statutes passed by Congress, TSCA is implemented by the Environmental Protection Agency (EPA). The Act confers comprehensive authority on EPA to evaluate and, if necessary, regulate the effects of chemicals on human health and the environment. Within EPA, TSCA has been implemented by the Office of Pesticide and Toxic Substances (OPTS), which is directed by an EPA Assistant Administrator.

TSCA applies to all "chemical substances" and "mixtures" manufactured, processed, distributed or used in the United States. However, excluded from the Act's coverage by Section 3(2) are chemicals whose specific uses are subject to preexisting federal laws like the Federal Food, Drug, and Cosmetic Act (FD&C Act) and the Federal Insecticide, Fungicide, and Rodenticide Act (FIFRA). In addition, while EPA's jurisdiction under TSCA is extremely broad, the statute reflects a Congressional intent to limit EPA's focus to those aspects of industrial chemicals that cannot be adequately controlled under other health and environmental statutes. Thus, as a practical matter, the Occupational Safety and Health Administration (OSHA) has remained primarily responsible for assuring the safe use of chemical substances and mixtures in workplaces; the Department of Transportation (DOT) has continued to exercise primary authority over the safe shipment of chemicals in commerce; and EPA itself has continued to regulate the discharge of chemicals into the environment principally under existing laws like the Clean Air Act, the Federal Water Pollution Control Act, and the Resource Conservation and Recovery Act.

Despite these limitations on EPA's activities, the Act affords substantial opportunities for the development of regulations requiring the evaluation and control of industrial chemicals. Areas of major concern under TSCA include the promulgation of testing requirements (Section 4); the review and control of new chemical substances (Section 5); restrictions on the manufacture and distribution of unsafe chemicals (Section 6); and the collection and analysis of information and data concerning the effects of chemicals on health and the environment (Section 8).

EPA has authority to require chemical labeling under Section 6 of TSCA. This provision empowers EPA to develop labeling requirements for particular chemicals or chemical classes (Section 6(a) ). It also specifically directs EPA to promulgate labeling requirements for polychlorinated biphenyls (PCBs) by prescribed deadlines (Section 6(e) ). While EPA has not yet formally exercised its labeling authority except in connection with PCBs, the development of hazard warnings under TSCA for individual chemicals or chemical classes remains a distinct possibility for the future.

Presented below is an analysis of EPA's general labeling authority under Section 6(a) and its special PCB labeling requirements under Section 6(e).

## EPA's GENERAL LABELING AUTHORITY UNDER SECTION 6(a)

Starting in 1979, EPA devoted substantial effort to developing proposed hazard warning labeling regulations for industrial chemicals shipped in commerce. These regulations, which would have applied to all chemical substances and mixtures subject to TSCA, were intended to complement OSHA hazard warning regulations applicable in the workplace. It was expected that OSHA regulations would assure the availability of information concerning chemical hazards in workplace settings where chemicals are used, while EPA regulations would provide for the dissemination of similar information as part of the labeling of chemicals shipped in commerce. Under the EPA phase of this program, the containers of industrial chemicals distributed commercially would be required to contain warning labels for acute hazards modeled on the voluntary standard for chemical hazard labeling adopted by the American National Standards Institute (ANSI). For chronic hazards, EPA intended to require warning labels on the containers of all chemical substances which the Agency had determined to be human or animal carcinogens.

EPA had virtually completed preparation of a notice of proposed rulemaking embodying the above approach when it abandoned its labeling activities in order to allow full responsibility for chemical labeling to be assumed by OSHA. While OSHA thereafter published a proposed labeling rule in the Federal Register (46 Fed. Reg. 4412), that proposal was subsequently revoked by the Reagan Administration in the face of widespread industry opposition. Since that time, OSHA has continued efforts to develop proposed labeling regulations, and a formal *Federal Register* notice embodying a revised OSHA approach has now been published.

The status of EPA's labeling efforts is still unclear. Although a recent report of the House Committee on Energy and Commerce urged EPA to resume its labeling activities under TSCA,[1] the Agency has not yet clarified whether it intends to pursue such a course. Industry attitudes towards a resumption of EPA's labeling efforts are mixed. On the one hand, there is a widespread belief that current labeling practices under the ANSI voluntary standard are adequate and a formal Federal labeling program is unnecessary. On the other hand, it is recognized that OSHA lacks statutory authority to mandate labeling requirements for chemicals shipped in commerce and that such requirements could only be imposed by EPA. Some observers have suggested that narrowly drawn EPA labeling regulations under TSCA may therefore be necessary to complement any OSHA requirements applicable in the workplace.

Although EPA's labeling activities were curtailed before the Agency issued a formal rulemaking proposal, numerous issues relating to EPA's labeling authority under TSCA were crystalized and discussed. Described below are the principal aspects of EPA's labeling authority under TSCA which would come into play if EPA decides to apply label requirements to individual chemicals or chemical classes.

### Definition of Unreasonable Risk

Under Section 6 of TSCA, EPA is empowered to take a broad spectrum of regulatory actions in order to reduce or eliminate risks to human health or the environment. These actions vary markedly in their severity and potential impact on manufacturers and processors. EPA may: (1) prohibit or limit the manufacturing, processing or distribution of a chemical; (2) prohibit or limit the manufacturing, processing or distribution of a chemical for a particular use or in excess of a specified concentration; (3) require that a chemical be marked with or accompanied by clear and adequate warnings and instructions; (4) prohibit or otherwise regulate any manner or method of a chemical's commercial use; (5) prohibit or otherwise regulate any manner or method of a chemical's disposal; or (6) require manufacturers or processors of a chemical to notify distributors or the general public of its risks and either replace the chemical with an alternative product or repurchase it from its users.

In choosing among these regulatory measures, Section 6(a) directs EPA to restrict the chemical only "to the extent necessary to protect adequately against [the] risk" it poses and to impose "the least burdensome requirements." This emphasis on minimizing the burdens of regulation is also reflected in Section 2(b)(3) of TSCA, under which EPA must exercise its authority "in such a manner as not to impede unduly or create unnecessary economic barriers to technological innovation."

These provisions have the effect of requiring EPA to consider label warnings and similar instructional measures as an alternative to banning or limiting the production of a chemical that it believes may be unsafe. As stated by Representative McCollister during the House debate on TSCA, "EPA could not ban a substance for a particular use if a labeling requirement would provide adequate protection." *Legislative History of Toxic Substances Control Act* 519 (1976). Thus, where EPA elects not to require label warnings but to impose more stringent requirements, it must be able to show that label warnings would not be adequate to protect exposed members of the population or the environment from harm.

Even where EPA seeks to impose labeling requirements, moreover, the Agency has the burden of demonstrating that the chemical or chemical class it seeks to regulate is capable of adverse human health or environmental effects. Under Section 6(a), before EPA can invoke any of the Section's regulatory mechanisms, including labeling requirements, it must determine that there is "a reasonable basis to conclude" that the chemical "presents or will present" an "unreasonable risk of injury to health or the environment."[2]

While TSCA does not specifically define the term "unreasonable risk," some guidance on its meaning is provided by Section 6(c). Under this provision, EPA must consider, and publish a statement with respect to, four factors before promulgating a rule under Section 6(a):

> (A) the effects of such substances or mixture on health and the magnitude of the exposure of human beings to such substance or mixture;
>
> (B) the effects of such substance or mixture on the environment and the magnitude of the exposure of the environment to such substance or mixture;
>
> (C) the benefits of such substance or mixture for various uses and the availability of substitutes for such uses; and
>
> (D) the reasonably ascertainable economic consequences of the rule, after consideration of the effect on the national economy, small business, technological innovation, the environment, and public health.

EPA's findings on these issues must, under Section 19(c) of TSCA, be "supported by substantial evidence in the rulemaking record . . . taken as a whole."

Taken together, the four factors specified by Section 6(c) require EPA to weight the potential health or environmental harm that regulatory action will prevent against the adverse economic and social impact of regulatory requirements. The first phase of this balancing process is specified by paragraphs (A) and (B), which direct EPA to assess the chemical's harmful effects on health and the environment and the magnitude of human and environmental exposure to the chemical. Paragraphs (C) and (D), by contrast, direct EPA to evaluate the burdens that regulating the chemical will place on industry and the public. The Agency must consider the chemical's "benefits" for "various uses" and "the availability of substitutes for such uses." It must also consider "the reasonably ascertainable economic consequences" of its proposed form of regulatory control.

The balancing process required by Section 6(c) suggests that the "unreasonableness" of a risk is to be judged in terms of the regulatory action by which EPA proposes to control that risk. If the rule's adverse economic consequences will be substantial but the chemical's potential for health or environment harm will be small, EPA could properly conclude that the risk associated with the chemical is "reasonable." On the other hand, where the proposed rule would not materially reduce the chemical's benefits but would prevent serious injury to health or the environment, the Agency could properly conclude that the chemical's risks are "unreasonable."

Confirming these principles is the legislative history of Section 6. The only draft version of TSCA which expressly defined "unreasonable risk" was H.R. 7548, 94th Cong., 1st Sess. (1976), which provided as follows:

> (23) "unreasonable risk" means any risk of a chemical substance to human health or the environment which outweighs the benefits of the use of the chemical substance to the consuming public or the environment.

No definition of "unreasonable risk" was included in the final House bill. However, the House Report makes it clear that the House endorsed the general approach of balancing a chemical's risks and benefits:

> During the hearings, a number of witnesses recommended that the bill include a definition of unreasonable risk. Because the determination of unreasonable risk involves a consideration of probability, severity, and similar factors which cannot be defined in precise terms and is not a factual determination but rather requires the exercise of judgment on the part of the person making it, the Committee did not attempt a definition of such risk. *In general, a determination that a risk associated with chemical substance or mixture is unreasonable involves balancing the probability that harm will occur and the magnitude and severity of that harm against the effect of proposed regulatory action on the availability to society of the benefits of the substance or mixture, taking into account the availability of substitutes for the substance or mixture which do not require regulation, and other adverse effects which such proposed action may have on society.* [H.R. Rep. No. 94-1341, 94th Cong. 2d Sess. 13-14 (1976) (emphasis supplied).]

Of particular note in this passage is the Committee's explicit recognition that an "unreasonable risk" determination involves balancing a chemical's potential for health or environmental harm against the "effect of proposed regulatory action" on the chemical's benefits and the "other adverse effects which such proposed action may have on society." The clear implication of this statement is that the "reasonableness" of a risk depends on the nature of the regulatory controls under consideration; when EPA proposes more stringent regulatory mechanisms such as a ban on production, it must demonstrate that a chemical poses a correspondingly greater threat to health or the environment.

To the same effect is another passage in the House Report, which discusses the different provisions of TSCA where the term "unreasonable risk" appears:

> Although the standard for defining the regulatory authority of the Administrator throughout the bill is unreasonable risk, the implementation of the standard will of necessity vary depending on the specific regulatory authority which the Administrator seeks to exercise. For example, a testing rule under section 4 will ordinarily not result in depriving the public of the benefits of a substance or mixture sub-

ject to the rule. This is because such a rule does not prohibit the manufacture, processing, etc., of existing substance or of mixtures. At the most a testing rule may, through section 5(d), delay the commercial availability of new substances and new uses of existing substances subject to the testing rule. Similarly, a requirement imposed under section 5(g) (regulation of new substances and significant new uses of substances pending the development of information) will only delay or restrict the availability of a substance subject to it until adequate health and safety data can be developed and evaluated.

*However, this is to be contrasted with the effect of the imposition of a requirement under section 6 on a substance. Such a requirement may remove a substance from the market or impose lesser restrictions on its availability and such a requirement is not of limited duration. This, the effect on society may be far-reaching. As a result regulatory effect will be of greater significance in a determination of unreasonable risk for purposes of section 6 than for a determination for purposes of section 4 or 5(g).* Conversely, with respect to section 4 or 5(g), because the regulatory effect of action taken under either of those sections is less than that of action taken under section 6, the requirements for a determination of unreasonable risk for purposes of section 4 or 5(g) are less demanding. [H.R. Rep. No. 94-1341, *supra,* at 14–15.]

This passage principally relates to EPA's greater burden of proof when proceeding under Section 6 as opposed to the other provisions of TSCA which require a showing of "unreasonable risk." Nevertheless, the Committee's reasoning is equally applicable to EPA's burden of proof when choosing among the various regulatory mechanisms prescribed by Section 6 itself. Where EPA proposes to "remove a substance from the market or impose lesser restrictions on its availability", the "effect on society may be far-reaching." Conversely, where EPA proposes less drastic measures like labeling, the "regulatory effect of [its] action . . . is less" and "the requirements for a determination of unreasonable risk . . . are less demanding." Thus, a risk which may be sufficiently "unreasonable" to justify labeling a chemical may be a "reasonable" risk where EPA proposes a ban or limitation on the chemical's production.

One consequence of this principle is that, once EPA has adopted labeling requirements for a chemical, it cannot impose more drastic requirements on that chemical on the basis of its original "unreasonable risk" determination. Rather, a new "unreasonable risk" determination concerning the chemical will be necessary. For this purpose, EPA must again analyze the evidence relating to the chemical's risks and benefits. Based on this examination, the Agency must then make the findings required by Section 6(c) in light of the further regulatory restrictions which it proposes to impose. Only if the Agency determines that the adverse economic consequences of such restrictions are outweighed by the protection they can afford to health or the environment would EPA be able to support a new "unreasonable risk" determination.

This approach is also mandated by those portions of Section 6(a) which permit EPA to impose restrictions on a chemical only to the "extent necessary to protect adequately against [the] risk" it poses and only if those restrictions represent "the least burdensome requirements" available. Under these provisions, EPA must be able to show that the regulatory mechanism it has selected represents the least drastic means of effectively controlling a chemical and that less burdensome approaches could not protect health or the environment to an acceptable extent.

Once EPA has determined that a given requirement (e.g., labeling) meets this standard, there would be a presumption that more drastic requirements (e.g., a ban on production) would be unduly "burdensome" and therefore "unnecessary" to control the chemical's risks. At the very least, a burden would now be on the Agency to show that its initial evaluation of the relative costs and benefits of regulatory action was erroneous, and more severe restrictions on the chemical are in fact "necessary." To make this showing, EPA would need to conduct a full reanalysis of the evidence concerning the chemical's health and environmental effects, exposure patterns, uses and benefits.

## Hearings

Before any labeling requirement can be imposed under Section 6, a hearing can be demanded pursuant to Sections 6(c)(2) and (3) of the Act on disputed issues of material fact. Thus, all of the statutory requirements are subject to close scrutiny through the hearing process. At such a hearing, live witness testimony could be presented by the various parties, and these witnesses would be subject to cross-examination and rebuttal on disputed factual issues.

## Adequate Protection

As noted above, Section 6(a) requires that EPA find that any label warning that it proposes is "necessary to protect adequately against such risk." This requirement has two practical consequences. First, it underscores that only real and substantial risks can be the subject of a required warning. To the extent that the risk is only theoretical or insubstantial, there could be no showing that the warning is "necessary" to provide "adequate" protection. Second, this statutory provision requires proof that the warning will in fact provide protection. EPA is thus without authority to require a warning absent evidence that it will be effective to protect against the risk involved.

## Least Burdensome Requirement

Section 6(a) provides that, in considering regulatory alternatives, EPA must choose "the least burdensome requirements." Under Section 6(a)(3), when EPA determines that some kind of warning meets the other statutory criteria, it has discretion to determine whether the substance or mixture should "be marked with or accompanied by clear and adequate warnings and instructions." Thus, in lieu of label warnings, EPA could instead require a wide variety of other educational efforts, such as instruction through employee programs, placards, employee manuals, and other techniques. EPA must analyze all of these alternatives and adopt the least burdensome requirement that will provide adequate protection against the unreasonable risk found to exist.

### Category of Chemicals

Sections 6(a) and (c) refer to particular chemical substances or mixtures, rather than to chemicals in general. Thus, except in unusual circumstances, Section 6 labeling rules would be limited to individual chemicals or mixtures which pose a specified risk.

Section 26(c) provides, however, that EPA may take action under the Act with respect to a category of chemical substances or mixtures, rather than individual chemical substances or mixtures, where such category is "suitable for classification as such for purposes of this Act." As examples of proper categories, the Act refers to groups of chemicals with similarities in (1) molecular structure, (2) physical, chemical, or biological properties, (3) use, or (4) mode of entry into the human body or into the environment. According to TSCA's legislative history, "in taking action under any provision of the [Act] respecting a category of chemical substances, the Administrator will not have to make the requisite finding for such action with respect to every chemical within the category." H.R. Rep. No. 94-1341, 94th Cong., 2d Sess. 61 (1976).

Because Section 6 is intended to apply primarily to individual chemical substances, regulation of entire categories should be limited to situations where EPA demonstrates that the categorization involved is appropriate to fulfill the objectives of Section 6. Under this approach, EPA's definition of a category will not be "suitable" to achieve the "purpose of the Act" unless EPA proves that the definition constitutes a rational grouping of chemicals under the Section 6 criteria for determining "unreasonable risk."

For example, as discussed above, Section 6(c) requires EPA to make findings concerning the health and environmental effects of the chemicals under consideration for regulation, the magnitude of human exposure to those chemicals, the chemicals' benefits for their various uses, and the eonomic consequences of the proposed regulatory action. The purpose of these findings is to permit a careful balancing of the harm that regulatory requirements will prevent against the adverse social and economic effects of those requirements, taking into account the actual conditions of use and exposure associated with the chemicals in question. Any "category" which indiscriminately grouped together different chemicals with dissimilar use and exposure patterns—without making any allowances for the differences among them—would be an "unsuitable" category since EPA could not meaningfully engage in the balancing process that Secton 6(c) requires. Thus, EPA's definition of a category for Section 6 purposes should incorporate all of the factors—i.e., potential for harm, actual exposure, use, and economic effects—that an "unreasonable risk" determination involves.

Whenever EPA seeks to require label warnings by category, it must initially justify its definition of the category as a rational grouping of chemicals for Section 6 purposes. It would then seem necessary for the Agency to permit an opportunity for interested persons to show that particular chemicals or mixtures that fall within the category are not properly subject to the proposed labeling requirement because the statutory criteria for determining "unreasonable risk" do not apply. Without this opportunity, EPA's actions would arguably be unauthorized since EPA would then be imposing requirements under Section 6(a) on individual members of a category that do not satisfy the Section 6 criteria for demonstrating an "unreasonable risk."

### Relation to Other Statutes

Section 6(c) requires EPA to determine whether a risk could be eliminated or reduced to a sufficient extent by actions taken under another Federal law or laws administered in whole or in part by EPA. In addition, Section 9(a) requires EPA to determine whether the risk may be prevented or reduced to a sufficient extent by action taken under other Federal laws. In the latter case, EPA must submit information about the risk to the other agency and ask for a response. Both the EPA report and the response of the other agency must be published in the *Federal Register*. If the other agency does not agree that a risk is presented, or undertakes itself to control the risk, EPA is precluded from taking action under Section 6 of TSCA. Finally, Section 9(d) requires coordination by EPA with other Federal agencies in order to impose "the least burdensome or duplicative requirements on those subject to the Act."

A number of other Federal health and safety statutes bear upon the requirement that EPA determine whether the laws administered by it or other Federal agencies adquately protect against the risk associated with a chemical. For example, OSHA protects against workplace hazards, DOT against transportation hazards, CPSC against household hazards, and the air and water statutes against both human and environmental hazards generally. EPA is obligated to examine each of these statutes to determine whether it is potentially applicable, its relationship to the specific risk at hand, and the possibility that it would be a better vehicle for controlling that risk than Section 6 of TSCA.

### Effective Date

Section 6(d) states that the effective date of any regulation requiring a label warning shall be specified in the regulation itself, and shall be "as soon as feasible." To determine whether a proposed effective date is "feasible", EPA is obligated to undertake a full inquiry into the economic and practical consequences of potential effective dates before it imposes the final effective date for its labeling regulation. Such an inquiry must encompass the economic value of the inventory of current labels that would have to be destroyed, the availability of printing resources to undertake all of the relabeling required, and other similar considerations.

### Required Testing

Section 4 of TSCA contains detailed criteria for determining when EPA may require industry to test specified chemicals. Both these criteria and TSCA's legislative history suggest that testing is to be highly selective and, for chemicals covered by the Act, will be the exception rather than the norm. Thus, it can be argued that, using its authority under Section 6, EPA has no power to require chemicals to be tested as a first step in complying with labeling requirements. Rather, EPA's labeling rules can only require warnings for those chemicals which have either been tested already or for which existing information and experience indicate that warnings are applicable.

Section 6(a)(4) of TSCA authorizes EPA to require manufacturers and processors of a substance or mixture which presents an unreasonable risk to "make and retain records of the processes used to manufacture or process such substance or mixture and monitor or conduct tests which are reasonable and necessary to assure compliance" with any regulatory requirements which are imposed. It is possible that EPA could invoke this provision as the basis for requiring testing in order to determine whether a chemical meets the Agency's criteria for warning requirements. In response to this position, it could be argued that Section 6(a)(4) only grants EPA limited authority to require quality assurance testing in those cases where its rules prohibit or limit the use of particular contaminants, impurities or other chemicals. A broader interpretation of this provision, it could be maintained, would allow EPA to circumvent the requirements of Section 4 and mandate across-the-board testing under Section 6.

### Confidentiality Protections

A legal issue which could receive careful attention in any EPA rulemaking proceeding involving labeling is the Agency's authority to require the chemical industry to utilize warning labels which disclose confidential information, including the precise chemical identity of substances or mixtures whose composition is considered to be a trade secret. Governing the resolution of this issue would be Section 14 of TSCA, which addresses the disclosure of confidential data under the Act.

Section 14(a) of TSCA prohibits EPA, except in certain limited circumstances described below, from disclosing information that falls within Exemption 4 of the Freedom of Information Act (FOIA), 5 U.S.C. § 552(b)(4). FOIA Exemption 4 in turn encompasses two categories of information: (1) trade secrets and (2) commercial or financial information that is privileged or confidential. Although Congress did not explicity define these terms in the FOIA, the definition of "trade secret" had been well established by judicial decision before FOIA was enacted, and a definition of confidential commercial information has since evolved through FOIA litigation. When TSCA was enacted, Congress was thus well aware of the judicially established definitions of the Exemption 4 categories. Indeed, the Act's legislative history clearly expresses Congress' intention that confidentiality determinations under Section 14(a) would be governed by the substantive criteria established in the case law interpreting FOIA Exemption 4. See S. Rep. No. 698, 94th Cong., 2d Sess. 76 (1976); H.R. Rep. No. 1341, 94th Cong., 2d Sess. 50 (1976).

The *Restatement of Torts*[3] broadly defines a "trade secret" as follows:

> A trade secret may consist of any formula, pattern, device or compilation of information which is used in one's business, and which gives him an opportunity to obtain an advantage over competitors who do not know or use it. It may be a formula for a chemical compound, a process of manufacturing, treating or preserving materials, a pattern for a machine or other device, or a list of customers. . . . .

This definition has been consistently applied in trade secret litigation[4] and was adopted by the Supreme Court for federal patent law.[5] Moreover, courts have explicitly applied the *Restatement* definition in cases arising under FOIA.[6]

As is clear from the *Restatement* definition itself, chemical formulas and process information are classic trade secrets because of their ongoing commercial value to innovators. However, the *Restatement* definition also extends to "any . . . compilation of information" that confers a competitive advantage upon those who possess it. To determine whether specific information constitutes a "trade secret" under this aspect of the definition, courts must consider the following factors set forth in the *Restatement:*

1. the extent to which the data is independently known to outsiders or is used by outsiders for similar purposes;

2. the extent to which it is known by insiders;

3. the extent of the measures taken by an owner to guard its secrecy;

4. value of the data to the owner and others, including the extent to which, if used in conduct of the business, it would confer a competitive advantage on said owner;

5. the amount of effort or money expended on developing the data; and

6. the ease or difficulty with which the data could properly be acquired or duplicated by others.

As noted above, FOIA Exemption 4 protects not only trade secrets but also confidential commercial and financial information. Commercial or financial information is "confidential" if its disclosure "is *likely* . . . to cause substantial harm to the competitive position of the person from whom the information was obtained." *National Parks and Conservation Association* v. *Morton,* 498 F.2d 765, 770 (D.C. Cir. 1974) (*National Parks I*) (emphasis added). This test does not require that harm be proven with mathematical precision. Rather, it looks generally to the commercial sensitivity of the information. For example, in *National Parks and Conservation Association* v. *Kleppe,* 547 F.2d 637, 683 (D.C. Cir. 1976) (*National Parks II*), the United States Court of Appeals for the District of Columbia stressed that a court (or agency) "need only exercise its judgment in view of the nature of the material sought and the competitive circumstances in which the [information submitters] do business . . . ." Under this standard, virtually any information could qualify for nondisclosure providing its competitive sensitivity could be adequately documented.

Because of the broad protection afforded by Section 14(a), EPA has limited authority to require warning labels that disclose the specific identity of particular chemicals or other confidential information. Nevertheless, it should be noted that Section 14 permits disclosure of confidential information under certain enumerated criteria, which might be met in a labeling proceeding under appropriate circumstances.

First, Section 14(a)(3) authorizes disclosure when EPA determines it is "necessary to protect health or the environment against unreasonable risk." Thus,

EPA could require information on warning labels where there is evidence that the chemicals in question present an "unreasonable risk" and disclosure of confidential information is necessary to protect against that risk. It should be noted, however, that this provision imposes a high standard on EPA. The Agency must show not merely that a chemical poses an "unreasonable risk" but that the inclusion of confidential information in warning labels represents the only effective mechanism for controlling that risk and other approaches are inadequate. EPA probably could not meet this standard where users of a chemical would be adequately informed about potential risks by a generic description of the chemical's identity or where a description of the hazard posed by the chemical and appropriate precautionary instructions are adequate to assure its safe use.

Yet another exception to the nondisclosure principle of Section 14(a) is contained in Section 14(a)(4), which permits the disclosure of confidential information "when relevant in any proceeding under this Act . . ." It is possible that EPA could argue that the promulgation of labeling requirements under Section 6 represents a "proceeding" within the meaning of this provision and that, accordingly, the Agency's final labeling rule can require the disclosure of confidential information where it is "relevant." Even if this rationale were to apply,[7] however, Section 14(a) would require EPA to disclose the information in question "in such manner as to preserve confidentiality to the extent practicable without impairing the proceeding." Thus, here as well, EPA would only be able to disclose confidential information if it determined that no other approach would be adequate to achieve its regulatory objectives. If releasing confidential information would have a detrimental commercial impact but would not contribute signficantly to public health protection, such release would be precluded.

## LABELING REQUIREMENTS FOR PCBs

### Statutory Provisions Relating to PCBs

Polychlorinated biphenyls ("PCBs") have been manufactured and used commercially for 50 years because of their chemical stability, fire resistance, and electrical resistance properties. PCBs are frequently used in electrical transformers and capacitors. However, concern has been expressed that PCBs may be toxic to humans and to wildlife. Because of these concerns, the major American manufacturer of PCBs limited its sales of PCBs after 1972 to manufacturers of transformers and capacitors and then in 1977 ceased all manufacture of PCBs and shipped the last of its inventory. Today, PCBs are produced in this country only as incidental byproducts of industrial chemical processes. There are known natural sources of PCBs.

Although TSCA is generally designed to cover the regulation of all chemical substances, Section 6(e) refers solely to the disposal, manufacture, processing, distribution, and use of PCBs. No other section of the Act addresses the regulation of a single class of chemicals in this fashion. The unusual attention that TSCA focusses on PCBs reflects the high priority that Congress assigned to PCB control. *See, e.g.,* H. Rep. No. 94-1341, *supra,* at 133-134.

Under Section 6(e)(1), within six months of the effective date of the Act, EPA is obligated to promulgate rules that prescribe methods for the disposal of PCBs and require PCBs to be marked with clear and adequate warnings, as well as with instructions with respect to their processing, distribution in commerce, use, or disposal. Section 6(e)(2) provides that, one year after the effective date of the Act, no person may manufacture, process, distribute in commerce, or use any PCB other than in a totally enclosed manner. Section 6(e)(3) provides that no person may manufacture any PCB after two years from the effective date of the Act, or process or distribute in commerce any PCB after two and one-half years after such date. The provisions of Sections 6(e)(1) and 6(e)(2) allow the Administrator to promulgate rules granting exemptions from the control measures mandated by Section 6(e).

### EPA's Disposal Regulations

EPA's first set of regulations to implement Section 6(e)—the so-called Disposal Regulations—set forth specific rules governing the disposal and marking of PCBs. These rules were proposed on May 24, 1977 (42 Fed. Reg. 26564) and issued in final form on February 17, 1978 (43 Fed. Reg. 7150). Covered by EPA's requirements were not only pure PCB compounds but mixtures containing at least 500 parts per million (ppm) PCBs and articles whose surfaces had been in contact with such PCB mixtures.

In the preamble to its proposed rule, EPA explained that its 500 ppm cutoff level was selected to differentiate between commercial products which generally are called PCBs and used as such and other commercial products in which PCBs may be produced as a side-effect of the manufacturing process. It was not regulating products in the latter category, EPA indicated, because both the economic effects of such regulation and its public health benefits were uncertain. 42 Fed. Reg. 26565. In the preamble to its final rule, EPA acknowledged that levels of PCB lower than 500 ppm could cause adverse health and environmental effects, and stated that the economic impact of regulation on products containing such PCB levels might be less substantial than at first appeared. 43 Fed. Reg. 7151. Nevertheless, the Agency announced that it would continue to use a 500 ppm cutoff level for disposal and marking requirements and would consider controlling lower PCB concentrations in its then-forthcoming regulations concerning PCB manufacture, processing use and distribution. 43 Fed. Reg. 7151.

The marking provisions of EPA's regulation were set forth in 40 C.F.R. § 761.20. Under these provisions, a variety of deadlines—ranging from July 1, 1978 to January 1, 1979—were prescribed for the marking of different PCB products. In addition, EPA devised different label statements whose application depended on the nature of the product involved.

Marking deadlines of July 11, 1978 were established for the following products: PCB containers, PCB transformers, PCB large high-voltage capacitors, equipment containing a PCB transformer or large high-voltage capacitor, PCB large low-voltage capacitors, electric motors using PCB coolants, hydraulic machinery using PCB hydraulic fluid, heat transfer systems using PCBs, and storage areas used to store PCBs for disposal. In addition, a deadline of October 1, 1978 was set for marking transport vehicles loaded with certain PCB containers, and a deadline of January 1, 1979 was established for marking certain additional transformers and large high-voltage capacitors.

Section 761.44 of EPA's regulatons established two formats for marking—a Large PCB Mark containing relatively detailed warnings concerning PCBs and a Small PCB Mark highlighting the need for caution in handling PCBs and the need to contact EPA for proper disposal information. The regulations prescribed the precise size of each mark and stated that the mark was to be sufficiently durable to equal or exceed the life of the equipment or container. Pursuant to Section 761.20(a)(5), the Large PCB Mark was to be used unless the article or equipment involved was too small to accommodate it; in this event the Small Mark was permissible.

Under Section 761.20, small capacitors, large low-voltage capacitors, and fluorescent light ballasts which do not contain PCBs were to be marked with a label statng "No PCBs." According to EPA, this marking requirement was designed to assist disposers in distinguishing between existing electrical equipment, which generally contains PCBs, and equipment manufactured in the future, which was expected not to contain PCBs.

Finally, EPA's regulation required that, after January 1, 1979, equipment containing PCB small capacitors was to be marked with the statement "This Equipment Contains PCB Capacitor(s)." In the preamble to its final rule, EPA stated that, even though such equipment was not subject to disposal requirements, it would be required to bear this marking "in order to discourage massive stockpiling of PCB articles and incorporation of the items and equipment indefinitely into the future." 43 Fed. Reg. 7152.

### EPA's Ban Regulations

On June 7, 1978, EPA issued proposed Ban Regulations to implement Sections 6(e)(2) and 6(e)(3). These regulations define "totally enclosed manner", authorize several non-totally enclosed uses, and set forth the procedures for obtaining exemptions from the Act's prohibitions. 43 Fed. Reg. 24801. As foreshadowed in the final Disposal Regulations, the final Ban Regulations (issued May 31, 1979) set 50 ppm as a cutoff for regulatory controls. 43 Fed. Reg. at 24813; 44 Fed. Reg. at 31543. In addition, the final regulations defined all intact electrical capacitors, electromagnets, and nonrailroad transformers as "totally enclosed", and thus automatically exempted them from the PCB ban. The final regulations also authorized 11 non-totally enclosed PCB uses to continue, including the servicing of totally enclosed uses. The continuation of these uses was based on EPA's analysis of the health and environmental effects of PCBs, the exposure to PCBs resulting from these activities, the availability of substitutes for the PCBs, and the economic impact of restricting those uses. Under the regulations, most authorizations for continued PCB uses were to extend until July 1, 1984, with some authorizations expiring sooner. Only the authorization for carbonless copy paper was permitted to continue indefinitely.

Among the features of EPA's rule was the application of all disposal and marking requirements included in EPA's February 1978 rule to PCB containers, articles, equipment and transport vehicles containing 50 ppm or greater PCBs. According to the Agency, this extension of the disposal and marking requirements was "essential to insure that PCB Items regulated under this rule was properly identified, handled and disposed of to minimize the potential risks of exposure to PCBs." 44 Fed. Reg. 31521. The Agency's rule set a deadline of October 1, 1979 for marking these additional PCB items. Id.; 40 C.F.R. § 761.20.

Nevertheless, EPA created a number of exceptions to the extension of marking requirements to items containing between 50 and 500 ppm PCBs. First, PCB-contaminated transformers were not required to be marked because of cost considerations; according to EPA, 35 million such transformers were in use and marking all of them would cost about $350 million. 44 Fed. Reg. 31521. In addition, EPA stated that persons who had petitioned for exemptions from the ban on PCB manufacture would not be required to label chemicals that contained less than 500 ppm PCB until EPA acted on their petitions. 44 Fed. Reg. 31522.

Finally, EPA's rule added a new paragraph to Section 761.20 requiring that marks be placed on the exterior of PCB items and transport vehicles so that the marks can be "easily read by persons inspecting or servicing the marked PCB Items or transport vehicles." According to the Agency, this addition was intended to correct an "oversight" in the original marking requirements. 44 Fed. Reg. 31522.

### Court Decisions Involving PCBs

Although no court challenge was made to EPA's Disposal Regulations, the Environmental Defense Fund (EDF) brought suit to challenge the Agency's Ban Regulations in the United States Court of Appeals for the District of Columbia Circuit. An opinion was issued by that Court on October 30, 1980. *Environmental Defense Fund* v. *EPA*, No. 79-1580. The court found that the administrative record lacked substantial evidence to support EPA's decision to exclude from regulation all materials containing concentrations of PCBs below 50 ppm. Accordingly, with respect to this issue, the Court held unlawful and set aside the challenged regulations and remanded them to EPA for further proceedings consistent with the opinion. In addition, however, the Court did find that there was substantial evidence in the administrative record to support the Agency's determination to allow continued use of the 11 nontotally enclosed uses. Accordingly, the Court upheld this aspect of EPA's regulations. The Agency subsequently sought, and obtained, a stay of the Court's mandate pending reconsideration and repromulgation of its PCB Ban Regulations. 46 Fed. Reg. 16090 (March 10, 1981).

The Court's decision did not specifically overturn the marking requirements in EPA's Ban Regulations. Thus, those requirements will remain in effect for the foreseeable future. It is possible, however, that EPA's reconsideration of its 50 ppm cutoff level could lead to a decision to extend marking requirements to PCB items containing lower PCB concentrations. This could substantially expand the scope of the marking requirements presently in effect.

### FOOTNOTES

1. H. Rep. No. 97-86, 97th Cong., 1st Sess. 9 (1981).
2. The term "unreasonable risk of injury" also appears in Sections 4 and 5 of TSCA.

For example, Section 4 of the Act authorizes EPA to promulgate rules which require the testing of particular substances or mixtures. Under Section 4(a), one of the criteria for promulgating such rules is whether the chemical "may present an unreasonable risk of injury to health or the environment." Under Section 5 of the Act, companies must notify EPA 90 days before they intend to manufacture a "new chemical substance" or manufacture or process an existing chemical substance for "a significant new use." Under Section 5(e)(1), upon receiving that notice, EPA may require further postponements in manufacture or processing if it determines that, without the development of additional information, the chemical in question "may present an unreasonable risk of injury to health or the environment."

3. *Restatement of Torts* § 757, Comment b at 5 (1939).
4. E.g., Cataphote Corp. v. Hudson, 422 F.2d 1290, 1293 (5th Cir. 1970); Sears, Roebuck & Co. v. 1-M Mfg. Co. 256 F.2d 517 519 (3d Cir. 1958); B. F. Gladding & Co. v. Scientific Anglers, Inc., 245 F.2d 722, 728 (6th Cir. 1957).
5. Kewanee Oil Co. v. Bicron Corp. 416 U.S. 470, 474-75 (1974).
6. Union Oil Co. v. FPC, 542 F.2d 1036, 1044 (9th Cir. 1976); Washington Research Project, Inc. v. HEW, 504 F.2d 238, 245 n.8 (D.C. Cir. 1974), *cert. denied,* 421 U.S. 965 (1975); Mobay Chem. Corp. v. Costle, 447 F. Supp. 811, 825 (W. D Mo. 1978), *appeal dismissed,* 439 U.S. 320, *rehearing denied,* 440 U.S. 940 (1979); Chevron Chem. Co. v. Costle, 443 F. Supp. 1024, 131 (N.D. Cal. 1978).
7. It is somewhat strained logic to regard the requirements of a final labeling rule as a "proceeding" within the meaning of Section 14(a)(4).

# 12

# Labeling Under the Resource Conservation and Recovery Act (RCRA)

**Robert M. Sussman**
*Covington and Burling*
*Washington, DC*

**Jennifer Machlin**
*Orrick, Herrington and Sutcliffe*
*San Francisco, CA*

## THE STATUTORY HAZARDOUS WASTE MANAGEMENT PROGRAM

Subtitle C of the Resource Conservation and Recovery Act of 1976, 42 U.S.C. §§ 6901 et seq., as amended (RCRA), requires the Environmental Protection Agency (EPA) to identify hazardous wastes and promulgate management standards and permit requirements for persons who generate, transport, store, treat or dispose of such wastes. Key provisions in this legislation are outlined below, accompanied by a brief summary of the regulations promulgated by EPA to date in its ongoing effort to implement those provisions. Those regulations incorporate by reference many of the Department of Transportation (DOT) rules that govern the labeling of hazardous waste intended for transport and transportation. The applicable DOT rules are analyzed at length elsewhere in this book. The discussion that follows therefore focuses on the RCRA provisions that define the persons subject to the RCRA labeling requirements and hence to the DOT labeling requirements.

### Identification of Hazardous Waste: RCRA §§ 1004(5), 1004(27) and 3001

A person who seeks to determine whether the RCRA regulations apply to him must first ascertain whether he handles waste defined as "hazardous" under the statute and under EPA's regulations. RCRA § 1004(27) defines "solid waste" as "discarded material" such as garbage, refuse, sludge from waste or water supply treatment plants or air pollution facilities, and gases from various industrial and community activities. Section 1004(5) defines "hazardous waste" as solid waste that may (i) cause or contribute to an increase in mortality or serious illness or (ii) pose a substantial hazard to human health or the environment when improperly managed. Section 3001 directs EPA to implement that definitional language by promulgating criteria for identifying hazardous waste and publishing lists of wastes deemed hazardous under those criteria. EPA's response to that directive is explained below.

**Determining Whether a Waste is "Solid" for Subtitle C Purposes:** EPA's definition of "solid waste" is set forth at 40 C.F.R. § 261.2. This definition expands upon the language of RCRC § 1004(27) by including (i) garbage, refuse, and sludge, regardless of their intended use, and (ii) "other waste material," identified by EPA as including certain spent materials and manufacturing by-products destined either for discard or for reclamation, recycling, or reuse. See 45 Fed. Reg. 33090-94; 33119.

This definition, in particular the provisions that include materials destined for beneficial uses rather than for discard, has been the subject of negotiations between the Agency and concerned industries since February 1981. The scope of the final regulatory definition will determine, to a large extent, the size of the regulated community. Certain materials are already excluded from EPA's definition, among them domestic sewage, sewage and other waste mixed in passing through a sewer system to a public treatment plant, discharges from industrial point sources and irrigation return flows, in-situ mining waste, and certain nuclear materials.

EPA's definitional provisions also specify that certain solid wastes are not "hazardous" for Subtitle C purposes. These wastes include household wastes (e.g., garbage, trash, and wastes in septic tanks) and certain wastes generated by growing crops, raising animals, mining, combustion of fossil fuels, and gas and oil production. See § 261.4(b). It should be noted that, while such wastes are not subject to the regulations promulgated under Subtitle C, they may be subject to the land disposal prohibitions incorporated in Subtitle D and implemented in proposed EPA regulations. See RCRA, Subtitle D, §§ 4004 & 4005; 45 Fed. Reg. 73340-72713; 46 Fed. Reg. 29064-29149.

**Determining Whether a Solid Waste is "Hazardous" for Subtitle C Purposes:** A person who handles a solid waste not within one of the regulatory exclusions must look to §§ 261.30-33 and 261.20-24 to determine whether that waste is "hazardous" and hence subject to the full panoply of Subtitle C regulations. Sections 261.30-33 contain EPA's hazardous waste lists; §§ 261.21-24 identify four characteristics (ignitibility, corrosivity, reactivity, and Extraction Procedure toxicity) that make a waste hazardous regardless of whether it is included on a hazardous waste list.

This initial determination is not always conclusive, however, First, even if a facility is generating a waste designated as hazardous on one of EPA's lists, EPA will "delist" the waste produced *by that particular facility* upon a showing that the waste does not meet EPA's listing criteria. See § 260.22 and EPA's preambles at 45 Fed. Reg. 33095 and 86543-48. Second, as indicated above, certain solid wastes are expressly excluded from EPA's definition of hazardous waste, among them household waste, waste related to agriculture, waste generated by fossil fuel burning, mining waste, and cement kiln dust waste. See RCRA § 3001, as amended; § 261.4(b), 45 Fed. Reg. 33120. Also excluded are hazardous waste residues in a container or container liner that meets EPA's definition of "empty." See § 261.7, 45 Fed. Reg. 78529. Third, hazardous waste generated in units that hold such waste only incidentally (product or raw material storage tanks, transport vehicles or vessels, and manufacturing process units) are exempted from certain Subtitle C regulations under certain conditions. § 261.4(c), 45 Fed. Reg. 72028.

The Subtitle C regulations, moreover, have been suspended with regard to owners and operators of (1) certain tanks used to treat or store wastewaters that are hazardous wastes and (2) neutralization tanks used to treat wastes that are hazardous only because corrosive. The Agency has proposed that such owners and operators be eligible for a "permit-by-rule," i.e., they may operate without actually applying for individual RCRA permits so long as their facilities meet the regulatory definitions of "wastewater treatment unit" or "elementary treatment unit." See 45 Fed. Reg. at 76074-76083. Finally, it should be noted that, in the event of spills of hazardous wastes or materials that become hazardous when spilled, actions taken to contain and treat the spills are not viewed as "treatment and storage of hazardous waste" for purposes of the Subtitle C regulations. 45 Fed. Reg. at 76626-76630.

### Standards Applicable to Persons Who Generate, Transport, Treat, Store, or Dispose of Hazardous Wastes: RCRA §§ 3002, 3003, and 3004

A person dealing with a "hazardous waste" as defined by the statute and by EPA must comply with the hazardous waste management regulations promulgated under Subtitle C. The regulations parallel the statute in identifying different, although overlapping, requirements for those who generate hazardous waste, those who transport it, and those who own or operate facilities to treat, store, or dispose of it. The most important of these requirements are summarized below; specific labeling requirements are discussed in detail in *RCRA Requirements for Labeling Hazardous Wastes* later in this chapter.

**Generators: RCRA § 3002; 40 C.F.R. Part 262:** RCRA § 3002 requires that EPA establish standards for hazardous waste generators that will ensure proper recordkeeping and reporting; the use of proper labels and containers; the furnishing of information on the composition of hazardous waste to persons transporting, treating, storing, or disposing of such waste; and the use of a manifest system to track shipments of hazardous waste. In particular, § 3002 directs EPA to establish requirements respecting "labeling practices for any containers used for the storage, transport, or disposal of such hazardous waste such as will identify accurately such waste." To implement this provision, EPA, in cooperation with the Department of Transportation (DOT), has developed special regulations that govern the shipping of hazardous waste. These regulations are covered in the section *RCRA Requirements for Labeling Hazardous Wastes* below.

Two categories of persons are excluded from EPA's generator requirements: "small quantity" generators and farmers disposing of pesticide residues that are hazardous waste. Under the small quantity exclusion, generators who produce less than 1,000 kilograms of hazardous waste in a calendar month are not subject to the full range of Subtitle C requirements. § 261.5, 45 Fed. Reg. 76623. The exclusion does not apply, however, to those who generate acutely toxic wastes; such wastes are regulated even if produced in quantities as small as one to one hundred kilograms. § 261.5(e). Farmers who use pesticides and are left with residues that constitute hazardous waste are excluded if they triple rinse empty pesticide containers and follow disposal instructions on the pesticide label. § 261.33, 45 Fed. Reg. at 33124; § 262.51, 45 Fed. Reg. at 33144.

A hazardous waste generator not covered by one of these exclusions is required to notify EPA and receive an identification number. Generators who ship hazardous waste off-site must comply with EPA's pretransport requirements, including packaging and labeling standards, and must prepare a manifest. §§ 262.20-23. Generators who accumulate hazardous waste on site for less than 90 days must comply with certain minimum requirements, discussed in the section *RCRA Requirements for Labeling Hazardous Waste*. Generators who accumulate waste on-site for more than 90 days before shipping it, or who treat, store, or dispose of hazardous waste on-site, must meet the notification requirements, apply for a RCRA permit to continue their activities, and comply with EPA's "interim status" standards, as explained in the section on *Owners and Operators* below.

**Transporters: RCRA § 3003; 40 C.F.R. Part 263:** RCRA § 3003 requires that EPA establish standards for hazardous waste transporters that "shall include but need not be limited to" specified recordkeeping, labeling, manifest, and destination requirements. These standards, set out at 40 C.F.R. Part 263, basically extend the generator management system to persons who transport waste off-site. Thus, they do not apply to on-site transportation of hazardous waste by generators or by owners or operators of permitted hazardous waste management facilities. See § 263.10(b).

The technical standards contained in the DOT regulations cited above apply to hazardous waste transporters as well as to generators. § 263.10(a). Aside from these standards, the transporter regulations are relatively straightforward. Transporters must notify EPA; comply with the manifest system; utilize special shipping papers for bulk hazardous waste shipments by rail or water; and comply with the clean-up requirements in §§ 263.30 and 263.32 if hazardous waste is discharged during transport.

**Owners and Operators of Facilities for Hazardous Waste Treatment, Storage, or Disposal: RCRA §§ 3004 & 3005; 40 C.F.R. Parts 264 & 265:** RCRA §§ 3004 and 3005 require EPA to promulgate regulations that govern the management of treatment, storage and disposal facilities (TSDFs) and establish a system for issuing RCRA permits to those facilities. The literal terms of § 3005(a) prohibit facility owners or operators from storing, treating, or disposing of hazardous waste after November 19, 1980 unless they have been issued a final RCRA permit. Anticipating an extended rulemaking process with regard to the TSDF standards, however, Congress in § 3005(e) created the concept of "interim status," a status that entitles TSDF owner/operators to stay in business pending EPA's final action on their permit applications. To achieve that status, TSDF owner/operators must meet certain requirements, including notifying EPA of their hazardous waste activities by the date specified in 40 C.F.R. § 122.22(a)(1) and filing Part A of the permit application by November 19, 1980 or as specified in § 122.22(a)(2) and (3). To maintain interim status pending final action on their applications, they must, as of November 19, 1981, comply with temporary management standards (40 C.F.R. part 265) that will remain in effect until EPA publishes management standards in final form (Part 264) and issues general status permits.

The interim status standards in Part 265 are meant to apply only to facilities in existence on November 19, 1980. RCRA § 3005(e). EPA has acknowledged that continued hazardous waste generation will necessitate the construction of new land disposal facilities before Part 264 takes effect, and has thus promulgated a special package of interim standards that apply solely to new land disposal facilities. 40 C.F.R. Part 267, 45 Fed. Reg. 12414 12433. These standards have been criticized as excessively vague; the Agency has indicated that more specific guidelines may be forthcoming. It appears likely, in any event, that the full spectrum of Part 264 design and performance standards will not apply retroactively to land disposal facilities constructed before promulgation of such standards.

A detailed examination of the interim status regulations is beyond the scope of this chapter. The circumstances under which TSDF owner/operators must comply with the generator pretransport procedures, including DOT labeling requirements, are described in the following section.

## RCRA REQUIREMENTS FOR LABELING HAZARDOUS WASTE

As indicated above, those who generate, transport, store, treat or dispose of hazardous waste must in certain situations comply with the DOT labeling, marking, and placarding requirements set forth at 49 C.F.R. Part 172. EPA's regulations cite other DOT hazardous waste packaging requirements as well (see, e.g., § 262.30 and note following § 263.10(a)); thus, the full range of DOT labeling requirements is covered in the brief summary that follows.

The following circumstances trigger the need to comply with various DOT regulations.

> 1. Generators must comply with regulations in Parts 172, 173, 178, and 179 before they transport hazardous waste or offer it for transportation off-site. They must also comply with the Part 172 labeling and marking requirements when they accumulate hazardous waste on-site for less than 90 days. EPA allows generators to accumulate wastes without a permit or interim status if, within 90 days after accumulation begins, the waste is shipped off-site to a designated facility or placed in an on-site facility with a permit or interim status. § 262.34, 45 Fed. Reg. 76626. EPA has augmented the DOT labeling requirements, for purposes of the 90-day rule, with its own requirement that the date on which accumulation begins be clearly marked on each container. § 262.34(a)(3).

2. Transporters must comply with the DOT regulations whenever they are carrying hazardous waste under circumstances that require the shipper to prepare a manifest, i.e., whenever the waste is transported off-site for treatment, storage, or disposal (as noted above, on-site transportation is excluded). In particular, transporters are subject to the Part 172 DOT labeling requirements, and the other DOT packaging requirements incorporated into the generator pretransport regulations, when they import hazardous waste from abroad or mix hazardous wastes of different DOT shipping descriptions by placing them in a single container. 40 C.F.R. § 263.10(c), 45 Fed. Reg. 33151.

3. Persons who own and operate hazardous waste management (i.e., treatment, storage, or disposal) facilities must observe all the regulations that apply to generators, including the pretransport DOT regulations, whenever they ship hazardous waste off-site, whether the waste in question was generated at their facility or received from an outside source. § 262.10(f), 45 Fed. Reg. 86970. EPA's 90-day rule for unpermitted temporary on-site storage may be invoked by facility owners or operators, however, only for purposes of accumulating hazardous waste which they have generated. Id.

The labeling, marking, and placarding requirements developed by DOT in cooperation with EPA that apply in the circumstances outlined above are summarized below. Special EPA requirements are included where relevant.

## Part 171: General Information

Section 171.2 forbids the offer or acceptance of a hazardous material for transportation within the United States unless it is packaged, marked, and labeled in accordance with DOT's regulations. Special requirements for import and export hazardous waste shipments, including the clear and legible display of required specification markings on the package, are set forth in § 171.12.

## Part 172: Hazardous Waste Communication Regulations

All persons shipping or transporting "hazardous waste" as defined under RCRA should be familiar with this Part. Only the most significant Part 172 regulations are discussed below.

**Marking Requirements:** Section 172.304 contains the following general standards for markings that appear on hazardous waste containers: (1) they must be durable, in English, and printed on the package or affixed to a label; (2) they must be displayed on a background of sharply contrasting color; (3) they must be unobscured by labels or attachments; and (4) they must be located away from other markings, such as advertising, that could reduce their effectiveness.

Section 172.308 precludes the use of abbreviations in markings, except in rare circumstances. Section 172.332 requires that identification numbers be displayed on orange panels or placards in accordance with detailed specifications; under § 172.338, the carrier must replace any panels or placards lost or destroyed during transportation as soon as practicable.

Under special EPA requirements, hazardous waste shippers must mark each shipping container of 110 gallons or less with this warning: "HAZARDOUS WASTE—Federal Law Prohibits Improper Disposal. If found, contact the nearest police or public safety authority or the U.S. Environmental Protection Agency. Generator's Name and Address _____ . Manifest Document Number _____ ." 40 C.F.R. § 262.32. This warning must be displayed in accordance with the terms of DOT regulation § 172.304, summarized above.

**Labeling Requirements:** These requirements *prohibit* the use of labels, in transporting materials *other than* hazardous waste, that might be confused with DOT hazardous waste labels. Other regulations call for multiple labeling in certain circumstances; set forth requirements for labeling mixed and consolidated packages; and describe proper label placement (usually near the marked shipping name) and specifications (e.g., durable and highly weather resistant). The regulations also contain numerous illustrations of the shape, wording, and graphic designs required for labels serving specific purposes.

**Placarding Requirements:** These are detailed at 49 C.F.R. §§ 172.500-558. Different requirements are specified for rail and highway transport, and for freight containers and tanks. In general, a placard must be "readily visible from the direction it faces"; securely attached to the transport vehicle, tank, or container; located away from obstructions such as ladders, pipes, doors, or tarpaulins; and have words horizontally displayed. Illustrations for the required shape and design of placards serving specified purposes are also included.

### Part 173: General Shipment and Packaging Requirements

This section contains a number of regulations aimed at specifically identified substances (e.g., in Subpart D, flammable, combustible, and pyrophoric liquids; in Subpart E, flammable solids, oxidizers, and organic peroxides) and must be consulted by any shipper subject to EPA's pretransport regulation.

Special labeling requirements are scattered throughout Part 173. Section 173.25 sets forth marking requirements for outside containers that hold packages in compliance with DOT regulations, and § 173.28 requires that shippers who intend to reuse previously emptied containers maintain the specified markings in a legible condition or reproduce them on a metal plate. The regulation further provides that markings and labels on hazardous waste containers must be thoroughly removed or obliterated before those containers are used to ship other articles.

### Parts 174-177: Shipment of Hazardous Waste by Rail, Air, Vessel, and Public Highway

Shippers and transporters using rail carriage should scrutinize the regulations that apply to the particular materials they are carrying (e.g., explosives, gases, flammable liquids or solids, oxidizers, poisonous, radioactive, or corrosive materials). Placarding is required not only for cars carrying hazardous waste but also for empty cars that previously carried such waste.

Persons transporting hazardous waste by any of the modes mentioned above must maintain an adequate supply of the labels and placards required under Part 172 and replace any that are lost or destroyed during transport as soon as possible. Replacement labels should be worded in accordance with information provided on the shipping papers that accompany the hazardous waste to be labeled.

Under the aircraft requirements, § 175.630 prohibits the carrying of packages marked "POISON" in the same cargo compartment with material marked as or known to be food for consumption by humans or animals. Under the public highway requirements, the Part 172 marking and placarding requirements are lifted in emergency situations, *if* the vehicle is escorted by a state or local government representative, the carrier has permission from DOT to proceed without a placard, or moving the vehicle is necessary to protect life or property.

## Parts 178-179: Shipping Container and Tank Specifications

The shipping regulations are broken down into different specifications for different types of containers (e.g., rubber drums, cylinders, metal barrels, wooden barrels, and so on). Specified DOT markings are designated for each type of container.

Tank builders are responsible for marking ranks with DOT specifications are required under Part 179. That marking constitutes a certification that the tank has been constructed to meet DOT standards, and the builder must inform anyone to whom the tank is transferred of standards not satisfied at the time of transfer.

This summary, as noted above, reflects only a sampling of the more important labeling, marking, and placarding requirements contained in the DOT regulations and incorporated by reference into EPA's RCRA regulations. It is essential that persons who are subject to the RCRA regulations because they are shipping hazardous waste off-site or transporting such waste become thoroughly familiar with the DOT regulations that cover the type of waste being shipped and mode of transportation being used.

# 13

# Labeling Requirements Administered by the Consumer Product Safety Commission (CPSC)

**Robert M. Sussman**
*Covington and Burling*
*Washington, DC*

The Consumer Product Safety Commission (CPSC or Commission) is an independent regulatory agency administered by a Chairman and four Commissioners. The CPSC was created by Congress in 1972. Its mission is to assure the safety of products used by consumers in the household environment. The principal source of the Commission's responsibilities and powers is the Consumer Product Safety Act (CPSA), enacted by Congress in 1972 when it created the Commission.[1] In addition, the Commission has been assigned responsibility for implementing various safety statutes that were previously administered by other federal agencies. These statutes include the Federal Hazardous Substances Act (FHSA), the Poison Prevention Packaging Act, the Flammable Fabrics Act, and the Refrigerator Safety Act.[2]

The Commissioner utilizes several regulatory mechanisms to control unsafe products, including prospective design standards, product recalls and product bans. A major additional tool available to the Commission is the imposition of labeling requirements. Such requirements have played a significant role in compelling disclosure of the acute and chronic health effects of chemicals used in household products. The CPSC's labeling activities have primarily occurred under two statutes: (1) the FHSA, which creates a detailed, largely self-executing labeling scheme for household products that present acute hazards, and (2) the CPSA, under which the Commission may utilize labeling requirements in exercising its broad rulemaking power to improve product safety.

Discussed below are the principal aspects of those provisions of the FHSA and the CPSA that govern the labeling of household products.

## LABELING UNDER THE FHSA

### Purposes and Overall Approach of the FHSA

The FHSA (originally entitled the Federal Hazardous Substances Labeling Act) was enacted in 1960 to require precautionary labeling on household products which are capable of causing substantial injury or illness as a result of foreseeable handling or use.[3] Until the CPSC's creation in 1972, the FHSA was implemented by the Food and Drug Administration (FDA).

According to the FHSA's legislative history, there were numerous common household products, such as furniture polish, bleaches and detergents, "containing poisonous or dangerous substances that lack adequate warning labels, and adequate identification of the dangerous substances." S. Rep. No. 1158, 86th Cong., 2d Sess. 1 (1960). Congress therefore decided that new legislation was needed "to require uniform labeling of household substances for household use." Id., 3. Such labeling, Congress expected, would serve a two-fold function: first, "warn the user of any hazard in the customary use of the product", and second, "in the case of an accident identify the household ingredient for the attending physician." Id.

The hazard definitions and labeling terminology contained in the FHSA closely resemble the voluntary labeling guidelines of the LAPI Committee of the Manufacturing Chemists Association (now the Chemical Manufacturers Association). The FHSA also bears many similarities to labeling legislation enacted and implemented by several states prior to 1960. The FHSA contains highly detailed criteria for determining which substances are "hazardous"; it then prescribes a series of specific warning statements for inclusion on these substances' labels. Any hazardous substance which is not labeled in accordance with these requirements will be deemed "misbranded" and will be subject to seizure and condemnation. In addition, even when properly labeled, certain products may be so unsafe that they can be banned from household use. The designation of such "banned hazardous substances" is to be accomplished primarily through regulations promulgated by the CPSC.

### Definition of "Hazardous Substance"

The term "hazardous substance" is defined in Section 2(f)(1)(A) to include any substance which is toxic, corrosive, an irritant, a strong sensitizer, flammable or combustible, or which generates pressure through decomposition, heat, or other means. Where a product fits into one of these six categories, it will be considered a "hazardous substance" if it is capable of causing "substantial personal injury or substantial illness during or as a proximate result of any customary or reasonably foreseeable handling or use, including reasonably foreseeable ingestion by children."

Augmented by CPSC regulations, the statutory text prescribes highly detailed, objective criteria for determining whether a substance poses one of the hazards covered by Section 2(f). Among these criteria are the following:

**Definition of "Toxic":** Section 2(g) of the FHSA defines the term "toxic" to include any substance "which has the capacity to produce personal injury or illness to man through ingestion, inhalation, or absorption through any body surface." At 16 C.F.R. § 1500.3(c)(2), CPSC regulations elaborate on this definition as follows:

> "Toxic" means any substance that produces death within 14 days in half or more than half of a group of:
>
> (i) White rats (each weighing between 200 and 300 grams) when a single dose of from 500 milligrams to 5 grams per kilogram of body weight is administered orally. Substances falling in the toxicity range between 500 milligrams and 5 grams per kilogram of body weight will be considered for exemption from some or all of the labeling requirements of the act, under § 1500.82, upon a showing that such labeling is not needed because of the physical form of the substances (solid, a thick plastic, emulsion, etc.), the size or closure of the container, human experience with the article, or any other relevant factors;
>
> (ii) White rats (each weighing between 200 and 300 grams) when an atmospheric concentration of more than 200 parts per million but not more than 20,000 parts per million by volume or gas or vapor, or more than 2 but not more than 200 milligrams per liter by volume of gas or vapor, or more than 2 but not more than 200 milligrams per liter by volume of mist or dust, is inhaled continuously for 1 hour or less, if such concentration is likely to be encountered by man when the substance is used in any reasonably foreseeable manner; and/or
>
> (iii) Rabbits (each weighing between 2.3 and 3.0 kilograms) when a dosage of more than 200 milligrams but not more than 2 grams per kilogram of body weight is administered by continuous contact with the bare skin for 24 hours by the method described in § 1500.40.
>
> The number of animals tested shall be sufficient to give a statistically significant result and shall be in conformity with good pharmacological practices.

**Definition of "Highly Toxic":** Section 2(h)(1) defines the term "highly toxic" to include any substance which:

> Produces death within fourteen days in half or more than half of a group of ten or more laboratory white rats each weighing between two hundred and three hundred grams, at a single dose of fifty milligrams or less per kilogram of body weight, when orally administered; or (b) produces death within fourteen days in half or more than half of a group of ten or more laboratory white rats each weighing between two hundred and three hundred grams, when inhaled continuously for a period of one hour or less at an atmospheric concentration of two hundred parts per million by volume or less of gas or vapor or two milligrams per liter by volume or less of mist or dust, provided such concentration is likely to be encountered by man when the substance is used in any reasonably foreseeable manner; or (c) produces death within fourteen days in half or more than half of a

group of ten or more rabbits tested in a dosage of two hundred milligrams or less per kilogram of body weight, when administered by
continuous contact with the bare skin for twenty-four hours or less.

At 16 C.F.R. § 1500.3(c)(1), CPSC regulations provide an alternative definition
of "highly toxic" intended to permit flexibility in the number of animals tested.

**Definition of "Corrosive":** Section 2(i) of the FHSA defines the term "corrosive" as any substance "which in contact with living tissue will cause destruction of tissue by chemical action." CPSC regulations elaborate on this definition by providing as follows:

> "Corrosive" means a substance that causes visible destruction or ir-
> reversible alterations in the tissue at the site of contact. A test for a
> corrosive substance is whether, by human experience, such tissue de-
> struction occurs at the site of application. A substance would be con-
> sidered corrosive to the skin if, when tested on the intact skin of the
> albino rabbit by the technique described in § 1500.41, the structure
> of the tissue at the site of contact is destroyed or changed irreversibly
> in 24 hours or less. [16 C.F.R. § 1500.3(c)(3).]

**Definition of "Irritant":** Section 2(j) defines "irritant" as any substance
which "on immediate, prolonged, or repeated contact with normal living tissue
will induce a local inflammatory reaction." CPSC regulations differentiate between a "primary irritant to the eye" and substances that are "irritants to the
eye or to mucous membranes." The regulations also prescribe test methods for
determining whether a substance falls into one of these categories. See 16
C.F.R. §§ 1500.41 and 1500.42.

**Definition of "Strong Sensitizer":** Under Section 2(k), a "strong sensitizer" is a substance

> which will cause on normal living tissue through an allergic or
> photodynamic process a hypersensitivity which becomes evident on
> reapplication of the same substance . . .

In contrast to other hazards for which labeling is automatically required, a
substance's sensitizing properties must be "designated as such" by the CPSC
before labeling requirements will arise. In making such a designation, the
Agency is required to consider "the frequency of occurrence and severity of the
reaction" and to "find that the substance has a significant potential for causing
hypersensitivity." The FHSA's legislative history stresses that few substances
will satisfy this standard:

> Some portion of the population is sensitive in one way or another to
> almost every article that enters the household, including foods and
> household soap. To require precautionary labeling on all such prod-
> ucts is not intended. Precautionary labeling would be required under
> this bill on any substance which affects a significant portion of the
> population and which may cause a strong or severe reaction, if after
> a finding by the [Commission] that the substance had a significant
> potential for causing hypersensitivity. [S. Rep. No. 1158, 86th Cong.,
> 2d Sess. 11 (1960).]

CPSC regulations elaborate on the statutory definition of "strong sensitizer" by specifically differentiating between allergic and photodynamic sensitization. According to 16 C.F.R. § 1500.3(c)(5), an allergic sensitization develops by means of an "antibody mechanism." A photodynamic sensitizer, by contrast, is defined as

> a substance that causes an alteration in the skin or mucus membranes in general or to the skin or mucus membrane at the site of contact so that when these areas are subsequently exposed to ordinary sunshine (or equivalent radiant energy) an inflamatory reaction will develop.

**Definitions of "Flammability" and "Combustibility":** Under Section 2(1), the terms "extremely flammable", "flammable", and "combustible" are to be defined by regulations issued by the CPSC. These regulations are to specify test methods that the Commission has found are generally applicable and are compatible with the definitions and test methods adopted by other federal agencies "involved in the regulation of flammable and combustible substances in storage, transportation and use."

In general, CPSC regulations provide that a substance is "extremely flammable" if it has a flash point at or below 20° F and "flammable" if it has a flash point of between 20° and 80° F. 16 C.F.R. § 1500.3(c)(6). Test methods for making these determinations are described in 16 C.F.R. § 1500.43 and 1500.44.

**Pressure-Generating Substances:** Filling an apparent gap in the statutory definitions, CPSC regulations provide that a substance that "generates pressure through decomposition, heat or other means" under Section 2(f)(1)(A) is a hazardous substance under the following circumstances:

(A) If it explodes when subjected to an electrical spark, percussion, or the flame of a burning paraffin candle for 5 seconds or less.

(B) If it expels the closure of its container, or bursts its container, when held at or below 130° F. for 2 days or less.

(C) If it erupts from its opened container at a temperature of 130° F. or less after having been held in the closed container at 130° F. for 2 days.

(D) If it comprises the contents of a self-pressurized container. [16 C.F.R. § 1500.3(c)(7)(i).]

**Significance of Human Experience:** In applying the above definitions, Section 2(h) provides that, wherever data derived from animal tests is in conflict with human experience, the latter shall take precedence. CPSC regulations expand on this principle, stating that "experience may show that an article is more or less toxic, irritant or corrosive to man than to test animals" and that, in such cases, "the human experience takes precedence." 16 C.F.R. § 1500.4.

**Testing of Mixtures:** Section 2(f) of the FHSA recognizes that a mixture may itself qualify as a "hazardous substance." Frequently, the question will arise whether a mixture should be tested separately to determine its hazardous properties or whether the manufacturer can make reasonable judgments based on the properties of its component substances. 16 C.F.R. § 1500.5 provides guidance on this issue. It states that the manufacturer must separately evaluate the "physical, chemical, and pharmacological characteristics of the mixture." The regulation then advises that it may not be possible "to reach a fully satisfactory decision" concerning the properties of a mixture based on information about its components or ingredients and that, in such circumstances, "the mixture itself should be tested."

**The Role of the "If" Clause:** Even if a product falls within the six hazard categories enumerated in the statute, it may not be a "hazardous substance" if, pursuant to Section 2(f)1(A), it will not cause substantial injury or substantial illness as a result of any reasonably foreseeable handling or use. Application of this second statutory criterion for a "hazardous substance"—which is frequently described as the "if" clause—depends on such factors as the physical nature of the product, the method of packaging, the extent of any injury to be anticipated, reports of human experience, and other similar considerations.

The two-fold nature of the definition of "hazardous substance", and its relationship to the basic purposes of the Act, are clearly articulated in the House Report:

> In order to come within the basic definition of "hazardous substance", a substance (or mixture) must meet two requirements. First it must be "toxic", "corrosive", and "irritant" a "strong sensitizer" designated as such by the Secretary, "flammable", or be a substance which "generates pressure through decomposition, heat, or other means." (All these terms, except the self-explanatory phrase "generates pressure through decomposition, heat, or other means", are defined in the bill.) Secondly, it must be a substance (or mixture) which—
>
>> may cause substantial personal injury or substantial illness during or as a proximate result of any customary or reasonable foreseeable handling or use, including reasonably foreseeable ingestion by children.
>
> The term "substantial", in the expression "substantial personal injury or substantial illness", should be read in the light of the purposes of the bill. On the one hand, it is not intended to impose the impracticable and self-defeating requirement of cautionary labeling against wholly insignificant or negligible illness or injury, such as the very temporary indisposition that a child might suffer from eating a piece of the standard type of toilet soap . . . . On the other hand, the term "substantial" is not intended to limit the requirement of cautionary labeling to situations in which the injury or illness to be guarded against would be severe or serious. [H.R. Rep. No. 1861, 86th Cong., 2d Sess. 5 (1960).]

The Senate Report expands on these objectives as follows:

> To be meaningful and accomplish the purpose of this bill, it is equally important that substance which, as packaged, present only a minor hazard not be required to display a series of precautionary

> statements, since the public might quickly learn to disregard the importance and significance of precautionary labeling. [S. Rep. No. 1158, 86th Cong., 2nd Sess. 2 (1960)l]

An example of the two-step approach required by the FHSA may be helpful in understanding the design of the Act. The term "irritant" is defined in Section 2(j) as:

> Any substance not corrosive . . . which on immediate, prolonged, or repeated contact with normal living tissue will induce a local inflammatory reaction.

The definition of this term is quite broad and would cause many products to be classified as irritants and thus to meet the first of the two requirements in Section 2(f)1(A). Nevertheless, none of these substances could be "hazardous" unless they also met the requirements of the "if" clause in that Section. This is illustrated in the House Report by a discussion of the application of the Act to ordinary water:

> "Irritant" refers to a substance (other than one defined as "corrosive") that will induce a local inflammatory reaction on immediate, prolonged, or repeated contact with normal living tissue. It is recognized that immersing the hands in ordinary water and other mild liquids for long periods of time may cause some minor transitory irritation of the skin. To make the substance "hazardous" within the meaning of the bill, however—so as to require a warning label—the substance must be one that, under conditions of customary or reasonably foreseeable handling or use, may cause a substantial "local inflammatory reaction", as that term is used by the medical profession. (See discussion above as to the meaning of "substantial.") [H.R. Rep. No. 1861, 86th Cong., 2d Sess. 7 (1960).]

Thus, even water may be an "irritant" within the meaning of the Act, but there can be no question that it is not "hazardous" because it does not pose a risk of *substantial* injury or illness from foreseeable handling or use. This illustration makes clear the importance of the "if" clause to the congressional purpose. Disregard of that clause, and an assumption that any product which fits into one of the six "preliminary screening" categories is necessarily hazardous, would lead to overlabeling of products and thus dissipate the cautionary impact of labeling that is appropriate.

In the 20-year history of the FHSA, there has been no litigation requiring judicial interpretation of the "if" clause. Judicial discussion of the definition of "hazardous substance", however, has consistently included *both* aspects of the statutory standard. See, e.g., *Springs Mills, Inc.* v. *CPSC,* 434 F. Supp. 416 (D.S. Car. 1977); *United States* v. *Chalaire,* 316 F. Supp. 543 (E.D. La. 1970); *United States* v. *7 Cases . . . Clacker Balls,* 253 F. Supp. 771 (S.D. Tex. 1966).

### Warning Requirements Prescribed Under the FSHA

Once a substance has been determined to be "hazardous" under Section 2(f), it must be labeled in accordance with the detailed directions contained in the statute. Failure to adhere to those directions will render the substance "misbranded" under Section 2(p) of the Act. To comply with the Act's detailed labeling provisions, the label of a "hazardous substance" must meet the following requirements:

**Identifying Information:** The label of a hazardous substance must state conspicuously the name and place of business of the manufacturer, packer, distributor or seller. In addition, it must disclose the identity of the hazardous substance or, where the substance is a mixture, the identity of each component which contributes substantially to the hazard. The statute embodies a clear preference for providing the "common or usual name" of the hazardous substance; the substance's chemical name need only be provided if no common usual name exists. By regulation, the Commission may permit or require the use of a recognized generic name. Such a generic name would appear appropriate where a more detailed description would either confuse product users or disclose information about the product's formulation that the manufacturer considers proprietary.

**Signal Word:** The label of a hazardous substance must also contain a "signal word" to attract the attention of product users. The statute prescribes the word "DANGER" for substances which are extremely flammable, corrosive, or highly toxic; highly toxic substances must also contain the word "poison." The labels of all other hazardous substances must contain the signal words "WARNING" or "CAUTION".

**Description of Hazard:** The label of a hazardous substance must provide an affirmative statement of the principal hazard or hazards presented. As examples of such statements, the statute specifies "Flammable", "Combustible", "Vapor Harmful", "Causes Burns", and "Absorbed Through Skin." Although the statutory language suggests that these statements are merely illustrative and that other statements can be substituted in the manufacturer's discretion, industry practice has led to a high degree of standardization in labeling terminology.

**Precautionary Measures:** The label of a hazardous substance must also describe the precautionary measures necessary to use the substance safely. According to the statute, instruction for first-aid treatment should be provided when necessary or appropriate, and there should also be instructions for the handling and storage of packages which require special care. In addition, the label must either state "Keep Out of the Reach of Children" or, if the article is intended for use by children, provide adequate directions for their protection.

**Presentation of Label Warnings:** Under Section 2(p)(2) of the Act, all required label statements must appear in a prominent location and must be presented in conspicuous and legible type contrasting by typography, layout or color with other label statements. Under Section 2(n), all required warning information must appear on any outside container or wrapper in which the substance is sold, and on all accompanying literature which contains directions for its use.

## Compliance With Statutory Labeling Requirements

For the most part, the labeling requirements of the FHSA are self-executing. The statute fully defines "misbranded hazardous substance", and thus imposes labeling requirements directly upon the manufacturer. Failure to meet those requirements on a product intended for interstate shipment constitutes an unlawful act; with minor exceptions, no administrative action is necessary to make the statute applicable to particular products.[4]

However, the self-executing character of the statute does not mean that the CPSC cannot issue regulations interpreting, and stating the positions it will apply in enforcement of, the various FHSA provisions. Sections 10(a) of the FHSA empowers the CPSC to prescribe regulations "for the efficient enforcement of this Act." Such regulations have been promulgated and are codified in 16 C.F.R. As noted above, these regulations set forth in considerable detail definitions and test methods for identifying substances that are "toxic", "corrosive", "irritants", etc. Because these regulations have been applied for several years without challenge, they possess a strong presumption of validity. The guidance they provide should simplify the task of the individual manufacturer in deciding whether his products fit into any of the "screening" categories in Section 2(f)1(A).

That initial determination, however, does not fully control whether labeling is required; for this purpose a further inquiry under the "if" clause is necessary. For example, a product may be "toxic" within the definition of that term in the Act and the regulations issued by the CPSC, but the manufacturer must still determine whether, in accordance with the "if" clause, the product may cause substantial personal injury or substantial illness as a result of foreseeable handling or use. In making this ultimate decision, the manufacturer is free to seek the advice of the CPSC if it wishes, but it is not required to do so, and if the CPSC disagrees with its conclusion, the question must be resolved by a court in an appropriate enforcement proceeding.

The heart of the Act's enforcement provisions is Section 4, which defines various "prohibited acts." Among these acts is the introduction, delivery for introduction, or receipt in interstate commerce of any "misbranded hazardous substance." When a prohibited act under Section 4 has been committed, the Commission can choose among several remedial provisions. Under Section 5, it can seek to impose criminal penalties, which may include both a fine and imprisonment. Under Section 6, it can seize any hazardous substance which is misbranded. Finally, under Section 8, it may seek an injunction against violations of the Act. Under Sections 5(b) a firm may defend against criminal charges by demonstrating that it received from the product's supplier a guarantee or undertaking "to the effect that a hazardous substance is not a misbranded hazardous substance." The receipt of such a guarantee or undertaking, however, will not protect a firm against product seizure under Section 6 or injunctive relief under Section 8.

## CPSC Authority to Designate Hazardous Substances by Regulation or Prescribe Special Labeling Requirements

While the FHSA's labeling requirements are primarily self-executing, the statute does authorize the Commission to issue implementing regulations in certain narrow but important circumstances.

First, Section 3(a) of the FHSA empowers the Commission, by regulation, to declare that specific articles qualify as "hazardous substances" under Section 2(f)1(A). This power may be exercised whenever, in the Commission's judgment, it will "promote the objectives of this Act by avoiding or resolving uncertainty as to its application . . . " In declaring that particular articles are "hazardous substances", the Commission must utilize the rulemaking procedures prescribed by Section 701(e)-(g) of the Federal Food, Drug, and Cosmetic Act, which provide for a formal trial-type administrative hearing. See *Pactra Industries, Inc.* v. *CPSC,* 555 F.2d 677 (9th Cir. 1977).

The rulemaking provisions of Section 3(a) were inserted in the FHSA at the insistence of the Secretary of HEW (who was initially charged with administering the Act) for the purpose of resolving uncertainties in applying the Act's labeling requirements. The reasoning of the Secretary was set forth in the following statement submitted to both congressional committees:

> *Declaratory regulations as to coverage.* It is apparent that, even with the above suggested clarifications, the application of the second part (i.e., the so-called "if" clause) of the basic definition of "hazardous substance" in the bill is so largely dependent on judgmental factors—e.g., what is "reasonably foreseeable"—that it will lead to considerable uncertainty and much costly litigation, with different courts and juries reaching different results, unless some mechanism for authoritatively resolving this uncertainty short of litigation is devised. We realize that on the one hand, in view of the broad sweep of the bill, and because of the constant development of new useful but hazardous substances suitable for household use, the inclusion of a statutory list of covered substances (in analogy to the list in the Federal Caustic Poison Act) or the limitation of coverage to substances listed by regulation would not be feasible. And while, on the other hand, we would prefer elimination of the "if" clause altogether from the point of facility of enforcement, we recognize that the inclusion of some such clause can be justified.
>
> It is feasible, however, and we strongly urge, that the committee include in the bill provisions deeming a substance to be hazardous where the Secretary by regulation declares it to be such upon the basis of a finding that it meets the requirements of the Bill's basic definition of "hazardous substance." The Secretary should be authorized to take such action whenever in his judgment this will promote the objectives of the bill by avoiding or resolving uncertainty. (The failure of the Secretary to take such action, of course, should not absolve anyone from the consequence of noncompliance with the labeling requirements of the bill in the case of a substance which is "hazardous" under the basic definition.) We would not object to making the issuance, amendment, or repeal of these declaratory regulations subject to procedural safeguards (with opportunity for administrative hearing and for judicial review on the basis of the hearing record) such as those contained in section 701 (e)-(g) of the Federal Food, Drug, and Cosmetic Act.[S. Rep. No. 1158, 86th Cong., 2d Sess. 24 (1960).]

As is evident from this statement, the Secretary was concerned about the factors of judgment entering into determinations under the "if" clause, and the confusion that might result, and therefore wished to have the authority to resolve any uncertainties about the Act's scope by declaring that particular substances met both requirements of Section 2(f)1(A) and therefore were "hazardous."

Despite the importance that was assigned to Section 3(a) when the FHSA was enacted, however, there has been little occasion to issue regulations designating particular articles as "hazardous substances." Under 16 C.F.R. § 1500.12, the only products so designated are charcoal briquettes and other forms of charcoal. Thus, it would appear that the CPSC (and previously FDA) has concluded that the "if" clause in the Act's definition of "hazardous substance" could be successfully applied by manufacturers to individual products without agency guidance and that regulations prescribing labeling requirements for particular products are unnecessary.

In addition to authorizing the Commission to promulgate regulations declaring particular articles "hazardous substances", Section 3(b) allows the Commission to issue regulations establishing additional label requirements for hazardous substances where "necessary for the protection of the public health and safety." Under Section 3(b), such additional labeling requirements are appropriate where the Commission finds that the customary label statements required under Section 2(p)1 are "not adequate for the protection of the public health and safety in view of the special hazard presented by any particular hazardous substance ... " Pursuant to Section 3(d), Commission regulations require special labeling for a number of products. These products, listed in 16 C.F.R. § 1600.14, include ethylene glycol, methyl alcohol, turpentine, benzene and other petroleum distillates, and firework devices.

## Banned Hazardous Substances

Under unusual circumstances, the FHSA provides for the outright banning of certain articles that qualify as "hazardous substances" under the Act's criteria.

First, under Section 2(q)(1)(A), any article will automatically be considered a "banned hazardous substance" if it is a toy or other article intended for use by children which either is a "hazardous substance itself" or "bears or contains a hazardous substance in such manner as to be susceptible of access by a child ... " Thus, in the case of products sold as toys or customarily used by children, precautionary labeling will be deemed inadequate to protect against harm; if the product has hazardous properties, its introduction into commerce is *per se* forbidden.

Where a product is not intended for use by children, it still may be banned, but the Commission will be required to meet a substantially more stringent standard. Under Section 2(q)(1), the CPSC may ban such a product only if it finds that, notwithstanding its cautionary labeling, "the degree or nature of the hazard involved in the presence or use of such substance in households is such that the objective of the protection of the public health and safety can be adequately served only by keeping such substance ... out of the channels of interstate commerce." In interpreting this provision, the courts have emphasized that the Commission must demonstrate an extremely high level of risk before an adequately labeled product can be banned. For example, in *Committee for Hand Gun Control* v. *CPSC,* 388 F. Supp. 216, 218 (D.D.C. 1975), the court observed that a ban on hand guns labeled in accordance with the Act was "unlikely" because of "the extensive findings that must be made to justify such a ban." As the court indicated, "the CPSC could not order a ban such as is proposed unless that it found that the hazard was very great and that no cautionary labeling would adequately protect the public." A similar conclusion was reached in *R.B. Jarts, Inc.* v. *Richardson*, 438 F.2d 846 (2d Cir. 1971).

16 C.F.R. § 1500.17 enumerates the products that the Commission has determined to be "banned hazardous substances" under Section 2(q)(1)(B) of the Act. Among these products are mixtures containing carbon tetrachloride, paints containing certain quantities of lead, garments containing asbestos, and self-pressurized products containing vinyl chloride monomer.

When the Commission invokes Section 2(q)(1)(B) and determines that a hazardous substance should be banned because precautionary labeling is inadequate to protect the public from harm, the procedures that the Commission must follow are prescribed by Section 2(q)(2). Under this provision, the Agency must utilize the formal trial-type rulemaking process outlined in Section 701(e)-(g) of the Federal Food, Drug, and Cosmetic Act. The only exception to this requirement is where the Commission "finds that the distribution for household use of the hazardous substance involved presents an imminent hazard to the public health." In this situation, the Commission may publish a Federal Register notice summarily declaring that the product is a "banned hazardous substance" and thereafter conduct the rulemaking proceeding required by statute.

Originally, the FHSA was silent on the procedures that the Commission was required to follow in determining that a hazardous substance intended for use by children is a "banned hazardous substance" within the meaning of Section 2(q)(10(A). In the absence of pertinent statutory provisions, the Commission took the position that it could make such determinations on an informal basis and then bring suit to seize, or enjoin the distribution of, violative products. One court, however, ruled that the Agency had to follow a different approach. In *Springs Mills, Inc.* v. *Consumer Product Safety Commission*, 434 F. Supp. 416 (D.S.C. 1977), the court held that the Commission could only determine that tris phosphate, an alleged carcinogen used in children's sleepwear, was a "banned hazardous substance" by first conducting a trial-type hearing.[5]

In the Consumer Product Safety Amendments of 1981, Congress directly addressed this issue by amending Section 3 of the FHSA to prescribe detailed procedures for classifying an article as a "banned hazardous substance." Under Section 3(f), the Commission must publish an advance notice of proposed rulemaking (ANPR) describing the regulatory options under consideration and inquiring whether an adequate voluntary standard might be developed to control the risk in question. If the Commission thereafter decides to proceed with rulemaking proceedings, Section 3(h) requires it to undertake a "regulatory analysis" which evaluates the relative costs and benefits of CPSC action and the adequacy of any voluntary safety standard that might be developed. Under Section 3(i), the Commission must then finalize its regulatory analysis by making various findings relating to the advantages and disadvantages of different regulatory actions, the relationship between the costs and benefits of the Commission's regulation, the absence of less burdensome but equally effective regulatory mechanisms, and the availability of adequate voluntary standards.

Under the previous version of Section 15 of the Act, all "banned hazardous substances" were subject to an elaborate repurchase procedure designed to remove them from commerce. Under this procedure, retailers had to repurchase the product from end-users; distributors then had to repurchase the product from retailers; and, finally, manufacturers had to repurchase the product from distributors. Although the Commission had promulgated detailed regulations governing the repurchase of banned hazardous substances,[6] this facet of the FHSA was confusing, costly and cumbersome.[7]

The Consumer Product Safety Amendments of 1981, however, improved matters by eliminating the repurchase scheme of the original statute and substituting streamlined procedures modeled on Section 15 of the CPSA. Under these new procedures, the Commission may require public notice of the hazards associated with a banned hazardous substance pursuant to Section 15(a) or repair, repurchase or refund remedies relating to the substance pursuant to Section 15(b). The statute provides that the Commission may not impose these remedies without affording affected manufacturers an opportunity for an adjudicative hearing.

### Regulation of Electrical, Mechanical and Thermal Hazards

The FHSA also empowers the CPSC to prescribe product safety standards for certain selected classes of products. Under Section 2(f)1(D), 15 U.S.C. § 1261(f)1(D), the CPSC may issue regulations determining the conditions under which toys or other articles intended for use by children will be deemed "hazardous substances" because they present "an electrical, mechanical, or thermal hazard." The terms "mechanical" and "thermal" hazard, in turn, apply to articles whose "design or manufacture presents an unreasonable risk of personal injury or illness" because of their mechanical or thermal properties. Section 2(s)-(t), 15 U.S.C. § 1261(s)-(t).

The FHSA does not define the term "unreasonable risk of injury." Nevertheless, the legislative history of the FHSA emphasizes that a determination of "unreasonable risk" involves balancing a product's potential harm against the economic burdens and other adverse effects of proposed regulatory requirements:

> In the definition of "mechanical hazards" the concept of unreasonable risk has been introduced to allow the Secretary greater discretion in determining which mechanical dangers are so great as to require banning. This discretion is beneficial to the manufacturer. There are numerous toys and other articles on the market which can cause personal injury because of their mechanical aspects but would probably not as a class be banned. Bicycles are one example. Before the Secretary banned such items he would consider such factors as the utility of the object, the degree of danger it presents, and the feasibility of designing out that danger. [S. Rep. No. 91-237, 91st Cong., 1st Sess. 7 (1969)]

The leading decision under the FHSA applying this concept of "unreasonable risk" is *Forester* v. *Consumer Product Safety Commission,* 559 F.2d 774 (D.C. Cir. 1977). This case involved the validity of CPSC regulations prescribing design and performance standards for bicycles and determining that, without compliance with these standards, such bicycles would be deemed to present a "mechanical hazard" within the meaning of the FHSA. The court defined the criteria for an "unreasonable risk" as follows:

> The Commission is, however, permitted by the FHSA to regulate only mechanical hazards that present "an unreasonable risk" of consumer injury. This means that the Commission must determine (1) that the risk posed by the hazard is an unreasonable one, and (2) that there is a sufficient nexus between the regulation and the hazard it is designed to prevent. The requirement that the risk be "unreasonable" necessarily involves a balancing test like that familiar in tort law: The regulation may issue if the severity of the injury that may result from the product, factored by the likelihood of the injury, offsets the harm the regulation itself imposes upon manufacturers and consumers. [559 F.2d at 789.]

The court then proceeded to apply this standard to various provisions of the Commission's regulation. As the following excerpts from its analysis indicate, the court explicitly recognized that the level of risk that the Commission is required to demonstrate is directly proportional to the burdens imposed by its standard and, where such burdens will be negligible, a minimal showing of risk will suffice:

> Forester assails § 15124(i), which requires that control cable ends be capped or treated to prevent unravelling, on the ground that the Commission has not shown that a "significant proportion of cyclist injuries" are caused by fraying control cable ends . . . We find this provision to rest upon a rational basis. *Although the injuries caused in this manner are minor, the means to be used for reducing the risk of injury is itself quite inexpensive and has not been shown to interfere in any way with the function of bicycles.* On these facts, it was within the discretion of the Commission to determine that the risk of injury from untreated control cable ends is "unreasonable." [559 F.2d at 791 (emphasis added).]

> \* \* \* \*

> When the product characteristics being regulated have no countervailing benefits at all, and the cost of eliminating the defects is slight, the burden of supporting the regulation is not heavy. [559 F.2d at 796.]

> \* \* \* \*

> Reflectors appear to provide a significant margin of added safety at a relatively small monetary cost and loss in bicycle efficiency. In view of the Commission's careful balancing of the relevant factors, we do not find this standard to be irrational. [559 F.2d at 798.]

Thus, a small risk may be "unreasonable" when the burdens of protecting against it will be minimal, while a larger risk may be "reasonable" when the mechanism proposed for controlling it will have serious adverse effects on manufacturers.

In addition to its regulations concerning bicycle safety, the Commission has utilized its authority under Section 2(f)1(D) to develop a variety of safety requirements for products used by children. These requirements include regulations governing sharp points and small parts (16 C.F.R. § § 1500.48-1500.53, 16 C.F.R. § § 1501); regulations governing electrically operated toys (16 C.F.R. § § 1505); and regulations governing cribs, rattles and pacifiers (16 C.F.R. § § 1508–1511).

### Preemption of State Regulation

As the FHSA's legislative history emphasizes, before the Act's enactment, numerous states had passed legislation regulating the labeling of products intended for household use. One goal of the FHSA, therefore, was to establish a uniform and comprehensive federal labeling scheme for products shipped in interstate commerce and to eliminate the possibility of inconsistent labeling requirements from state to state. S. Rep. No. 1158, 86th Cong. 2d Sess. 3 (1960). This goal is embodied in Section 18 of the FHSA, which addresses the Act's effect on state law. Under Section 18(b)(1)(A), where a substance is subject to a labeling requirement under Section 2(p) or 3(b) of the Act, no state may establish or continue to enforce a cautionary labeling requirement applicable to that substance and designed to protect against the same risk of injury unless the requirement has the same effect as the provisions of the FHSA.[8] Thus, for all practical purposes, enactment of the FHSA has precluded states from adopting and implementing their own labeling programs for household products.

## CONSUMER PRODUCT SAFETY ACT

While labeling requirements may be prescribed under the CPSA, it is not primarily a labeling statute. Rather, it gives the Commission broad authority to ban, recall or prescribe safety standards for a wide variety of household products. Under the FHSA, by contrast, the Commission's primary focus is on labeling; while other regulatory mechanisms are available, the Commission must meet a very high standard for invoking them and they can be applied only to certain narrow classes of products. Moreover, while the labeling requirements prescribed by the FHSA are largely defined by the statute itself, the Commission has greater latitude under the CPSA to tailor labeling requirements to individual products. In this additional respect, the Commission has greater flexibility and freedom under the CPSA than under the FHSA.

The CPSA and the FHSA are extremely similar to their coverage. The FHSA generally applies to products "intended, or packaged in a form suitable for use in the household or by children . . ." (Section 2(p) ). The CPSA applies to "consumer products," a term defined in Section 3(a)(1) to cover articles produced or distributed for use by consumers in or around a household or residence. Because many products are subject to both the CPSA and FHSA, it is frequently necessary for the Commission to choose between invoking one statute or the other. Section 30(d) of the CPSA provides standards for making this decision. Under the provision, a risk of injury which could be "eliminated or reduced to a sufficient extent by action" under the FHSA may be regulated under the CPSA "only if the Commission by rule finds that it is in the public interest" to do so.

Three provisions of the CPSA have a bearing on the labeling responsibilities of chemical manufacturers: (1) Section 7, which authorizes the Commission to prescribe safety standards which may include labeling requirements; (2) Section 27(e), which the Commission views as an additional source of labeling authority; and (3) Section 15, under which an inadequately labeled product already distributed in commerce can be subjected to recall, replacement, notification and other remedies.

A more detailed discussion of these provisions is presented below.

### Safety Standards and Bans

Under Section 7(a)(1) of the CPSA, the Commission may promulgate "consumer product safety standards" that enhance product safety by prescribing (1) requirements concerning product performance, or (2) requirements that a consumer product be marked with or accompanied by clear and adequate warnings or instructions.[9] The Consumer Product Safety Amendments of 1981 stress Congress' preference for the adoption of voluntary safety standards in lieu of mandatory federal requirements. In its amended form, Section 7(b) provides that the Commission "shall rely upon voluntary consumer product safety standards ... whenever compliance with such voluntary standards would eliminate or adequately reduce the risk of injury addressed and it is likely there will be substantial compliance with such voluntary standards."

In the case of unusually hazardous products, the Commission is authorized by Section 8 of the CPSA to declare a product ban. The precondition for such a ban is a finding that the product presents an unreasonable risk of injury and that no feasible consumer product safety standard would protect the public from the hazard involved.

**Procedures:** The development of consumer product safety standards and bans is governed by Section 9 of the CPSA. In 1981, Congress significantly reworked the procedures prescribed by this provision to place greater emphasis on the encouragement of voluntary standard-development activities and to increase the role of cost-benefit analysis in CPSC decision-making.

Section 9(a) requires the CPSC to commence a standard development proceeding by publishing an ANPR which identifies the product and risk of injury that are of concern to the Commission, the adequacy of any applicable voluntary standards, and the regulatory alternatives under consideration. The ANPR must also inquire about the willingness of industry or other groups to develop an appropriate voluntary standard if such a standard does not exist already. If the Commission determines that existing or prospective voluntary standards will adequately protect consumers, it must terminate rulemaking proceedings.

Section 9(c) provides that, where the Commission concludes that voluntary standard-development activities will be unsatisfactory, it may propose a rule adopting a mandatory safety standard. The Commission's proposed rule must discuss the status of voluntary standard-development activities and evaluate the benefits and costs of the proposed mandatory standard and all alternative regulatory approaches.

Under Section 9(f)(2), the Commission's final rule must be accompanied by a "regulatory impact analysis" which describes the potential costs and benefits associated with the rule, analyzes the costs and benefits of alternative approaches that the Commission has considered and rejected, and discusses all significant issues raised during the public comment period. Pursuant to Section 9(g)(1), the rule must specify an effective date which does not exceed 180 days from the rule's promulgation unless the Commission finds that a delayed effective date is in the public interest.

As they apply to the promulgation of safety standards and bans, the rulemaking procedures authorized by Section 9(d)(2) incorporate and slightly enlarge upon the Administrative Procedure Act's provisions for informal rulemaking: rules must be published in the form of a proposal, comments (both oral and written) must be permitted and, after the comment period, final rules must be adopted or the proposal withdrawn.

**The Criteria for Requiring a Safety Standard:** The text and legislative history of the CPSA establish that the Commission must adhere to two criteria in promulgating a consumer product safety standard: (1) the consumer injury the standard will prevent must outweigh the standard's adverse effect on other forms of consumer welfare; and (2) the Commission must adopt the regulatory requirement that will impose the smallest possible burdens while still affording effective protection to consumers.

In conjunction with Section 7, Section 9 sets forth the substantive criteria for promulgating a safety standard. Section 7(a) provides that any requirement of a standard shall "be reasonably necessary to prevent or reduce an unreasonable risk of injury associated with" the product to which it applies. Section 9(f)(3) repeats this test in slightly different language, stating that the Commission is precluded from promulgating a "consumer product safety rule"[10] unless it finds that "the rule (including its effective date) is reasonably necessary to eliminate or reduce an unreasonable risk of injury associated with such product" *and* "that the promulgation of the rule is in the public interest." Section 9(f)(3) also requires the CPSC to make three additional findings: (1) that any existing voluntary standards would not afford adequate protection against the unreasonable risk that the Commission has identified; (2) that the benefits expected from the rule bear a reasonable relationship to its costs; and (3) that the rule imposes the least burdensome requirement which prevents or adequately reduces the risk of injury that the CPSC is seeking to address.

Section 9(f)(1) identifies four issues that the Commission must address in determining whether a proposed standard is justified:

(A) the degree and nature of the risk of injury the rule is designed to eliminate or reduce;

(B) the appoximate number of consumer products, or type of classes thereof, subject to such rule;

(C) the need of the public for the consumer products subject to such rule, and the probable effect of such rule upon the utility, cost, or availability of such products to meet such need; and

(D) any means of achieving the objective of the order while minimizing adverse effects on competition or disruption or dislocation of manufacturing and other commercial practices consistent with the public health and safety.

The first two issues involve the "benefits" of a standard—e.g., the seriousness of the consumer injury it purports to prevent (death, serious illness, minor burns, etc.), the number of people who might be exposed to that injury, and the likely effectiveness of the standard in reducing the injury's occurrence.

The second two issues are directed at a measurement of the standard's "costs." (C) is aimed at the direct burdens which the standard may impose on consumers. In light of the consumer need which the product satisfies, the Commission is directed to determine to what extent the standard may increase product cost or decrease product availability and utility. (D), on the other hand, is aimed at the indirect burdens the standard may impose on consumers. It encompasses the standard's adverse effects on "competition" (perhaps by driving smaller concerns out of business) and on "dislocation of manufacturing and other commercial practices" (perhaps by creating unemployment or requiring reconstructions of plant and equipment).

The legislative history of the original CPSA confirms the Congressional desire to balance a proposed standard's benefits to consumer safety against its drawbacks in other areas of consumer welfare. S. 3419, the Senate version of the Act, included an express definition of "unreasonable risk of injury":

> "Unreasonable risk of injury presented by a consumer product" means that degree of risk which the Commission determines is incompatible with the public health and safety either because the degree of anticipated injury or the frequency of such injury, or both, is unwarranted because—(A) the degree of anticipated injury or the frequency of such injury can be reduced without affecting the performance or availability of the consumer product but the effect on such performacne or availability is justified when measured against the degree of anticipated injury or the frequency of such injury.

The Report of the Commerce Committee contains an exhaustive explanation of this definition:

> The definition of "unreasonable risk of injury presented by a consumer product" in paragraph (8) of section 101 highlights those factors which are relevant when drawing the line between risks which are reasonable and those which are unreasonable. The definition sets forth a balancing test which emphasizes the primacy of health safety factors. Two particular measures of public health and safety are to be considered: (1) the degree of *anticipated* injury, and (2) the frequency of such injury. In those situations where either the degree of anticipated injury or the frequency of such injury can be reduced without affecting the "performance" or "availability" of that class of consumer product, then almost any risk capable of consumer product injury becomes unwarranted. When "performance" or "availability" are affected, then a balancing of competing interests must be undertaken. [S. Rep. No. 92–749, 92 Cong., 2nd Sess., 14–15 (1972)]

A definition of "unreasonable risk" was deliberately omitted from H.R. 15003, the bill passed by the House, and this omission was carried forward into the final version of the Act by the Conference Committee. The House's reasons for omitting a definition of unreasonable risk, however, do not reflect a rejection of the Senate's cost-benefit approach. As the House Committee on Interstate and Foreign Commerce explained:

> Your committee has not included a definition of "unreasonable hazards" within this bill. Protection against unreasonable risks is central to many Federal and State safety statutes and the courts have had broad experience in interpreting the term's meaning and application. *It is generally expected that the determination of unreasonable hazard will involve the Commission in balancing the probability that risk will result in harm and the gravity of such harm against the effect on the product's utility, cost and availability to the consumer. An unreasonable hazard is clearly one which can be prevented or reduced without affecting the product's utility, cost, or availability; or one in which the effect on the product's utility, cost or availability is outweighed by the need to protect the public from the hazard associated with the product.* There should be no implication, however, that in arriving at its determination the Commission would be required to conduct and complete a cost-benefit analysis prior to promulgating standards under this act. Of course, no standard would be expected to impose added costs or inconvenience to the consumer unless there is reasonable assurance that the frequency or severity of injuries or illnesses will be reduced. [H. Rep. No. 92-1153, 92nd Cong., 2d Sess., 33 (1972) (emphasis added).]

**Judicial Decisions:** A significant number of judicial decisions have construed the "unreasonable risk" standard in Sections 7 and 9 of the CPSA. These decisions provide considerable insight into the criteria that the Commission must observe in promulgating safety standards which include labeling requirements. While the judicial decisions construing Sections 7 and 9 all predate the 1981 amendments, the basic approach they adopt was expressly reaffirmed by Congress when it recently modified the CPSA.

*Aqua Slide 'N' Drive* v. *Consumer Product Safety,* 569 F.2d 831 (5th Cir. 1978), is of special interest because it involved labeling. In *Aqua Slide,* the court set aside warning sign requirements in a swimming pool slide standard on the basis that they were not, as required by statute, "reasonably necessary to eliminate or reduce an unreasonable risk of injury." These signs were designed to reduce injury from sliding into a swimming pool head-first. The Commission had justified the signs on the ground that they "may achieve" a reduction in injury and that it "felt" the signs would be a "positive step." However, the court found this justification for the "benefits" of the standard to be insufficient. Such benefits, it held, must be based on "empirical evidence to show that the signs will work," not on the Commission's "inferences." 569 F.2d at 842.

The court also found that the Commission failed to assess adequately the burdens of the signs, including their adverse effect on consumer purchases of slides. As the concurring opinion explained this aspect of the court's holding:

But the Commission is required to do more than determine whether there are many benefits to its regulation. Congress required it to consider the economic costs. *In this case, those costs are the effects on the manufacturers, and the effects on consumers who will be frightened away from purchasing pool slides.* The Commission's determination that benefits exist is not the only conclusion that must be supported by substantial evidence. Most importantly, the benefits and the costs must have such support. This is the argument pressed by Aqua Slide. The benefits from these signs have no reasonable relationship to the costs they will impose . . . These signs are not so innocuous as to be presumed "inexpensive." With no evidence on the cost side of the ledger, the Commission's cost-benefit analysis is without substantial evidence for support. [569 F.2d at 845 (emphasis supplied).]

In summarizing the balancing test that the CPSC had neglected to apply, the court stated:

In evaluating the "reasonable necessity" for a standard, the Commission has a duty to take a hard look, not only at the nature and severity of the risk, but also at the potential the standard has for reducing the severity or frequency of the injury, and the effect the standard would have on the utility, cost or availability of the product. In this case, the Commission neglected that duty. [569 F.2d at 844.]

In reaching this conclusion, the court emphasized the remoteness of the risk that the Commission was attempting to reduce:

*The risk of paraplegia from swimming pool slides, however, is extremely remote. More than 350,000 slides are in use, yet the Commission could find no more than 11 instances of paraplegia over a six-year period.* According to Institute figures, the risk, for slide users, is about one in 10 million, less than the risk an average person has of being killed by lightning. App. 583. The standard faces an initial difficulty because it is not easy to predict where paraplegia will next occur, and *to burden all slide manufacturers, users, and owners with requirements that will only benefit a very few, is questionable.* [569 F.2d at 840 (emphasis added).]

In view of the "infrequency of the risk", the court determined that the Commission had a special burden to demonstrate that the warning·would achieve measurable benefits. This burden, the court indicated, could only be satisfied by producing evidence of the warning's efficacy in reducing the risk. As the court said, "[w]hile it is no doubt rational to assume the warning signs would be heeded, mere rationality is not enough." 569 F.2d at 841.

The final consideration that the court emphasized was the Commission's failure to consider the adverse effect of warnings on the public's perception of swimming pool slides and, hence, its reduced willingness to continue purchasing those slides. As the court explained, the extremely low risk associated with swimming pool slide use had been totally overlooked by the agency:

> In this case, the prime disadvantage to which Aqua Slide points is the warning's effect on the availability of the slides. Because the Commission did not test the signs, it provided little evidence of whether the signs were so explicit and shocking in their portrayal of the risk of paralysis as to constitute an unwarranted deterrent to the marketing of slides, and hence, their availability to users. *The record provides only scant assurance that purchasers would not be so alarmed by the warning signs that they would unnecessarily abstain. The signs do not indicate paralysis is a one in 10 million risk.* [569 F.2d at 842 (emphasis added).]

Thus, the court recognized that where the risk of harm is remote, warnings which unduly alarm consumers will not be "reasonably necessary."

Shedding additional light on these issues is the Fifth Circuit's recent decision in *Southland Mower Co. v. CPSC,* 619 F.2d 499 (5th Cir. 1980). In that case, the court overturned portions of the CPSC's safety standard for power lawnmowers which required a foot-probe test for the mower's rear area. To justify this requirement, the Commission had relied on a documented instance of serious injury due to a mower's discharge chute, but the court held that this one incident did not provide the requisite "[s]ubstantial evidence that such injury is *significantly* likely to occur . . . " 619 F.2d at 509 (emphasis added). Recognizing that the Commission's safety standard was intended to prevent the amputation of toes and that "the seriousness of these injuries cannot be gainsaid", the court nevertheless rejected the proposition that "any risk, however remote, is unreasonable" within the meaning of the CPSA. *Id.*

At the same time, however, the court in *Southland* upheld the portion of the CPSC' standard which prescribed labeling requirements for certain types of lawnmowers. Pursuant to these requirements, reel-type and rotary-power mowers were required to bear a warning stating "DANGER, KEEP HANDS AND FEET AWAY" and a pictorial representation of a blade-like object cutting into a finger. In upholding this requirement, the court determined that the "Commission properly concluded that there was a need for labeling as a safety measure" since other forms of protection would not be "fail-safe." 619 F.2d at 522. The court also determined that the reasonableness of the labeling requirement was supported by the record. As the court noted, the Commission had tested various warning labels and determined that the cautionary statement it required was likely to be effective. It had also determined that the labeling requirement would impose little additional cost since the industry's voluntary standard already called for warning labels.

The court also differentiated between the Commission's approach to lawnmower labeling and the warning signs overturned in *Aqua Slide.* In the case before it, the court stressed, the warnings' "contents are not shocking or gruesomely explicit and would not pose an unwarranted deterrent to potential purchasers of lawnmowers." The court then explained the significance of these factors as follows:

> Our own examination of the record in *Aqua Slide* disclosed some evidence that the signs, which did not indicate that the risk of paralysis was only one in ten million, were so alarming that they might well impose high costs on the industry and on potential purchasers of

swimming pool slides. *Id.* Since the proof that the signs could reduce the risk of injury was weak, an adequate analysis of buyer reaction to the alarming signs had not been performed, and the Commission's analysis of the economic impact of the entire standard was otherwise unreliable, we ruled that substantial evidence did not support the crucial finding that the "relationship between the advantages and disadvantages of the signs is reasonable" and that consequently the reasonable necessity of the warning signs had not been demonstrated. 569 F.2d at 841. In the present case, an unreasonable risk of blade-contact injury has been shown, and the warning label clearly does not threaten to impose costs on lawn mower manufacturers and consumers like those that were at issue in *Aqua Slide.* On this record, we hold that the labeling requirement is reasonably necessary to reduce an unreasonable risk of injury. [619 F.2d at 522.]

Based on the *Aqua Slide* and *Southland* decisions, it is possible to identify certain general principles that will govern the development of labeling requirements under Section 7 of the CPSA. Where the Commission proposes to require warnings as part of a safety standard, it must first demonstrate that the warnings address a real and significant, rather than a theoretical or minor, risk. It must then demonstrate that warnings represent a necessary measure for reducing the risk of injury and that other measures being taken are inadequate for this purpose. Next, it must show that the warnings are likely to be effective in reducing the probability of injury and that they accurately disclose the nature of the risk in question. Finally, the Commission must investigate whether warning labels will diminish sales of the regulated product or otherwise impose costs on manufactures and, if so, whether the burdens imposed by the warnings are outweighed by their benefits to consumers. Only if all of these requirements are met can the Commission conclude that its labeling requirement is "reasonably necessary" to reduce or eliminate an "unreasonable risk of injury."

## Labeling Requirements Promulgated Under Section 27(e)

Prior Commission labeling requirements have been based not merely on the standard-development provisions of Section 7 of the CPSA but on Section 27(e) of the Act as well. Industry representatives have argued that the Commission's labeling authority under Section 27(e) is extremely limited and that the Agency cannot use Section 27(e) as a statutory basis for warning requirements. While two courts have ruled on this issue to date, the precise scope of Section 27(e) has not yet been clarified.

Section 27(e) provides that "[t]he Commission may by rule require any manufacturer of consumer products to provide . . . such performance and technical data related to performance and safety as may be required . . . " Having obtained such data from manufacturers, the Commission may then require manufacturers to provide appropriate "notification" of the data "to prospective purchasers" of the products involved.

Certain obvious differences exist between Section 27(e) and Section 7 which, from the Commission's standpoint, make the latter a significantly more attractive vehicle for promulgating labeling requirements. First, under Section

27(e), the Commission need not observe the elaborate procedures and stringent deadlines that, under Section 7 and 9 of the CPSA, govern the development of consumer product safety rules. While Section 27 requires the Commission to proceed by promulgating regulations, it does not specify the procedures that the Agency must employ. Thus, the Commission is free to conduct an informal notice-and-comment rulemaking in accordance with the Administrative Procedure Act. See 5 U.S.C. § 533. Equally important, when proceeding under Section 27(e), the Commission need not meet Section 7's criteria for developing a safety standard—i.e., that the standard is "reasonably necessary to prevent or reduce an unreasonable risk of injury." Rather, Section 27(e) merely requires the Commission to show that any notification scheme it adopts is "necessary to carry out the purposes" of the CPSA. Satisfying this test will normally be far easier than satisfying the "unreasonable risk" standard of Sections 7 and 9.

Based on these considerations, industry litigants have argued that Section 7 is the *only* provision of the CPSA that authorizes the promulgation of warning requirements and other forms of precautionary labeling and that Commission efforts to use Section 27(e) for this purpose represent an attempt to circumvent clear Congressional intent. In *Southland Mower,* the Fifth Circuit upheld this position, ruling that Section 27(e) did not authorize the Commission to require a warning label on lawnmowers relating to the dangers of direct contact with blades.[11] In reaching this conclusion, the court employed the following reasoning:

> We agree with OPEI that section 27(e) does not authorize promulgation of the warning label requirement. Notification of the hazard of blade-contact injury by means of a graphic design and the exclamation, "Danger, Keep Hands and Feet Away," does not fit within the "plain, obvious, and rational meaning" of "performance and technical data" that the Commission may order manufacturers to provide to prospective purchasers and to a product's first consumer under section 27(e). *See Old Colony Railroad* v. *Commissioner,* 284 U.S. 552, 560, 52 S. Ct. 211, 213, 76 L. Ed. 484 (1932) (in interpreting statutory language, "the plain, obvious and rational meaning of a statute is to be preferred to any curious, narrow, hidden sense"). [619 F.2d at 521.]

The court acknowledged that the Commission could properly order manufacturers to furnish consumers with information concerning lawnmower rotation speed under Section 27(e) but that a warning label simply did not "communicate performance and technical data" within the scope of the statutory language. Id.

A different result, however, was reached in *United States* v. *Falcon Safety Products* (D.N.J. January 27, 1981). In that case, the District Court upheld the Commission's use of Section 27(e) to require a warning for all aerosolized consumer products containing chlorofluorocarbon propellants. The Commission's warning, which was similar to warnings adopted by the Food and Drug Administration and the Environmental Protection Agency for products within their jurisdictions, stated as follows:

WARNING—Contains a Chloroflurocarbon That May Harm the Public Health and Environment by Reducing Ozone in the Upper Atmosphere.[12]

In its decision, the court noted that the Commission had proceeded under Section 27(e) because standard development proceedings under Section 7 would be too lengthy and cumbersome and there was a compelling interest in developing a warning requirement for chloroflurocarbon propellants as soon as possible. The court also noted that the stringent protections of Section 7 and 9 were not always necessary for labeling requirements, which imposed relatively limited burdens on industry. Finally, the court emphasized the need to defer to the CPSC's own interpretation of its governing statute. According to the court, these factors persuaded it that Section 27(e) authorized the warning requirements under challenge.

Although the court in *Falcon Safety Products* acknowledged the *Southland Mower* decision, it made no effort to reconcile its approach with that of the Fifth Circuit. Moreover, viewed in conjunction, the two decisions appear to be directly conflicting. Accordingly, further judicial guidance will be necessary to clarify the precise scope of the Commission's labeling authority under Section 27(e). In the meantime, the Commission has displayed its continued willingness to use Section 27(e) as the basis for labeling requirements.[13]

### Reporting Under Section 15

To date, the labeling issues addressed by the Commission have principally arisen under Sections 7 and 27(e). Nevertheless, manufacturers should recognize that Section 15 of the CPSA also has important implications for product labeling.

Section 15, unlike Section 7 and 9, seeks to provide protection against dangerous products already manufactured and in the hands of consumers. The Section authorizes the Commission to take two measures against such products. The first measure, authorized by subsection (c), is to require notification of the dangers which the product may present. This notification may be directed to the general public (presumably in the form of radio or television broadcasts or newspaper announcements), to manufacturers, distributors and retailers of the product, or to all identifiable purchasers and users of the product.

The second (and considerably more burdensome) measure, authorized under subsection (d), is to require that the dangerous product be repaired or replaced or its purchase price refunded. Once the Commission has determined that repair, replacement or refund is necessary, the persons (manufacturers, retailers or distributors) it has designated to take action may elect which of these remedies will be pursued. To insure that repair, replacement or refund is satisfactorily carried out, the Commission may require the advance submission of a corrective action plan for its approval.

Paragraph (d)(1) of Section 15 provides that consumers who avail themselves of repair, replacement or refund cannot be directly charged therefor and will receive reimbursement for any reasonable and foreseeable expenses which these remedies may entail. Paragraph (d)(2) permits the costs of notification or repair, replacement or refund to be allocated to other manufacturers and distributors of the products involved. The precondition for requiring this sharing of costs is that it be "in the public interest."

The principal requirement for both notification under subsection (c) and repair, replacement or refund under subsection (d) is a determination that the product in question presents a "substantial product hazard." Section 15(a) defines a "substantial product hazard" as:

(1) a failure to comply with an applicable consumer product safety rule which creates a substantial risk of injury to the public, or

(2) a product defect which (because of the pattern of the defect, the number of defective products distributed in commerce, the severity of the risk, or otherwise) creates a substantial risk of injury to the public.

The meaning of "failure to comply with an applicable consumer product safety rule" is self-explanatory; in the labeling context, failure to implement warning requirements imposed under Section 7 would potentially trigger Section 15 and, if a substantial risk of injury were created, give rise to a duty to repair, replace, or refund the purchase price of the products involved. The concept of "defectiveness" under Section 15 is somewhat more involved and is discussed more fully below.

Perhaps the most important provision of Section 15 is subsection (b), under which manufacturers, distributors or retailers of products that may present a "substantial product hazard" are required to submit reports to the Commission. Section 15(b) provides that companies must "immediately" inform the Commission when they obtain information "which reasonably supports the conclusion" that a product that they manufacture, distribute or retail either "fails to comply with an applicable consumer product safety rule" or "contains a defect which could create a substantial product hazard." The only exception to this requirement is where the company involved "has actual knowledge that the Commission has been adequately informed of such defect or failure to comply."

In practice, these reporting provisions have been the focal point for the Commission's activities under Section 15. The reporting obligation created under the section is framed in extremely broad terms, and failures to report are potentially punishable by civil and criminal penalties under Sections 20 and 21 of the CPSA. Accordingly, since its creation in 1972, the Commission has received an extremely large number of "substantial risk" reports from industry. Upon their receipt, these reports have been evaluated by the Commission's staff, which generally requests follow-up information and urges the company involved to take appropriate remedial action.[14] A very large number of Section 15(b) reports have resulted in some form of negotiated corrective action, including public notice of possible hazards and recalls of the products involved.

In determining whether a possible safety problem is reportable under Section 15, the first question is whether a product "defect" may exist. The Act itself does not define this term. Nevertheless, the legislative history suggests that Congress intended the concept of "defectiveness" under the CPSA to be at least as broad as the comparable concept in products liability law. See 117 Cong. Rec. 9930 (June 21, 1972). Thus, it appears that Section 15 would encompass both manufacture and design defects. It also appears that the Section would encompass violations of the duty to warn, as defined by Section 402A of the *Restatement of Torts* and pertinent judicial decisions. For this reason, a manufacturer who has not utilized adequate precautionary labeling to acquaint consumers with the possible risks of product use may be producing a "defective" product under Section 15.

CPSC regulations stress that the term "defect" should be construed extremely broadly under the CPSA and extends at least as far as the concept of "defect" in products liability law. See 16 C.F.R. § 1115.4. Significantly, Commission regulations state that a defect can occur in a product's "packaging, warnings, and/or instructions." According to the Commission, "a consumer product may contain a defect if the instructions for assembly or use could allow the product, otherwise safely designed and manufactured, to present a risk of injury."

A product "defect" will only be reportable under Section 15 if it "could create a substantial product hazard." Thus, inadequate labeling or other product shortcomings will not be reportable unless they result in an increased risk of injury. It should be noted, however, that clear and convincing evidence of a potential hazard is unnecessary; the statute requires reporting wherever a product defect "could create" a substantial risk of injury.

The Act specifies three criteria for determining whether a risk of injury is "substantial": the pattern of defect, the number of defective products distributed in commerce, and the severity of the risk. These criteria suggest that the overriding inquiry in measuring substantiality must be twofold: first, how certain is it that the product will cause injury and, second, how extensive is that injury likely to be? Discussing the definition of substantial risk, the CPSA's legislative history states:

> [T]his definition looks to the extent of public exposure to the hazard. A few defective products will not normally provide a basis for compelling notification under this Section. [H. Rep. No. 92-153, 92nd Cong., 2d Sess., 42 (1972).]

Plainly, a product that is defective in design will be at a serious disadvantage under Section 15—each individual product would contain a defect. Nevertheless, the number of defective products alone will not be decisive. The statute requires that equal weight be given to the "severity of the risk" which the defect creates.

While the Act specifies no criteria for measuring the severity of risk, two considerations seem logically relevant. First, what is the probability that the product defect will in fact injure consumers? In other words, what fraction of the consumers who are exposed to the product will be injured? Second, in those cases where the product will cause injury, what type of injury will occur? Will that injury be death, serious bodily harm or merely minor injury? In applying these considerations to specific products, the product's historical safety record will undoubtedly be relevant, as will an analysis of the conditions which must be present in order for the product defect to result in injury. In the latter connection, it would be important to examine the extent to which injury is tied to such factors as type, duration and intensity of use, external conditions like heat, moisture or physical stress, manner of product installation, and consumer ability to detect and correct product malfunctions before they cause harm. Assuming these factors demonstrated that actual injury would be extremely infrequent, a strong argument could be made that any risk that exists is not "severe." Nevertheless, because the statute expressly mentions "the number of products in distribution," what may be a slight risk for a product manufactured in small quantities will become steadily more severe when the product reaches increasing numbers of consumers.

In addition to amplifying on the statutory criteria, Commission regulations stress that information indicating that a product has caused "death or grievous bodily injury" is normally reportable. See 16 C.F.R. § 1115.12(c). The regulations also indicate that firms should carefully evaluate such additional information as product liability suits and consumer complaints. Despite the additional guidance provided by Commission regulations, however, the determination whether a risk is "substantial" remains very much a matter of informed judgment for the company involved.

## FOOTNOTES

1. Pub. L. 92-573, 86 Stat. 1207, 15 U.S.C. § 2051 et seq. Major amendments to the CPSA were enacted in 1981.
2. Authority to administer these laws was transferred to the Commission by Section 30 of the CPSA, 15 U.S.C. § 2079.
3. The FHSA was amended by the Child Protection Act of 1966, the Child Protection & Toy Safety Act of 1969 and the Poison Prevention Packaging Act of 1970. For the most part, the purpose of these amendments was to strengthen the provisions of the FHSA which apply to toys and other products used by children.
4. The only exceptions are substances that are "strong sensitizers" or "radioactive."
5. In its Federal Register notice regarding Tris, the Commission relied on portions of the FHSA's legislative history indicating that hazardous substances intended for use by children are banned by the language of the statute itself. 434 F. Supp. at 424. It is, however, open to question whether the FHSA's definition of "hazardous substance" was ever intended to be applied to chemicals asserted to be "toxic" not because of acute health effects, but because of long-term chronic effects like carcinogenicity.

   After the district court decision, the CPSC announced that it intended to take action against Tris-treated garments by bringing enforcement actions against individual manufacturers. The use of seizure proceedings for this purpose, without prior agency rulemaking, was upheld in *United States v. Articles of Hazardous Substance*, 588 F.2d 39 (4th Cir. 1978), which held that the hearing required by due process could properly be conducted by the district court in the condemnation action itself.
6. 16 C.F.R. § § 1500.202, 1500.203.
7. It was, however, settled that the repurchase process could only be initiated by a Commission determination that a product is a "banned hazardous substance," and not by a private lawsuit seeking a judicial declaration to that effect. *Riegel Textile Corporation* v. *Celanese Corporation*, (2nd Cir. May 6, 1981).

8. The Act gives the states somewhat greater latitude to ban or otherwise restrict hazardous substances. Under Section 18(b)(1)(B), the issuance of a regulation under Section 2(q) banning or regulating the safety of a product will not preclude states from promulgating requirements that provide a "higher degree of protection" where the state has sought and received CPSC authorization to proceed with its own regulatory requirements.

9. Prior to 1981, the CPSA permitted the Commission to prescribe requirements relating to product composition, contents, design or packaging. The Consumer Product Safety Amendments of 1981, however, narrowed the Commission's authority, directing the Commission to promulgate standards "expressed in terms of performance requirements." Section 7(a)(1). The Congressional policy in favor of performance-oriented standards derives in part from a 1980 Fifth Circuit decision emphasizing the disadvantages of CPSC-prescribed design requirements. *Southland Mower Co.* v. *Consumer Product Safety Commission,* 619 F.2d 499, 515 (5th Cir. 1980).

10. Defined in Section 3(a)(2) of the Act as either a safety standard or a product ban.

11. As described above, however, the Court did agree with the Commission that its warning requirements were justified under the "unreasonable risk" criteria of Section 7.

12. 16 C.F.R., Part 1401.

13. On June 10, 1980, the Commission proposed a rule under Section 27(e) that would have required manufacturers of urea-formaldehyde foam insulation to alert purchasers and prospective purchasers to certain possible acute health effects. 45 Fed. Reg. 39434. The Commission's rule was never adopted because the Agency subsequently decided to propose a ban on urea-formaldehyde insulation based on evidence indicating that formaldehyde may be carcinogenic. The Commission finalized that ban in a rule published on April 2, 1982. 47 Fed. Reg. 14366. In the preamble to the final rule, the Commission explained why warnings relating to the carcinogenic hazards presented by urea-formaldehyde would not be sufficient to protect consumers. 47 Fed. Reg. 14401.

14. The Commission has promulgated elaborate and detailed regulatons governing the submission of "substantial risk" reports. See 16 C.F.R., Part 1115. These regulations also discuss the submission of follow-up information and the negotiation of voluntary corrective action.

# 14

# Labeling in Transportation

**John E. Gillick**
*Kirby, Gillick, Schwartz and Tuohey, P.C.*
*Washington, D.C.*

## INTRODUCTION

The Department of Transportation's hazardous materials transportation regulatory program is, by any measure, massive in scope. Comprising over 1,000 pages in the *Code of Federal Regulations,* the program prescribes "cradle-to-grave" requirements for the preparation, shipping and carriage of materials deemed by the regulations to be "hazardous" when transported.

Given the breadth of the program and the relatively narrow focus of this chapter, it is simply not possible to discuss each of the program's individual requirements in detail. Instead, this chapter provides highlights of the program's development over the past decade, a brief overview of compliance with the program's general requirements, a far more detailed discussion of the aspect of the program that is conceptually thought of as "labeling" but actually encompasses the "labeling", "marking", and "placarding" requirements of the regulations, and, finally, a discussion of the likely evolution of the program during the 1980s.

## DEVELOPMENT OF THE TRANSPORTATION DEPARTMENT'S HAZARDOUS MATERIALS TRANSPORTATION REGULATORY PROGRAM OVER THE LAST DECADE

The decade of the 1970s was marked by significant changes in both the nature and scope of the Department of Transportation's hazardous materials regulatory program. While an extensive discussion of each of these developments is beyond the scope of this chapter, the following overview of the major modifications of the program during this period, with particular emphasis on the modifications pertaining to the "labeling" requirements, is designed to place these hazard communication requirements in context.

By way of background, the Department of Transportation was created in 1967, and it was not until then that the authority and responsibility for regulating the shipment and transportation of hazardous materials at the federal level, which previously had been divided among the Interstate Commerce Commission, the Federal Aviation Agency, and the Coast Guard, was consolidated into one Cabinet level department. Unfortunately, while the hazardous materials authority was consolidated in the Department, the responsibility for assuring safety in air, rail, pipeline, and motor carrier transportation was delegated by the Act which established the DOT to the various modal administrations created within DOT. As a first step towards enabling shippers and carriers to rely upon a cohesive set of regulations in preparing, shipping, and transporting materials, however, the Secretary established a Hazardous Materials Regulations Board. The Board was composed of representatives of the various modal administrations as well as the DOT Assistant Secretary for Research and Technology, who was designated Chairman. Its function was to handle all matters relating to the regulations for the shipment and transportation of hazardous materials, although the regulations themselves remained scattered throughout the Code of Federal Regulations.

Almost immediately after its formation, the Board set out on an ambitious course of action to make extensive revisions in the regulatory program, with particular emphasis on "casting the regulations in general terms and eliminating much of the detail" as well as initiating rulemaking with respect to classification and labels, handling and stowing, placards and emergency procedures, and packaging. While, as often has happened in this program, the Board's intentions were not matched by its ability to accomplish these modifications in short order, substantial progress was nevertheless achieved in several areas during the 1970s. From the standpoint of the "labeling" requirements, the most notable development occurred in the following areas:

- Establishment of a Hazard Information System
- Consolidation of the various individual regulatory programs into a single set of regulations (an effort "aided and abetted" by the passage of the Hazardous Materials Transportation Act in 1975)
- Evolution of an improved environment for emergency response capability
- Expansion of the program's scope so as to include "environmental and health effects" materials
- Enhancement of the compatibility of the DOT program with international programs

**Hazard Information System**

Turning initially to the development of a "Hazard Information System," this proposal was advanced in 1972 to provide for more complete identification and communication of the hazards presented by materials during their transportation. The rationale for the proposal was that the then-effective communications requirements of the regulations were not generally addressed to more than one hazard, did not in all instances require disclosure of the presence of hazardous materials in transport vehicles, were not addressed to the different hazard characteristics of a mixed load of hazardous materials, failed to provide sufficient information whereby fire fighting and other emergency response personnel could achieve adequate immediate information to handle emergency situations, and were inconsistent in their application to the different transportation modes. Among the requirements proposed at that time by the Board were the assignment of hazard information numbers to each hazardous material; designation of each label by class marking, rather than color, in the list of hazardous materials; various shipping paper requirements; inclusion of the hazard information number on labels; several new placarding requirements (including the establishment of a distinction between large and small packages so as to establish a break point where labeling would stop and placarding would begin, detailed requirements and provisions for attaching placards to transport vehicles, and specifications for the new placards); and the consolidation of the hazardous materials communications regulations.

The first element of this proposal to be adopted was revised labeling requirements, patterned after the United Nations Labeling System. Although the Board had proposed adoption of the UN system as early as 1968, the subsequent inclusion of the proposed labeling system as part of the "Hazardous Information System" proposal delayed its implementation until 1973, when the Board, in reaction to the mandated use of the UN labels by many foreign countries in early 1973, was compelled to adopt this aspect of the information system independent of the remaining proposals.

In 1976, the Board finally issued a massive modification of the regulations that, among other things, adopted many of the elements of the proposed hazard information system, including a uniform vehicle placarding system (discussed in the section *Placarding* later in this chapter), a uniform marking system (discussed in section *Marking*), and several modifications of the requirements for labeling (discussed in section *Labeling*).

### Consolidation of Regulations

As indicated earlier, the regulations concerning the transportation of hazardous materials by air, water and surface transportation had, historically, been set forth in three different volumes of the *Code of Federal Regulations*. In an effort to introduce as much uniformity as possible in the regulations, the Department proposed, in early 1974, to consolidate these regulations, while maintaining the differing requirements designed to deal with the particularized circumstances presented by air and water transportation.

As this regulatory consolidation effort was moving ahead, the Congress, in reaction to various hazardous materials accidents that had occurred in the early 1970s, and frustrated by the gaps in the legal basis for various aspects of the regulatory program as well as the lack of a centralized authority for assuring consistent regulatory treatment for the transportation of these materials, enacted the Hazardous Materials Transportation Act in early 1975. The principal thrust of the Act was to grant broad regulatory and enforcement authority to the Secretary of Transportation, *not* the modal administrations, "to protect the Nation adequately against the risks to life and property inherent in the transportation of hazardous materials in commerce." Among the most significant provisions of the new Act, which today provides the statutory basis for the regulatory program, was the authority to designate materials as being hazardous during transportation; regulate materials so designated; establish criteria for the handling of hazardous materials; require the registration of persons involved in each phase of hazardous materials transportation; grant exemptions from the regulations in carefully prescribed circumstances; provide regulations for the transportation of radioactive materials on passenger-carrying aircraft; seek civil and criminal penalties and injunctive relief; and, most importantly from the standpoint of maintaining a program national in scope, preempt inconsistent state and local requirements.

The most immediate consequence of the enactment of this legislation was the replacement, in July 1975, of the Hazardous Materials Regulations Board by the Materials Transportation Bureau. Thus, as incredible as it may seem, the authority to regulate hazardous materials in transportation was, for the first time since the Department was created, unified in one agency with a single administrator.

Prodded by the enactment of the legislation, the Bureau, in early 1976, consolidated the previously separate air, water and surface regulations into one volume of the *Code of Federal Regulations.*[1] Among the most significant of the "consolidation" modifications were those resulting from the incorporation of the U.S. Coast Guard and Federal Aviation Regulations into the overall Hazardous Materials Transportation Regulations.

In this regard, one of the most important elements of this consolidation was the creation of a regulatory mechanism whereby the application of the regulations to each of the modes of transportation would be clearly established. This posed no problem for those materials already included in the regulation's existing classifications. As to other materials, which had previously been described as "hazardous articles" when transported by water, "other restricted articles" when transported by air, or partially "exempt" materials when transported by highway or rail, it was necessary to make the regulations applicable only to transportation by air or water (or both), due to the kinds of potential hazards presented when transported by those modes. This was accomplished in the regulations by the creation of an "Other Regulated Material" (ORM) category which was, in turn, divided into the following four subcategories to identify the types of materials covered and to establish the basis for the application of the various regulations to them when transported by air, highway, rail, or water:[2]

> *An ORM-A material* is a substance which has an anesthetic, irritating, noxious, toxic, or other similar property and which can cause extreme annoyance or discomfort to passengers and driver or crew in the event of leakage during transportation;
>
> *An ORM-B material* is a substance capable of causing significant damage to a transport vehicle or vessel from leakage during transportation;

*An ORM-C material* is a substance which has other inherent characteristics not described as an ORM-A or ORM-B but which make it unsuitable for shipment, unless properly identified and prepared for transportation; and

*An ORM-D material* is a material that is classed as a flammable liquid, corrosive material, flammable compressed gas, flammable solid, oxidizing material, or organic peroxide, that, due to its limited quantity in a package, may be described and shipped as an ORM material. Shippers of ORM-D materials would be required to comply with the regulations pertaining to such materials regardless of the transportation mode utilized. However, no carrier operating regulations apply, such as the carrying of shipping papers, unless the ORM-D materials are to be transported by aircraft.

Over the next five years, the three most significant developments with an impact on the program's labeling requirements occurred in the areas of emergency response capability, expansion of the regulatory program so as to include specified "environmental and health effects" materials, and authorization of the use of UN shipping descriptions and identification numbers for certain hazardous materials in lieu of the descriptions otherwise required by the regulations—each of which were ultimately adopted as final regulations in what has become known as DOT's "Super Docket", published in May 1980.

### Emergency Response Capability

Turning initially to enhanced emergency response capability, the Department had, for several years, been working on various systems to improve the capability of emergency personnel to identify hazardous materials quickly, to ensure the accurate transmission of information to and from the scenes of accidents involving hazardous materials, and to enable such personnel to gain quick access to immediate response infomation from a guidebook prepared by the Deparment. As a means of accomplishing these objectives, the amendments, discussed in greater detail in the section *Marking,* required the display of identification numbers on shipping papers and packages in association with proper shipping names and the display of identification numbers on orange panels or placards affixed to portable tanks, cargo tanks, and tank cars, as well as the inclusion of more specific identification for poisonous materials.

### Environmental and Health Effects Materials

Prior to the expansion of the scope of the program to include "environmental and health effects" materials, the primary focus of the Department's hazardous materials transportation regulatory program had been upon the properties of materials that pose substantial potential hazards to persons or property from acute exposures. Consequently, the program had been directed primarily towards controlling the handling of such materials in transportation.

During the 1970s, however, various public and private organizations and environmental agencies urged the Department to consider establishing transportation controls to deal with materials which were not, or were only partially, regulated by the Department at that time. Additionally, several federal and state regulatory programs were enacted during the 1970s which addressed the health or environmental effects of various materials.

In response to these developments, DOT issued an advance notice of proposed rulemaking in late 1976 soliciting comments on whether new or additional controls were necessary for classes of materials presenting certain hazards to human health and to the environment which were not then subject to the existing Hazardous Materials Regulations. Included among the issues upon which the Department solicited comments were—what sort of human health and environmental effects should be considered, what criteria should be used to ascertain effects and identify materials (including the possible use of lists of materials identifed by other agencies as having adverse environmental or health effects), whether modifications to existing DOT hazardous material classifications or establishment of new classes would best accommodate the identified environmental and health effects materials, and what type of transportation controls may be needed for these materials (including the possible use of performance standards rather than specification standards in the establishment of packaging controls).

In the final regulations in the "Super Docket," the Department significantly expanded the scope of the program by promulgating final regulations (1) addressing the transportation of "hazardous wastes", and (2) providing for the identification of "hazardous substances" during transportation as well as the reporting of discharges of such substances into or on navigable waters.

The "hazardous waste" modifications, which were developed in conjunction with EPA's establishment of a national hazardous waste program under the Resource Conservation and Recovery Act,[3] are designed to assure that hazardous waste materials are properly identified for transportation and such wastes ultimately are delivered to predetermined designated facilities through implementation of certain record-handling requirements. Significantly, with the exception of the one-time requirement for a carrier identification number and the ongoing requirement for the cleanup of spills or discharges, EPA has indicated that a carrier that complies with the DOT requirements will be considered in compliance with the otherwise applicable EPA regulations for transporters of hazardous wastes.

The "hazardous substance" modifications, on the other hand, are designed to provide for the identification of substances designated as being hazardous by the EPA under Section 311 of the Clean Waste Act, when a "reportable quantity" of such a material is contained in a package so that appropriate action may be taken if the substance should be discharged into the navigable waters of the United States. This goal is to be accomplished by requiring that the name of the hazardous substance and the letters "RQ" be placed on shipping papers and marked on packages in association with the descriptions for hazardous substances, as well as including a reporting requirement for discharges of hazardous substances into or upon the navigable waters or adjoining shorelines.

## UN Shipping Descriptions and Identification Numbers

Finally, the "Super Docket" amendments authorized the use of United Nation's shipping descriptions and identification numbers for certain hazardous materials in place of the descriptions required by existing DOT regulations. Those amendments were intended to facilitate the international transportation of hazardous materials and minimize the economic burdens imposed on shippers by the multiplicity of package markings and shipping paper descriptions required for compliance with both domestic and international requirements. In response to requests from many surface carriers and carrier associations subsequent to the issuance of the final rule, however, the Department decided to limit the use of the descriptions in the Optional Table to international shipments involving transportation by vessels in anticipation of efforts by these carriers to accommodate these international transportation descriptions in surface transportation as well.

In summary, the 1970s were the most significant decade in the history of the hazardous materials regulatory program, as substantial progress was made towards developing a more coherent and comprehensive set of regulations. Among the most important changes in that program were the modifications to the hazard communication regulations, which are the subject of the remainder of this chapter.

## CONCEPTUAL OVERVIEW OF COMPLIANCE WITH THE HAZARDOUS MATERIALS REGULATORY PROGRAM

Although a detailed analysis of the myriad requirements contained in the hazardous materials transportation regulations is beyond the scope of this chapter, an understanding of the operation of those regulations and their basic requirements is a necessary predicate for a full understanding of the labeling regulations. Toward that end, the following discussion sets forth the step-by-step approach for preparing a shipment for transportation as contemplated by the regulations.

*First,* the composition and characteristics of the material to be shipped must be determined. This is a critically important step in the process since the entire regulatory scheme is premised upon an accurate determination of the precise properties of the material to be shipped and, based on this determination, the proper classification of the material under the regulations. In this regard, Chapter 2 discusses various sources of information that can be used to assist in the identification of a material's particular properties.

*Second,* after identifying the characteristics of the material, the material must be classified in accordance with the regulations, and the proper DOT shipping name must be chosen.[4]

If the material to be shipped is specified by name in the Hazardous Materials Table (Section 172.101), the classification and name selection process is relatively simple, since the proper name for purposes of the regulations is the name set forth in the Table and the hazard class for that particular material is listed in column 3 of the Table. Thus, for example, "Nitrogen tetroxide, liquid" is listed specifically in the Table, and is classified as a "Poison A" material for purposes of the regulations.

If, on the other hand, the material is not named in the Table, it must be evaluated from a technical standpoint to determne its proper classification. In this regard, the following classes of materials are presently included within the scope of the regulatory program and are described in the referenced portion of Part 173 of the regulations:

| Classification | Part 173 Reference[5] |
|---|---|
| Explosives (solid and liquid) | Subpart C |
| Flammables (solid, liquid and gaseous) | Subparts D and E |
| Combustible liquids | Subpart D |
| Oxidizers | Subpart E |
| Organic peroxides | Subpart E |
| Corrosive materials (solid and liquid) | Subpart F |
| Compressed gases | Subpart G |
| Poisonous materials (solid, liquid and gaseous) | Subpart H |
| Etiologic agents | Subpart H |
| Radioactive materials (solid, liquid and gaseous) | Subpart H |
| Other regulated materials, ORM-A, B, C, D and E | Subparts J–O |

In making these classification determinations, the regulations provide that, in general,[6] a hazardous material having more than one hazard must be classed according to the following order of hazards:

(1) Radioactive material

(2) Poison A

(3) Flammable gas

(4) Non-flammable gas

(5) Flammable liquid

(6) Oxidizer

(7) Flammable solid

(8) Corrosive material (liquid)

(9) Poison B

(10) Corrosive material (solid)

(11) Irritating material

(12) Combustible liquid (in containers having capacities exceeding 110 gallons)

(13) ORM-B

(14) ORM-A

(15) Combustible liquid (in containers having capacities of 110 gallons or less)

(16) ORM-E[7]

Once having determined the proper classification for the material, the proper shipping name, i.e., those names shown in Roman type, must be selected from the Table. In that regard, the regulations state that, if an appropriate technical name is not shown in the Table (e.g., "Nitrogen tetroxide, liquid" as referenced earlier), selection of a proper shipping name must be made from the general descriptions or n.o.s.[8] entries corresponding to the specific hazard class of the material being shipped. The name that most appropriately describes the material must be used. Thus, for example, an alcohol not listed by name in the Table must be shipped as "Alcohol, n.o.s." rather than "Flammable liquid, n.o.s." In addition, the regulations provide that some mixtures may be more appropriately described according to their application, such as "compound, cleaning, liquid" or "compound rust removing," rather than by n.o.s. entry, such as "Corrosive liquid, n.o.s.".

At this point, it is particularly important to note that the regulations operate to forbid the transportation of certain materials by either all or certain modes of transportation on the basis that such materials are simply too dangerous to be transported. Thus, for example, "Ammonium nitrite" is forbidden for all forms of transportation,[9] whereas "Amyl trichlorosilane" is forbidden on passenger-carrying aircraft, railcars, and passenger vessels.[10] Additional prohibitions are found in Part 173 of the regulations,[11] and columns 6 and 7 of the Table establish quantity limitations for one package.

*Third,* the correct packaging for the material must be selected. This is accomplished by, once again, referring to the Hazardous Materials Table, which, in column 5, lists the section of Part 173 describing the packaging requirements for the material with that given shipping name as well as the availability of any exceptions from the packaging requirements. Using "Nitrogen tetroxide, liquid" as an example once again, column 5 of the Table indicates that the packaging requirements for this material are found in Section 173.336, and that no packaging exceptions are available.

*Fourth,* the shipment must be labeled, marked, and placarded in the manner described in the section Compliance with the Program's Labeling Requirements of this chapter.

*Fifth,* the shipping papers must be completed for the shipment as specified in the regulations. The detailed shipping paper requirements, as well as the exceptions from these requirements that are available for specified ORM-A, B, C, or D shipments moving by certain transportation modes, are found in Subpart C of Part 172 of the regulations (49 C.F.R. §§ 172.200-.205).[12]

*Sixth,* the special requirements applicable to the particular mode or modes of transportation for the given shipment involved must be reviewed[13] so as to assure that these requirements are met, including, for example, those that impose particular quantity limitations on transportation by certain modes, additional shipping paper requirements, and differing requirements for passenger, as opposed to cargo-only aircraft and vessels.[14]

Finally, care must be taken to comply with any requirements imposed by individual carriers or group of carriers[15] as well as any international requirements that may be applicable to exports.

## COMPLIANCE WITH THE PROGRAM'S LABELING REQUIREMENTS

Although generically referred to as "labeling", the regulatory requirements relating to the communication of the hazard or hazards presented by a particular shipment are actually divided into three discrete, but carefully interrelated, subcategories of requirements denominated "labeling", "marking" and "placarding". From a conceptual standpoint, each of these three regulatory "mini-programs" is designed to communicate a particular level of information concerning the shipment to one or more potential audiences, and the regulatory requirements, discussed extensively in the following sections, reflect these complementary communication objectives.

### Labels

Under the DOT regulatory scheme, labels are square or diamond (square-on-point) hazard warnings that must be affixed to a shipment for the purpose of communicating the particular hazard presented by the shipment during transportation.[16] More than any of the other hazard communication requirements, the labeling regulations contain numerous exceptions, as well as a multiplicity of additional requirements for various packagings, materials, and modes of transportation, each of which is explained after discussion of the general labeling requirements.

**General Labeling Requirements, Prohibitions and Exceptions:** Packages, overpacks and freight containers containing hazardous materials must generally[17] be labeled with the label prescribed for the material contained in the shipment as specified in column 4 of the Hazardous Materials Table (when authorized).[18] Thus, by way of example, acetone shipments must, according to the Hazardus Materials Table, be labeled with a flammable liquid label, acrylic acid shipments must be labeled with a corrosive label, and liquid tetraethyl lead must be labeled with a poison label.

In general, labels must be durable and weather resistant, meet the size, color, and content specifications set forth in Sections 172.407–172.450, must be printed on, placed on, or affixed to, the surface of the package near the marked proper shipping name,[19] be affixed to a background of contrasting color (or have a dotted or solid line outer border), and not be obscured by markings or other attachments.

*Samples*—A package containing a material (other than a forbidden material (see Section 173.21) or an explosive[20]) for which a reasonable doubt exists as to its class and labeling requirements and for which a sample must be transported for laboratory analysis may be labeled according to the shipper's tentative class assignment based upon the defining criteria in the regulations, the hazard precedence prescribed in Section 173.2,[21] and the shipper's knowledge of the material.

*Multiple Labeling, and Mixed and Consolidated Packaging*—Certain materials are deemed by the regulations to present such a risk, because of their multiple hazards, that they must bear multiple labels as specified in Table 14.1.

## Table 14.1: Hazard Classes and Required Labeling

| Hazard Classes | Required Labeling |
|---|---|
| An explosive A, poison A, or radioactive material | Must be labeled as required for each class |
| A poison B liquid that also meets the definition of a flammable liquid | Must be labeled POISON and FLAMMABLE LIQUID |
| An oxidizer, flammable solid or flammable liquid that also meets the definition of a poison B | Must be labeled POISON as well as OXIDIZER, FLAMMABLE SOLID or FLAMMABLE LIQUID |
| A flammable solid that also meets the definition of a water reactive material | Must have both the FLAMMABLE SOLID and DANGEROUS WHEN WET labels affixed |
| A corrosive material that also meets the definition of a poison B* | Must be labeled CORROSIVE and POISON** |
| A poison B that also meets the definition of a corrosive material* | Must be labed POISON and CORROSIVE |
| A flammable liquid that also meets the definition of a corrosive material* | Must be labeled FLAMMABLE LIQUID and CORROSIVE |
| A flammable solid that also meets the definition of a corrosive material* | Must be labeled FLAMMABLE SOLID and CORROSIVE |
| An oxidizer that also meets the definition of a corrosive material* | Must be labeled OXIDIZER and CORROSIVE |

*This multiple labeling requirement is not required prior to July 1, 1983.
**This multiple labeling requirement does not apply, however, to a material that would cause death due to corrosive destruction of tissue rather than by systemic poisoning.

When hazardous materials having different hazard classes are packed within the same packaging or within the same outside container or overpack (so-called "mixed packaging"), the packaging, outside container or overpack must be labeled as required for each class of hazardous material contained therein. When two or more packages containing compatible hazardous materials are placed within the same container or overpack (so-called "consolidated packaging"), the outside container or overpack must also be labeled as required for each class of hazardous material contained therein.

In the event that two or more labels are required as a consequence of the multiple labeling, or mixed or consolidated packaging requirements, the labels must be displayed or affixed next to each other.

*Prohibited Labels*—Packages must not be labeled with a label unless the package contains a material that is a hazardous material and the label represents a hazard of the hazardous material in the package. A package which bears any marking or labeling which by its color, design, or shape could be confused with or conflict with a label prescribed by the regulations may not be offered for transportation or transported by a carrier.

These restrictions do not apply, however, to packages labeled in conformance with (1) any United Nations recommendation, including the class number (see Section 172.407), in the document entitled "Transport of Dangerous Goods (1970)", (2) the Intergovernmental Maritime Consultative Organization requirements, including the class number (see, once again, Section 172.407), in the document entitled "International Maritime Dangerous Goods Code", or (3) the International Civil Aviation Organization Technical Instructions for the Safe Transport of Dangerous Goods by Air.[22]

*Generally Available Exceptions*—As mentioned earlier, the regulations provide numerous labeling exceptions.[23] Thus, a label is *not* required for—

- A cylinder containing a compressed gas classified as flammable or nonflammable that is carried by a private or contract motor carrier, not overpacked, and durably and legibly marked in accordance with the requirements of Appendix A of Compressed Gas Association Pamphlet C-7,

- Military ammunition shipped by, for, or to the U.S. Department of Defense (DOD) when in freight containerload, carload or truckload shipments, if loaded and unloaded by the shipper or DOD,

- A package containing a hazardous material other than ammunition that is loaded and unloaded under the supervision of DOD personnel, and escorted by DOD personnel in a separate vehicle,

- A compressed gas cylinder permanently mounted in or on a transport vehicle,

- A portable tank which is placarded in accordance with Section 172.514 of the regulations,

- A freight container having a volume of 640 cubic feet or more which is subject to Section 172.512 (which establishes a placarding requirement for such a container),

- A package containing a material classed as ORM-A, B, C, D or E if that package does not contain any other material classed as a hazardous material that requires labeling,

- A package containing a combustible liquid,

- A package of low specific activity radioactive material, when being transported in a transport vehicle assigned for the sole use of the consignor under Section 173.392(b),

- A cargo tank or tank car other than a multi-unit tank car tank, and

- A package for which labeling is not required under the conditions set forth in the hazardous materials regulations.

**Labeling Exceptions for "Limited Quantities" of Hazardous Materials:** In addition to the just described generally available labeling exceptions, the regulations also provide exceptions from the labeling requirements (*except* when offered for transportation by air) for "limited quantities" of the following materials as defined in the referenced regulations:

| Material | Regulatory Provision Delineating the "Limited Quantity" Exception Available |
|---|---|
| Flammable liquids | Section 173.118 |
| Combustible liquids | Section 173.118 (a) |
| Flammable solids | Section 173.153 (a) |
| Oxidizers | Section 173.153 (b) (1) |
| Organic peroxides | Section 173.153 (b) (2) |
| Corrosive materials | Section 173.244 |
| Compressed gasses | Section 173.306 |

**Special Labeling Requirements for Large Packages, Specified Freight Containers, Metal Drums and Barrels, Multi-Unit Tank Car Tanks, Portable Tanks, and Empty Packaging:** In addition to the generally applicable labeling requirements, the regulations provide the following special labeling requirements for the following packages and containers:

1. When labeling is required, the labels must be displayed on at least two sides or two ends (excluding the bottom) of–

    Each package having a volume of 64 cubic feet or more,

    Each freight container having a volume of at least 64 cubic feet but less than 640 cubic feet (except when placard in accordance with Section 172.512 (b) )[24] with one of each of the appropriate labels being displayed on or near the closure, and

    Each portable tank having a rated capacity of less than 1,000 gallons, except when placarded in accordance with Section 172.514(a);

2. In order to warn those opening a metal drum that internal pressure exists, metal barrels or drums containing a flammable liquid having a vapor pressure between 16 and 40 psia at 100°F. must have a BUNG label (see Section 173.119(i)) in addition to a FLAMMABLE LIQUID label;

3. A DOT specification 106 or 110 tank (i.e., multi-unit tank car tanks) must be labeled on each end for the hazardous material it contains; and

4. Finally, an "empty" packaging having a capacity of 110 gallons or less that has not been cleaned and purged of *all* residue must be labeled as required when it previously contained a greater quantity of hazardous material.

**Special Labeling Requirements for Etiologic Agents, Oxygen, Chlorine, and Radioactive Materials:** Special labeling requirements exist for etiologic agents, oxygen, chlorine, and radioactive materials, certain of which operate to impose additional requirements while others establish exceptions from otherwise applicable requirements.

Each package containing an *Etiologic Agent*[25] (except a diagnostic specimen[26] or a biological product[27]) must be labeled as prescribed by the regulations of the Department of Health and Human Services found at 42 C.F.R. ;SS 72.25(c) (4). For export shipments, if use of the POISON label is required by the regulations of another country, it may be used in addition to the ETIOLOGIC AGENT label.

For a package containing *Oxygen,* the word "OXYGEN" may be used in the place of the word "OXIDIZER" on the OXIDIZER label, provided the letter size and color for OXYGEN are the same as those required for OXIDIZER.

For a package containing *Chlorine,* the word "CHLORINE" may be used in the place of the word "POISON" on the POISON label, provided the letter size and color for CHLORINE are the same as those required for POISON. Additionally, a CHLORINE label may be used in place of the NON-FLAMMABLE GAS and POISON labels otherwise required for Chlorine.

Extensive specific labeling requirements apply to the shipment of *Radioactive Materials.* Thus, every package of radioactive materials must be labeled, unless specifically excepted from labeling as being either (a) a "limited quantity" of a radioactive material or a manufactured article containing radioactive materials as specified in Section 173.391, or (b) a "low specific activity" radioactive material transported in a transport vehicle (other than by aircraft) assigned for the sole use of the consignor and otherwise meeting the requirements of Section 173.392 of the regulation.

The labels specified for radioactive materials are as follows–

> A RADIOACTIVE WHITE-I label must be affixed to each package measuring 0.5 millirem or less per hour at each point on the external surface of the package, provided the package is not a Fissile Class II or III, or does not contain a "large quantity" of radioactive material, as defined in Section 173.389 of the regulations

> A RADIOACTIVE YELLOW-II label must be affixed to each (1) package measuring more than 0.5 but not more than 50 millirem per hour at each point, and not exceeding 1.0 millirem per hour at three feet from each point on the external surface of the package, and (2) Fissile Class II package having a transport index of 1.0 or less, and

> A RADIOACTIVE YELLOW-III label must be affixed to each package which measures more than 50 millirem per hour at each point or exceeds 1.0 millirem per hour at three feet from each point on the external surface, is a Fissile Class III, or contains a "large quantity" of radioactive material as defined in Section 173.389 of the regulations.[28]

Each package containing a radioactive material must have labels displayed on at least two sides or two ends (excluding the bottom). In addition, each package containing a radioactive material that also meets the definition of one or more additional hazards must be labeled both as a radioactive material (as required by Section 172.403) and for each additional hazard. Thus, for example, packages containing the solid nitrates of uranium or thorium must be labeled RADIOACTIVE and OXIDIZER, and packages containing nitric acid solutions of radioactive material must be labeled RADIOACTIVE and CORROSIVE.

The following items of information must also be entered in the blank spaces on the RADIOACTIVE label by legible printing using a durable weather resistant means of marking:

> "Contents"–the name of the radionuclides as taken from the listing of radionuclides in Section 173.390 (symbols which conform to established radiation protection terminology are authorized, i.e., $^{99}$Mo, $^{60}$Co, etc.). For mixtures of radionuclides, the most restrictive radionuclides on the basis of radiotoxicity must be listed as space on the label allows.

> "Number of curies"–units shall be expressed in appropriate curie units, i.e., curies (Ci), millicuries (mCi) or microcuries ($\mu$Ci) (abbreviations are authorized). For a fissile material, the weight in grams or kilograms of the fissile radioisotope also may be inserted.

> The "Transport index".[29]

Finally, any packaging or accessory which has been used for a shipment of radioactive materials and which contains residual internal radioactive contamination must, when shipped as empty, have an EMPTY label affixed to the packaging.

**Additional Requirements for Air Transportation:** Finally, the regulations specify certain additional labeling requirements for air transportation. Thus, a package meeting the definition of a magnetized material (see Section 173.1020 of the regulations) must have a MAGNETIZED MATERIAL label in order to assure that the material is loaded on an aircraft in a manner that will not interfere with the aircraft's compass. In addition, the regulations preclude the carriage of certain materials on passenger carrying aircraft, and, consequently, a package containing a hazardous material authorized only on cargo aircraft must have a CARGO AIRCRAFT ONLY label to assure that this restriction is complied with.

### Marking

Marking, as that term is used in the DOT hazard information scheme, refers to the application of the descriptive name, instructions, cautions, weight, identification numbers, or specification marks, or any combination thereof, on the outside containers of hazardous materials. The purpose of these requirements is to communicate specific (and, with certain materials, rather extensive) information concerning the particular shipment to those persons handling it, as well as to emergency response personnel that may be called upon to deal with any incident involving the container. Consequently, the marking requirements have been individually crafted by the DOT so as to provide the appropriate level of information for a particular shipment of a given material.

Other than the markings that are required for specification packaging (which are beyond the scope of this chapter),[30] the regulations prescribe marking requirements for all outside containers, and certain additional requirements for portable tanks, cargo tanks, tank cars and multi-unit tank car tanks, specified hazardous materials, and exports by water, each of which is considered in turn.

**General Marking Requirements for Packaging:** As a general matter, any markings required by the regulations must be–

- Durable, in English, and printed on or affixed to the surface of a package or, when authorized or required, on a label, tag or sign,

- Displayed on a background of sharply contrasting color,

- Unobscured by any other label or attachment, and

- Located away from any other marking that could substantially reduce the marking's effectiveness.

Each person who offers a hazardous material for transportation in a packaging having a rated capacity of 110 gallons or less is required, in general, to mark the package with the proper shipping name[31] and, effective July 1, 1983, the identification number (including the description, preceded by either "UN" or "NA" as appropriate) assigned to the material[32] in either the Hazardous Materials Table (Section 172.101) or, when authorized, the Optional Hazardous Materials Table (Section 172.102).[33]

Each such package must also be marked with the name and address of the consignee or consignor, *except* when the package is transported by highway and will not be transferred from one motor carrier to another, *or* is part of a carload lot, truckload lot, or freight containerload, and the entire contents of the rail car, truck or freight container are tendered from one consignor to one consignee.

Each package having an inside packaging containing liquid hazardous materials[34] must be packed with closures upward and legibly marked "THIS SIDE UP" or "THIS END UP" indicating the upward position of the inside packaging.[35] This requirement does *not*, however, apply to either (1) cylinders of liquified compressed gas and specification containers 6D, 37M, 37P, and 21P, except as otherwise prescribed in Part 173, or (2) limited quantities of flammable liquids packed in inside packagings of one quart or less, prepared in accordance with Sections 173.118(a) and 173.1200(a) (1), when shipped other than by air,[36] and, if shipped by air, when packed with sufficient absorption material between the inner and outer packagings to completely absorb the liquid contents.

If an authorized package is shipped in an outside container as provided in

Section 173.25, the outside container must be marked with the proper shipping name for each material contained therein and marked "INSIDE PACKAGES COMPLY WITH PRESCRIBED SPECIFICATIONS", unless the shipping name and specification markings on the inside packaging are visible.[37]

Finally, packaging that contains the residue and/or vapor of a hazardous material must be marked as required when it previously contained a greater quantity of the hazardous material. The word "waste" need not be included in the proper shipping name, however, if the package bears the EPA marking prescribed by 40 C.F.R. § 262.32.[38]

**Additional Marking Requirements for Certain Hazardous Materials:** The regulations also require certain additional markings for packages containing radioactive materials, "Other Regulated Materials" (ORM), and hazardous substances and wastes.

Turning initially to *radioactive materials,* each package of such materials in excess of 110 pounds must have its gross weight marked "plainly and durably" on the outside of the package. In addition, each package of radioactive materials which conforms to the requirements for Type A (see Sections 173.389 (j) and 173.398 (b) ) or Type B (see Sections 173.389 (k) and 173.398 (c) ) packaging must be plainly and durably marked "TYPE A" or "TYPE B", as appropriate, in letters at least ½ inch high. Each package of radioactive material destined for export shipment must also be marked "USA" in conjunction with the specification marking or other package certificate identification.

Insofar as "Other Related Materials (ORM)" are concerned, each packaging having a rated capacity of 110 gallons or less and containing a material classed as ORM-A, B, C, D or E must be "plainly, durably, and legibly" marked on at least one side or end with the appropriate ORM designation (see Section 172.316 (a) (1) - (7) ) immediately following or below the proper shipping name of the material and placed within a rectangle that is approximately ¼ inch larger on each side than the designation.[39] This marking, ORM-A, B, C, D or E, is the certification by the person offering the package for transportation that the material is properly described, classed, packaged, marked and labeled (when required) and in proper condition for transportation according to the applicable regulations.[40]

As discussed previously, the transportation of previously unregulated hazardous wastes and substances became subject to the DOT regulations in November 1980. In that regard, *hazardous wastes,* in addition to meeting the otherwise applicable DOT requirements, must also meet the following EPA requirements:

1. Hazardous wastes in containers of 1·10 gallons or less which are to be transported off-site must be marked as follows:

   HAZARDOUS WASTE—Federal Laws Prohibit Improper Disposal. If found, please contact the nearest police or public safety authority or the U.S. Environmental Protection Agency.

Generator's Name and Address————————————

————————————————————————————————

Manifest Document Number————————————
(40 C.F.R. § 262.32)

2. Hazardous waste which is accumulated on-site without a permit must be in containers marked with the date upon which the period of accumulation began (see 40 C.F.R. § 262.34).

Finally, packages containing *hazardous substances* and having a capacity of 110 gallons or less must, effective July 1, 1983, have:

1. The name or names of such hazardous substance constituents as shown in Section 172.101[41] entered in association with the proper shipping name (if the proper shipping name for a mixture or solution that is a hazardous substance does not identify the constituents making it a hazardous substance); and

2. The letters RQ displayed in association with the proper shipping name.

**Additional Marking Requirements for Portable Tanks, Cargo Tanks, Tank Cars and Multi-Unit Tank Car Tanks:** In addition to the more general marking requirements for packages, the regulations prescribe the following specific requirements for portable tanks, cargo tanks, tank cars and multi-unit tank car tanks.

*Portable Tanks*—Turning initially to a portable tank, such a tank must be marked with—[42]

• The name of the owner or, if leased, the lessee,

• The proper shipping name of the material on two opposing sides,

• The identification number of the material specified in the Hazardous Materials Table or Optional Hazardous Materials Table (when authorized, as discussed earlier) on each side and each end if the tank has a capacity of 1,000 gallons or more, or on two opposing sides in association with the proper shipping name if the tank has a capacity of less than 1,000 gallons,[43] and

• The identification number, effective November 1, 1981, on either an orange display panel (see Section 172.332) or on the DOT placard[44] (but *not* a POISON GAS, RADIOACTIVE or EXPLOSIVES placard) as described in Section 172.334.[45]

A portable tank marked with the name or identification number of a hazardous material may not be used to transport any other material unless the marking is removed, or changed to identify the hazardous material in the portable tank, whichever is appropiate. Additionally, each marked portable tank must remain marked unless it is filled with a material not subject to the regulations, or sufficiently cleaned of residue and purged of vapor to remove any potential hazard.

*Cargo Tanks*—A cargo tank must, in general,[46] be marked with the proper shipping name of the material displayed on each end and each side, and a cargo tank transporting flammable or nonflammable compressed gasses must be marked, on each end and each side, with either the proper shipping name of the gas or an appropriate common name for the material such as "Refrigerant Gas". In addition, each specification MC 330 and MC 331 cargo tank must be appropiately marked "QT" or "NQT" to indicate whether it is constructed of quenched and tempered steel (QT) or not (NQT), with these markings being placed near the specification identification plate indicators.

A cargo tank must also be marked with the appropriate indentification number on each side and the rear end. As with a portable tank, the identification number must also be placed on orange display placards, DOT placards (except POISON GAS, RADIOACTIVE or EXPLOSIVES placards), or a white square-on-point configuration with the same dimensions as a placard. Identification numbers are *not* required, however—

- On a cargo tank containing only gasoline, if the tank is marked "Gasoline" on each side and rear in letters no less than 2 inches high or placarded in accordance with Section 172.542 (c),

- On a cargo tank containing only fuel oil, if the cargo tank is marked "Fuel Oil" on each side and rear in letters no less than 2 inches high or placarded in accordance with Section 172.544 (c),

- For different liquid distillate fuels, including gasoline, in a compartmented cargo tank or tank car, if the identification number displayed is for the liquid distillate fuel having the lowest flash point,

- On nurse tanks meeting the provisions of Section 173.315 (m), and

- As with a portable tank, on the ends of a tank car having more than one compartment which is transporting more than one hazardous material (but the identification numbers must be displayed on the sides in the same sequence as the compartments containing the materials they identify).

Unless the cargo tank is already properly marked, a person offering a motor carrier a hazardous material for transportation in a cargo tank shall, prior to or at the time the material is offered for transportation, provide the motor carrier with the required identification numbers on placards or shall affix orange panels containing the required identification numbers, and a person offering a cargo tank containing a hazardous material for transportation other than by motor carrier shall affix the required identification numbers on panels or placards prior to tendering the cargo tank for shipment.

Finally, as with a portable tank, a marked cargo tank cannot be used to transport any other material unless the marking is either removed or changed to identify the hazardous material then in the tank, and must remain marked unless reloaded with a material not subject to the regulations or sufficiently cleaned of residue and purged of vapor to remove any potential hazard.

*Tank Cars*—Unless previously marked or excepted, a tank car must be marked on each side, when required by Part 173 or 179 of the regulations with—

- The proper shipping name of the material or the common name authorized in the regulations for the material (such as, for example, "Refrigerant Gas") marked on each side and each end,

- The identification number specified in the Hazardous Materials Table or Optional Hazardous Materials Table (when authorized),[47] and

- The identification number, as with other tanks, on orange display panels, DOT placards (except POISON GAS, RADIOACTIVE or EXPLOSIVE placards), or, as appropriate, on white square-on-point configuration which is the same size as a placard.

Finally, exceptions from the identification number making requirement are provided for compartmented tank cars (no identification numbers on car's ends) and distillate fuels (including gasoline) transported in a compartmented tank car if the identification number is displayed for the distillate fuel having the lowest flash point.

*Multi-Unit Tank Car Tanks*—A multi-unit tank car tank containing hazardous materials, unless previously marked or otherwise excepted, must be marked on opposing sides, in letters at least two inches high, with the proper shipping name specified for the material in the Hazardous Materials Table or Optional Hazardous Materials Table (when authorized) or an authorized common name, and, beginning July 1, 1983, the identification number specified for the material in either of these tables. A motor vehicle or rail car used to transport a multi-unit tank car tank must be marked on each side and each end with the proper identification number. If a multi-unit tank car tank contains chlorine, marking of the name "Chlorine" is not required when the CHLORINE label is used as provided in Section 172.405(b). Finally, as with a tank car, a marked multi-unit tank car tank cannot be used to transport another material unless the marking is removed or changed to identify another material then contained in the tank, and must remain marked unless reloaded with a nonregulated material or sufficiently cleaned of residue and purged of vapor to remove any potential hazard.

**Export Shipments by Water:** In order to enhance safety in the transportation of water shipments, the regulations retain a requirement found initially in the Coast Guard regulations. Each package of hazardous material for export by water and described by a n.o.s. entry in the Hazardous Materials Table or Optional Hazardous Materials Table (when authorized) must have the technical name or names of the material added in parenthesis immediately following the proper shipping name—e.g., Corrosive liquid, n.o.s. (Caprylyl chloride). For a mixture of two or more hazardous materials, the technical name of at least the two components predominantly contributing to the hazard or hazards of the mixture must be added in parenthesis immediately following the proper shipping name.

## Placarding

The final category of hazard communication requirements, placarding, refers to the placing of square-on-point hazard warnings on motor vehicles, rail cars, freight containers, portable tanks, cargo tanks, and tank cars. The principal purpose of the present placarding requirements, not formulated on a uniform basis during the mid-1970s, is to communicate the contents of the trailer, container, or rail car to the public, carrier employee[48] and, in the event of an accident, to emergency response personnel.

The general specifications for placards are found in Section 172.519 of the regulations, the color tolerances are set forth in Appendix A of Part 172 of the regulations, the dimensional specifications are found in Appendix B to Part 172, and the additional specifications for each of the individual placards are found in Sections 172.521-.558.[49]

In addition to conforming to these size, color, and content requirements, placards must be—

- Made of plastic, metal or other material that is equal to or better in strength and durability than the tag-board specified in the regulations,
- Securely attached or placed in a holder,[50]
- Located clear of appurtenances and devices such as ladders, pipes, doors and tarpaulins,
- Located so that dirt and water is not directed to it from the wheels of the transport vehicle,
- Located at least three (3) inches from any marking that could substantially reduce its effectiveness,
- Placed so that the words on the placard read horizontally from left to right,
- Readily visible, on either a motor vehicle or a rail car, from the direction a placard faces except from the direction of another motor vehicle or rail car to which it is coupled,[51] and
- Maintained by the carrier so as to assure their legibility, color and visibility.

**General Placarding Requirements:** Except as specified below, each motor vehicle, rail car, freight container, portable tank, and cargo tank containing any quantity of a hazardous material must be placarded on each end and each side with the type of placards specified in Tables 14.2 and 14.3. No placard is required, however, for[52]—

- Etiologic agents,
- Hazardous materials classed as ORM-A, B, C, D, or E,
- Hazardous materials authorized to be shipped as "limited quantities" (discussed *supra*) when identified as such on shipping papers in accordance with Section 172.203(b) of the regulations,
- A motor vehicle, or a freight container if transported by highway only, containing less than 1,000 pounds (aggregate gross weight) of one or more materials covered by Table 14.3, above,[53] or
- A rail car loaded with freight containers or motor vehicles when *each* freight container or motor vehicle contains less than 1,000 pounds (aggregate gross weight) of one or more materials covered by Table 14.3.[53]

A freight container, motor vehicle, or rail car, which contains two or more classes of materials requiring different placards as specified in Table 14.3, may be placarded DANGEROUS in place of the separate placarding specified for each of those classes of material specified in Table 14.3.[54] When 5,000 pounds

Table 14.2

| If the motor vehicle, rail car, or freight container contains a material classed (described) as — | The motor vehicle, rail car, or freight container must be placarded on each side and each end — |
|---|---|
| Class A explosives | EXPLOSIVES A[1] |
| Class B explosives | EXPLOSIVES B[2] |
| Poison A | POISON GAS[1] |
| Flammable solid (DANGEROUS WHEN WET label only) | FLAMMABLE SOLID W[3] |
| Radioactive material | RADIOACTIVE[4,5] |
| Radioactive material: | |
| Uranium hexafluoride, fissile (containing more than 0.7% $U^{235}$) | RADIOACTIVE[4] and CORROSIVE[6] |
| Uranium hexafluoride, low specific activity (containing 0.7% or less $U^{235}$) | RADIOACTIVE[4,5] and CORROSIVE[6] |

[1] See § 172.510 (a).
[2] An EXPLOSIVES B placard is not required if the freight container, motor vehicle, or rail car contains class A explosives and is placarded EXPLOSIVES A as required.
[3] A FLAMMABLE SOLID "W" placard is required only when the DANGEROUS WHEN WET label is specified in § 172.101 for a material classed as a flammable solid.
[4] Applies only to any quantity of packages bearing the RADIOACTIVE YELLOW-III label (see § 172.403).
[5] See §§ 173.389 (c) and 173.389 (o), for full-load shipments of radioactive materials meeting the definition of low specific activity when transported pursuant to § 173.392 (b).
[6] A CORROSIVE placard is not required for shipments of less than 1,000 pounds gross weight.

Table 14.3

| If the motor vehicle, rail car, or freight container contains a material classed (described) as — | The motor vehicle, rail car, or freight container must be placarded on each side and each end — |
|---|---|
| Class C explosives | DANGEROUS[1,9] |
| Blasting agents | BLASTING AGENTS[9,10] |
| Nonflammable gas | NONFLAMMABLE GAS[8] |
| Nonflammable gas (chlorine) | CHLORINE[7] |
| Nonflammable gas (fluorine) | POISON |
| Nonflammable gas (oxygen, pressurized liquid) | OXYGEN[2] |
| Flammable gas | FLAMMABLE GAS[8] |
| Combustible liquid | COMBUSTIBLE LIQUID[3,4] |
| Flammable liquid | FLAMMABLE |
| Flammable solid | FLAMMABLE SOLID[5] |
| Oxidizer | OXIDIZER[9,10] |
| Organic peroxide | ORGANIC PEROXIDE |
| Poison B | POISON |
| Corrosive material | CORROSIVE[6] |
| Irritating material | DANGEROUS |

[1] Applies only to a Class C explosive required to be labeled with an EXPLOSIVE C label.

[2] OXYGEN placards may be used to identify liquefied pressurized oxygen contained in a manner so it does not meet the definition in § 173.300.

[3] A COMBUSTIBLE placard is required only when a material classed as a combustible liquid is transported in a packaging having a rated capacity of more than 110 gallons, a cargo tank, or a tank car.

[4] A FLAMMABLE placard may be used on a cargo tank and a portable tank during transportation by highway and water.

[5] Except when offered for transportation by water, a FLAMMABLE placard may be displayed in place of a FLAMMABLE SOLID placard except when a DANGEROUS WHEN WET label is specified for the material in § 172.101 (see Table 14.2).

[6] See § 173.245 (b) for authorized exemptions.

[7] A CHLORINE placard is required only for a packaging having a rated capacity of more than 110 gallons; the NONFLAMMABLE GAS placard for packaging having a rated capacity of 110 gallons or less.

[8] A NONFLAMMABLE GAS placard is not required on a motor vehicle displaying a FLAMMABLE GAS placard.

[9] BLASTING AGENTS, OXIDIZER and DANGEROUS placards need not be displayed if a freight container, motor vehicle, or rail car also contains Class A or Class B explosives and is placarded EXPLOSIVES A or EXPLOSIVES B as required.

[10] Except for shipments by water, OXIDIZER placards need not be displayed if a freight container, motor vehicle or rail car also contains blasting agents and is placarded BLASTING AGENTS as required.

or more of one class of material is loaded therein at one loading facility, however, the placard specified for that class in Table 14.3 must be applied.[55]

In addition to specifying the placarding that is authorized, the regulations contain several placarding prohibitions. In that regard, placards are not to be affixed or displayed on a portable tank, freight container, motor vehicle or rail car, unless the material being offered or transported is a hazardous material, and the placard represents a hazard of that hazardous material. In addition, no person may affix or display any sign or other device on a motor vehicle, rail car, portable tank, or freight container, that by its color, design, shape or content could be *confused* with any placard prescribed in the regulations. It is important to note that neither of these prohibitions applies to portable tanks, freight containers, motor vehicles or rail cars which, in addition to any placards required by the DOT regulations, may also be placarded in conformance with the International Maritime Consultative Organization code.

**Special Placarding Requirements for Cargo and Portable Tanks, Tank Cars, and Freight Containers:** In addition to these more general placarding responsibilities, special requirements exist for the placarding of cargo and portable tanks, tank cars, and freight containers.

Turning initially to *cargo tanks* and *portable tanks,* each person who offers a cargo tank or a portable tank containing a hazardous material for transportation is responsible for assuring that the proper placards are provided to the motor carrier (in the case of a shipment by motor carrier), and are actually affixed to the tank in the case of a shipment by rail.[56] It is also critically important to note that each cargo tank and portable tank that must be placarded when it contains a hazardous material must also *remain placarded* when it is empty unless it is (1) reloaded with a material that is not subject to the regulations, or (2) sufficiently cleaned and purged of vapors to remove any potential hazard.

As with cargo and portable tanks, *tank cars* containing a hazardous material which requires placards must be appropriately placarded on each end and each side. Significantly, however, once that tank car has been emptied, it must be placarded with an EMPTY placard that corresponds to the placard that was required for the material the tank car last contained, unless the tank car either last contained a combustible liquid or has, as with cargo and portable tanks, been (1) reloaded with a material that is not subject to the regulations, or (2) sufficiently cleaned of residue and purged of vapor to remove any potential hazard.

Finally, *freight containers* with capacities of 640 cubic feet or more and containing a hazardous material must be appropriately placarded on each side, except placarding is not required if the container is to be transported (a) by highway and contains less than 1,000 pounds (aggregate gross weight) of one or more materials set forth in Table 14.3 or (b) by highway or rail for delivery to a consignee immediately following an air or water shipment when the container contains, once again, less than 1,000 pounds (aggregate gross weight) of one or more materials set forth in Table 14.3.

Freight containers with capacities of less than 640 cubic feet need not be placarded when containing hazardous materials if the container is labeled in accordance with Subpart E of Part 172 and the mode of transportation is other than by air. When air transportation is involved, such containers must be placarded with one placard, unless the container is labeled in accordance with Section 172.406(e)(3), or contains radioactive materials requiring the RADIOACTIVE YELLOW-III label and is placarded with one RADIOACTIVE placard and is labeled in accordance with Section 172.406(e).

**Additional Requirements for Highway and Rail Transportation:** Persons offering a motor carrier a hazardous material for transportation by highway shall provide to the motor carrier the required placards for the material being offered prior to or at the same time the material is offered for transportation, unless the carrier's motor vehicle is already placarded for the particular material. No motor carrier may transport a hazardous material in a motor vehicle unless the placards required for the hazardous material are affixed thereto as required in the regulaions. Finally, effective January 1, 1982, each motor vehicle used to transport a package of "large quantity radioactive materials," as defined in Section 173.389(b) of the regulations must have the required RADIOACTIVE warning placard placed on a square background as described in Section 172.527[57]

Persons offering a hazardous material for transportation by rail must affix the appropriate placards to the rail car containing the material, unless the placards already displayed on motor vehicles, transport containers, or portable tanks that are on a rail car comply with the rail placarding requirements. As with motor carriers, no rail carrier may accept a rail car containing a hazardous material for transportation unless the placards for the hazardous material are affixed thereto.

Furthermore, each EXPLOSIVE A, POISON GAS, and POISON GAS-EMPTY placard affixed to a rail car must be placed on a square background in the manner specified in Section 172.527. Each domed tank car containing a flammable liquid having a vapor pressure exceeding 16 psi at 100°F (37.8°C) must have a DOME placard affixed thereto as specified in Section 173.119(h).

Finally, each freight container, motor vehicle, and rail car containing lading that has been fumigated or treated with poisonous liquid, solid, or gas, and that is offered for transportation by rail must have a placard (as specified in Section 173.426) indicating that the lading has been fumigated or treated.

## HAZARDOUS MATERIALS TRANSPORTATION REGULATION IN THE 1980s

As with any federal regulatory program that must simultaneously be responsive to the vagaries of politics, evolution of technology, growth in state and local government involvement, and explosion in international trade, it is simply not possible to predict the precise manner in which the Transportation Department's hazardous materials regulatory program will evolve over the next decade. On the other hand, consideration of the various efforts that have begun as well as anticipation of the issues that are likely to arise during this time period indicates that the following areas will form the nucleus of the Department's regulatory agenda for the 1980s.

First, in what may be the most revealing exposition of the program's likely evolution in the 1980s, the Department recently published a 10-year plan, set forth in Appendix A, for the review of the agency's rules to determine if they serve a beneficial purpose relative to their cost of compliance, looking towards the elimination of those requirements that are no longer considered essential and reviewing those that are retained for clarity.

While the history of this regulatory program is littered with unkept deadlines, the plan is, nevertheless, instructive as to the Department's relative priorities for reviewing its regulations. In that regard, it is significant to note that the plan indicates the Department does not intend to review the hazardous materials communications regulations until 1990. Consequently, it seems clear that the Department, after directing a substantial amount of time and regulatory effort during the 1970s towards the modification of those communication regulations, has determined to devote its regulatory resources to other areas for the greater part of this decade. This is not to say, of course, that there will not be modifications to the "labeling" regulations in order to accomodate developments in these other areas, but it appears that the Department does not intend, at least at this time, to reinvent that "regulatory wheel" for many years to come.

While this review process will, no doubt, occupy a significant portion of the Department's energies, it appears that hazardous materials regulation at the federal, state, local and international levels will require a substantial amount of attention by the Department in the 1980s.

Turning initially to activity at the federal level, the recently enacted Comprehensive Environmental Response, Compensation, and Liability Act of 1980 (the so-called "Superfund" legislation) requires that any "hazardous substance[58]" designated under that Act must, within 90 days of such designation, be listed as a hazardous material under the Hazardous Materials Transportation Act. This assures the Transportation Department's ongoing involvement in the regulation of materials with adverse "environmental and health effects" (as discussed in the Introduction to this chapter). Similarly, the Occupational Safety and Health Administration is presently considering adoption of labeling requirements for the workplace, once again requiring the Department's involvement to assure that whatever program OSHA may ultimately adopt is both compatible and consistent with the Department's transportation program.[59]

At the state and local level, the ever-increasing interest on the part of state and local governments in the transportation of hazardous materials (particularly the carriage of radioactive materials)[60] makes it even more important that the operation of any such state or local governmental program is consistent with the Transportation Department's regulatory program. As mentioned briefly earlier, Section 112 of the Hazardous Materials Transportation Act provides a rather elaborate process for involving the Department in such determinations,[61] and it is anticipated that this mechanism will be used frequently throughout the decade in view of the ongoing interest in such regulation at the state and local level.

Finally, and perhaps most importantly given the ever-increasing flow of goods between the United States and foreign countries, there will be ongoing efforts to coordinate more closely the transportation of hazardous materials between various countries. As discussed earlier, the Transportation Department has made several recent amendments to its regulations that are designed to accommodate international requirements and methodologies, and is presently evaluating several additional modifications which can be expected to continue and expand in the future as U.S. companies expand their markets around the world.

In summary, the Transporation Department's principal focus during the 1970s was upon consolidating and substantively improving the quality of its regulations. While this latter objective will, and should, be an important goal of the Department in this decade, the most important challenge facing the Department over the next several years will be to coordinate the continuing development of the hazardous materials transportation program with related activities at the federal, state and local, and international levels—a challenge that may well determine the efficacy of the Transportation Department's program in the 1980s.

## APPENDIX A: SCHEDULE FOR REVIEW OF HAZARDOUS MATERIALS REGULATIONS BY THE DEPARTMENT OF TRANSPORTATION

| Hazardous Materials Regulation | Date of Review |
|---|---|
| Part 178. Review of packaging manufacturing regulations. Consideration will be given to adoption of performance oriented standards consistent with those contained in United Nations Recommendations for small packagings | 1981* |
| Part 173. Review of shipper regulations pertaining to selection and use of packagings for various chemicals | 1982 |
| Part 173. Review of regulations pertaining to packaging and shipping of explosives | 1983 |
| Part 173. Review of regulations pertaining to packaging and shipping of compressed gases | 1984 |
| Part 177. Review of carrier operations regulations pertaining to carriage of hazardous materials by motor vehicle | 1985 |
| Part 174. Review of carrier operations regulations pertaining to carriage of hazardous materials by railroad | 1986 |
| Part 176. Review of carrier operations regulations pertaining to carriage of hazardous materials by vessel | 1987 |
| Part 175. Review of carrier operations regulations pertaining to carriage of hazardous materials aboard aircraft | 1988 |
| Part 173. Review of regulations pertaining to packaging and shipments of radioactive materials | 1989 |
| Part 172. Review of regulations pertaining to hazardous materials communications | 1990 |

*This review was not initiated until 1982.

Source:  49 Fed. Reg. 33409 (June 29, 1981).

## APPENDIX B: GLOSSARY

The following list of terms used in the hazardous materials regulations is provided for the convenience of the reader.

*Cargo tank* means any tank permanently attached to or forming a part of any motor vehicle or any bulk liquid or compressed gas packaging not permanently attached to any motor vehicle which by reason of its size, construction, or attachment to a motor vehicle, is loaded or unloaded without being removed from the motor vehicle. Any packaging fabricated under specifications for cylinders is not a cargo tank.

*Consumer commodity* means a material that is packaged and distributed in a form intended or suitable for sale through retail sales agencies or instrumentalities for consumption by individuals for purposes of personal care or household use. This term also includes drugs and medicines.

*Cylinder* means a pressure vessel designed for pressures higher than 40 psia and having a circular cross section. It does not include a portable tank, multi-unit tank car tank, cargo tank, or tank car.

*Flash point* means the minimum temperature at which a substance gives off flammable vapors which in contact with spark or flame will ignite.

*Freight container* means a reusable container having a volume of 64 cubic feet or more, designed and constructed to permit being lifted with its contents intact and intended primarily for containment of packages (in unit form) during transportation.

*Hazardous material* means a substance or material which has been determined by the Secretary of Transportation to be capable of posing an unreasonable risk to health, safety and property when transported in commerce, and which has been so designated.

*Hazardous substance* for the purposes of the DOT regulations, means a material, and its mixtures or solutions, that is identified by the letter "E" in Column 1 of the table to § 172.101 when offered for transportation in one package, or in one transport vehicle if not packaged, and when the quantity of the material therein equals or exceeds the reportable quantity (RQ). This definition does not apply to petroleum products that are lubricants or fuels; or to a mixture or solution containing a material identified by the letter "E" in Column 1 of the table to § 172.101 if it is in a concentration less than that shown in the following table based on the reportable quantity (RQ) specified for the materials in Column 2 of the table to § 172.101:

| RQ (lb) | RQ (kg) | Concentration by Weight % | ppm |
|---|---|---|---|
| 5,000 | 2,270 | 10 | 100,000 |
| 1,000 | 454 | 2 | 20,000 |
| 100 | 45.4 | 0.2 | 2,000 |
| 10 | 4.54 | 0.02 | 200 |
| 1 | 0.45 | 0.002 | 20 |

*Hazardous waste,* for the purpose of the DOT regulations, means any material that is subject to the hazardous waste manifest requirements of the EPA specified in 40 C.F.R. Part 262 or would be subject to these requirements absent an interim authorization to a state under 40 C.F.R. Part 123, Subpart F.

*Limited quantity,* when specified as such in a section applicable to a particular material with the exception of Poison B materials, means the maximum amount of a hazardous material for which there is a specific labeling and packaging exception.

*Marking* means applying the descriptive name, instructions, cautions, weight or specification marks or combination thereof required by the regulations to be placed upon outside containers of hazardous materials.

*Mixture* means a material composed of more than one chemical compound or element.

*Motor vehicle* includes a vehicle, machine, tractor, trailer, or semi-trailer, or any combination thereof, propelled or drawn by mechanical power and used upon the highways in the transportation of passengers or property. It does not include a vehicle, locomotive, or car operated exclusively on a rail or rails, or a trolley bus operated by electric power derived from a fixed overhead wire, furnishing local passenger transportation similar to street-railway service.

*Outside container* means the outermost enclosure used in transporting a hazardous material other than a freight container.

*Overpack* means an enclosure not intended for reuse that is used by a single consignor to consolidate two or more packages for convenience in handling.

*Package* or *Outside Package* means a packaging plus its contents.

*Packaging* means the assembly of one or more containers and any other components necessary to assure compliance with the minimum packaging requirements of the DOT regulations and includes containers (other than freight containers or overpacks), portable tanks, cargo tanks, tank cars, and multi-unit tank car tanks.

*Passenger-carrying aircraft* means an aircraft that carries any person other than a crew-member, company employee, an authorized representative of the United States, or a person accompanying the shipment.

*Passenger vessel* means - (1) a vessel subject to any of the requirements of the International Convention for the Safety of Life at Sea, 1960, which carries more than 12 passengers; (2) a cargo vessel documented under the laws of the United States and not subject to the Convention, which carries more than 16 passengers; (3) a cargo vessel of any foreign nation that extends reciprocal privileges and is not subject to the Convention and which carries more than 16 passengers; and (4) a vessel engaged in a ferry operation and which carries passengers.

*Portable tank* means any packaging (except a cylinder having a 1,000 pound or less water capacity) over 110 U.S. gallons capacity and designed primarily to be loaded into or on or temporarily attached to a transport vehicle or ship, and equipped with skids, mounting or accessories to facilitate handling of the tank by mechanical means. It does not include any cargo tank, tank car tank, tank of the DOT-106A or 110A type, or trailers carrying 3AX, 3AAX, or 3T cylinders.

*Rail freight car* means a car designed to carry freight or nonpassenger personnel by rail, and includes a box car, flat car, gondola car, hopper car, tank car, and occupied caboose.

*Technical name* means a recognized chemical name currently used in scientific and technical handbooks, journals and texts. Generic descriptions authorized for use as technical names are, Organic phosphate compound, Organic phosphorus compound, Organic phosphate compound mixture, Organic phosphorus compound mixture, Methyl parathion.

*Transport vehicle* means a motor vehicle or rail car used for the transportation of cargo by any mode. Each cargo-carrying body (trailer, railroad freight car, etc.) is a separate transport vehicle.

*Vessel* includes every description of watercraft, used or capable of being used as a means of transportation on the water.

## FOOTNOTES

1. 49 C.F.R. Parts 171-179 [Part 171–General Information, Regulations, and Definitions; Part 172–Hazardous Materials Table and Hazardous Materials Communications Regulations; Part 173–Shippers—General Requirements for Shipments and Packagings; Part 174–Carriage by Rail; Part 175–Carriage by Aircraft; Part 176–Carriage by Vessel; Part 177–Carriage by Public Highway; Part 178–Shipping Container Specifications; Part 179 Specifications for Tank Cars.]
2. The operation of these ORM classifications vis-a-vis the "labeling" requirements is described in the section *Compliance with the Program's Labeling Requirements.*
3. The labeling requirements under this Act are discussed in great detail in Chapter 12.
4. Chapter 4 describes this classification process in detail.
5. Part 173 of the regulations, among other things, defines hazardous materials for transportation purposes and prescribes certain requirements to be observed in preparing them for shipment by air, highway, rail or water, or any combination of these modes of transportation.
6. This requirement does not apply to a material specifically identified in the Table, certain explosives, blasting agents, and etiologic agents, and all organic peroxides as set forth in Section 173.2(b) of the regulations.
7. ORM-E is a material not included in any other hazard class, but is subject to the regulations, and includes hazardous "waste" and "substances".
8. The term "n.o.s." means "not otherwise specified".

9. This prohibition is contained in column 3 of the Hazardous Materials Table.

10. The prohibitions concerning transportation by passenger carrying aircraft or railcar are found in column 6 of the Table, while the prohibitions concerning passenger vessels are found in column 7(b) of the Table.

11. By way of example, Section 173.21 lists various forbidden materials, and Section 173.51 lists forbidden explosives.

12. As a consequence of the recent expansion of the program to include "hazardous wastes", the preparation of a hazardous waste manifest for the transportation of such materials has been included as an additional shipping paper requirement.

13. These special requirements are found throughout Parts 172 and 173 of the regulations, and, accordingly, the regulations must be consulted carefully to assure compliance with these provisions.

14. Parts 174-77 contain specific requirements relating to the carriage of hazardous materials by rail, aircraft, vessel and public highway, and impose important regulatory requirements on the carriers of hazardous materials.

15. For example, the International Air Transport Association, an organization whose membership is composed of international airlines, has its own tariff that sets forth various requirements for the international transportation of hazardous materials by air.

16. The regulations setting forth the labeling requirements are found in Subpart E of Part 172 of the regulations (49 C.F.R. § § 172.400-.450). A chart setting forth the authorized D.O.T. labels is found in Appendix A to this chapter.

17. As noted above and discussed below, the regulations provide certain exceptions from the otherwise applicable labeling requirements.

18. Except for packages not requiring a label, when the proper shipping name marked on a package is selected from the Optional Hazardous Materials Table (Section 172.102) and it does not appear in the Hazardous Materials Table (Section 172.101), the package must be labeled as provided in the Optional Table.

19. Labels may, however, be printed on or placed on a securely affixed tag, or may be affixed by other suitable means to a package that contains no radioactive material and which either (a) has dimensions less than those of the required label, (b) is a compressed gas cylinder, or (c) is a package which has such an irregular surface that a label cannot be satisfactorily affixed.

20. Special procedures for the transportation of "new" explosives, including labeling requrirements, are provided in Section 173.86(d) of the regulations.

21. As discussed earlier in this chapter, Section 173.2 sets forth a mechanism, subject to certain exceptions, that provides for the classification of materials having more than one hazard in accordance with an established order of hazards.

22. In addition, except for Class A and Class B explosives and radioactive materials, the Intergovernmental Maritime Consultative Organization label specifications may be used as long as the labels are in English and meet the color tolerances of Section 172.407(d).

23. These exceptions do not apply to the CARGO AIRCRAFT ONLY label or the MAGNETIZED MATERIAL label, discussed below.

24. Placarding may not, however, be used instead of labeling on a package containing radioactive material.

25. An "etiologic agent" is defined as a viable microorganism, or its toxin, which causes or may cause human disease, and is limited to those agents listed in 42 C.F.R. § 72.25(c) of the regulations of the Department of Health and Human Services. An exception is provided, however, for cultures of etiologic agents of 50 milliliters (1.666 fluid ounces) or less total quantity in one outside package.

26. A "diagnostic specimen" means any human or animal material including, but not limited to, excreta, secreta, blood (and its components), tissue, and tissue fluids, being shipped for purposes of diagnosis.

27. A "biological product" means a material prepared and manufactured in accordance with the provisions of 9 C.F.R. Part 102 (licensed veterinary biological products), 21 C.F.R. Part 601 (licensing), 21 C.F.R. Part 312.1 (conditions for exception of new drugs for investigational use), 9 C.F.R. Part 103 (biological products for experimental treatment of animals), or 21 C.F.R. § 312.9 (new drugs for investigational use in laboratory research animals or in vitro tests), and which, in accordance with these provisions, may be shipped in interstate commerce.

28. Radioactive materials labels required by the regulations in effect prior to November 20, 1980, may continue in effect until July 1, 1983.

29. The transport index is a calculation, described in Section 173.89(i) of the regulations, which is used to determine the degree of control to be exercised by the carrier during transportation.

30. The marking requirements for specification packaging are found in Parts 178 and 179 of the regulations. The purpose of such marking is to act as a certification that a particular packaging meets the specifications as marked.

31. Abbreviations may not be used in a proper shipping name marking except that "W" and "W/O" may be used as abbreviations for "with" and "without" for marking descriptions of ammunition, and "ORM" may be used in place of the words "Other Regulated Material". Additionally, when it has been determined by the shipper that a package has been previously marked as required for the material it contains, it need not be remarked.

32. As noted earlier, use of the identification number serves the dual purpose in an emergency of (a) assuring the proper identification and (b) providing a reference for proper use of the "Emergency Response Guidebook", published by the D.O.T.

33. Use of the Optional Hazardous Materials Table for purposes of describing a material under the regulations is presently authorized only for international shipments involving transportation by vessels.

34. It should also be noted that the specific requirements for certain other materials, found in Part 173 of the regulations, require that additional markings be placed on packaging. Thus, for example, Section 173.277 requires that hypochlorite solutions in Specification 37P steel drums must be marked "KEEP THIS END UP".

35. The regulations also indicate that an arrow symbol indicating "THIS WAY UP" as specified in ANSI (American National Standards Institute) MH6.11968 should be used in addition to the required marking. Additional arrows for purposes other than indicating proper package orientation may not, however, be displayed on a package containing a liquid hazardous material.

36. As indicated in the labeling discussion, the regulations provide certain exceptions from the labeling and packaging requirements for specified flammable liquids shipped in "limited quantities". In order to qualify for such an exemption, the flash point of the particular material must be marked on the outside package. *See* Section 173.118(a) and (b).

37. Packages required to be marked "THIS SIDE UP" or "THIS END UP" must be packed in the outside container with their filling holes up and the outside container marked "THIS SIDE UP" or "THIS END UP" to indicate the upward position of closures.

38. The marking requirements set forth in the EPA's regulations are discussed in Chapters 11-13.

39. If the ORM-D marking, including the proper shipping name, cannot be affixed on the package surface, it may be on an attached tag.

40. This certification does not, however, satisfy the requirement for a certificate on a shipping paper when required by Subpart C of Part 172 of the regulations.

41. This requirement also applies when descriptions from the Optional Hazardous Materials Table in Sectin 172.102 are used.

42. These required markings must be at least two inches in height.

43. Identification numbers are not required on the ends of a portable tank with more than one compartment if hazardous materials having different identification numbers are being transported therein, but the identification numbers on the sides of the tank should be displayed in the same sequence as the compartments containing the materials they identify. If the identification number marking is not visible, a transport vehicle, or freight container used to transport a portable tank must be marked on each side and each end with the appropriate identification number.

44. For hazardous materials not requiring placarding, identification numbers may be displayed on a white square-on-point configuration having the same outside dimensions as those prescribed for authorized placards.

45. If more than one of the identification number markings on the placards or orange panels that are required to be displayed are lost or destroyed during transportation, the carrier is required to replace all the missing identification numbers as soon as practicable. In such circumstances, the numerals may be entered legibly by hand using an indelible marking material.

46. The lettering for all marking on cargo tanks must be at least two inches in height.

47. The letters in this marking must be at least 4 inches high, have at least a ⅝ inch stroke, and be at least ¾ of an inch apart.

48. In this regard, the regulations concerning carriage by rail have detailed requirements relating to placarded cars, including the inclusion of additional information on shipping papers, the giving of notice of rail cars placarded EXPLOSIVE A or POISON GAS, as well as the handling of placarded cars.

49. A chart setting forth the various placards is presented at the front and back of this book.

50. The regulations permit a placard to be hinged, provided the required format, color, and legibility of the placard are maintained.

51. The required placarding of the front of a motor vehicle may be on the front of a truck-tractor instead of or in addition to the placarding on the front of the cargo body to which a truck-tractor is attached.

52. In addition, the regulations provide that any packaging having a capacity of 110 gallons or less that contains only the residue of a hazardous material covered by Table 14.3 need *not* be included in determining the applicability of the placarding requirements.

53. These placarding exceptions do not, however, apply to portable tanks, cargo tanks or tank cars, or to transportation by air or water.

54. Use of the DANGEROUS placard in this manner does *not* apply, however, to a portable tank, cargo tank, or tank car.

55. It should be remembered, however, that these placarding exceptions relate only to materials found in Table 14.3 but *not* Table 14.2 (Class A and B explosives, Poison A, specified Flammable solids, and specified Radioactive materials). Placards *must* be applied for each material in Table 14.2

56. A portable tank havng a rated capacity of less than 1,000 gallons, if placarded instead of beng labeled as authorized in Section 172.406(e)(4), need be placarded on only two oppooiste sides.

57. In January 1981, the D.O.T. adopted regulations establishing routing and driver training requirements for highway carriers of large quantity packages of radioactive materials. The placarding requirement is designed to assist enforcement per-

sonnel in distinguishing between "large quantity" shipments and other placarded shipments for which preferred routing is not required.

58. The term "hazardous substance" is broadly defined so as to include:

(A) any substance designated pursuant to Section 311(b)(2)(A) of the Federal Water Pollution Control Act, (B) any element, compound, mixture, solution, or substance designated pursuant to section 102 of this Act, (C) any hazardous waste having the characteristics identified under or listed pursuant to section 3001 of the Solid Waste Disposal Act (but not including any waste the regulation of which under the Solid Waste Disposal Act has been suspended by Act of Congress), (D) any toxic pollutant listed under section 307(a) of the Federal Water Pollution Control Act, (E) any hazardous air pollutant listed under section 112 of the Clean Air Act, and (F) any imminently hazardous chemical substance or mixture with respect to which the Administrator has taken action pursuant to section 7 of the Toxic Substances Control Act.

59. In this regard, the Transportation Department is, itself, developing a regulatory proposal that, if adopted, would be designed to reduce radiation levels to which transportation workers are exposed.

60. As indicated in a recent survey of state and local activity prepared by the Oak Ridge National Laboratory and entitled "Compilation of Legislation Relevant to Transportation of Hazardous Materials (1981)", a total of more than fifty-five state and local governments (including 27 States) regulate the transportation of hazardous materials. See Marten, "Regulation of the Transportation of Hazardous Materials: A Critique and a Proposal", *5 Harv. Envir. L. Rev.* 345, 354 (1981).

61. Under Section 112, any requirement of a State or political subdivision which is inconsistent with any requirement set forth in the Act or its regulations is preempted, unless the Secretary determines, upon application of the appropriate State agency, that such requirement affords an equal or greater level of protection to the public than is afforded by the Act or regulations and does not unreasonably burden commerce. In this latter regard, the Department has issued implementing regulations which establish procedures for seeking such a "non-preemption" determination. See 49 C.F.R. § § 107.201-.225.

# 15

# Occupational Safety and Health Administration (OSHA)—Labeling in the Workplace

**Flo H. Ryer**
*U.S. Environmental Protection Agency*
*Washington, DC*

## INTRODUCTION

When Congress passed the Occupational Safety and Health Act in 1970, "to assure safe and healthful working conditions for working men and women," the Secretary of Labor was given the responsibility for implementing the Act.

In the Act, the Congress directed the Secretary of Labor to pursue to inform employees; to "prescribe the use of labels or other appropriate forms of warning" (Section 6(b)(7)). Responsibility for the implementation of the Act was delegated to the newly-formed Occupational Safety and Health Administration (OSHA) under the Department of Labor. OSHA's involvement in the generic approach "to insure that employees are apprised of all hazards to which they are exposed" (Section 6(b)(7)) and to identify all toxic and hazardous chemicals in the workplace, formally began in 1974.

The Standards Advisory Committee on Hazardous Materials Labeling was established under section 7(b) of the OSHA Act to develop guidelines for the implementation of section 6(b)(7) of the Act with respect to hazardous materials. On June 6, 1975, the Committee submitted its final report which identified issues and recommended guidelines for categorizing and ranking chemical hazards. Labels, material safety data sheets (MSDS), and training programs were also prescribed.

The National Institute for Occupational Safety and Health (NIOSH) published a criteria document in 1974, which recommended a standard to OSHA. The document entitled "A Recommended Standard . . . An Identification System for Occupationally Hazardous Materials," included provisions for labels and material safety data sheets.

In 1976, Congressman Andrew Maguire (New Jersey) and the Health Research Group petitioned OSHA to issue a standard to require the labeling of all workplace chemicals. The House of Representatives Committee on Government Operations in 1976 and 1977 recommended that OSHA should enforce the health provisions of the OSH Act by requiring manufacturers to disclose any toxic ingredients in their products, and by requiring employers to disclose this information to workers.

Many of the initial OSHA standards already requires signs, tags, markings or labels designed to warn employees of a wide variety of safety and health hazards to which they might be exposed. Under Section (6)(a) of the Act, the Secretary was directed to adopt "any national consensus standard, and any established Federal standard, unless he determines that the promulgation of such a standard would not result in improved safety or health for specifically designated employees." As a result of this direction, in 1971, OSHA incorporated many voluntary consensus standards, such as those of the National Fire Protection Association (NFPA), the American National Standards Institute (ANSI), and the National Electric Code (NEC), into its safety and health standards for General Industry, 29 CFR 1910. Many of these standards required signs, tags, markings or labels to warn employees of hazards to which they might be exposed.

## LABELING REQUIREMENTS IN OSHA SAFETY STANDARDS

Some OSHA safety standards require signs, tags or markings to provide information on means of egress, exits, fire extinguishers, compressed gas cylinders and standpipe and hose fire-fighting equipment.

Certain load markings also are mandatory for cranes, derricks, powered industrial trucks and powered platforms.

Other standards call for the indentification of containers or systems for potentially hazardous materials such as bulk oxygen, gaseous and liquid hydrogen, and anhydrous ammonia. Hazard warning information may also appear on the container or system. For example, a label for a gaseous or liquid hydrogen system would require the following text:

HYDROGEN

FLAMMABLE GAS

NO SMOKING

NO OPEN FLAMES

Standards involving flammable and combustible liquids contain various requirements for signs and labels. For example, a "NO SMOKING" sign must be displayed in an area where spraying is done with organic peroxides, and where dip tanks containing flammable or combustible liquids are located. Also, a storage cabinet for flammable or combustible liquids must be labeled conspicuously with the warning:

"FLAMMABLE — KEEP FIRE AWAY."

The OSHA standards for ionizing and nonionizing radiation also contain specific requirements for caution signs and labels. The familiar magenta and yellow symbol, along with appropriate "CAUTION" warnings, must be used to indicate areas of radioactivity or containers of radioactive materials where ionizing radiation may present a hazard. The nonionizing radiation standard has adopted the use of the red, aluminum and black symbol from ANSI to warn of radio-frequency radiation hazards.

In additon, OSHA adopted two standards from ANSI which contain specifications for some of the signs, tags, markings and labels required in other standards, such as those already mentioned. The OSHA standard 1910.144 entitled "Safety Color Code for Marking Physical Hazards" sets forth appropriate situations in which the colors red and yellow should be used for hazard warnings. The standard entitled "Specification for Accident Prevention Signs and Tags," 1910.145, applies to the design, application and use of signs or symbols for "Danger", "Caution," safety instructions, slow-moving vehicles and biological hazards. Specifications for certain accident prevention tags are also set forth. These include tags for "Danger," "Caution," "Do not start," "Out of order" and biological hazards.

## LABELING REQUIREMENTS IN OSHA HEALTH STANDARDS

There are very specific signs and label requirements in the OSHA health standards which regulate exposures to individual toxic substances. These requirements were not adopted from voluntary consensus standards, and are very specific to the type of hazard presented by the substance.

The OSHA authority to promulgate labeling requirements is given in Section 6(b)(7) of the Act which states:

> Any standard promulgated under this subsection shall prescribe the use of labels or other appropriate forms of warning as are necessary to insure that employees are apprised of all hazards to which they are exposed, relevant symptoms and appropriate emergency treatment, and proper conditions and precautions of safe use or exposure. Where appropriate, such standard shall also prescribe suitable protective equipment and control or technological procedures to be used in connection with such hazards. . . .

Each health standard promulgated by OSHA since 1971. except for 1910.1002, Coal Tar Pitch Volatiles, contains a requirement for the posting of appropriate signs in specific locations where the substance may pose a problem to workers. Most of these same standards also contain requirements for the labeling of containers.

### Asbestos

The Asbestos standard, 1910.1001, requires caution signs and labels warning of possible lung disease. In addition to size and location specifications, the following legend is required on all signs for areas of potential or actual asbestos overexposure:

ASBESTOS
DUST HAZARD
Avoid Breathing Dust
Wear Assigned Protective Equipment
Do Not Remain in Area Unless Your Work
Requires It
Breathing Asbestos Dust May Be Hazardous
to Your Health

Caution labels must be affixed to all materials containing asbestos, including employees' contaminated protective clothing, or to their containers. This requirement is waived if the asbestos fibers have been modified in some way, and the possibility of employee overexposure to asbestos does not exist. The label must be readily visible and legible, and shall state:

CAUTION
Contains Asbestos Fibers
Avoid Creating Dust
Breathing Asbestos Dust May Cause
Serious Bodily Harm

**Carcinogens**

The next group of health standards containing the requirements for signs and labels is the fourteen carcinogens (now thirteen), 1910.1003 to 1910.1016. All of these requirements are identical. At entrances to regulated areas, signs must be posted containing the text:

CANCER-SUSPECT AGENT
AUTHORIZED PERSONNEL ONLY

At entrances to regulated areas, where an employee could directly have contact with the carcinogen, e.g., during maintenance or repair operations, the following sign must be posted:

CANCER-SUSPECT AGENT EXPOSED IN
THIS AREA
IMPERVIOUS SUIT INCLUDING GLOVES,
BOOTS, AND AIR-SUPPLIED HOOD
REQUIRED AT ALL TIMES
AUTHORIZED PERSONNEL ONLY

Other appropriate signs and instructions must be posted at the entrance to, and exit from a regulated area to inform employees of procedures to follow in entering and leaving the regulated area.

Labeling requirements are also designed to warn employees of possible exposure to a carcinogen. Containers of the carcinogen which are accessible to, and handled only by, authorized or trained personnel may be labeled with a generic or proprietary name and the percent of carcinogen present. Containers of the carcinogen, or of clothing or equipment contaminated with the carcinogen, which are accessible to, or handled by, personnel who are not authorized or trained, must be labeled with the full chemical name of the carcinogen and its Chemical Abstracts Service Registry (CAS) number. All containers must have the words "CANCER-SUSPECT AGENT" adjacent to the contents identification. In addition, any mixture which contains a carcinogen and which has corrosive or irritating properties shall be labeled with suitable hazard warning statements noting, if applicable, any particularly sensitive or affected parts of the body.

### Vinyl Chloride

The next OSHA health standard, 1910.1017, Vinly Chloride, contains numerous signs and label requirements. Again, as for the 13 carcinogens, entrances to regulated areas must be posted with signs warning:

CANCER-SUSPECT
AUTHORIZED PERSONNEL ONLY

Any areas containing hazardous operations, that is, where a release of vinyl chloride gas or liquid might be possible, must be posted with signs warning:

CANCER-SUSPECT AGENT IN THIS AREA
PROTECTIVE EQUIPMENT REQUIRED
AUTHORIZED PERSONNEL ONLY

There are also specific content requirements for labels. Containers of polyvinyl chloride resin waste from reactors, or any other wastes contaminated with vinyl chloride, shall be legibly labeled:

CONTAMINATED WITH VINYL CHLORIDE
CANCER-SUSPECT AGENT

All containers of polyvinyl chloride (PVC) shall be legibly labeled:

POLYVINYL CHLORIDE (OR TRADE NAME)
CONTAINS
VINYL CHLORIDE
VINYL CHLORIDE IS A CANCER-SUSPECT
AGENT

All containers of vinyl chloride shall be legibly labeled:

VINYL CHLORIDE
EXTREMELY FLAMMABLE GAS
UNDER PRESSURE
CANCER-SUSPECT AGENT

### Arsenic

The Inorganic Arsenic Standard, 1910.1018, again contains provisions for warning employees of a possible cancer hazard. The employer must post signs with the following legend at all regulated areas:

DANGER
INORGANIC ARSENIC
CANCER HAZARD
AUTHORIZED PERSONNEL ONLY
NO SMOKING OR EATING
RESPIRATOR REQUIRED

In addition, the employer shall apply precautionary labels to all shipping and storage containers of inorganic arsenic, and to products containing inorganic arsenic, where the possibility of airborne exposure to arsenic exists. The labels must read:

DANGER
CONTAINS INORGANIC ARSENIC
CANCER HAZARD
HARMFUL IF INHALED OR SWALLOWED
USE ONLY WITH ADEQUATE VENTILATION
OR RESPIRATORY PROTECTION

## Lead

Due to the unique problems presented in the workplace by lead, the sign and label requirements in OSHA's Lead standard, 1910.1025, differ significantly from those already discussed. The following sign must appear at each work area where the OSHA Permissible Exposure Limit (PEL) for lead is exceeded:

WARNING
LEAD WORK AREA
POISON
NO SMOKING OR EATING

In addition, containers of protective clothing which are contaminated with lead must bear labels with the following legend:

CAUTION: CLOTHING CONTAMINATED WITH LEAD.
DO NOT REMOVE DUST BY BLOWING OR SHAKING.
DISPOSE OF LEAD CONTAMINATED WASH WATER
IN ACCORDANCE WITH APPLICABLE LOCAL,
STATE OR FEDERAL REGULATIONS.

## Benzene

The Benzene standard, 1910.1028, contains requirements for signs and labels. Signs shall be posted in all regulated areas, and shall bear this legend:

DANGER
BENZENE
CANCER HAZARD
FLAMMABLE—NO SMOKING
AUTHORIZED PERSONNEL ONLY
RESPIRATOR REQUIRED

Caution labels must be affixed to all containers of benzene, and of products containing benzene, and must read:

CAUTION
CONTAINS BENZENE
CANCER HAZARD

In addition, the employer shall assure that these labels remain affixed when the products are sold, distributed or otherwise leave the workplace.

## Coke Oven Emissions

The next standard containing sign and label requirements is Coke Oven Emissions, 1910.1029. The employer must post signs in regulated areas which bear the legend:

DANGER
CANCER HAZARD
AUTHORIZED PERSONNEL ONLY
NO SMOKING OR EATING

In areas where the OSHA PEL for Coke Oven Emissions is exceeded signs must be posted which read:

DANGER
RESPIRATOR REQUIRED

Protective clothing which is contaminated with coke oven emissions must be placed in containers. These containers must be labeled as follows:

CAUTION
CLOTHING CONTAMINATED WITH COKE EMISSIONS
DO NOT REMOVE DUST BY BLOWING OR SHAKING

## Cotton Dust

The OSHA Cotton Dust Standard, 1910.1043, contains sign requirements specifically warning employees of the possibility of contracting "brown lung" disease from overexposure to cotton dust. The employer must post the following sign in each work area where the OSHA PEL for cotton dust is exceeded:

WARNING
COTTON DUST WORK AREA
MAY CAUSE ACUTE OR DELAYED LUNG INJURY
(BYSSINOSIS)
RESPIRATORS REQUIRED IN THIS AREA

Similarly, the standard for Exposure to Cotton Dust in Cotton Gins, 1010.1046, the following sign must be posted in each work area where there is potential employee exposure to cotton dust.

WARNING
COTTON DUST WORK AREA. MAY CAUSE
ACUTE OR DELAYED LUNG INJURY
(BYSINOSSIS)

### 1,2-Dibromo-3-chloropropane and Acrylonitrile

Two other standards, 1,2-Dibromo-3-chloropropane (DBCP—and Acrylonitrile (AN), 1910.1044 and 1910.1045, respectively, contain similar requirements for signs and labels. All regulated areas for DBCP must be posted with a sign bearing the legend:

DANGER
1,2-Dibromo-3-chloropropane
(insert appropriate trade or common names)
CANCER HAZARD
AUTHORIZED PERSONNEL ONLY
RESPIRATOR REQUIRED

Similarly, all workplaces, where acrylonitrile concentrations exceed the OSHA PELs of 2 parts per million (ppm) as an 8-hour time-weighted average (TWA) or 10 ppm over any fifteen-minute period, shall be posted with a sign which indicates:

DANGER
ACRYLONITRILE (AN)
CANCER HAZARD
AUTHORIZED PERSONNEL ONLY
RESPIRATORS MAY BE REQUIRED

Containers of DBCP, or products containng DBCP, must be labeled with the following warning:

DANGER
1,2-Dibromo-3-Chloropropane
CANCER HAZARD

Similarly, containers of liquid AN and AN-based materials must be labeled:

DANGER
CONTAINS ACRYLONITRILE (AN)
CANCER HAZARD

### CANCER POLICY

On January 22, 1980, OSHA issued its Cancer Policy, 29 CFR 1990, the Identification, Classification and Regulation of Potential Carcinogens. This cancer policy, which will directly influence the promulgation of future OSHA health standards for carcinogens, contains provisions for signs and labels.

There are two model standards as part of the policy: one for an Emergency Temporary Standard, and the other for a Permanent Standard. These standards contain provisions for signs and labels to warn employees of the hazard for Category I potential carcinogens.

Signs must be posted to indicate all workplaces where the carcinogen may be present or where employee exposures exceed the action level. The required legend is:

DANGER
(insert appropriate trade or common name)
CANCER HAZARD
AUTHORIZED PERSONNEL ONLY

Labels must be affixed to all containers of the carcinogen, and of products containing the carcinogen, and require a similar legend:

DANGER
CONTAINS (name of carcinogen)
CANCER HAZARD

The labels must remain affixed to the containers when the products are sold, distributed, or otherwise leave the workplace.

In addition, both model standards contain the provision whereby the employer must label, or otherwise inform employees who may contact waste material containing the carcinogen, of the contents of such materials. This provision applies also be contaminated protective clothing.


## THE NEED FOR LABELING

The preamble to the Cancer Policy proposal sets forth OSHA's reasons including provisions for signs and labels in health standards as follows:

> OSHA believes that it is important and indeed section 6(b)(7) of the Act requires, that appropriate forms of warning, as necessary, be used to apprise employees of the hazards to which they are exposed in the course of their employment. OSHA believes, as a matter of policy, that employees should be given the opportunity to make informed decisions as to whether to work at a job under the particular working conditions extant. Furthermore, OSHA believes that when the control of potential safety and health problems involves the cooperation of employees, the success of such a program is highly dependent upon the employee's understanding of the hazards attendant to that job.

OSHA believes that sign-posting and labeling are effective means of informing employeers of the hazards to which they may be exposed. During the Cancer Policy hearing, employees who worked around signs which indicated that vinyl chloride was a "cancer-suspect agent" testified that they find sign-posting very helpful. It reminds them of what atmosphere they are in, they can act accordingly, and they know what obligation the company is under in handling this product. Other testimony indicated that, very often, labeling on containers served as the only source of information about its contents, which in turn was essential to assessing the potential for employee exposure or appropriate handling procedures. During the hearing for Access to Employee Exposure and Medical Records, testimony again indicated that workers are not often informed by their employers of the identities of substances with which they work. As a result, without adequate substance identification, many workers are exposed routinely to toxic substances, are unaware of the hazards, and are incapable of protecting themselves or ensuring that their employers provide them with adequate protection.

Thus, OSHA believes that the general requirement to label containers constitutes responsible industry practice. For this reason OSHA is presently in the process of promulgating a generic labeling standard.

## HISTORY OF THE PROPOSED LABELING STANDARD

As stated earlier, OSHA's involvement in the identification of toxic and hazardous chemicals in the workplace began as long ago as 1971 with adoption of the initial standards. Then, to briefly review, a Standards Advisory Committee for Hazardous Materials Labeling was established in 1974. The Committee forwarded its final report to OSHA in June, 1975. In the report, the Committee recommended guidelines for categorizing and ranking chemical hazards. Labels, data sheets and training programs were also prescribed. Congressman Andrew Maguire (New Jersey) and the Health Research Group petitioned OSHA, in 1976, to issue a standard requiring the labeling of all workplace chemicals. The House of Representatives' Committee on Government Operations in 1976 and 1977 reported that OSHA should require manufacturers to disclose any toxic ingredients in their product and to make this information available to workers.

Therefore, on January 28, 1977, OSHA published an advance notice of proposed rulemaking on chemical labeling in the *Federal Register* (42 FR 5372). The notice requested public comment as to what provisions should be contained in the regulation to ensure that employees are apprised of the hazards to which they are exposed. Eighty-one comments were received from a variety of federal, state and local government agencies, trade associations, businesses, and labor organizations.

On September 27, 1979, the Public Citizen Health Research Group, the Philadelphia Area Project on Occupational Safety and Health (PHILAPOSH), and United States Congressman Andrew Maguire filed suit in District Court against the Secretary of Labor. The suit sought an order directing OSHA to promulgate rules requiring employers to apprise employees and employee representatives of the identity of all potentially toxic materials and harmful physical agents to which they may be or may have been exposed in the workplace.

On October 17, 1979, the Chemical Manufacturers Association (CMA) filed for intervention in the suit, alleging that any "court-ordered deadline may not allow the Secretary (of Labor) sufficient time (a) to consider the extent of his authority to require notification in all cases, and (b) to weigh the competing, but nevertheless valid, considerations of employee knowledge, on the one hand, and the employer's legitimate trade secrets, on the other hand." The Court determined that the Secretary could take whatever time is necessary to promulgate an appropriate chemical substance identification standard.

The Interagency Regulatory Liaison Group (IRLG), which consisted of representatives from OSHA, Environmental Protection Agency (EPA), Food and Drug Administration (FDA), Consumer Product Safety Commission (CPSC), and the Federal Nutrition Service of the Department of Agriculture, was established to coordinate federal activity and to reform the regulatory process regarding the protection of workers and public health.

The IRLG Labeling Regulations Task Force held meetings to consider OSHA's proposed labeling regulation along with a similar rulemaking effort currently under development at EPA. The purpose of this task force was to ensure that the two rules appropriately augmented each other, so that burdens on employers having to comply with both regulations would be minimized.

## PROVISIONS OF THE PROPOSED LABELING STANDARD(S)

Briefly, the provisions of the early OSHA 1980 draft proposed standard for Chemical Substance Identification applied to all hazardous or toxic substances and mixtures which are:

1. Listed in the Environmental Protection Agency Toxic Substances Control Act Inventory (EPA's TSCA Inventory) or in the National Institutes for Occupational Safety and Health Registry of Toxic Effects of Chemical Substances (NIOSH's RTECS), and

2. Present in the workplace where workers may be exposed under normal conditions of use.

The provisions included in OSHA's 1980 draft proposal include:

1. Substance-employee identification lists for each work area

2. Labels for containers of regulated substances

3. Materials Safety Data Sheets (MSDS), to be obtained if available

4. Record preservation of substance-employee identification lists and material safety data sheets.

At that time, OSHA stated the belief that "this 'Chemical Substance Identification' Standard, when promulgated, will definitely aid in the protection of the health and safety of all workers routinely exposed to toxic and hazardous substances in the workplace."

Finally, on January 16, 1981, OSHA published, in the *Federal Register,* a proposed chemical labeling standard entitled "Hazard Identification." The standard was heavy with specification language. It emphasized identification of chemicals and covered all manufacturers, importers and repackagers of all chemical products. It required specific search and evaluation of all relevant studies, had extensive and specific labeling requirements, did not allow for existing systems of labeling and included the labeling of pipes. It made no allowance for trade secrets, required the provision of MSDSs if they were available and mandated a certification for "no hazard." It specified hazard evaluation records to be kept for three years and employee access to MSDSs; and it made no provision for education and training. The initial (start-up) cost was estimated to be between $2.6 and $3.0 billion, depending on the amount of testing conducted to meet the certification requirement.

Two weeks after publication in the *Federal Register*, the standard was recalled by the Department of Labor for consideration of regulatory alternatives. Meanwhile a spokesman for the Office of Management and Budget (OMB) was quoted as saying "We do have some problems with the costs and benefits of the labeling proposal." OSHA addressed the question of costs and benefits in the *Federal Register*, January 13, 1982. OSHA stated that cost estimates vary widely, depending on the provisions and gave a range from $581.5 million to $7.6 billion. However, OSHA emphasized that their "new proposal is likely to fall near the bottom of that range," whereas the standard proposed last January was expected to be highly costly.

More than one year after withdrawal of the Hazard Identification proposed standard, the new proposed standard (standard) entitled "Hazard Communication," was published in the *Federal Register*, Volume 47, No. 5, Friday, March 19, 1982. It proposes that chemical manufacturers assess the hazards of chemicals which they produce and that all employers in SIC Codes 20 through 39 (Division D, Standard Industrial Classification Manual) provide information to their employees about the hazardous chemicals which they use by means of a hazard communication program, labels, placards, material safety data sheets, and information and training.

The standard applies to "any chemical which is known to be present in the workplace in such a manner that employees may be exposed under normal conditions of use or in a foreseeable emergency." It also applies to any mixture which is comprised of at least one (1) per cent (by weight or volume) of any chemical determined to be hazardous unless the mixture has been evaluated as a whole and the data indicates it is not hazardous.

When "employee protection necessitates disclosure of hazardous chemicals comprising less than one (1) per cent (by weight or volume) of a mixture, the Assistant Secretary may lower or eliminate this concentration exemption by a rulemaking notice in the *Federal Register*."

The standard does not apply to chemicals being developed and used only in research laboratories or to chemicals which are foods, drugs, cosmetics or tobacco products intended for personal consumption by employees while in the workplace.

When the standard becomes effective the chemical manufacturer will be required to evaluate the chemicals produced in his/her workplace to determine if they are hazardous. He will be given guidelines to follow in making this evaluation. Also, each employer must develop and implement a hazard communication program for his/her workplace which meets the criteria specified in the standard for labels and placards, MSDSs and employee information and training.

The standard requires that the employer ensure that each container of hazardous chemicals in the workplace is labeled, tagged or marked with the identity of the hazardous chemical(s) and the hazard warnings.

When stationary containers in the work area have similar contents and hazards, the employer may post signs or placards to convey the required information rather than affixing labels to each individual container.

When containers leave the workplace the employer is responsible for labeling, tagging or marking them. The containers should have the following information:

> Identity of the hazardous chemical(s),
>
> Hazard warnings, and
>
> Name, address and telephone number of the manufacturer.

An exception is that the employer is not required to label containers of ten gallons (37.8 liters) or less in volume, if the contents were transferred from labeled containers, and are intended only for the immediate use of the transferee.

Each employer shall obtain or develop MSDS for each hazardous chemical which he/she produces or uses. Each MSDS shall reflect the information contained in the sources consulted by the chemical manufacturer in his/her hazard determination.

The MSDS must contain at least the following information:

- The chemical and common name(s), CAS Number(s) and the identity used on the label for all hazardous ingredients which comprise greater than one (1) percent of the chemical (except as provided by paragraph (g) of this section on trade secrets);
- Physical and chemical characteristics of the hazardous chemical (such as vapor pressure, flash point);
- The physical hazards of the hazardous chemical, including the potential for fire, explosion, and reactivity;
- Known acute and chronic health effects of exposure to the hazardous chemical, including signs and symptoms of exposure, and medical conditions which may be aggravated by exposure to the chemical;
- The primary route(s) of entry and permissible exposure limit (for those hazardous chemicals for whch OSHA has promulgated a permissible exposure limit);

- Emergency and first aid procedures
- Precautions for safe handling and use, including appropriate hygienic practices, procedures for decontaminating equipment prior to performing repairs and maintenance, and procedures for clean-up of leaks or spills;
- The date of preparation of the MSDS or the last change to it; and
- The name, address and telephone number of the manufacturer preparing the sheet.

Also recommended to be included in the MSDS are engineering controls, work practices, personnel protection equipment, and emergency and first aid procedures. Blank spaces on existing MSDS will be considered to indicate that information was sought but not found. It is the chemical manufacturer's responsibility to ensure that this interpretation is accurate.

If the employer becomes aware of any information which is both new and significant regarding the health hazard of a chemical, he is required to add this to the MSDS within a reasonable period of time.

Chemical manufacturers are required to ensure that manufacturing purchasers of hazardous chemicals are provided an appropriate MSDS with their initial shipment, and with the first shipment after the MSDS is updated. If the MSDS is not provided with the shipment, the purchasing manufacturing employer shall obtain one from the chemical manufacturer as soon as possible.

The employer must maintain copies of the required MSDS for each hazardous chemical in the workplace, and shall ensure that they are readily accessible to employees exposed to the hazardous chemicals.

The employer shall provide employees with information and training on hazardous chemicals in the workplace at the time of their initial assignment, and whenever a new hazardous chemical is introduced into their work area.

The standard states insofar as trade secrets are concerned, "an employer may withhold the precise chemical name of a chemical if:

- The employer can substantiate that it is a trade secret;
- The chemical is not a carcinogen, mutagen, teratogen, or a cause of significant irreversible damage to human organs or body systems for which there is a need to know the precise chemical name;
- The chemical is identified by a generic chemical classification which would provide useful information to a health professional;
- All other information on the properties and effects of the chemical required by this is contained in the material safety data sheet;

- The material safety data sheet indicates which category of information is being withheld on trade secret grounds; and

- The withheld information is provided on a confidential basis to a treating physician who states in writing (except in an emergency situation) that a patient's health problems may be the result of occupational exposure. A statement to this effect with the name of the manufacturer and an emergency telephone number shall be included in the material safety data sheet.

- To the extent that names of trade secret chemicals are disclosed, the employer may condition employee, designated representative, and downstream employer access to such information upon acceptance of a reasonable confidentiality agreement. The agreement may restrict use of the information to health purposes, prohibit disclosure of the information to anyone other than a treating physician without the consent of the originating employer, and provide for compensation or other legally appropriate relief for any competitive harm which results from a breach of the agreement.

## Effective Dates

Employers are to be in compliance with the standard within the following time periods:

| Employer | Pure Substances | Mixtures |
|---|---|---|
| Chemical manufacturers: | | |
| More than 250 employees | 1 year | 2 years |
| 25 to 250 employees | 1½ years | 2½ years |
| Fewer than 25 employees | 2 years | 3 years |
| Other employers | 3½ years | 3½ years |

## Appendices in the Standard

There are two appendices in the standard which many employers may find helpful. Appendix A is a discussion of Health Hazards and Appendix B contains the Hazard Determination Guidelines and Sources. Appendix A states that the employer shall, when assessing the health hazard potential of a chemical for purposes of compliance with this standard, "consider the scientifically well-established evidence of any type of health effect which may occur in any body system of his/her employees." The employer is encouraged to consult Appendix B for sources of information to assist him/her in conducting the hazard evaluation.

## SUMMARY

The OSH Act directs the Secretary of Labor (OSHA) to "prescribe the use of labels or other appropriate forms of warning" and to ensure that employees are apprised of all hazards to which they are exposed" Section 6(b)(7).

Many of the initial standards adopted by OSHA in 1971 required signs, tags, markings or labels to warn employees of hazards to which they might be exposed.

There are specific signs and label requirements contained within the OSHA health standards promulgated since 1971. These requirements are very specific as to the type of hazard presented by the substance.

A Standards Advisory Committee on Hazardous Materials labeling was established under section 7(b) of the Act to develop guidelines for the implementation of section 6(b)(7). The Committee submitted its final report recommending guidelines for categorization and ranking chemical hazards on June 6, 1975. Labels, MSDS and training programs were also prescribed in the report. In 1976 and again in 1977 the House of Representatives Committee on Government Operations reported that OSHA should require manufacturers to disclose any toxic ingredients in their products and make this information available to workers.

On January 28, 1977, OSHA published an advance notice of proposed rulemaking on chemical labeling in the *Federal Register*. Comments were requested and eighty-one were received from a variety of federal state and local government agencies, trade associations, businesses, and labor organizations. In general, there was support for the concept of a labeling standard.

On September 27, 1979, a suit was brought to direct OSHA to promulgate labeling requirements. The suit was intervened October 17, 1979. The Court determined that the Secretary of Labor should take whatever time was necessary to promulgate an appropriate labeling standard.

Finally, January 16, 1981, after several drafts of a labeling standard, OSHA published a proposed standard entitled "Hazard Identification." The standard was withdrawn two weeks later and OSHA addressed the question of costs and benefits in the *Federal Register,* January 13, 1982.

On March 19, 1982, a new proposed standard entitled "Hazard Communication," was published by OSHA in the *Federal Register*. It differs from the Hazard Identification standard in that it emphasizes "Communication" rather then "Identification." It is more performance oriented and less specific. The coverage is more limited and the evaluation less extensive. It allows for trade secrets and provides for performance-oriented education and training. In this standard only chemical manufacturers evaluate the hazards whereas in the earlier proposal all employers were required to evaluate hazards.

The comment period for the proposed OSHA Hazard Communication standard closed on May 18, 1982. The hearings were held throughout the country on June 15, 1982.

The final standard was published in the *Federal Register* on Friday, November 25, 1983. It is a performance oriented standard and reflects the comments received during the hearings and comment period. While there is general agreement that such a standard is necessary, many have differed sharply over the specifics of the standard.

The standard as published declares Federal preemption as one purpose of the standard, and opinion is divided over whether there will be challenges to OSHA's stand.

Hazard Communication is the first major regulatory proposal that OSHA has published under this Administration. It is deserving of much attention, careful scrutiny, serious review and an awareness of all it entails.

This new standard is given in the Appendix starting on page 452.

# Part IV

# Industry Standards and Practice

In this last section there is a discussion of the ANSI Guide to Precautionary Labeling of Hazardous Chemicals. The Standard, sponsored by CMA has its roots in MCA's original LAPI Manual, first published in 1938, and subsequently revised. It remains the best overall guide for developing chemical labels.

The ANSI Guide uses the basic DOT classification scheme with few modifications; one is, however, of major importance. An Appendix creates a standard of care for the labeling of serious chronic hazards. Three general classes are established, (1) carcinogen, (2) reproductive toxicants and (3) other serious chronic effects.

In a departure from previous label standards, there is no requirement for the use of signal words based solely upon statements of serious chronic hazards. The gravity of the warning is considered sufficient to ensure attention.

This section also includes a review of the NFPA Identification System. This color-keyed system is in use within the chemical industry for in-plant identification—blue for health, red for flammability, and yellow for reactivity (instability). A fourth square is reserved for special or unusual hazards, as for example, radioactivity or an especially violent reaction with water. These four squares are arranged to form a larger square on point. The "top" (red) is always flammability, the "right" (yellow) is always reactivity (instability), the "left" (blue) is always Health, and the "bottom" special hazards.

The definitions in the NFPA System include the products of combustion as part of their scope. This differs from most other systems. It is a worthwhile and useful refinement to keep in mind when preparing instructions for the "In Case of Fire" section of a label. NFPA, incidentally, publishes a wide range of fire related data, including their extremely useful manual of hazardous chemical reactions. A complete list of these publications can be obtained from NFPA.

*403*

These two systems are, and should be, viewed as complementary. The NFPA System, utilizing color keyed graphics, has a high recognition rate with excellent retention. It is simple, colorful, and easily understood with little training. The ANSI System, relying upon text, is able to convey more information and greater detail. A comprehensive hazard communication system would utilize both systems, as well as include a material safety data sheet program.

An example of a combined system is the NIOSH developed, "An Identification System For Occupationally Hazardous Materials" recommended standard. This report was finally published in 1974. It remains one of the best examples of a comprehensive visual hazard communication system.

As pointed out, the NIOSH System draws heavily upon NFPA-704 and CMA-LAPI for its conceptual basis. As in NFPA-704, red represents flammability, blue stands for health, and yellow for reactivity. While the definitions and degrees of hazard vary somewhat from their antecedent systems, they are, in general, compatible. The statements of hazard and precautionary language are very similar to those used in other systems.

As in the NFPA 704 System, the signal words "Danger", "Warning", "Caution" (used in other systems to indicate level or degree of hazard), are replaced by a numerical 0–4 rating system to indicate *degree* of hazard while the color or color background defines the *kind* of hazard involved.

The NIOSH Standard includes a consideration of chronic health effects. These effects include carcinogenicity, mutagenicity, teratogenicity, as well as long and short term disease or bodily injury caused by exposure to workplace chemicals.

This Standard consists of three major parts, placards, labels, and MSDSs. The placard is intended for area labeling and should be recognizable from a distance. Placards may contain only the hazard color symbols. Labels are intended to usually combine text and color hazard symbols for use principally on containers. MSDS are to be readily available within the workplace. While the Standard does discuss the need for training, it does not offer a specific program.

The ASTM's proposed Safety, Signs and Colors Z 535.2 Standard is also discussed. This system relies upon color, shape, and signal word to alert workers to workplace hazards. These distinctive shapes: oval, truncated diamond, and a rectangle are used with the signal words, Danger, Warning, Caution. Red and black combine with "Danger" on an oval to indicate the highest level of hazard; orange and black are used with the truncated diamond on which appears "Warning"; and "Caution" appears in a rectangle in black and yellow.

The National Paint and Coating Association (NPCA) Label Guide and the Hazardous Materials Identification System (HMIS) are then presented. HMIS is the most comprehensive in-plant hazard communication program developed as a single total integrated system.

Conceptually, it appears to draw upon NFPA-NIOSH concepts, and enlarges these systems by adding a specific training program and a number of other features. HMIS includes labels for containers, tags for piping, placards for tanks, symbols for personal protection, a rating system for raw materials, an MSDS program, an audio-visual training program; and as reinforcement for these elements, a wallet card, wall posters, and an attractive and highly readable employee handout. NPCA even includes a sample letter to suppliers requesting the data needed to implement portions of the system. HMIS is a good example of the benefits trade association members can receive from a well run organization.

Lastly, materials safety data sheets, their advantages and faults and their place within a hazard labeling/communication system, are discussed.

# 16

# American National Standards Institute (ANSI) Guide to Precautionary Labeling of Hazardous Chemicals

**Jay A. Young**
*Consultant*
*Silver Spring, MD*

Labels are constructed, built from selected words, phrases, and symbols, arranged within an area on the surface of a container. A label is intended to convey information describing the inherent nature, suitable use, and appropriate treatment (handling, storage, disposal) of the containers' contents. This information is prepared by the manufacturer and directed to the person who will use, or handle, the material described by the label. Ordinarily, the manufacturer has no other means to directly inform those persons about the contents at point of use. That is, a container label is usually the only communication device available to a manufacturer for directly informing the person who will use or handle, the manufactured material.

## PRECAUTIONARY LABELS

Precautionary labels apply to hazardous chemicals. As used here, "hazard" and its derivatives (such as "hazardous") refer to a potential to cause harm. Risk refers to the probability that harm will in fact ensue. The likelihood or probability of causing harm can be reduced by taking appropriate precautions. Thus, a hazardous chemical need not cause harm if the precautions that are taken in its use have sufficiently reduced the risk. This is essentially the "Expected Value" concept applied to the industrial handling of chemicals. The information precautionary labels convey is intended to alter the users behavior so as to reduce the likelihood of harm that might otherwise ensue during the reasonably foreseeable use, transportation, storage, and disposal of the contents. Considering the large number of hazardous chemicals in commerce and the variety of properties of these chemicals and their mixtures, the need for precautionary information to be conveyed to users is obvious.

Perhaps not as obvious is the label practitioner's requirement that to the degree possible all precautionary labels for hazardous chemicals should share certain common characteristics. This demand derives from communication theory which holds that a common understanding is necessary for good communication. These elements include:

> *Readability:* Esoteric or unfamiliar words may not convey the intended information. Simple language is preferred.
>
> *Similarity of meaning:* The same words, phrases, and symbols used on different labels should have the same meaning and significance. Different words should not be used to convey the same meaning or significance.
>
> *Legibility:* The printing should be easily read.
>
> *Permanence:* Printing ink, paper/film and the adhesive used should be resistant to the adverse effects of reasonably expectable weathering.
>
> *Arrangement:* The relative type sizes and placement of the various words, phrases, and symbols should be such as to invite attention first to the most important, or most likely to be severe, hazard.
>
> *Avoidance of "over-labeling":* Only reasonably foreseeable hazards and the associated reductions of risk should be addressed. Thus, although ammonia, a gas, will burn, CPSC has taken the position that is is not ordinarily desirable to identify this hazard on the label of aqueous ammonia cleaning solutions used in many households. They apparently reason that since the level of free ammonia is so low, this hazard will not cause harm in the reasonably foreseeable use of household products. If a label identifies a hazard that is not consistent with the users common sense and experience, that person will tend to ignore both the information related to the inapplicable hazard *and* that related to those hazards that *can* occur if precautions are not taken. Material safety data sheets are used for fuller explanations of unusual hazards and the label directs the user to read the MSDS before using the product.

## HISTORICAL DEVELOPMENT

The development of present chemical labeling has evolved principally during the last fifty years. Precautionary labeling in the United States really began in 1927 with the passage of the Federal Caustic Poison Act and the attendant regulations by the Food and Drug Administration. Currently, the Federal

Hazardous Substances Act as amended supercedes the earlier legislation; this Act is administered by the Consumer Products Safety Commission at present. Other Acts also include precautionary labeling provisions, for example the Federal Insecticide, Fungicide and Rodenticide Act, administered by the Environmental Protection Agency. The Occupational Safety and Health Act, administered by the Occupational Safety and Health Administration contains provisions for precautionary labeling. Other label requirements of principal interest to the chemical industry are also contained in the Resource Conservation and Recovery Act, The Toxic Substance Control Act, and the Transportation Act.

In general however, the precautionary labeling of *Industrial* hazardous chemicals has remained a voluntary activity, sponsored, endorsed, and promoted by the chemical industry for over half of a century. This activity has largely centered in the Chemical Manufacturers Association, formerly known as Manufacturing Chemists Association, although other trade associations have been active. Label recommendations for gases have been published by the Compressed Gas Institute. The National Fire Protection Association has published and developed a special label system for chemicals involved in fires, and the National Paint and Coatings Association has published a Labeling Guide for the Paint Industry and an in-plant Hazardous Materials Information System, HMIS.

In the decade, 1930–1940, following the passage of the Federal Caustic Poison Act, further activity included the development of drafts of Volatile Poisons Bills which were never passed by Congress. These deliberations enhanced the recognition of the importance of precautionary labeling as a menas of reducing risks from hazardous chemicals and renewed the awareness of the chemical industry to its responsibilities in this area. As a consequence, the "Chemical Products Agreement Committee" appointed by the Surgeon General developed the voluntary "Surgeon General's Agreements" subscribed to by the U.S. chemical industry in mid-decade. This agreement included texts for the precautionary labeling of certain specific hazardous chemicals. These standards remained in effect until 1952 when the Surgeon General agreed to their replacement by the broader and more widely applicable precautionary labeling practices voluntarily developed by the Labels and Precautionary Information (LAPI) Committee of the Manufacturing Chemists Association. This Committee was disbanded in 1978 during a period when the Manufacturing Chemists Association reorganized its internal structure and changed its name to Chemical Manufacturers Association. Former LAPI members were the nucleus for the formation of the currently existing professional organization dealing with the labeling of chemicals, the American Conference on Chemical Labeling, Inc.

In 1946, the Labels and Precautionary Information Committee published the first voluntary precautionary labeling standard for hazardous chemicals, "Guide to Precautionary Labeling of Hazardous Chemicals", familiarly known as the "LAPI Manual" or "Manual L-1". This publication during the life time of its seven successive editions, 1946 through 1970, was recognized throughout the world as the standard for precautionary labeling. Its principles were applied, for example, in the development of the "Paint Industry Labeling Guide" now in its second edition (1979) with supplement (1981) (National Paint and Coatings Association, Washington, DC).

The principles from the LAPI Manual were also applied in the development of State and Federal regulations for precautionary labeling. Several states, perhaps a dozen, have adopted the LAPI Guide as a State standard. New Jersey's Safety Notice #2 is one example.

In 1976 through the efforts of the LAPI Committee, chaired by Charles J. O'Connor, the LAPI Manual was itself replaced by the American National Standards Institute (ANSI) voluntary standard, "American National Standard for the Precautionary Labeling of Hazardous Industrial Chemicals, ANSI Z129.1-1976" (American National Standards Institute, New York, NY).

## ANSI Z129.1

As is clear from the foregoing, precautionary labeling is a developing activity. Almost as soon as ANSI Z129.1-1976 was printed, discussions for the next edition, originally scheduled for publication in 1979, were begun. The second edition of ANSI Z129.1 was published in early 1983. The draft of the second edition was approved September 16, 1982. CMA is acting as the secretariat for this standard.

The second edition consists of an introduction, four sections, and two appendices. (The appendices present examples and are *not* part of the consensus document.) Section 1 identifies the scope of the standard and section 2 lists definitions for particular terms. Thus, paraphrasing, "adequate ventilation" refers to a condition in which air contaminant concentrations are below levels that cause injury or illness, or, that the vapors of flammable liquids are well below the lower flammable limit. A "toxic chemical" has an oral $LD_{50}$ for albino rats greater than 50 mg/kg but not greater than 500 mg/kg, or a 24 hr. skin contact $LD_{50}$ for albino rabbits more than 200 mg/kg but not more than 1000 mg/kg, or an inhalation $LC_{50}$ for albino rats more than 200 ppm but not more than 2000 ppm of gas or vapor or more than 2 mg/$\ell$ but not more than 20 mg/$\ell$ of dust or mist, provided that such exposures are reasonably likely to be encountered by humans in their use of the chemical.

Section 3 deals with General Requirements, Section 4 with the Selection of Precautionary Label Text, Appendix A with illustrative examples, and Appendix B with the Labeling of Serious Chronic Hazards.

## APPENDIX B

Appendix B is not a part of the American National Standard for the Precautionary Labeling of Hazardous Industrial Chemicals, Z129.1-1982, but is included for information purposes only. Because of the widespread interest in chronic hazards, it is reproduced below.

## Labeling Serious Chronic Hazards

### B.1 General

Until definitions and test protocols for chronic effects are established, precautionary label texts for serious chronic effects should be used when there is generally accepted, well-established evidence that such serious chronic effects exist or when required by law.

A chronic effect is an effect from exposure to a hazardous chemical resulting in either of the following:

(1) a persistent illness or injury that develops over time from a single exposure, or

(2) a persistent illness or injury that develops from prolonged or repeated exposure under conditions that do not produce the effect from a single exposure.

### B.2 Labeling Carcinogens

When there is generally accepted, well-established evidence that a chemical is known to cause cancer in humans, it should be labeled with the following statements of hazard or equivalent:

CANCER HAZARD
OVEREXPOSURE MAY CREATE CANCER RISK
CANCER SUSPECT AGENT (as required by Government
Regulation)

When there is generally accepted, well-established evidence that a chemical is known to cause cancer in animals and not known to cause cancer in humans, but where the animal data suggest that the chemicals would probably cause cancer in humans, it should be labeled with the following statements of hazard or equivalent:

CANCER HAZARD BASED ON TESTS WITH
LABORATORY ANIMALS
OVEREXPOSURE MAY CREATE CANCER RISK

When there is generally accepted, well-established evidence that a chemical is known to cause cancer in animals, but not known to cause cancer in humans, it should be labeled with the following statements of hazard or equivalent:

POSSIBLE CANCER HAZARD BASED ON TESTS WITH
LABORATORY ANIMALS
OVEREXPOSURE MAY CREATE CANCER RISK

### B.3 Labeling Reproductive Toxicants

When there is generally accepted, well-established evidence that a chemical is known to be a reproductive toxicant it should be labeled with the following statement of hazard or equivalent:

POSSIBLE REPRODUCTIVE HAZARD
OVEREXPOSURE MAY CAUSE (FEMALE, MALE)
REPRODUCTIVE DISORDER(S)[1]

### B.4 Labeling Other Serious Chronic Effects

When there is generally accepted, well-established evidence that a chemical causes a serious chronic effect other than carcinogenicity or reproductive toxicity the chemical should be labeled with one of the following statements of hazard or equivalent:

OVEREXPOSURE MAY CAUSE (Specify the organ[s])
DAMAGE[2]
OVEREXPOSURE MAY CAUSE (Specify the
disease[s])[3]

### B.5 Labeling Mixtures

If a mixture contains more than one chemical capable of causing a serious chronic effect, the potential effects of each should be combined in one statement of hazard.

### B.6 Signal Words Applicable to Serious Chronic Effects

When a combination of acute and serious chronic effects exists for a chemical, the signal word required for the acute hazards should be used.

### B.7 Precautionary Measures Applicable to Serious Chronic Effects

Statements of hazard for serious chronic effects necessarily require appropriate precautionary measures for avoidance of exposure. The precautionary measures shown in Table 1 for the most severe hazards of contact, inhalation and skin absorption should be used for protection against serious chronic effects. Additional useful precautionary measures may be found in 4.8.

### B.8 Placement of Statements of Hazard for Serious Chronic Effects

Statements of hazard for serious chronic effects should be placed after (below) any statements of hazard for which the signal word DANGER is required by Section 4.1 and before (above) any statement of hazard for which the signal words WARNING or CAUTION are required by Section 4.1.

Statements of hazard for carcinogenicity or reproductive toxicity should be placed before (above) any statements of hazard pertaining to other serious chronic effects.

HAZARDS:

Known to Cause Cancer in Humans
Irritant, Eye and Skin

```
CANCER HAZARD
OVEREXPOSURE MAY CREATE CANCER RISK
WARNING: CAUSES IRRITATION

Do not get in eyes, on skin, on clothing.
Do not breathe (dust, vapor, mist, gas.)*
Keep container closed.
Use only with adequate ventilation.
Wash thoroughly after handling.

FIRST AID: In case of contact, immediately flush eyes
with plenty of water for at least 15 minutes. Call a
physician. Flush skin with water. (Wash clothing
before reuse.)**
```

*Select applicable word or words.
**Use phrase when appropriate.

HAZARDS:

Corrosive, Eye and Skin
Known to Cause Cancer in Laboratory
Animals but not Known to Cause
Cancer in Humans

```
DANGER! CAUSES BURNS
POSSIBLE CANCER HAZARD BASED ON TESTS
WITH LABORATORY ANIMALS
OVEREXPOSURE MAY CREATE CANCER RISK

Do not get in eyes, on skin, on clothing.
Do not breathe (dust, vapor, mist, gas.)*
Keep container closed.
Use with adequate ventilation.
Wash thoroughly after handling.

FIRST AID: In case of contact, immediately flush eyes
or skin with plenty of water for at least 15 minutes
while removing contaminated clothing and shoes. Call
a physician. Wash clothing before reuse. (Destroy
contaminated shoes.)** (Thoroughly clean shoes before
reuse.)**
```

*Select applicable word or words.
**Use phrase when appropriate.

HAZARD:

Known to Cause Cancer in Humans

```
 _____
|                                                                |
|               CANCER HAZARD                                    |
|               OVEREXPOSURE MAY CREATE CANCER RISK              |
|                                                                |
|   Do not get in eyes, on skin, on clothing.                    |
|   Do not breathe (dust, vapor, mist, gas.)*                    |
|   Keep container closed.                                       |
|   Use only with adequate ventilation.                          |
|   Wash thoroughly after handling.                              |
|_____|
```

*Select applicable word or words.

HAZARD:

Known to Cause Male Sterility in Humans

```
 _____
|                                                                |
|               REPRODUCTIVE HAZARD                              |
|               OVEREXPOSURE MAY CAUSE MALE REPRODUCTIVE         |
|               DISORDER                                         |
|                                                                |
|   Do not get in eyes, on skin, on clothing.                    |
|   Do not breathe (dust, vapor, mist, gas.)*                    |
|   Keep container closed.                                       |
|   Use only with adequate ventilation.                          |
|   Wash thoroughly after handling.                              |
|_____|
```

*Select applicable word or words.

## Footnotes to Appendix B

1. Add the phrase "BASED ON TEST WITH LABORATORY ANIMALS" if only animal data are available to substantiate the chronic effect.
2. Ibid.
3. Ibid.

## CONSTRUCTION AND JUDGMENT

A cursory reading of ANSI Z129.1 may suggest that the preparation of a label for a hazardous chemical is a straight-forward procedure. This is not the case; labels are constructed, not prepared in a routine manner by selecting a set of words and phrases from a menu-like document such as ANSI Z129.1. A better analogy would involve a formulation chemist who is developing a new product. This typically requires balancing a number of different factors. The chemist must be concerned with stability, solvents, compatability, consumer acceptability, cost, effectiveness, shelf life, odor and so on. *Judgment* is required.

If the foreseeable use of a corrosive chemical may involve the generation of a mist, is "Avoid breathing mist" an appropriate precautionary measure, or would a stronger direction be better? Should a respirator be recommended? If so, should a specific respirator be identified? Perhaps engineering controls should be recommended. When such specific recommendations are made, have you accepted an unreasonable level of liability, or have you diminished your liability? *Judgment* is required. If it is reasonably foreseeable that a nonhazardous chemical can be put to a surreptitious use, with the result that a different and hazardous chemical is produced, should that consequence be addressed on the label, or would such mention perhaps encourage such undesirable use?

Should a precautionary label address a foreseeable hazard that arises when a chemical is used in a plausible but unintended way—perhaps as one ingredient in a three component mixture that could result in a hazardous product: For example, carbon, intended and labeled for use in an air purifying system *can* be mixed with potassium nitrate and sulfur to form "Black Powder", a common explosive. Should a label address a foreseeable consequence that is minimally harmful but is extremely unpleasant? To what extent should a label-constructor depend upon specialized knowledge possessed by an intended user of a chemical? Clearly, *judgment* is required.

Those new to label construction can benefit from consultation with an experienced general labeler. There are a number of professional and technical consultants who specialize in labeling and hazard communication systems. Many are members of the ACCL. These consultants are primarily chemists, chemical engineers, toxicologists and industrial health physicians. Most have multidisciplinary backgrounds, often combining chemistry and toxicology. There are also a number of lawyers and law firms who have developed special expertise in label preparation and review as part of their practice.

# 17

## Other Recognized Labeling Standards

### Charles J. O'Connor
*Greens Farms, CT*

Five other special purpose labeling systems should be mentioned in a handbook on the subject of precautionary labeling. These are the National Fire Protection Association (NFPA) 704 system, the National Institute for Occupational Safety & Health (NIOSH) system, the American Society for Testing and Materials (ASTM) proposal, the NPCA Paint Label Guide, the NPCA Hazardous Materials Information System (HMIS) and ANSI adjunct systems by the J. T. Baker Chemical Company and the Fisher Scientific Company.

### NATIONAL FIRE PROTECTION ASSOCIATION—IDENTIFICATION SYSTEM NFPA 704 FIRE HAZARDS OF MATERIALS—1980

The 704 system is designed to convey information on Health, Fire and Reactivity effects to firemen fighting a fire that involves chemicals.

This standard was first tentatively adopted in 1960, officially adopted by the National Fire Protection Association in 1961, and revised in 1964, 1966, 1975 and 1980. It grew out of development work begun in 1952 by the NFPA sectional committee on Classification, Labeling, and Properties of Flammable Liquids. As originally conceived, this standard is to safeguard the lives of those who are concerned with fires and fire control and prevention, in industrial plants or storage locations. It is particularly useful where the fire hazards of materials may not be readily apparent.

The scope of NFPA 704 applies to manufacturing, storage, and use of hazardous materials. It is concerned primarily with short term exposure for the areas of health, fire, reactivity, and related hazards under fire or similar emergency conditions. It is not recommended for small containers used in laboratories, for transportation purposes or for use by the general public. It *does* apply to industrial and institutional facilities.

The utility of the 704 system in preventing harm to fireman cannot be overemphasized; it is an excellent method for conveying summary information rapidly to those with a need to know. Unfortunately, this system has at times, been misused. Persons with limited or no fire fighting experience have assigned their own subjective ratings to chemicals and to mixtures of chemicals in a nonstandardized manner. The assignment of these numbers are a critical part of the system and must be done by experienced professionals, preferably in concert with NFPA. The NFPA 704 placard is designed to be used to identify areas or rooms that contain hazardous chemicals or, on containers large enough to hold significant amounts of hazardous material and thus with ample space available for the application of colored squares or numbers that will be visible from some distance away. The counsel of an experienced fire fighter should always be sought before using the NFPA 704 system (contact NFPA for the names of such professionals in your area). The use of 704 placards for identifying rooms or areas must be accompanied by a periodic review to assure that no important changes have taken place in the kinds of chemicals stored or used in such rooms or areas.

The system is simple and easy to understand. It identifies the *kind* of hazard by color and the *level* of hazard by number (0-4). A "0" indicates little or no unusual hazard in a fire situation; a "4" indicates the highest degree of hazard; intermediate numbers signify corresponding intermediate degrees of hazard. This simple categorization is enhanced by the arrangement of the symbols (see NFPA Chapter 6 reproduced below) to form a diamond, or square on point. The "right" is always yellow and only refers to "reactivity"; the "left" is always blue and only refers to "health"; the "top" is always red and only refers to "flammability"; the "bottom" square is reserved for identifying those materials that may cause special problems or require special fire fighting techniques. Materials that in contact with water demonstrate unusual reactivity are identified with the letter W with a horizontal line through the center. Phosphorus pentasulfide, methyl trichlorosilane, lithium, sodium hydride, and potassium peroxide are examples of materials that require such a W, placed in the bottom square.

Materials which are oxidizers are identified by the letters OX. This hazard symbol would be placed in the bottom square for such materials as potassium permanganate, potassium persulfate, dibenzoyl peroxide, and liquid oxygen.

Materials that possess radioactivity hazards are identified with the standard radioactivity symbol, placed in the bottom square.

## Kind of Hazard

**Health:** NFPA 704 recognizes essentially three basic hazard classes, health, reactivity, and fire. In general, the health hazard class refers to the capacity of a material to cause injury from contact with or absorption into the body. In this classification scheme, only the health hazard arising out of an inherent property of the material is considered. Injury from the heat of a fire or the force of an explosion or violent reaction is not included in health evaluation.

The NFPA 704 system recognizes two sources of health hazard: one is the toxic products of combustion or the toxic products of decomposition under fire conditions; the second source is from the inherent properties of the material. The standard principally contemplates a single exposure which may vary from a few seconds to an hour. The level of physical effort which fire fighting or other emergency conditions require often intensifies the effects of exposure. Exposure is defined as inhalation, skin contact, oral ingestion and skin absorption.

While the emphasis in the 1980 revision of this Standard in defining health hazard is principally upon acute effects, the fire-fighting community *is* seriously concerned with chronic health effects. Epidemiologic studies demonstrate that firemen pay a significant health toll in protecting society from the devastation of uncontrolled fire. Thus the residual injury that may result from exposure to toxic chemicals or their decomposition and/or combustion products should be a consideration when evaluating health hazard.

**Reactivity:** Reactivity refers to the susceptibility of materials to release energy. This class includes materials which are capable of rapid release of energy by themselves, either by polymerization or self-reaction. It also includes materials which will react violently with water or other extinguishing agents, either violently erupting or exploding. This class is meant to include chemicals which will react violently when in contact with common materials; although this is usually limited to reactions with water, individual assessments should be made.

Materials which are readily capable of detonation or of explosive decomposition by mechanical or thermal shock at normal temperature or pressure or at elevated temperature should be classed as reactive.

Many chemicals that are stable at room temperature, will, if confined, undergo either detonation or explosion under fire conditions. By contrast, stable materials have the capacity to resist changes in their chemical composition despite exposure to air, water, and heat under fire conditions.

**Flammability:** Flammability hazard is concerned with the ease with which materials can be ignited and continue to burn. A major consideration is the rate of burning. Clouds of fine combustible dusts, for example, burn so rapidly that they have the force and effect of an explosion. There are various criteria which have been developed to identify flammable materials. Flash points, fire points, and autoignition temperature are 3 common measures of flammability. Flame propagation and the explosive or flammable range are measures commonly used for gases, vapors and air-suspended fine combustible dusts. Pyroforic materials are included in this class.

While no single measure is sufficient for all purposes, the flash point of a material is a good general measure of flammability. It is often the first estimator of flammability developed for a chemical, and the most commonly available measure found in literature.

### Degree of Hazard

**Flammability:** The 704 System ranks level or degree of hazard by assigning a number, 0, 1, 2, 3 or 4. Zero represents materials that offer little or no hazard beyond that of ordinary combustible material. A zero level flammability hazard includes materials which will not burn in air when exposed to a temperature of 1500°F for a period of five minutes. Chloroform, carbon tetrachloride, zirconium tetrachloride, sulfuric acid and sodium hydroxide are among the chemicals that are assigned a zero rating for flammability.

Sodium hydrosulfite, paraformaldehyde, maleic anhydride and arsenic trisulfide are all assigned a flammability rating of one (1). In general, this means materials must be preheated before ignition occurs. It includes liquids, solids, and semisolids which have a flash point above 200°F.

Degree two (2) includes liquids with a flash point above 100°F but below 200°F, and solids and semisolids which easily release flammable vapors. Proprionic anhydride, ortho toluidine, methacrylic acid and dimethyl sulfate are examples of this class.

Diethyl zinc, diethylamine, hydrazine, methyl parathion in xylene and sodium hydride are in degree three (3). This degree includes liquids with a flash point below 73°F and a boiling point at or above 100°F, and liquids with a flash point at or above 73°F but below 100°F. It includes: coarse dusts which will burn rapidly, but do not form explosive mixtures with air; as well as materials which burn with extreme rapidity often by reason of self-contained oxygen.

Degree four (4) includes liquids with a flash point below 73°F and a boiling point below 100°F. It also incudes materials which rapidly vaporize under normal conditions and which are combustible. Various gases and fine combustible dusts are in this class. Acetylene, acetaldehyde, calcium carbide, cyanogen, ethylamine, and nickel catalyst are among the chemicals in degree four (4).

**Health:** The NFPA 704 System includes five (5) degrees of health, which are reproduced below. Degree zero (0) includes materials such as red phosphorus, diethyl zinc, magnesium and dilauroyl peroxide.

Degree one (1) representative chemicals are acetylene, calcium carbide, dibenzoyl peroxide, hydroxylamine and ethene.

Degree two (2) has been assigned to such chemicals as nickel catalyst, acetaldehyde, proprionic anhydride, diethylamine, and paraformaldehyde.

Degree three (3) includes among others, acrolein, sodium hydride, ethylamine, carbon tetrachloride and hydrazine.

Health hazard degree four (4), the most dangerous, is characterized by such chemicals as cyanogen, dimethyl sulfate, methyl parathion in xylene, beryllium and hydrogen flouride.

**Reactivity:** Degree zero (0) consists of materials which are stable under fire conditions and do not react with water. Arsenic trisulfide, beryllium, carbon tetrachloride and hydrogen sulfide are in degree zero (0).

Degree one (1) includes aluminum, ethyl ether, dimethyl sulfate, sodium hydroxide and stannic chloride. The materials in this rank often become unstable at elevated temperatures or pressures and may react with water, but not violently.

Degree two (2) materials are normally unstable and may readily undergo violent chemical change, but do not detonate. It includes chemicals which react violently with water. Cyanogen, ethene, sodium hydride, calcium carbide and acetaldehyde are chemicals classed as degree two (2).

Degree three (3) chemicals include acetylene, diethyl zinc, bromine trifluoride and chloropicrin. This rank includes materials which will detonate but require a strong initiating source, and those chemicals which react explosively with water.

Finally, reactivity degree four (4) chemicals, the most dangerous group, are materials which are easily capable of detonation at normal temperatures, or may undergo explosive decomposition or reaction under normal conditions. Dibenzoyl peroxide, picric acid, tertiary butyl hydroperoxide, ethyl nitrite, and trinitrobenzene are representative of degree four (4) chemicals.

### Spatial and Color Arrangement

Chapter Six and Appendix A of the 1980 NFPA 704 Standard are reproduced below. Appendix A provides a good recap of the general system. One should obtain a copy of the complete 704 System from NFPA before attempting to implement this label system. Contact Mr. Martin F. Henry, Secretary, Committee on Fire Hazards of Materials, NFPA, Battery March Park, Quincy, MA 02269, for further information on this system.

### NIOSH—AN IDENTIFICATION SYSTEM FOR OCCUPATIONALLY HAZARDOUS MATERIAL

Richard J. Lewis, Sr., Division of Technical Service, The Industrial Hygiene Branch of The National Institute for Occupational Safety & Health, had primary responsibility for the development of this recommended Standard. Although Lewis led the project, input was received from General Motors Corporation, Olin Corporation and Shell Oil Company, the Michigan Department of Health, The University of Missouri, and of course various offices and divisions of NIOSH. The work draws heavy upon NFPA's 704 system and CMA's LAPI Guide to Precautionary Labeling. The recommended colors follow NFPA, as do the three major hazard classes: health, reactivity, and flammability.

A major departure occurs in the NIOSH definition of "health class". NIOSH is concerned with both acute and low level long term exposure, perhaps for a lifetime, that a worker may experience, primarily occurring under normal work conditions. NIOSH is not directly concerned with the decomposition or combustion products (as NFPA is) of chemicals unless they constitute the normal exposure of a worker. There is greater concern with chronic effects.

Thus the NIOSH criteria for health include substances which are carcinogens, teratogens, mutagens; or which may cause significant irreversible health effects. The criteria also capture acute health effects, based primarily upon $LD_{50}$ and $LC_{50}$ acute toxicity values. These values range from an oral $LD_{50}$ of less than 1 mg/kg to an $LD_{50}$ of 5000 mg/kg; $LC_{50}$ values between less than 1 ppm and 10,000 ppm; and dermal $LD_{50}$ values between less than 1 mg/kg to 2,800 mg/kg.

The health, reactivity and flammability hazard classes are divided into 5 levels and assigned a number from 0 to 4, as in the NFPA 704 system. This system, thus, utilizes *kind* of hazard and *level* of hazard NFPA concepts.

However, NIOSH has provided, based primarily upon the CMA label guide, recommended label statements for health, fire, reactivity, disposal, and first aid. The NIOSH system proposes a label that includes a numeric NFPA-type

## Chapter 6    Identification of Materials by Hazard Signal System

**6-1**    One of the systems delineated in the following illustrations shall be used for the implementation of this standard.

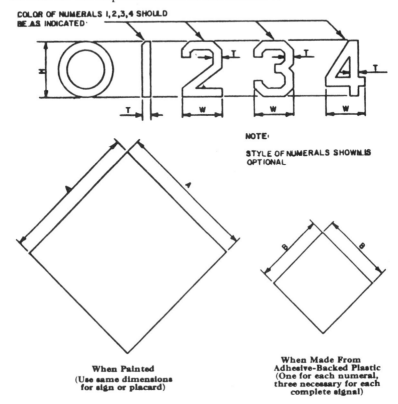

**Minimum Dimensions of White Background for Signals**
(White Background is Optional)

| Size of Signals | | | | | |
|---|---|---|---|---|---|
| H | W | T | A | B |
| 1 | 0.7 | 5/32 | 2½ | 1¼ |
| 2 | 1.4 | 5/16 | 5 | 2½ |
| 3 | 2.1 | 15/32 | 7½ | 3¾ |
| 4 | 2.8 | 5/8 | 10 | 5 |
| 6 | 4.2 | 15/16 | 15 | 7½ |

IDENTIFICATION OF
MATERIALS BY
HAZARD SIGNAL
DIMENSIONS

All Dimensions Given in Inches

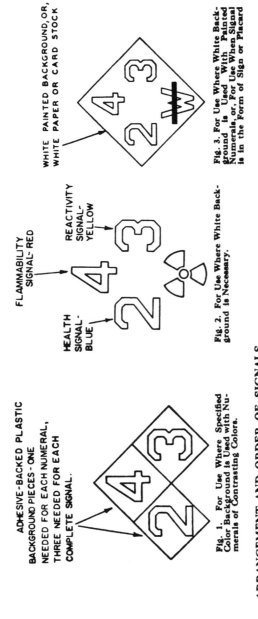

ADHESIVE-BACKED PLASTIC BACKGROUND PIECES - ONE NEEDED FOR EACH NUMERAL, THREE NEEDED FOR EACH COMPLETE SIGNAL.

Fig. 1. For Use Where Specified Color Background is Used with Numerals of Contrasting Colors.

HEALTH SIGNAL-BLUE

FLAMMABILITY SIGNAL-RED

REACTIVITY SIGNAL-YELLOW

Fig. 2. For Use Where White Background is Necessary.

WHITE PAINTED BACKGROUND, OR, WHITE PAPER OR CARD STOCK

Fig. 3. For Use Where White Background is Used With Painted Numerals, or, For Use When Signal is in the Form of Sign or Placard

IDENTIFICATION OF MATERIALS BY HAZARD SIGNAL ARRANGEMENT

ARRANGEMENT AND ORDER OF SIGNALS — OPTIONAL FORM OF APPLICATION

NOTE:
This shows the correct spatial arrangement and order of signals used for identification of materials by hazard

| Distance at Which Signals Must be Legible | Minimum Size of Signals Required |
|---|---|
| 50 feet | 1″ |
| 75 feet | 2″ |
| 100 feet | 3″ |
| 200 feet | 4″ |
| 300 feet | 6″ |

**Fig. 4. Storage Tank**

## Appendix A

*This Appendix is not a part of the requirements of this NFPA document . . . but is included for information purposes only.*

This is a system for the identification of hazards to life and health of people in the prevention and control of fires and explosions in the manufacture and storage of materials.

The bases for identification are the physical properties and characteristics of materials that are known or can be determined by standard methods. Technical terms, expressions, trade names, etc., are purposely avoided as this system is concerned only with the identification of the involved hazard from a standpoint of safety.

The explanatory material in this Appendix is to assist users of this guide, particularly the person who assigns the degree of hazard in each category.

| Identification of Health Hazard Color Code: BLUE | | Identification of Flammability Color Code: RED | | Identification of Reactivity (Stability) Color Code: YELLOW | |
|---|---|---|---|---|---|
| Type of Possible Injury | | Susceptibility of Materials to Burning | | Susceptibility to Release of Energy | |
| Signal | | Signal | | Signal | |
| 4 | Materials which on very short exposure could cause death or major residual injury, even though prompt medical treatment were given. | 4 | Materials which will rapidly or completely vaporize at atmospheric pressure and normal ambient temperature, or which are readily dispersed in air and which will burn readily. | 4 | Materials which in themselves are readily capable of detonation or of explosive decomposition or reaction at normal temperatures and pressures. |
| 3 | Materials which on short exposure could cause serious temporary or residual injury even though prompt medical treatment were given. | 3 | Liquids and solids that can be ignited under almost all ambient temperature conditions. | 3 | Materials which in themselves are capable of detonation or explosive reaction but require a strong initiating source or which must be heated under confinement before initiation or which react explosively with water. |
| 2 | Materials which on intense or continued exposure could cause temporary incapacitation or possible residual injury unless prompt medical treatment is given. | 2 | Materials that must be moderately heated or exposed to relatively high ambient temperatures before ignition can occur. | 2 | Materials which in themselves are normally unstable and readily undergo violent chemical change but do not detonate. Also materials which may react violently with water or which may form potentially explosive mixtures with water. |
| 1 | Materials which on exposure would cause irritation but only minor residual injury even if no treatment is given. | 1 | Materials that must be preheated before ignition can occur. | 1 | Materials which in themselves are normally stable, but which can become unstable at elevated temperatures and pressures or which may react with water with some release of energy but not violently. |
| 0 | Materials which on exposure under fire conditions would offer no hazard beyond that of ordinary combustible material. | 0 | Materials that will not burn. | 0 | Materials which in themselves are normally stable, even under fire exposure conditions, and which are not reactive with water. |

ranking and the typical precautionary statement found on most CMA-LAPI labels.

This standard requires labels for all containers, placards for areas and buildings and a MSDS available in the work place. It briefly refers to hazard training for employees but does not present a program to accomplish this portion of the standard. In that sense this standard falls short of being a complete communication system. In spite of this, it remains a good workable compromise between a hazard alert system such as NFPA 704 and the LAPI-based ANSI system which relies primarily upon worker-oriented label statements.

Representative tables are reproduced below. The NIOSH Standard can be obtained from The National Technical Information Service (NTIS) by writing to:

National Technical Information Service
U.S. Department of Commerce
5285 Port Royal Rd.
Springfield, VA 22161.

RELATIVE TOXICITY RATING FOR HAZARDOUS MATERIALS
(Human Exposure by Any Route)

| Rating | Key Words | Acute (Single Exposure, immediate or delayed effects) | Chronic (Repeated Exposure) |
|---|---|---|---|
| 4 | EXTREME HEALTH HAZARD | Death | Death* |
| 3 | HIGH HEALTH HAZARD | Major temporary or permanent injury May threaten life | Major permanent injury (Includes mutagens and teratogens) |
| 2 | MODERATE HEALTH HAZARD | Minor temporary or permanent injury** (Includes nonlife threatening substances which sensitize the majority of exposed workers) | Minor temporary or permanent injury (Includes skin carcinogens) |
| 1 | SLIGHT HEALTH HAZARD | Minor injury readily reversible** | Minor injury readily reversible |
| 0 | NO SIGNIFICANT HEALTH HAZARD | Materials which produce toxic effects only under the most unusual conditions or by overwhelming dosage. | |

*Includes substances which bear a significant relationship to the development of cancer in man, but excluding the common varieties of skin cancer.
**Allergens are rated according to their sensitizing potential rather than the severity of an allergic reaction upon reexposure to a substance by a sensitized worker.

RELATIVE ACUTE TOXICITY CRITERIA

| Rating | Key Words | LD50 Single Oral Dose: Rats mg/kg | LC50 Inhalation Vapor Exposure: Rats ppm | LD50-Skin Rabbits: mg/kg |
|---|---|---|---|---|
| | | less than or equal to | less than or equal to | less than or equal to |
| 4 | EXTREMELY HAZARDOUS | 1 | 10 | 5 |
| 3 | HIGHLY HAZARDOUS | 50 | 100 | 43 |
| 2 | MODERATELY HAZARDOUS | 500 | 1,000 | 340 |
| 1 | SLIGHTLY HAZARDOUS | 5,000 | 10,000 | 2,800 |
| 0 | NO SIGNIFICANT HAZARD | 5,000 or greater | 10,000 or greater | 2,800 or greater |

RELATIVE FLAMMABILITY CRITERIA
FOR LIQUIDS HAVING A FLASH POINT

| Numerical Rating | Key Terms | Flash Point F   (C) |
|---|---|---|
| 4 | EXTREMELY FLAMMABLE | below 73 (22.8) |
| 3 | HIGHLY FLAMMABLE | at or above 73 (22.8) but below 100 (37.8) |
| 2 | MODERATELY COMBUSTIBLE | at or above 100 (37.8) but below 200 (93.4) |
| 1 | SLIGHTLY COMBUSTIBLE | at or above 200 (93.4) |
| 0 | NONCOMBUSTIBLE | greater than 1,500 (815) |

EXAMPLE LABEL

| 4 | Extreme Health Hazard |
|---|---|
| 4 | Extremely Flammable |
| 2 | Moderately Reactive |

Fatal if swallowed, inhaled, or absorbed through the skin.

Causes severe eye burns.

Protect from all sources of ignition.

Subject to violent polymerization.

Do not breathe vapor or get in eyes, on skin, on clothing.

When possibility of contact exists:
Wear full neoprene suit, rubber boots, rubber gloves, and self-contained breathing apparatus.

Avoid contact with acid, organic compounds, or water.

FIRST AID                    CALL A PHYSICIAN AS SOON AS POSSIBLE

Immediately upon exposure flush skin and eyes with water for 15 minutes while removing contaminated clothing and shoes. Wash clothing before reuse. Discard contaminated shoes.

Refer to Data Sheet on file.

## ASTM Z535.2 PROPOSAL

The American Society for Testing and Materials' proposal is currently under discussion within the Z535.2 committee. It is presently designed to be used as a safety sign, not only in an industrial environment, but perhaps for general use. The system consists of three elements. The first, *words*-"Danger", "Warning", "Caution", the second, *shape of the sign*-an oval, a truncated diamond, and a rectangle; and the *third,* colors-red, orange, yellow and black. These are not independent elements, however.

"Danger" is always used with the oval, "Warning" is always used with the truncated diamond and "Caution" is always used with the rectangle. Red and black are the colors always associated with "Danger" and oval; orange and black are used only with "Warning", on a truncated diamond background; "Caution" is always associated with yellow and black on a rectangle.

This proposal draws on the DOT and NFPA concepts that utilize signs or label shapes with fixed colors for various hazard classes. The ASTM proposal uses the three LAPI signal words, "Danger", "Warning", and "Caution" to represent level of hazard, but in a change from other systems *also* uses color to indicate level of hazard. This system, thus, reinforces level of hazard by fixed shape, word and color.

Several advantages are claimed for this general warning sign system; it is informative to foreign workers, the illiterate (including the very young), the color-blind, and the visually impaired.

There is little doubt that with some training, this would be an easy system to learn, retain and use. This is also true, however, of the NFPA system, the ANSI system and the NIOSH system. In distinction, ASTM is designed to be a *hazard alert* sign system, especially visible and informative at a distance. It is supposed to put a reader on notice of the existence of a hazard and hopefully cause him to adopt a cautious attitude and behavior.

## NATIONAL PAINT AND COATINGS ASSOCIATION

### The Paint Industry Labeling Guide

First published by the National Paint and Coatings Association in 1972, the current 1979 Second Edition with "Supplement No. One", 1981, of the *Paint Industry Labeling Guide* applies and extends the principles of ANSI 129.1-1976 to the products of the paint industry as these are used by both industrial customers and by home consumers.

In 1953, the Association developed suggestions for labeling based principally on various State laws pertaining to the labeling of hazardous substances. This effort to promote uniformity in hazardous labeling language was in part responsible for the enactment of the Federal Hazardous Substances (Labeling) Act. In 1962 the Association published a pamphlet, "Precautionary Labels", which was supplemented in 1964 and again in 1965.

In its present form, the *Paint Industry Labeling Guide,* more than 200 pages in length, provides labeling and related information to the paint and allied products industry. It consists of nine sections. The first two, on household-consumer products and industrial products, include discussions on the basic requirements of relevant laws and regulations for precautionary labeling of representative types of household and industrial paints and allied products. Several examples of specimen labels are provided. These are clearly derived from ANSI Z129.1-1976 principles, as is shown by Figure 17.1.

The remaining seven sections treat labeling-related matters not addressed directly in the ANSI standard. These are: products classed as pesticides; state and local laws and regulations; miscellaneous matters-the Poison Prevention Packaging Act, the Lead Based Paint Poisoning Prevention Act, and the Consumer Product Safety Act; Occupational Safety and Health; transportation (currently superceded by a separate NPCA publication, "A Guide to the Shipment of Paints and Coatings"); weights and measures; and consumer information labeling for coatings. A tenth section is reserved for a treatment of pro-

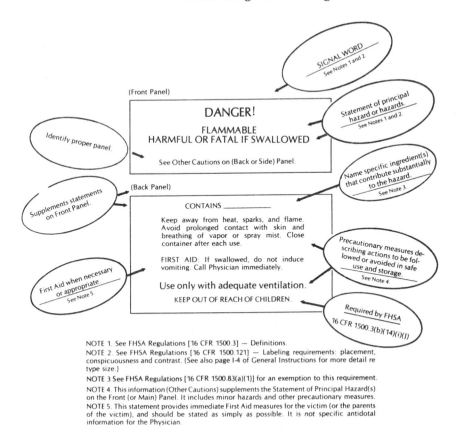

NOTE 1. See FHSA Regulations [16 CFR 1500.3] — Definitions.
NOTE 2. See FHSA Regulations [16 CFR 1500.121] — Labeling requirements: placement, conspicuousness and contrast. (See also page I-4 of General Instructions for more detail re type size.)
NOTE 3 See FHSA Regulations [16 CFR 1500.83(a)(1)] for an exemption to this requirement.
NOTE 4. This information (Other Cautions) supplements the Statement of Principal Hazard(s) on the Front (or Main) Panel. It includes minor hazards and other precautionary measures.
NOTE 5. This statement provides immediate First Aid measures for the victim (or the parents of the victim), and should be stated as simply as possible. It is *not* specific antidotal information for the Physician.

Figure 17.1: Sample label format, from NPCA Paint Industry Labeling Guide, reproduced with permission.

spective rulemaking by the Office of Toxic Substances, EPA, related to Section 6(a) of the Toxic Substances Control Act—as its provisions may apply to the labeling of paints and allied products.

### Hazardous Materials Identification System

On April 30, 1981, NPCA sent a comprehensive implementation starter kit to each of its members. This manual is the most complete hazard communication system available today. It is identified as NPCA in-plant Hazardous Materials Identification System (HMIS).

While this system is directed to the paint industry, it is applicable to all industries that use hazardous materials. It is an excellent example of the benefits to be attained by membership in this small but effective trade association. HMIS consists of a label system with symbols, for containers; placards for tanks; tags for piping; a MSDS program; and a section devoted to explicit employee training. The employee training program includes an audio-visual script for instructional presentation. NPCA has also prepared, in semi-comic book form, an employee handout entitled *Forewarned is Forearmed!* This easy to read attractive pamphlet is further supported by a varnished wallet card (2-1/2" × 3-1/2") which explains the HMI System, and includes reproductions of the symbols used to indicate personal protection. NPCA completes the training program by providing wall posters (22" × 28") that reinforce these other elements.

The label system employed by the HMI System is a derivative of the NFPA and NIOSH concepts; ratings, from one (1) to five (5) are assigned for health, flammability, and reactivity. The NFPA-NIOSH colors are also used: blue for health, red for flammability, and yellow for reactivity.

The definitions which form the basis for assigning numerical ratings for reactivity and flammability are similar to the NIOSH and NFPA concepts. A significant difference occurs, however, in the health definitions. NPCA uses the Health Hazard Ratings System developed by Henry F. Smyth in cooperation with the Medical Department, Union Carbide Corporation. This system enlarges previous concepts to specifically include chronic exposure and chronic effects. So called supplemental effects such as photosensitization, acne-like eruptions, eye burn without pain, and metal fume fever among others are also identified.

Summary portions of the HMI System are reproduced below.

Contact Patrick J. Hurd at NPCA, Inc., 1500 Rhode Island Avenue, NW, Washington, DC 20005, or telephone (202) 462-6272, for further information on either HMIS or the Paint Industry Labeling Guide.

# PART I: Introduction

## A. Why a Hazardous Materials Identification System

In recent years we have seen a heightened interest in occupational health and safety in the United States. One result has been a need for employers to inform their workers of the hazards associated with the performance of the employee's job. A number of states, as well as the federal government, require employers to inform their workers about on the job hazards to their health and safety.

Many of these regulations, primarily those written for a specific material, impose requirements for labelling and placarding. Such requirements may be appropriate for a workplace using only a few substances, or for extremely hazardous materials. However, the blanket application of such requirements to the paint manufacturing industry, which can use upward of 1500 separate and distinct raw materials, is often inappropriate. Therefore, the problem exists for a paint manufacturer to identify the hazard associated with the use of this diverse list of raw materials for his employee.

The hazards may range from those with severe immediate impact, such as acute overexposure to solvent fumes, or drying of the skin from solvent exposure; to very subtle health effects such as sensitizations from isocyanates. Long term, or chronic, effects such as cancer from selected solvents or pigments must also be included.

In undertaking such an identification, the paint manufacturer must achieve the seemingly opposite chores of accurately and concisely communicating hazard information without so inundating the employee that the hazard warning is lost.

In order to assist NPCA member companies in communicating such health and safety information to their workers. NPCA through the Occupational Health and Safety Task Force (and in conjunction with the Canadian Paint and Coatings Association—CPCA) has developed a Hazardous Materials Identification System (HMIS) for use by the paint manufacturers. The system allows the paint manufacturing industry to quickly and concisely inform their employees about the variety of hazards presented by the large number of raw materials in the plant.

It is important to note that this system can stand on its own as a means of communication, but is developed to be compatible with other systems calling for specific instructions or precautionary language. Therefore, it can be used in conjunction with existing hazard identification programs, and

where identification of extremely hazardous materials are mandated by law.

The HMIS as developed by NPCA and CPCA is based on systems developed by Inmont Canada Limited and PPG Industries Inc. Personnel from these companies were of great assistance in the development of this system, and we gratefully acknowledge that assistance.

## B. The Purpose of the Manual

This manual is a comprehensive explanation of the HMIS for paint manufacturers and is designed to assist them in the implementation of this system in their paint manufacturing facilities. In addition, it includes recommendations for worker education and training programs which would ensure that they are thoroughly educated in the use of the system.

The completion of the identification elements of the system and their deployment throughout the plant and on raw material packages is the responsibility of the paint manufacturer. The raw material suppliers to the paint industry will develop information on the hazards of the specific raw material and supply it to the paint manufacturer so that he can complete the HMIS element (label, placard, etc.)

NPCA plans to negotiate with raw material suppliers to have the labels pre-printed or pre-affixed to raw material packages. We will keep NPCA members apprised of this activity.

## C. Concept of the System

The HMIS has been designed so that employers and their employees can quickly and easily identify the hazards associated with the use of a material in the paint manufacturing process. These three hazards will be identified by this system:

**Health**

**Flammability**

**Reactivity**

Hazards will be assessed by a five-tiered scale which ranges from 0 to 4. A "0" hazard denotes a minimal hazard, while a "4" hazard indicates a severe hazard. In addition, the system tells the employee the proper personal protective equipment to be worn while using a material under specific conditions. The degree of hazard for each of the three types of hazard is explained in table 1.

has the potential to present a high degree of hazard. Further, the health hazard rating can utilize an asterisk to differentiate between a chronic (resulting from long term exposure) hazard or an acute hazard (resulting from a one time or short period of exposure.) This option is explained under Appendix A.

## D. Use of Personal Protective Equipment

NPCA urges the use of engineering controls and administrative procedures as the primary means of limiting worker exposure to hazardous materials. Most member companies support this idea, but because of the unique processes employed by this industry, personal protective equipment must be considered in a paint manufacturing facility.

Unlike many other industrial operations, the batch nature of paint manufacturing necessitates a flexibility of worker and equipment. Such flexibility means that workers cannot always be adequately protected through the use of engineering and administrative controls. Therefore, some degree of personal protective equipment is needed.

For this reason, the Hazardous Materials Identification System also informs workers as to the proper protective equipment to be used.

## E. Danger vs Hazard

One of the most important aspects of the HMIS is that the system informs the worker about the hazardous properties associated with a raw material. While the raw materials (indeed all materials) possess **inherent dangerous properties,** among these toxicity, flammability and reactivity, the availability of that material to fulfill the danger potential determines its **hazard.** This dangerous property can become hazardous through the way it is used, the form in which it exists, or its proximity to the worker.

As an example, a drum of acetone possesses inherent flammability characteristics. These flammability characteristics are a danger wherever the acetone is present. If a person is standing directly beside a drum of acetone, that acetone is not only a **danger,** but a **hazard** to that person. However, if the drum of acetone is two miles from the person, it continues to be a flammability danger but it is no longer a flammability hazard to that person.

Another example would be a pigment which causes adverse health effects when it is inhaled into the lungs. The chemical or physical properties responsible for the adverse health effects are inherent dangerous properties. Therefore, a worker opening a bag of this dry powdered pigment is directly exposed to these dangerous properties because the pigment may be inhaled. Hence, the pigment presents a specific health hazard.

Now consider the same pigment having been completely wetted and incorporated into an alkyd semi-gloss enamel. A worker opening that can of paint, while as close to the pigment as the worker opening a bag of the pigment, is no longer exposed to the dangerous property of the pigment. Since in its present form, it is completely wetted in a paint, the pigment cannot be inhaled by the worker. Therefore, for this **dangerous** property, the pigment is no longer a **hazardous** material.

As will be explained later, the ratings will often be based on the inherent properties of the material.

# PART II: Implementing the System

## A. Compatibility with Other Health and and Safety Programs

The system, as described in the first portion of this manual, is the basic Hazardous Materials Identification System. As noted earlier, it is a flexible system which may stand on its own as a means of worker hazard identification and is also compatible with other existing plant education and training programs. Legislative or regulatory provisions that call for the special identification of extremely hazardous materials (such as carcinogens) can easily be incorporated into the HMIS, too.

However, as many NPCA members may wish to customize the system to better suit their own operations, Appendix A includes a list of options that can be used either singly or in groups. This will allow "custom building" to a particular company's needs.

## B. Description of the Elements of the System

As described earlier, the system consists of a number of visual "elements" (labels, placards, tags, etc.) for communicating with the worker. This visual communication system is based on a four-part, four-color square (or five-part) four-color square, if the company chooses to include either the raw material codes or the specific chemical's name on the identification elements. These squares will appear on or nearby the raw material package in order to communicate to the worker the hazards associated with that raw material. The specifics of these elements are as follows:

### 1. The Label
The most widely used element will be a label affixed to the raw material package. Presently NPCA has designed the label in two sizes. These are an 8" x 8" label and a 2" x 2" label. The large label is for use on raw material packages, such as 55 gallon drums, pigment bags, fiber drums and five gallon containers.

The smaller label is for use on one gallon containers, pint containers of extremely hazardous material and specialty items such as pre-batching cards, etc.

The dimensions and the colors for the labels are included in Appendix B. You will need this information to order labels from a printer.

These labels are supplied only with the written designation for the type of hazard and protective equipment. The completion of the numerical ratings and the protective equipment symbol is the responsibility of the paint manufacturer as described later in this manual.

### 2. The Placard
Placards are of limited use, but can be used on such equipment as tanks, or storage vessels. NPCA has therefore developed a 15" x 15" self-adhesive placard to be used in such situations.

### 3. Tags for Piping and Spigots
Often a manufacturer will find it necessary to affix an HMIS element to piping or spigots from which solvent or other materials, such as resins are drawn. Tags may be used to highlight special hazards in areas congested with pipes where labelling each pipe would be confusing.

In order to assist those who are labelling pipes, NPCA has developed a 1-1/2" by 1-1/2" tag that can be affixed to piping. This tag can be produced in either a single use cardboard format or a reusable plastic format. Due to its small size, abbreviations are used to describe the various hazards and personal protective equipment.

### 4. Other Sizes and Elements
Anyone wishing to design and use an identification element in a different size, but based on the specifications offered under Appendix B, is welcome to do so. However, the manufacturer should keep in mind that the system relies upon the visual identification of hazards and protective equipment. Therefore, the label and the information on it must be easily seen by the workers.

## C. Information to Go on the Identification Elements

### 1. The Ratings
Before you can begin implementation of the HMIS, you must obtain the ratings assigned to the various raw materials by their supplier. NPCA has developed and distributed to raw material suppliers in the United States a Rating Protocol through which they can make the appropriate ratings assignments.

Since a raw material supplier is the most knowledgable source regarding the materials that are marketed by his firm, he is the best person to undertake the job of rating them. However, you may find it necessary to have a purchasing agent request that a raw material supplier assign hazard ratings to the raw materials used by your company.

equipment is the protective equipment required under the following assumed conditions:

1. Direct worker contact with the material is possible.
2. The material is used routinely; equipment is not for emergency or misuse situations.
3. The material will be used in the absence of engineering controls or protective measures.

Based on these assumptions and identified hazardous properties of the raw material (such as those that appear in Section V of the MSDS), the supplier will assign the "maximum protective equipment" required. This maximum protective equipment recommendation will match the recommendation on the MSDS contained under the personal protective equipment section. The task then falls to the paint manufacturer to evaluate his suppliers recommendation in light of other protective measures within his plant. This includes such things as engineering controls and administrative control procedures. Then the paint manufacturer must assign the proper protective equipment for his workers.

## E. Means for Assigning the Personal Protective Equipment

Many paint manufacturers already require the use of certain personal protective equipment in various areas of their plants or under specific circumstances. Protective measures which are already in place, can serve as the basis for the personal protective measures required by the HMIS. However, if a manufacturer wishes to use the suppliers maximum personal protective equipment recommen-

dation, he should examine the specific use of the material within his plants.

For example:

Consider a raw material having a maximum protective equipment requirement of "F". "F" calls for gloves, safety glasses, synthetic apron, and a dust respirator.

However, the paint manufacturer, through measurement of airborne levels of the specific material, has determined that the material is controlled well below recommended or required airborne levels. Thus some regulated airborne levels are in compliance, or judged of a low hazard potential and there is no need for the use of a dust respirator. Therefore, the appropriate protective equipment would be gloves, safety glasses and synthetic apron. This corresponds with a personal protective designation of "C". A manufacturer would then put "C" on the element. This is known as the "derived personal protective equipment."

NPCA would like to remind paint manufacturers that the needs of different work sites within a plant will vary. Personal protection equipment assignments should be determined by the specific needs of a plant or work site rather than the establishment of one company wide standard.

NPCA strongly urges paint manufacturers to adopt a company wide personal protection equipment code. However, this code should be deployed on the basis of the specific needs of the individual plants. No two plants are exactly alike. The measures required to adequately protect your employees are certain to vary as a result of different processes or engineering controls in separate plants. As a minimum though, most paint operations require the wearing of safety glasses in manufacturing areas.

# PART III: Deployment of Elements

Now that the paint manufacturer has the hazard ratings for all of the raw materials and has assigned the personal protective equipment requirements for the specific plant and worksite, we will address the "nuts and bolts" of implementing the system.

## A. Overall Responsibility for the Hazardous Materials Identification System

The responsibility for the implementation of the Hazardous Materials Identification System should be assigned to one individual by the plant manager.

This will either be a plant engineer, an industrial hygienist, the plant manager or a supervisor.

## B. Inventory of the Ratings and Personal Protective Equipment

### 1. Maintenance of a Notebook

The simplest means for maintaining the codes is to keep a notebook listing all of the health and safety ratings and the personal protective equipment requirements. This could be the responsibility of the health and safety coordinator, the plant manager, the chief safety and health officer, or the department responsible for health and safety. In setting up such a notebook, we would recommend including the headings shown below.

1. Raw Material Code
2. Trade Name
3. Chemical Name
4. Supplier
5. Ratings (H, F, R)
6. Maximum Personal Protection
7. Derived Personal Protection

This list should be continually updated. A sample page appears in appendix D.

### 2. Integration into Existing Computer Raw Material Inventories.

For those companies using a computer to inventory raw materials, we recommend that the HMIS information be incorporated into the program for each raw material. The ratings and the required protective equipment will take eight characters on the printout. They will consist of the letters "H, F, R and P" to denote health, flammability, reactivity and personal protective equipment, followed by the specific hazard rating and protective equipment designation. For example:

H1F2R0PG relates to a material with a health hazard rating of one, flammability rating of two,

reactivity rating of zero and personal protective equipment G, calling for safety glasses, gloves, and a vapor respirator.

In addition, if the computer is used to generate batch cards this information should be printed on the batch card.

## C. Identification Elements

### 1. Labels

The following are recommended instructions for the use of the HMIS labels.

#### a. Securing Labels

The labels as received from the printer (or printed by an individual manufacturer) should contain the written designation (health, flammability, reactivity, or personal protective equipment), but the actual ratings and designation for personal protective equipment will not be on the label. It is the responsibility of the paint manufacturer to complete the individual information on each raw material.

We might note here that there may be some raw material suppliers who are affixing or pre-printing the HMIS label element to their raw material packages, if so, as discussed later, there are specific ways to deal with this.

#### b. Entry of the HMIS Information on the Label

Most companies using a hazardous materials identification system have found that the best way to affix HMIS labels is to assign a person in the receiving department the responsibility for inventorying and affixing the HMIS labels. This can be accomplished in two ways, as described in c and d.

#### c. Completion of the HMIS Label Upon Receipt of the Raw Material

The first option for entering the HMIS information on the visual element is to complete it by hand at the time it is affixed to the raw material package. The person assigned to perform this task will obtain the necessary information from either the invoice on the raw material or from a code book that contains the hazard ratings and proper personal protective equipment designation for each raw material. Then he will complete the label by writing the proper hazard ratings and the proper (derived) personal protective equipment designation directly on the HMIS label with an indelible felt-tipped pen. The label is then placed on the raw material package.

tags should be the responsibility of the health and safety coordinator. If the piping might contain several different raw materials during the workday, the first line supervisor should be the person responsible for making sure that the proper HMIS tag element is attached to the piping.

These tags, like the placards. should be stored in the area occupied by the health and safety coordinator and distributed by same.

### 8. Other Identification Means

#### a. Batch Tickets

As mentioned earlier, anyone using computer printouts for batch tickets should incorporate the hazard ratings and required personal protective equipment into the computer program for each individual raw material. Then when the batch ticket is printed by the computer, the HMIS information will be included on the batch ticket.

For those companies not using a computer to generate batch tickets. the person preparing the batch ticket should have ready access to the notebook containing the HMIS information. Then when the batch ticket is prepared, the information may be entered on the ticket in the manner described earlier. That is, the following designations, H for health, F for flammability, R for reactivity and P for personal protective equipment, should be used as prefixes with the specific numbers and letter for personal protection following. For example:

A material having a health rating of 2, a flammability rating of 1, a reactivity rating of 1, and personal protective requirements of F, would appear on the batch card as: H2F1R1PF.

# PART IV: Education and Training Program

The HMIS labels, tags and placards are of little use if the workers do not know the meaning of the rating numbers and letter codes that appear on them. Therefore, NPCA has developed a complete education and training program to explain the HMIS to the employee. NPCA will make available to member companies, at a nominal cost, the complete education and training program. This program includes an audio-visual presentation and handout material.

## A. Audio Visual Presentation

This presentation can be given by either the person responsible for health and safety within the plant or by supervisory personnel. It is a flexible program that may be presented by persons of varying levels of knowledge. A copy of the script may be found in Appendix F.

## B. The Wall Poster

Along with this manual you received a copy of the HMIS wall poster. This poster should be placed in highly visible locations throughout the plant. The receiving area, open wall space in work sites, employee lounges and hallways are all appropriate. The wall poster is meant to reinforce the HMIS

**WALL POSTER**

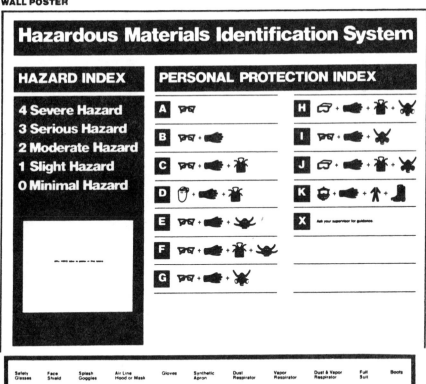

training program and serve as a back up for employees to use to check a rating or equipment assignment. Each poster should have an actual HMIS label affixed to it over the white box on the lower left hand side.

Artwork for printing the HMIS posters has also been supplied to you with this manual. The black and white version of the poster is all of the artwork that you need to print as many posters as you desire. You must also give the printer the following specifications. The poster is printed on 18 point stock, white carolina coated (one side) with PMS 362 (green) ink. It measures 22" x 28".

## C. The Wallet Card

The sample wallet card that accompanied this manual may also be used as artwork to print additional wallet cards. To do that you will need to give the printer the following information: The card measures 2-1/2" x 3-1/2" with rounded corners (this reduces wear on the card). It is printed on 12 point cardboard, coated on two sides. It should be printed with one color ink and varnished.

## D. Conduct of Training

This training program consists of a set of slides and a script and/or a taped narration on a cassette. The program can be presented to the employees with the instructor advancing the slides and either reading the copy or playing the taped narration.

An employee handout in comic book form, which corresponds to the audio/visual program, has also been prepared by NPCA. It includes a short quiz and a form that may be removed from the handout and placed in an employee's permanent file to record his training in the HMIS.

The handout, like the wallet card, may be used as artwork to print additional copies. Tell the printer to use coated stock suitable for reproduction. The handout measures 8-1/2" x 11", has 12 pages and is printed on both sides.

## WALLET CARD

Front

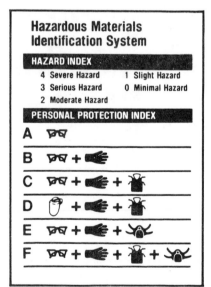

Back

# APPENDIX B: Dimensions and Use of the Various HMIS Elements

## A. Introduction

The various visual elements are the key to successful use of the HMIS. This appendix describes in detail the various visual elements and the means for deploying them throughout the plant. It also gives you the dimensions of the different elements.

## B. Label Configuration

The basic HMIS element can be in either a four-part or five-part configuration. The use of these configurations is described below.

In all illustrations "A" = the described dimension of the sides of the square element.

### 1. Four-Part Element Configuration

This is the basic configuration and is shown in figure B-1. The element is divided into four parts, the top division is blue and rates health hazards, the second division is red and rates flammability hazards, the third division is yellow and rates reactivity hazards and the fourth division is white and assigns personal protection equipment. This is the basic design element giving maximum space on a specific element.

### 2. Five-Part Element Configuration

In addition, NPCA has designed a five-part element which is illustrated below. This five-part element is identical to the four-part element described above, except that a second (blank) white division has been added at the top of the element with corresponding narrowing of the other four sections.

This white division at the top has one of three purposes:

1. To allow a manufacturer space to identify the raw material through code, or product name.
2. To allow a paint manufacturer space to identify the chemical identity of the specific raw material **where** that requirement is mandated by legislation regulation.
3. A combination of 1 and 2.

This five-part configuration is illustrated in figure B-2.

A manufacturer may of course use both the four-part and five-part element in the same operation, as the information conveyed for health, flammability and reactivity hazards and for Personal Protection are the same on both variations.

**Figure B-1**

### Proportions of the Hazardous Materials Identification System Label

Illustrated below are the proportions of the basic HMIS label. For the 50 lb. bag and 55 gallon drum A = 6" (15.25 cm) and 1/4A = 1.5" (3.813 cm). Other sizes should be scaled up or down according to these proportions.

**Figure B-2**

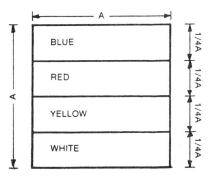

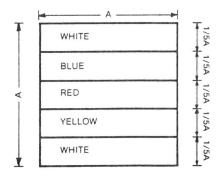

Many manufacturers will find the five-part element design the best choice for placards designed according to option 2 or 3 above. This design assures that the raw material is identified on the placard reducing the chance of putting the wrong placard on the tank or process equipment.

## F. Tags

The tag element of the HMIS can be used in a permanent or semi-permanent form. Due to its small dimensions (shown below) we recommend the use of the following abbreviations: Flammability—Flam., Reactivity—React., and Personal Protection Equipment—p.p.e.

Additionally, the tag element can be prepared as a reusable plastic tag or as a one use cardboard tag.

## G. Other Means of Identification

In addition, if a manufacturer so desires the HMIS element can be painted directly on a process vessel, equipment or other vessels that are to be re-used for various raw materials. This would allow for either a permanent or temporary entry of the HMIS information on the specific vessel.

**Figure B-5**

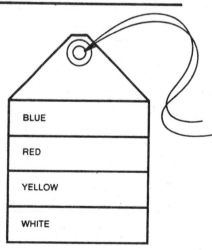

BLUE

RED

YELLOW

WHITE

# APPENDIX C: Protocol for Assigning Hazard Ratings

### Introduction

In order to generate hazard ratings as part of the NPCA Hazardous Materials Identification System we have requested that suppliers to the paint manufacturing industry undertake to generate the health, flammability and reactivity ratings, and assign a "maximum personal protection" designation to each raw material. However, we realize that there are many paint manufacturers who either wish to assign hazard ratings to the raw materials on their own, or wish to know the means whereby the hazard ratings are assigned. Therefore, we have included the exact rating protocol that was given to the raw material suppliers. That protocol is included in this appendix.

## ASSIGNMENT OF THE RATINGS

## A. Considerations for the Assignment of Ratings

### 1. Hazard vs Danger

One of the important concepts associated with the development of the system is the fact that the system is a "hazardous" materials identification system. While a material possesses inherent dangerous properties (toxicity, flammability, reactivity) the availability of the material to fulfill this property, through its use or physical form, determines its hazard.

Since the supplier doing the rating will be unable to ascertain all conditions associated with the use of this raw material in the paint industry, a large portion of the hazard assessment will fall upon the paint manufacturer. This assessment will be done by the paint manufacturer when he assigns specific protective equipment to be used. This assignment will be based on the paint manufacturing operations and use of the raw material.

Therefore, the assignment of a rating to the raw materials by the supplier will be based, **in large part,** on the inherent dangerous properties of the material. However, the ratings process should take into account the hazardous situation relating to the form in which the raw material is supplied, and hence its ability to fulfill the specific dangerous property or properties. Several examples of the way in which the form will influence the development of a hazard rating are given in this appendix as annex 1.

### 2. Recommended Criteria

NPCA has developed recommended criteria for establishing hazard ratings for a material. This criteria is based on systems developed by other experts, which have been widely utilized. This recommendation does not preclude the use of other criteria in assigning the ratings. However, any other criteria used **must be** scientifically sound.

### 3. Material to be Rated

The ratings are to be developed on the raw material as it is **packaged** and received. This means that if it is a mixture or contains significant trace substances, these components should be acknowledged, and their properties used in the development of the hazard rating.

In the development of the ratings the following conditions should be assumed:

1) Direct worker contact with the material is possible;
2) It is available for absorption into the system (based on hazard evaluation);
3) The material will be used in the absence of any controls or protective measures;
4) The exposure will occur on an 8 hour a day schedule.

In assigning the ratings to a specific raw material, any tests done on the entire material, as marketed, are of course the most valid, and should be used in the assignment of the hazard ratings. While this may have been done relative to the flammability and reactivity properties, there is a good chance it was not done relative to health. Therefore, the specific comments are offered in the development of a hazard rating.

### 4. References

In order to assist you in undertaking these ratings, and to supplement specific information on the raw material in question, we are including a list of references later in this appendix in annex 2.

### 5. Rating Sheet

To develop the rating in a concise, orderly fashion we have prepared "classification" or "rating" sheets. Copies of these sheets are shown in annex

box checked in any section determines the overall health hazard rating.

If other information is utilized to accomplish a rating, it should be entered in the lined space provided under the proper heading.

A complete explanation of this system is given under annex 2.

In addition, to guide persons making a determination of the health hazard rating, NPCA offers the following points to be considered.

### 3. Absence of Specific Data

In specifically using the Smyth System, there is likely to be an absence of specific test data necessary to completely fill out this section. Or, under the strict utilization of the protocol, the only information available would result in a lower rating than is known to be proper.

In order to properly address this, the person doing the rating has several options open to him:

    1) Toxicity of an Analogous Material or Structure;

    2) Other Test Data;

    3) Internal Control Procedures.

### 4. Toxicology of an Analogous Structure or Material

This is the least desirable of the methods cited above, as the use of an analogous material or structure is frought with problems.

However, in some case analogies can be drawn, such as in the association between methyl-n-butyl ketone and hexane. Both materials are six carbon, straight chain hydrocarbons which have been indicted for causing nerve disorders.

In addition, it has been noted that often a material can be, or has been, determined to be a metabolite produced during the body's detoxification mechanism for another (analogous) material. If sufficient information is available to demonstrate that the body's treatment of the material, that is the identified metabolite, would be the same as that for the analogous material (before and after the specific point of metabolism), this might also be a means of establishing a health hazard rating.

We again stress that extreme care should be taken in using the toxicology of an analogous material to arrive at a health hazard rating. However, if such use is made, complete the health hazard portion of the form being used for the analogous material and enter the analogous material cited in the supplemental portion.

Additionally, if portions of the form are completed for the specific substance being rated and other portions are completed for an analogous material that results in a different health hazard rating, place an asterisk after the specific toxic property and make note of same in the supplemental information section.

### 5. Other Test Data

The manufacturer of the raw material may have conducted tests on the material that does not fit the Smyth criteria. Or, you may have found literature that indicates that these tests have been conducted. If this test data is available, it may be used to assign a health hazard rating on a 0 to 4 scale. This should only be done in the absence of information sufficient to complete the specific criteria headings under health hazard.

If this method is used, make a reference of the test under the proper heading on the Rating Sheet in the space provided under the appropriate heading.

### 6. Internal Control Procedures

Often a manufacturer of a raw material has established internal control procedures for the handling of the material during manufacture and packaging. These may include specific personal protective equipment requirements, internally generated Permissible Exposure Levels etc., and are usually reported on the suppliers MSDS.

This can be most valuable in assigning a hazard rating. If the rating is based on internal controls and/or internal procedure, reference same in the space provided.

## C. Flammability

The flammability rating protocol is taken directly from the latest definition offered by the National Fire Protection Association (NFPA) 704 system. These are rather straightforward physical constants and most manufacturers have determined 704 ratings for their raw materials. However, to assist you, the specific NFPA reference is cited in annex 2.

## D. Reactivity

Like the flammability rating protocol, the reactivity protocol is taken from NFPA 704. If you have not previously determined the NFPA reactivity rating for your raw materials, the references cited in annex 2 can be used to determine the reactivity.

## E. Assignment of Personal Protective Equipment

One of the most important aspects of this system is the assignment of the proper personal protective equipment to be worn by the worker having contact with a raw material. In section III of the manual, instructions are given to the paint manufacturer for the determination of this information

# APPENDIX C: Annex 1: Examples: Rating Raw Materials that are Mixtures

## Examples: Rating Raw Materials that are Mixtures

Included in this appendix are three examples of the rating process for theoretical raw materials that exists as mixtures. These examples illustrate the following situations:

1. A raw material consisting of a mixture of components with similar hazardous properties.
2. A raw material containing a trace contaminant (less than 1%) that would result in a higher hazard rating due to the contaminant.
3. A raw material containing a highly dangerous (toxic) component whose form results in a lower hazard rating than expected.

## EXAMPLE #1

A raw material consisting of a mixture of components of similar hazardous properties.

### Theoretical Raw Material

A Solvent Blend Consisting of:

| Methyl n-Butyl Ketone | 30% |
| Toluene | 70% |

The individual ratings on the components are:

| Material | Health | Flammability | Reactivity |
|---|---|---|---|
| Methyl n-Butyl Ketone | 3 | 3 | 2 |
| Toluene | 2 | 3 | 0 |

Since both components are present in substantial quantities all hazardous properties are capable of fulfilling their potential and the rating would be the highest available in each situation (in this case all belong to Mn-Bk).

Therefore the overall rating for the solvent blend would be:

| Health | 3 |
| Flammability | 3 |
| Reactivity | 2 |

## EXAMPLE #2

A raw material with a trace contaminant (less than 1%) that would result in a substantially higher hazard rating.

### Theoretical Raw Material

An epoxy resin consisting of:

| Resin | 70.00% |
| Acetone | 29.52% |
| Epichlorohydrin | 0.48% |

The hazard ratings for the individual materials would be:

| Material | Health | Flammability | Reactivity |
|---|---|---|---|
| Resin | 0 | 0 | 0 |
| Acetone | 1 | 3 | 0 |
| Epichlorohydrin | 4 | 3 | 1 |

### Discussion

The low amounts of epichlorohydrin (less than 1%) tend to negate its use as a consideration in assigning a hazard. However, it has been suggested that epichlorohydrin has carcinogenic properties and the material has a recommended ACGIH TLV of 5 ppm. Although possessing different evaporation rates, consider the volatilizing epichlorohydrin and acetone. If within a unit time period, the acetone were controlled to 1000 ppm (the current ACGIH TLV). The epichlorohydrin has the potential of reaching a maximum airborne level of around 16 ppm, three times the recommended TLV. Therefore, the presence of the epichlorohydrin does pose a serious threat, and should be considered in the assignment of ratings.

However, while it should be noted that the health rating of the epichlorohydrin will influence the overall rating, the other hazards for epichlorohydrin would not be a factor due to the dilution of the epichlorohydrin by the other components. It would not be expected to realize its hazardous properties. Therefore the rating for this material is:

| Health | 4 |
| Flammability | 3 |
| Reactivity | 0 |

## EXAMPLE #3

A raw material containing a highly dangerous (toxic) component whose form results in a lower hazard rating than expected.

### Theoretical Raw Material

A tinting paste consisting of:

| Lead Chromate ($PbCrO_4$) | 30% |
| Medium Oil Linseed Alkyd | 40% |
| Mineral Spirits | 30% |

The individual components (rated individually) would have the following hazard ratings:

| Material | Health | Flammability | Reactivity |
|---|---|---|---|
| Lead Chromate | 4 | 0 | 0 |
| Linseed Alkyd (Solids) | 0 | 0 | 0 |
| Mineral Spirits | 1 | 2 | 0 |

## ANSI ADJUNCT SYSTEMS

### The J. T. Baker Chemical Company Saf-T-Data™ System

The Saf-T-Data system is a pictorial-numeric-color system for identifying hazards, describing precautionary measures, and specifying storage restrictions. It is intended to be used as an addition to the ANSI Z129.1 system (see Chapter 16).

Two horizontal rows of symbols are used. The upper row is a series of four numbers, the lower a set of stylized pictorials, and both are superposed on a colored background. The whole is placed above the signal word, statements of hazard, precautionary measures, etc. of the ANSI Z129.1 label. The four numbers in the top row range from 0 through 4, as in the NFPA system, and are each captioned as signifying the degree of health, flammability, reactivity, and contact hazard, respetively. Zero signifies no significant hazard, 4 signifies extremely hazardous, 1, 2 and 3 intermediate degrees of hazard. The stylized pictures in the lower row, each briefly captioned, describe at a glance appropriate precautionary measures, such as eye and face protection devices, proper gloves, respirators, full colored background suggests separated storage locations, red for flammables, yellow for reactive substances, white for corrosives, blue for toxic chemicals requiring secure storage, and orange signifying not suitably characterized by any of the foregoing.

Essentially, the numbers and pictorials duplicate the signal word, statements of hazard and precautionary measures of the ANSI label—with the difference that they can be comprehended as a whole, without reading. The colored background is either a solid block of color, or, for red, yellow and white, a diagonal striped block. Thus, chemicals that burn readily are identified with either a solid red or a red and white striped background. Flammable liquids, for example, carry a solid red background block whereas sodium metal requires a red and white striped background (sodium should not be stored with flammable liquids in a sprinklered storage location). Strong acids carry a white background; strong bases a white and gray striped background since these two should not be stored together. Oxidizers carry a yellow background, reducing agents a yellow and white striped background. Although the degree of hazard numbering is derived from the NFPA system, only the flammability numbers are in general identical with the NFPA assigned flammability numbers. The NFPA health and reactivity numbers refer to hazards anticipated in a fire or other emergency situation, whereas the Saf-T-Data™ health, reactivity and contact numbers (NFPA does not rate contact hazards separately) refer to hazards in handling in unexceptional useage such as in a laboratory or the industrial workplace.

A bold-face question mark symbol in the Saf-T-Data™ system signifies that the degree of hazard is not known and that the chemical should be handled as though it were extremely hazardous. An octagonal shaped "Stop sign" symbol is used when appropriate precautionary measures cannot be described by pictorials. The Stop sign also carries the wording "Exceptional hazard; read the material safety data sheet."

### The Fisher Scientific Company Chem-Alert™ system

A somewhat similar pictorial-numeric-color system has been announced by the Fisher Scientific Company as their Chem-Alert™ system. This system uses the NFPA 704 numbering concepts as designed for fire and other emergency situations. It incorporates a slightly different color scheme for overall hazard and uses simpler pictorials to describe precautionary measures. It also is intended as an adjunct to the ANSI Z129.1 system.

# 18

# Material Safety Data Sheets

**Jay A. Young**
*Consultant*
*Silver Spring, MD*

A material safety data sheet (MSDS) is usually a printed document of one or more pages conveying cautionary information and data about a hazardous chemical or mixture. It is intended primarily for use by supervisory personnel as an outline and source of data for topics to be discussed when supervisors directly instruct employees in the work place on the precautions to be followed and the reasons therefor in using, storing, or otherwise handling the material. An MSDS is also useful to industrial hygienists, occupational health personnel, and physicians as a guide for controlling exposures and for treatment in case of over-exposure.

Because it is physically separate from a chemical container, an MSDS does not replace a label. In some cases, the "container" while physically present may not be readily apparent to an employee. For example, the "container" may be a pipe with an on/off valve. The hazardous chemical was delivered from the manufacturer in a tank truck and transferred at the work place to a storage tank—to which the valved pipe is connected. Chemical manufacturers cannot enter a customer's premises to give precautionary information to, nor to enforce precautionary practices by, a customer's employees. In instances where the manufacturer cannot communicate by an affixed precautionary label, an MSDS backed by employee training is the best way to convey the information to those who need it. Some chemicals by their nature can pose hazards to third parties. For example, cement is mixed with sand, aggregate, water to form mortar which hardens into concrete. Concrete is susceptible to weathering; freeze-thaw cycling can severely weaken concrete originally made with "too much" water. When the proper amount of water is used and the correct mixing procedures have been followed, the mortar will have very small bubbles of air entrained in the mixture; the resulting concrete will withstand severe weathering conditions. Thus, it would be prudent that an MSDS for cement would identify this hazard and describe in detail how to mix the mortar in order to reduce this risk.

Most chemical manufacturers furnish safety data sheets routinely with shipments to customers, whether or not the materials are labeled. At present, there is no regulatory requirement to do so (except for certain maritime applications, see below), although a current draft of a to be proposed OSHA regulation may eventually so require.

Two material safety data sheets for borax are shown in Figures 18.1 and 18.2. Figure 18.1 is an example of a currently used borax MSDS with only the manufacturer's identity changed. It is an "OSHA Form 20" required by OSHA regulations[1] to be furnished for hazardous materials in ship repair, ship building, and ship breaking. As is evident by comparing Figures 18.1 and 18.2, a completed OSHA Form 20 may be minimally useful in conveying precautionary information.

While Figure 18.1 is illustrative, it is based on a composite of features extracted from current examples of useful, informative, and up to date MSDSs from a variety of sources, including manufacturers, commercial and governmental users. Judging from such examples, an informative MSDS should at least include all of the following characteristics, that are appropriate.

It should be attractively printed with the information arranged in topical order so as to invite attention and efficiently convey the proffered information.

That information, not necessarily in the order given here, should include these subjects:

1. *The proper chemical name* of the hazardous chemical or of the hazardous component(s) of the mixture, if the concentration is sufficient to present a hazard. The trade name or code name, clearly identified as such, and common names or synonyms for the proper chemical name should be included.

2. *Manufacturer identification;* name, address, and *emergency telephone number.*

3. *Summary statement;* a tabulation of the recommended exposure limit and toxic dose data, with a general physical description and briefly stated information about fire hazards, unique storage hazards, decomposition hazards, reactivity and incompatibility hazards, disposal hazards, etc., as applicable.

4. *Hazard identification;* each reasonably foreseeable hazard (even though briefly mentioned in item 3) is identified, fire and explosion, corrosive, temperature or other instability, reactivity, impact sensitivity, incompatibility, toxicity, etc., with toxicological effects further identified as acute or chronic and typical symptoms of over exposure stated.

U.S. DEPARTMENT OF LABOR

Occupational Safety and Health Administration

Form Approved
OMB No. 44-R1387

# MATERIAL SAFETY DATA SHEET

Required under USDL Safety and Health Regulations for Ship Repairing,
Shipbuilding, and Shipbreaking (29 CFR 1915, 1916, 1917)

## SECTION I

| MANUFACTURER'S NAME | EMERGENCY TELEPHONE NO. |
|---|---|
| Aceb XL Chemical Company | 999/999-9999 |

ADDRESS *(Number, Street, City, State, and ZIP Code)*
123 Any Ave., Anytown, ZZ  99999

| CHEMICAL NAME AND SYNONYMS | TRADE NAME AND SYNONYMS |
|---|---|
| Sodium tetraborate decahydrate | Borax |

| CHEMICAL FAMILY | FORMULA |
|---|---|
| Sodium borate | $Na_2B_4O_7 \cdot 10 H_2O$ |

## SECTION II - HAZARDOUS INGREDIENTS

| PAINTS, PRESERVATIVES, & SOLVENTS | % | TLV (Units) | ALLOYS AND METALLIC COATINGS | % | TLV (Units) |
|---|---|---|---|---|---|
| PIGMENTS | | | BASE METAL | | |
| CATALYST | | | ALLOYS | | |
| VEHICLE    Does not apply | | | METALLIC COATINGS    Does not apply | | |
| SOLVENTS | | | FILLER METAL PLUS COATING OR CORE FLUX | | |
| ADDITIVES | | | OTHERS | | |
| OTHERS | | | | | |

| HAZARDOUS MIXTURES OF OTHER LIQUIDS, SOLIDS, OR GASES | % | TLV (Units) |
|---|---|---|
| | | |
| | | |
| | | |
| | | |

## SECTION III - PHYSICAL DATA

| BOILING POINT (°F.)    Does not apply | SPECIFIC GRAVITY ($H_2O=1$)    1.73 |
|---|---|
| VAPOR PRESSURE (mm Hg.)    Does not apply | PERCENT, VOLATILE BY VOLUME (%)    none |
| VAPOR DENSITY (AIR=1)    Does not apply | EVAPORATION RATE (_____ =1)    Does not apply |
| SOLUBILITY IN WATER    Moderate | |

APPEARANCE AND ODOR    White, odorless solid

## SECTION IV - FIRE AND EXPLOSION HAZARD DATA

| FLASH POINT (Method used)    Does not apply | FLAMMABLE LIMITS    Does not apply | Lel | Uel |
|---|---|---|---|

EXTINGUISHING MEDIA    None, Product inherent fire retardant

SPECIAL FIRE FIGHTING PROCEDURES    None

UNUSUAL FIRE AND EXPLOSION HAZARDS    None

**Figure 18.1:** An OSHA Form 20 for borax. (a) Front
side. (b) Back side.

## SECTION V · HEALTH HAZARD DATA

THRESHOLD LIMIT VALUE 5 mg/M$^3$ (Amer. Conf. Gov't Ind. Hygenists, 1981)

EFFECTS OF OVEREXPOSURE Minor skin irritant; moderate eye irritant

EMERGENCY AND FIRST AID PROCEDURES Wash with water

## SECTION VI · REACTIVITY DATA

| STABILITY | UNSTABLE | | CONDITIONS TO AVOID None | |
|---|---|---|---|---|
| | STABLE | XX | | |

INCOMPATABILITY *(Materials to avoid)* None

HAZARDOUS DECOMPOSITION PRODUCTS None

| HAZARDOUS POLYMERIZATION | MAY OCCUR | | CONDITIONS TO AVOID None | |
|---|---|---|---|---|
| | WILL NOT OCCUR | XX | | |

## SECTION VII · SPILL OR LEAK PROCEDURES

STEPS TO BE TAKEN IN CASE MATERIAL IS RELEASED OR SPILLED

Standard disposal procedures — presents no health hazard

WASTE DISPOSAL METHOD

Standard disposal procedures — presents no health hazard

## SECTION VIII · SPECIAL PROTECTION INFORMATION

RESPIRATORY PROTECTION *(Specify type)* No specific protection required

| VENTILATION | LOCAL EXHAUST Normal | SPECIAL |
|---|---|---|
| | MECHANICAL *(General)* | OTHER |

PROTECTIVE GLOVES Not needed   EYE PROTECTION Avoid eye contact

OTHER PROTECTIVE EQUIPMENT None

## SECTION IX · SPECIAL PRECAUTIONS

PRECAUTIONS TO BE TAKEN IN HANDLING AND STORING None needed

OTHER PRECAUTIONS None needed

Form OSHA-20
Rev. May 72

MATERIAL SAFETY DATA SHEET: BORAX

Manufacturer:
ABee Que Chemicals
321 Any Ave.
Thiscity, VV  99999

Emergency Telephone
 (111) 111-1111

Formula:  $Na_2B_4O_7 \cdot 10\ H_2O$

Synonyms: None

Chemical name: Sodium tetraborate decahydrate

TLV:  5 mg/$M^3$, ACGIH - 1981

Toxic dose: oral-human adult $LD_{lo}$ 709 mg/kg

oral-human infant $LD_{lo}$ 1000 mg/kg

General description: Borax is a white solid; it cannot burn; it is mildly toxic -- exposure should
be avoided.  It is not considered to be a reactive chemical.  Borax is used
in water solution with detergents in "cleaning compounds"; it is also used as
a flux -- to dissolve metal oxides, etc. -- in brazing and welding.

| HAZARDS | PRECAUTIONS | IN CASE OF, DO THIS |
|---|---|---|
| **Acute exposures** | | |
| Eye: redness, pain | Close fitting goggles | Copious water for 15 min., take to physician |
| Skin: no effect or mild irritation | Gloves | Wash with soap and water |
| Inhalation: cough, nausea, vomiting | Adequate ventilation or respiratory protection | Fresh air, rest, physician's attention |
| Ingestion: nausea, vomiting, diarrhea, muscular spasms | No food, tobacco in area, wash before leaving area | Wash out mouth, drink water, call physician |
| **Chronic exposures** | | |
| Skin: dryness, eczema | Gloves | Wash with soap and water, avoid occasion for further exposure |
| Ingestion: gastritis | No food, tobacco in area, wash before leaving area | Physician's attention |

REACTIVITY: Generally stable, reactions with other chemicals not markedly exo- or endo- thermic; not self-reactive; does not decompose spontaneously.

SPILLS:  Sweep or scoop up solid, wash area with water drained to catch basin.  Dike solutions and transfer to tank, wash area as above.  Notify authorities.

STORAGE:  Cool, dry preferable but not essential; however avoid excess humidity.

DISPOSAL:  If recycling or re-working not possible, disposal in controlled land fill may be required; check with local authorities.

PHYSICAL PROPERTIES:

| | |
|---|---|
| Melting point: | 140 - 390° F. (apparent, while losing water of hydration) 167° F.  (anhydrous) |
| Boiling point: | 608° F. |
| Density (relative to water = 1.0) | 1.7 |
| Water solubility: | 1.7 oz./gal. at 32°F.; 2 lbs./gal. at 150°F. |
| Molecular weight: | 381.4 |
| Color: | White, technical grades may be off-white |
| Sensitivity to light: | Unaffected |

NOTES: Sometimes used in water solution as an antiseptic.  This use is NOT recommended.

The information in this material safety data sheet should be provided to all who will use, handle, store, transport, or otherwise be exposed to borax.  This information has been prepared for the guidance of plant engineering, operations and management and for persons working with or handling borax.  ABee Que Chemicals believes this information to be reliable and up to date as of July, 1982 but it makes no warranty that it is.

In particular, DO NOT RELY ON THE INFORMATION PRESENTED HERE AFTER JULY, 1983.

ABQCHEM-5M-7882

**Figure 18.2**: Material safety data sheet for borax.

5. *Precautionary information* for each identified hazard, along with instructions on what to do if the precautions fail, or were not taken. This section should include first aid as appropriate, and instructions to a physician.

6. *Spill, leak and disposal* treatment, including personal protective equipment and other precautions, if needed.

7. *Fire fighting procedures* that are effective, including recommended personal protective equipment if appropriate.

8. *Storage conditions* that reduce hazards, including reference to previously cited incompatibilities, if any. If storage life is limited, probable life is stated, identified as an estimate.

9. *Other information* such as special packaging requirements, or unique hazardous properties not described in the foregoing—such as third party hazards.

10. *The publication date* of the MSDS and the date when recent developments might be expected to render the proffered information significantly incomplete.

MSDSs should cite the sources for the information; almost all available MSDS carry a disclaimer; some also show a copy of the precautionary label.

## FOOTNOTE

1. Title 29, Code of Federal Regulations, parts 1915.57, 1916.57, and 1917.57. These regulations require the use of the form shown, or an "essentially similar" form. An essentially similar form may be different in only two ways: Section II can be modified to better fit the hazardous material described and the space provided for answers to non-applicable items may be reduced to that necessary for entering "n.a." provided that the sequential ordering of these items is not changed.

# Appendix

## OSHA 1983 Standard

Subpart Z of Part 1910 of Title 29 of the Code of Federal Regulations (CFR) is hereby amended by adding a new s1910.1200 to read as follows:

### s1910.1200 HAZARD COMMUNICATION

(a) *Purpose.* The purpose of this section is to ensure that the hazards of all chemicals produced or imported by chemical manufacturers or importers are evaluated, and that information concerning their hazards is transmitted to affected employers and employees within the manufacturing sector. This transmittal of information is to be accomplished by means of comprehensive hazard communication programs, which are to include container labeling and other forms of warning, material safety data sheets and employee training. This occupational safety and health standard is intended to address comprehensively the issue of evaluating and communicating chemical hazards to employees in the manufacturing sector, and to preempt any state law pertaining to this subject. Any state which desires to assume responsibility in this area may only do so under the provisions of s18 of the Occupational Safety and Health Act (29 U.S.C. 651 et. seq.) which deals with state jurisdiction and state plans.

(b) *Scope and application.*

(1) This section requires chemical manufacturers or importers to assess the hazards of chemicals which they produce or import, and all employers in SIC Codes 20 through 39 (Division D, Standard Industrial Classification Manual) to provide information to their employees about the hazardous chemicals to which they are exposed, by means of a hazard communication program, labels and other forms of warning, material safety data sheets, and information and training. In addition, this section requires distributors to transmit the required information to employers in SIC Codes 20-39.

(2) This section applies to any chemical which is known to be present in the workplace in such a manner that employees may be exposed under normal conditions of use or in a foreseeable emergency.

*452*

(3) This section applies to laboratories only as follows:

    (i) Employers shall ensure that labels on incoming containers of hazardous chemicals are not removed or defaced;

    (ii) Employers shall maintain any material safety data sheets that are received with incoming shipments of hazardous chemicals, and ensure that they are readily accessible to laboratory employees; and,

    (iii) Employers shall ensure that laboratory employees are apprised of the hazards of the chemicals in their workplaces in accordance with paragraph (h) of this section.

(4) This section does not require labeling of the following chemicals:

    (i) Any pesticide as such term is defined in the Federal Insecticide, Fungicide, and Rodenticide Act (7 U.S.C. ss136 et seq.), when subject to the labeling requirements of that Act and labeling regulations issued under that Act by the Environmental Protection Agency;

    (ii) Any food, food additive, color additive, drug, or cosmetic, including materials intended for use as ingredients in such products (e.g., flavors and fragrances), as such terms are defined in the Federal Food, Drug, and Cosmetic Act (21 U.S.C. ss301 et seq.) and regulations issued under that Act, when they are subject to the labeling requirements of that Act and labeling regulations issued under that Act by the Food and Drug Administration;

    (iii) Any distilled spirits (beverage alcohols), wine, or malt beverage intended for nonindustrial use, as such terms are defined in the Federal Alcohol Administration Act (27 U.S.C. s201 et seq.) and regulations issued under that Act, when subject to the labeling requirements of that Act and labeling regulations issued under that Act by the Bureau of Alcohol, Tobacco, and Firearms; and,

    (iv) Any consumer product or hazardous substance as those terms are defined in the Consumer Product Safety Act (15 U.S.C. ss2051 et seq.) and Federal Hazardous Substances Act (15 U.S.C. ss1261 et seq.) respectively, when subject to a consumer product safety standard or labeling requirement of those Acts, or regulations issued under those Acts by the Consumer Product Safety Commission.

(5) This section does not apply to:

    (i) Any hazardous waste as such term is defined by the Solid Waste Disposal Act, as amended by the Resource Conservation and Recovery Act of 1976, as amended (42 U.S.C. ss6901 et seq.), when subject to regulations issued under that Act by the Environmental Protection Agency;

    (ii) Tobacco or tobacco products;

    (iii) Wood or wood products;

    (iv) Articles; and,

    (v) Foods, drugs, or cosmetics intended for personal consumption by employees while in the workplace.

(c) *Definitions.*

"Article" means a manufactured item

> (i) which is formed to a specific shape or design during manufacture,
>
> (ii) which has end use function(s) dependent in whole or in part upon its shape or design during end use, and
>
> (iii) which does not release, or otherwise result in exposure to, a hazardous chemical under normal conditions of use.

"Assistant Secretary" means the Assistant Secretary of Labor for Occupational Safety and Health, U.S. Department of Labor, or designee.

"Chemical" means any element, chemical compound or mixture of elements and/or compounds.

"Chemical manufacturer" means an employer in SIC Codes 20 through 39 with a workplace where chemical(s) are produced for use or distribution.

"Chemical name" means the scientific designation of a chemical in accordance with the nomenclature system developed by the International Union of Pure and Applied Chemistry (IUPAC) or the Chemical Abstracts Service (CAS) rules of nomenclature, or a name which will clearly identify the chemical for the purpose of conducting a hazard evaluation.

"Combustible liquid" means any liquid having a flashpoint at or above 100°F (37.8°C), but below 200° F (93.3° C), except any mixture having components with flashpoints of 200°F (93.3°C), or higher, the total volume of which make up 99 percent or more of the total volume of the mixture.

"Common name" means any designation or identification such as code name, code number, trade name, brand name or generic name used to identify a chemical other than by its chemical name.

"Compressed gas" means:

> (i) A gas or mixture of gases having, in a container, an absolute pressure exceeding 40 psi at 70°F (21.1°C); or
>
> (ii) A gas or mixture of gases having, in a container, an absolute pressure exceeding 104 psi at 130°F (54.4°C) regardless of the pressure at 70°F (21.1°C); or
>
> (iii) A liquid having a vapor pressure exceeding 40 psi at 100°F (37.8°C) as determined by ASTM D-323-72.

"Container" means any bag, barrel, bottle, box, can, cylinder, drum, reaction vessel, storage tank, or the like that contains a hazardous chemical. For purposes of this section, pipes or piping systems are not considered to be containers.

"Designated representative" means any individual or organization to whom an employee gives written authorization to exercise such employee's rights under this section. A recognized or certified collective bargaining agent shall be treated automatically as a designated representative without regard to written employee authorization.

"Director" means the Director, National Institute for Occupational Safety and Health, U.S. Department of Health and Human Services, or designee.

"Distributor" means a business, other than a chemical manufacturer or importer, which supplies hazardous chemicals to other distributors or to manufacturing purchasers.

"Employee" means a worker employed by an employer in a workplace in SIC Codes 20 through 39 who may be exposed to hazardous chemicals under normal operating conditions or foreseeable emergencies, including, but not limited to production workers, line supervisors, and repair or maintenance personnel. Office workers, grounds maintenance personnel, security personnel or non-resident management are generally not included, unless their job performance routinely involves potential exposure to hazardous chemicals.

"Employer" means a person engaged in a business within SIC Codes 20 through 39 where chemicals are either used, or are produced for use or distribution.

"Explosive" means a chemical that causes a sudden, almost instantaneous release of pressure, gas, and heat when subjected to sudden shock, pressure, or high temperature.

"Exposure" or "exposed" means that an employee is subjected to a hazardous chemical in the course of employment through any route of entry (inhalation, ingestion, skin contact or absorption, etc.), and includes potential (e.g. accidental or possible) exposure.

"Flammable" means a chemical that falls into one of the following categories:

    (i) "Aerosol, flammable" means an aerosol that, when tested by the method described in 16 CFR 1500.45, yields a flame projection exceeding 18 inches at full valve opening, or a flashback (a flame extending back to the valve) at any degree of valve opening;

    (ii) "Gas, flammable" means:

        (A) A gas that, at ambient temperature and pressure, forms a flammable mixture with air at a concentration of thirteen (13) percent by volume or less; or

        (B) A gas that at ambient temperature and pressure, forms a range of flammable mixtures with air wider than twelve (12) percent by volume, regardless of the lower limit;

    (iii) "Liquid, flammable" means any liquid having a flashpoint below 100°F (37.8°C), except any mixture having components with flashpoints of 100°F (37.8°C) or higher, the total of which make up 99 percent or more of the total volume of the mixture.

    (iv) "Solid, flammable" means a solid, other than a blasting agent or explosive as defined in s1910.109(a), that is liable to cause fire through friction, absorption of moisture, spontaneous chemical change, or retained heat from manufacturing or processing, or which can be ignited readily and when ignited burns so vigorously and persistently as to create a serious hazard. A chemical shall be considered to be a flammable solid if, when tested by the method described in 16 CFR 1500.44, it ignites and burns with a self-sustained flame at a rate greater than one-tenth of an inch per second along its major axis.

"Flashpoint" means the minimum temperature at which a liquid gives off a vapor in sufficient concentration to ignite when tested as follows:

    (i) Tagliabue Closed Tester (See American National Standard Method of Test for Flash Point by Tag Closed Tester, Z11.24-1979 (ASTM D 56-79))-for liquids with a viscosity of less than 45 Saybolt Universal Seconds (SUS) at 100°F (37.8°C), that do not contain suspended solids and do not have a tendency to form a surface film under test; or

(ii) Pensky-Martens Closed Tester (see American National Standard Method of Test for Flash Point by Pensky-Martens Closed Tester, Z11.7-1979 (ASTM D 93-79))-for liquids with a viscosity equal to or greater than 45 SUS at 100°F (37.8°C), or that contain suspended solids, or that have a tendency to form a surface film under test; or

(iii) Setaflash Closed Tester (see American National Standard Method of Test for Flash Point by Setaflash Closed Tester (ASTM D 3278-78)).

Organic peroxides, which undergo autoaccelerating thermal decomposition, are excluded from any of the flashpoint determination methods specified above.

"Foreseeable emergency" means any potential occurrence such as, but not limited to, equipment failure, rupture of containers, or failure of control equipment which could result in an uncontrolled release of a hazardous chemical into the workplace.

"Hazard warning" means any words, pictures, symbols, or combination thereof appearing on a label or other appropriate form of warning which convey the hazards of the chemical(s) in the container(s).

"Hazardous chemical" means any chemical which is a physical hazard or a health hazard.

"Health hazard" means a chemical for which there is statistically significant evidence based on at least one study conducted in accordance with established scientific principles that acute or chronic health effects may occur in exposed employees. The term "health hazard" includes chemicals which are carcinogens, toxic or highly toxic agents, reproductive toxins, irritants, corrosives, sensitizers, hepatotoxins, nephrotoxins, neurotoxins, agents which act on the hematopoietic system, and agents which damage the lungs, skin, eyes, or mucous membranes. Appendix A provides further definitions and explanations of the scope of health hazards covered by this section, and Appendix B describes the criteria to be used to determine whether or not a chemical is to be considered hazardous for purposes of this standard.

"Identity" means any chemical or common name which is indicated on the material safety data sheet (MSDS) for the chemical. The identity used shall permit cross-references to be made among the required list of hazardous chemicals, the label and the MSDS.

"Immediate use" means that the hazardous chemical will be under the control of and used only by the person who transfers it from a labeled container and only within the work shift in which it is transferred.

"Importer" means the first business with employees within the Customs Territory of the United States which receives hazardous chemicals produced in other countries, for the purpose of supplying them to distributors or manufacturing purchasers within the United States.

"Label" means any written, printed, or graphic material displayed on or affixed to containers of hazardous chemicals.

"Manufacturing purchaser" means an employer with a workplace classified in SIC Codes 20 through 39 who purchases a hazardous chemical for use within that workplace.

"Material safety data sheet (MSDS)" means written or printed material concerning a hazardous chemical which is prepared in accordance with paragraph (g) of this section.

"Mixture" means any combination of two or more chemicals if the combination is not, in whole or in part, the result of a chemical reaction.

"Organic peroxide" means an organic compound that contains the bivalent -O-O- structure and which may be considered to be a structural derivative of hydrogen peroxide where one or both of the hydrogen atoms has been replaced by an organic radical.

"Oxidizer" means a chemical other than a blasting agent or explosive as defined in s1910.109(a), that initiates or promotes combustion in other materials, thereby causing fire either of itself or through the release of oxygen or other gases.

"Physical hazard" means a chemical for which there is scientifically valid evidence that it is a combustible liquid, a compressed gas, explosive, flammable, an organic peroxide, an oxidizer, pyrophoric, unstable (reactive) or water-reactive.

"Produce" means to manufacture, process, formulate, or repackage.

"Pyrophoric" means a chemical that will ignite spontaneously in air at a temperature of 130°F (54.5°C) or below.

"Responsible party" means someone who can provide additional information on the hazardous chemical and appropriate emergency procedures, if necessary.

"Specific chemical identity" means the chemical name, Chemical Abstracts Service (CAS) Registry Number, or any other information that reveals the precise chemical designation of the substance.

"Trade secret" means any confidential formula, pattern, process, device, information or compilation of information (including chemical name or other unique chemical identifier) that is used in an employer's business, and that gives the employer an opportunity to obtain an advantage over competitors who do not know or use it.

"Unstable (reactive)" means a chemical which in the pure state, or as produced or transported, will vigorously polymerize, decompose, condense, or will become self-reactive under conditions of shocks, pressure or temperature.

"Use" means to package, handle, react, or transfer.

"Water-reactive" means a chemical that reacts with water to release a gas that is either flammable or presents a health hazard.

"Work area" means a room or defined space in a workplace where hazardous chemicals are produced or used, and where employees are present.

"Workplace" means an establishment at one geographical location containing one or more work areas.

(d) *Hazard determination.*

    (1) Chemical manufacturers and importers shall evaluate chemicals produced in their workplaces or imported by them to determine if they are hazardous. Employers are not required to evaluate chemicals unless they choose not to rely on the evaluation performed by the chemical manufacturer or importer for the chemical to satisfy this requirement.

    (2) Chemical manufacturers, importers or employers evaluating chemicals shall identify and consider the available scientific evidence concerning such hazards. For health hazards, evidence which is statistically significant and which is based on at least one positive study conducted in accordance with established scientific principles is considered to be sufficient to estab-

lish a hazardous effect if the results of the study meet the definitions of health hazards in this section. Appendix A shall be consulted for the scope of health hazards covered, and Appendix B shall be consulted for the criteria to be followed with respect to the completeness of the evaluation, and the data to be reported.

(3) The chemical manufacturer, importer or employer evaluating chemicals shall treat the following sources as establishing that the chemicals listed in them are hazardous:

   (i) 29 CFR Part 1910, Subpart Z, Toxic and Hazardous Substances, Occupational Safety and Health Administration (OSHA); or,

   (ii) *Threshold Limit Values for Chemical Substances and Physical Agents in the Work Environment,* American Conference of Governmental Industrial Hygienists (ACGIH) (latest edition).

The chemical manufacturer, importer, or employer is still responsible for evaluating the hazards associated with the chemicals in these source lists in accordance with the requirements of this standard.

(4) Chemical manufacturers, importers and employers evaluating chemicals shall treat the following sources as establishing that a chemical is a carcinogen or potential carcinogen for hazard communication purposes:

   (i) National Toxicology Program (NTP), *Annual Report on Carcinogens* (latest edition);

   (ii) International Agency for Research on Cancer (IARC) *Monographs* (latest editions); or

   (iii) 29 CFR Part 1910, Subpart Z, Toxic and Hazardous Substances, Occupational Safety and Health Administration.

Note: The *Registry of Toxic Effects of Chemical Substances* published by the National Institute for Occupational Safety and Health indicates whether a chemical has been found by NTP or IARC to be a potential carcinogen.

(5) The chemical manufacturer, importer or employer shall determine the hazards of mixtures of chemicals as follows:

   (i) If a mixture has been tested as a whole to determine its hazards, the results of such testing shall be used to determine whether the mixture is hazardous;

   (ii) If a mixture has not been tested as a whole to determine whether the mixture is a health hazard, the mixture shall be assumed to present the same health hazards as do the components which comprise one percent (by weight or volume) or greater of the mixture, except that the mixture shall be assumed to present a carcinogenic hazard if it contains a component in concentrations of 0.1 percent or greater which is considered to be a carcinogen under paragraph (d)(4) of this section;

   (iii) If a mixture has not been tested as a whole to determine whether the mixture is a physical hazard, the chemical manufacturer, importer, or employer may use whatever scientifically valid data is available to evaluate the physical hazard potential of the mixture; and,

   (iv) If the employer has evidence to indicate that a component present in the mixture in concentrations of less than one percent (or in the case

container of hazardous chemicals leaving the workplace is labeled, tagged, or marked in accordance with this section in a manner which does not conflict with the requirements of the Hazardous Materials Transportation Act (18 U.S.C. ss1801 et seq.) and regulations issued under that Act by the Department of Transportation.

(3) If the hazardous chemical is regulated by OSHA in a substance-specific health standard, the chemical manufacturer, importer, distributor or employer shall ensure that the labels or other forms of warning used are in accordance with the requirements of that standard.

(4) Except as provided in paragraphs (f)(5) and (f)(6) the employer shall ensure that each container of hazardous chemicals in the workplace is labeled, tagged or marked with the following information:

    (i) Identity of the hazardous chemical(s) contained therein; and

    (ii) Appropriate hazard warnings.

(5) The employer may use signs, placards, process sheets, batch tickets, operating procedures, or other such written materials in lieu of affixing labels to individual stationary process containers, as long as the alternative method identifies the containers to which it is applicable and conveys the information required by paragraph (f)(4) of this section to be on a label. The written materials shall be readily accessible to the employees in their work area throughout each work shift.

(6) The employer is not required to label portable containers into which hazardous chemicals are transferred from labeled containers, and which are intended only for the immediate use of the employee who performs the transfer.

(7) The employer shall not remove or deface existing labels on incoming containers of hazardous chemicals, unless the container is immediately marked with the required information.

(8) The employer shall ensure that labels or other forms of warning are legible, in English, and prominently displayed on the container, or readily available in the work area throughout each work shift. Employers having employees who speak other languages may add the information in their language to the material presented, as long as the information is presented in English as well.

(9) The chemical manufacturer, importer, distributor or employer need not affix new labels to comply with this section if existing labels already convey the required information.

(g) *Material safety data sheets.*

    (1) Chemical manufacturers and importers shall obtain or develop a material safety data sheet for each hazardous chemical they produce or import. Employers shall have a material safety data sheet for each hazardous chemical which they use.

    (2) Each material safety data sheet shall be in English and shall contain at least the following information:

        (i) The identity used on the label, and, except as provided for in paragraph (i) of this section on trade secrets:

            (A) If the hazardous chemical is a single substance, its chemical and common name(s);

of carcinogens, less than 0.1 percent) could be released in
tions which would exceed an established OSHA permi:
sure limit or ACGIH Threshold Limit Value, or coul<
health hazard to employees in those concentrations, 1
shall be assumed to present the same hazard.

(6) Chemical manufacturers, importers, or employers evaluatin
shall describe in writing the procedures they use to determine
of the chemical they evaluate. The written procedures are to be
able, upon request, to employees, their designated representat
sistant Secretary and the Director. The written description ma
rated into the written hazard communication program rec
paragraph (e) of this section.

(e) *Written hazard communication program.*

(1) Employers shall develop and implement a written hazard co
program for their workplaces which at least describes how
specified in paragraphs (f), (g), and (h) of this section for lab
forms of warning, material safety data sheets, and employe
and training will be met, and which also includes the follo'

(i) A list of the hazardous chemicals known to be present
tity that is referenced on the appropriate material saf
(the list may be compiled for the workplace as a whole
ual work areas);

(ii) The methods the employer will use to inform employ
ards of non-routine tasks (for example, the cleaning
sels), and the hazards associated with chemicals
unlabeled pipes in their work areas; and,

(iii) The methods the employer will use to inform any con
ers with employees working in the employer's workp
ardous chemicals their employees may be exposed
forming their work, and any suggestions for approp:
measures.

(2) The employer may rely on an existing hazard communicat
comply with these requirements, provided that it meets th
lished in this paragraph (e).

(3) The employer shall make the written hazard communi
available, upon request, to employees, their designated repi
Assistant Secretary and the Director, in accordance with tl
of 29 CFR 1910.20 (e).

(f) *Labels and other forms of warning.*

(1) The chemical manufacturer, importer, or distributor shall
container of hazardous chemicals leaving the workplace i
or marked with the following information:

(i) Identity of the hazardous chemical(s):

(ii) Appropriate hazard warnings; and

(iii) Name and address of the chemical manufacturer, i
responsible party.

(2) Chemical manufacturers, importers, or distributors shall

(B) If the hazardous chemical is a mixture which has been tested as a whole to determine its hazards, the chemical and common name(s) of the ingredients which contribute to these known hazards, and the common name(s) of the mixture itself; or,

(C) If the hazardous chemical is a mixture which has not been tested as a whole:

   (1) The chemical and common name(s) of all ingredients which have been determined to be health hazards, and which comprise 1% or greater of the composition, except that chemicals identified as carcinogens under paragraph (d)(4) of this section shall be listed if the concentrations are 0.1% or greater; and,

   (2) The chemical and common name(s) of all ingredients which have been determined to present a physical hazard when present in the mixture;

(ii) Physical and chemical characteristics of the hazardous chemical (such as vapor pressure, flash point);

(iii) The physical hazards of the hazardous chemical, including the potential for fire, explosion, and reactivity;

(iv) The health hazards of the hazardous chemical, including signs and symptoms of exposure, and any medical conditions which are generally recognized as being aggravated by exposure to the chemical;

(v) The primary route(s) of entry;

(vi) The OSHA permissible exposure limit, ACGIH Threshold Limit Value, and any other exposure limit used or recommended by the chemical manufacturer, importer, or employer preparing the material safety data sheet, where available;

(vii) Whether the hazardous chemical is listed in the National Toxicology Program (NTP) *Annual Report on Carcinogens* (latest edition) or has been found to be a potential carcinogen in the International Agency for Research on Cancer (IARC) *Monographs* (latest editions), or by OSHA;

(viii) Any generally applicable precautions for safe handling and use which are known to the chemical manufacturer, importer or employer preparing the material safety data sheet, including appropriate hygienic practices, protective measures during repair and maintenance of contaminated equipment, and procedures for clean-up of spills and leaks;

(ix) Any generally applicable control measures which are known to the chemical manufacturer, importer or employer preparing the material safety data sheet, such as appropriate engineering controls, work practices, or personal protective equipment;

(x) Emergency and first aid procedures;

(xi) The date of preparation of the material safety data sheet or the last change to it; and,

(xii) The name, address and telephone number of the chemical manufacturer, importer, employer or other responsible party preparing or distributing the material safety data sheet, who can provide additional information on the hazardous chemical and appropriate emergency procedures, if necessary.

(3) If no relevant information is found for any given category on the material safety data sheet, the chemical manufacturer, importer or employer preparing the material safety data sheet shall mark it to indicate that no applicable information was found.

(4) Where complex mixtures have similar hazards and contents (i.e. the chemical ingredients are essentially the same, but the specific composition varies from mixture to mixture), the chemical manufacturer, importer or employer may prepare one material safety data sheet to apply to all of these similar mixtures.

(5) The chemical manufacturer, importer, or employer preparing the material safety data sheet shall ensure that the information recorded accurately reflects the scientific evidence used in making the hazard determination. If the chemical manufacturer, importer or employer becomes newly aware of any significant information regarding the hazards of a chemical, or ways to protect against the hazards, this new information shall be added to the material safety data sheet within three months. If the chemical is not currently being produced or imported the chemical manufacturer or importer shall add the information to the material safety date sheet before the chemical is introduced into the workplace again.

(6) Chemical manufacturers or importers shall ensure that distributors and manufacturing purchasers of hazardous chemicals are provided an appropriate material safety data sheet with their initial shipment, and with the first shipment after a material safety data sheet is updated. The chemical manufacturer or importer shall either provide material safety data sheets with the shipped containers or send them to the manufacturing purchaser prior to or at the time of the shipment. If the material safety data sheet is not provided with the shipment, the manufacturing purchaser shall obtain one from the chemical manufacturer, importer, or distributor as soon as possible.

(7) Distributors shall ensure that material safety data sheets, and updated information, are provided to other distributors and manufacturing purchasers of hazardous chemicals.

(8) The employer shall maintain copies of the required material safety data sheets for each hazardous chemical in the workplace, and shall ensure that they are readily accessible during each work shift to employees when they are in their work area(s).

(9) Material safety data sheets may be kept in any form, including operating procedures, and may be designed to cover groups of hazardous chemicals in a work area where it may be more appropriate to address the hazards of a process rather than individual hazardous chemicals. However, the employer shall ensure that in all cases the required information is provided for each hazardous chemical, and is readily accessible during each work shift to employees when they are in their work area(s).

(10) Material safety data sheets shall also be made readily available, upon request, to designated representatives and to the Assistant Secretary, in accordance with the requirements of 29 CFR 1910.20 (e). The Director shall also be given access to material safety data sheets in the same manner.

of carcinogens, less than 0.1 percent) could be released in concentrations which would exceed an established OSHA permissible exposure limit or ACGIH Threshold Limit Value, or could present a health hazard to employees in those concentrations, the mixture shall be assumed to present the same hazard.

(6) Chemical manufacturers, importers, or employers evaluating chemicals shall describe in writing the procedures they use to determine the hazards of the chemical they evaluate. The written procedures are to be made available, upon request, to employees, their designated representatives, the Assistant Secretary and the Director. The written description may be incorporated into the written hazard communication program required under paragraph (e) of this section.

(e) *Written hazard communication program.*

(1) Employers shall develop and implement a written hazard communication program for their workplaces which at least describes how the criteria specified in paragraphs (f), (g), and (h) of this section for labels and other forms of warning, material safety data sheets, and employee information and training will be met, and which also includes the following:

   (i) A list of the hazardous chemicals known to be present using an identity that is referenced on the appropriate material safety data sheet (the list may be compiled for the workplace as a whole or for individual work areas);

   (ii) The methods the employer will use to inform employees of the hazards of non-routine tasks (for example, the cleaning of reactor vessels), and the hazards associated with chemicals contained in unlabeled pipes in their work areas; and,

   (iii) The methods the employer will use to inform any contractor employers with employees working in the employer's workplace of the hazardous chemicals their employees may be exposed to while performing their work, and any suggestions for appropriate protective measures.

(2) The employer may rely on an existing hazard communication program to comply with these requirements, provided that it meets the criteria established in this paragraph (e).

(3) The employer shall make the written hazard communication program available, upon request, to employees, their designated representatives, the Assistant Secretary and the Director, in accordance with the requirements of 29 CFR 1910.20 (e).

(f) *Labels and other forms of warning.*

(1) The chemical manufacturer, importer, or distributor shall ensure that each container of hazardous chemicals leaving the workplace is labeled, tagged or marked with the following information:

   (i) Identity of the hazardous chemical(s):

   (ii) Appropriate hazard warnings; and

   (iii) Name and address of the chemical manufacturer, importer, or other responsible party.

(2) Chemical manufacturers, importers, or distributors shall ensure that each

container of hazardous chemicals leaving the workplace is labeled, tagged, or marked in accordance with this section in a manner which does not conflict with the requirements of the Hazardous Materials Transportation Act (18 U.S.C. ss1801 et seq.) and regulations issued under that Act by the Department of Transportation.

(3) If the hazardous chemical is regulated by OSHA in a substance-specific health standard, the chemical manufacturer, importer, distributor or employer shall ensure that the labels or other forms of warning used are in accordance with the requirements of that standard.

(4) Except as provided in paragraphs (f)(5) and (f)(6) the employer shall ensure that each container of hazardous chemicals in the workplace is labeled, tagged or marked with the following information:

(i) Identity of the hazardous chemical(s) contained therein; and

(ii) Appropriate hazard warnings.

(5) The employer may use signs, placards, process sheets, batch tickets, operating procedures, or other such written materials in lieu of affixing labels to individual stationary process containers, as long as the alternative method identifies the containers to which it is applicable and conveys the information required by paragraph (f)(4) of this section to be on a label. The written materials shall be readily accessible to the employees in their work area throughout each work shift.

(6) The employer is not required to label portable containers into which hazardous chemicals are transferred from labeled containers, and which are intended only for the immediate use of the employee who performs the transfer.

(7) The employer shall not remove or deface existing labels on incoming containers of hazardous chemicals, unless the container is immediately marked with the required information.

(8) The employer shall ensure that labels or other forms of warning are legible, in English, and prominently displayed on the container, or readily available in the work area throughout each work shift. Employers having employees who speak other languages may add the information in their language to the material presented, as long as the information is presented in English as well.

(9) The chemical manufacturer, importer, distributor or employer need not affix new labels to comply with this section if existing labels already convey the required information.

(g) *Material safety data sheets.*

(1) Chemical manufacturers and importers shall obtain or develop a material safety data sheet for each hazardous chemical they produce or import. Employers shall have a material safety data sheet for each hazardous chemical which they use.

(2) Each material safety data sheet shall be in English and shall contain at least the following information:

(i) The identity used on the label, and, except as provided for in paragraph (i) of this section on trade secrets:

(A) If the hazardous chemical is a single substance, its chemical and common name(s);

(h) *Employee information and training.* Employers shall provide employees with information and training on hazardous chemicals in their work area at the time of their initial assignment, and whenever a new hazard is introduced into their work area.

　(1) *Information.* Employees shall be informed of:

　　(i) The requirements of this section;

　　(ii) Any operations in their work area where hazardous chemicals are present; and,

　　(iii) The location and availability of the written hazard communication program, including the required list(s) of hazardous chemicals, and material safety data sheets required by this section.

　(2) *Training.* Employee training shall include at least:

　　(i) Methods and observations that may be used to detect the presence or release of a hazardous chemical in the work area (such as monitoring conducted by the employer, continuous monitoring devices, visual appearance or odor of hazardous chemicals when being released, etc.);

　　(ii) The physical and health hazards of the chemicals in the work area;

　　(iii) The measures employees can take to protect themselves from these hazards, including specific procedures the employer has implemented to protect employees from exposure to hazardous chemicals, such as appropriate work practices, emergency procedures, and personal protective equipment to be used; and,

　　(iv) The details of the hazard communication program developed by the employer, including an explanation of the labeling system and the material safety data sheet, and how employees can obtain and use the appropriate hazard information.

(i) *Trade secrets.*

　(1) The chemical manufacturer, importer or employer may withhold the specific chemical identity, including the chemical name and other specific identification of a hazardous chemical, from the material safety data sheet, provided that:

　　(i) The claim that the information withheld is a trade secret can be supported;

　　(ii) Information contained in the material safety data sheet concerning the properties and effects of the hazardous chemical is disclosed;

　　(iii) The material safety data sheet indicates that the specific chemical identity is being withheld as a trade secret; and,

　　(iv) The specific chemical identity is made available to health professionals, in accordance with the applicable provisions of this paragraph.

　(2) Where a treating physician or nurse determines that a medical emergency exists and the specific chemical identity of a hazardous chemical is necessary for emergency or first-aid treatment, the chemical manufacturer, importer, or employer shall immediately disclose the specific chemical identity of a trade secret chemical to that treating physician or nurse,

regardless of the existence of a written statement of need or a confidentiality agreement. The chemical manufacturer, importer, or employer may require a written statement of need and confidentiality agreement, in accordance with the provisions of paragraphs (i)(3) and (4) of this section, as soon as circumstances permit.

(3) In non-emergency situations, a chemical manufacturer, importer, or employer shall, upon request, disclose a specific chemical identity, otherwise permitted to be withheld under paragraph (i)(1) of this section, to a health professional (i.e. physician, industrial hygienist, toxicologist, or epidemiologist) providing medical or other occupational health services to exposed employee(s) if:

  (i) The request is in writing;

  (ii) The request describes with reasonable detail one or more of the following occupational health needs for the information:

   (A) To assess the hazards of the chemicals to which employees will be exposed;

   (B) To conduct or assess sampling of the workplace atmosphere to determine employee exposure levels;

   (C) To conduct pre-assignment or periodic medical surveillance of exposed employees;

   (D) To provide medical treatment to exposed employees;

   (E) To select or assess appropriate personal protective equipment for exposed employees;

   (F) To design or assess engineering controls or other protective measures for exposed employees; and,

   (G) To conduct studies to determine the health effects of exposure.

  (iii) The request explains in detail why the disclosure of the specific chemical identity is essential and that, in lieu thereof, the disclosure of the following information would not enable the health professional to provide the occupational health services described in paragraph (ii) of this section:

   (A) The properties and effects of the chemical;

   (B) Measures for controlling workers' exposure to the chemical;

   (C) Methods of monitoring and analyzing worker exposure to the chemical; and,

   (D) Methods of diagnosing and treating harmful exposures to the chemical;

  (iv) The request includes a description of the procedures to be used to maintain the confidentiality of the disclosed information; and,

  (v) The health professional, and the employer or contractor of the health professional's services (i.e., downstream employer, labor organization, or individual employer), agree in a written confidentiality agreement that the health professional will not use the trade secret information for any purpose other than the health need(s) asserted and agree not to release the information under any circumstances other than to OSHA, as provided in paragraph (i)(6) of this section, except as authorized by the terms of the agreement or by the chemical manufacturer, importer, or employer.

(4) The confidentiality agreement authorized by paragraph (i)(3)(iv) of this section:

   (i) May restrict the use of the information to the health purposes indicated in the written statement of need;

   (ii) May provide for appropriate legal remedies in the event of a breach of the agreement, including stipulation of a reasonable pre-estimate of likely damages; and,

   (iii) May not include requirements for the posting of a penalty bond.

(5) Nothing in this standard is meant to preclude the parties from pursuing non-contractual remedies to the extent permitted by law.

(6) If the health professional receiving the trade secret information decides that there is a need to disclose it to OSHA, the chemical manufacturer, importer, or employer who provided the information shall be informed by the health professional prior to, or at the same time as, such disclosure.

(7) If the chemical manufacturer, importer, or employer denies a written request for disclosure of a specific chemical identity, the denial must:

   (i) Be provided to the health professional within thirty days of the request;

   (ii) Be in writing;

   (iii) Include evidence to support the claim that the specific chemical identity is a trade secret;

   (iv) State the specific reasons why the request is being denied; and,

   (v) Explain in detail how alternative information may satisfy the specific medical or occupational health need without revealing the specific chemical identity.

(8) The health professional whose request for information is denied under paragraph (i)(3) of this section may refer the request and the written denial of the request to OSHA for consideration.

(9) When a health professional refers the denial to OSHA under paragraph (i)(8) of this section, OSHA shall consider the evidence to determine if:

   (i) The chemical manufacturer, importer, or employer has supported the claim that the specific chemical identity is a trade secret;

   (ii) The health professional has supported the claim that there is a medical or occupational health need for the information; and,

   (iii) The health professional has demonstrated adequate means to protect the confidentiality.

(10)   (i) If OSHA determines that the specific chemical identity requested under paragraph (i)(3) of this section is not a *bona fide* trade secret, or that it is a trade secret but the requesting health professional has a legitimate medical or occupational health need for the information, has executed a written confidentiality agreement, and has shown adequate means to protect the confidentiality of the information, the chemical manufacturer, importer, or employer will be subject to citation by OSHA.

(ii) If a chemical manufacturer, importer, or employer demonstrates to OSHA that the execution of a confidentiality agreement would not provide sufficient protection against the potential harm from the unauthorized disclosure of a trade secret specific chemical identity, the Assistant Secretary may issue such orders or impose such additional limitations or conditions upon the disclosure of the requested chemical information as may be appropriate to assure that the occupational health services are provided without an undue risk of harm to the chemical manufacturer, importer, or employer.

(11) If, following the issuance of a citation and any protective orders, the chemical manufacturer, importer, or employer continues to withhold the information, the matter is referrable to the Occupational Safety and Health Review Commission for enforcement of the citation. In accordance with Commission rules, the Administrative Law Judge may review the citation and supporting documentation *in camera* or issue appropriate protective orders.

(12) Nothwithstanding the existence of a trade secret claim, a chemical manufacturer, importer, or employer shall, upon request, disclose to the Assistant Secretary any information which this section requires the chemical manufacturer, importer, or employer to make available. Where there is a trade secret claim, such claim shall be made no later than at the time the information is provided to the Assistant Secretary so that suitable determinations of trade secret status can be made and the necessary protections can be implemented.

(13) Nothing in this paragraph shall be construed as requiring the disclosure under any circumstances of process or percentage of mixture information which is trade secret.

(j) *Effective dates.* Employers shall be in compliance with this section within the following time periods:

(1) Chemical manufacturers and importers shall label containers of hazardous chemicals leaving their workplaces, and provide material safety data sheets with initial shipments by November 25, 1985.

(2) Distributors shall be in compliance with all provisions of this section applicable to them by November 25, 1985.

(3) Employers shall be in compliance with all provisions of this section by May 25, 1986, including initial training for all current employees.

## APPENDIX A TO s1910.1200: HEALTH HAZARD DEFINITIONS (Mandatory)

Although safety hazards related to the physical characteristics of a chemical can be objectively defined in terms of testing requirements (e.g., flammability), health hazard definitions are less precise and more subjective. Health hazards may cause measurable changes in the body—such as decreased pulmonary function. These changes are generally indicated by the occurrence of signs and symptoms in the exposed employees—such as shortness of breath, a non-measurable, subjective feeling. Employees exposed to such hazards must be apprised of both the change in body function and the signs and symptoms that may occur to signal that change.

The determination of occupational health hazards is complicated by the fact that many of the effects or signs and symptoms occur commonly in non-occupationally exposed populations, so that effects of exposure are difficult to separate from normally occurring ill-nesses. Occasionally, a substance causes an effect that is rarely seen in the population at large, such as angiosarcomas caused by vinyl chloride exposure, thus making it easier to ascertain that the occupational exposure was the primary causative factor. More often, however, the effects are common, such as lung cancer. The situation is further compli-cated by the fact that most chemicals have not been adequately tested to determine their health hazard potential, and data do not exist to substantiate these effects.

There have been many attempts to categorize effects and to define them in various ways. Generally, the terms "acute" and "chronic" are used to delineate between effects on the basis of severity or duration. "Acute" effects usually occur rapidly as a result of short-term exposures, and are of short duration. "Chronic" effects generally occur as a result of long-term exposure, and are of long duration.

The acute effects referred to most frequently are those defined by the American Na-tional Standards Institute (ANSI) standard for Precautionary Labeling of Hazardous In-dustrial Chemicals (Z129.1-1982)—irritation, corrosivity, sensitization and lethal dose. Although these are important health effects, they do not adequately cover the considera-ble range of acute effects which may occur as a result of occupational exposure, such as, for example, narcosis.

Similarly, the term chronic effect is often used to cover only carcinogenicity, teratogenicity, and mutagenicity. These effects are obviously a concern in the workplace, but again, do not adequately cover the area of chronic effects, excluding, for example, blood dyscrasias (such as anemia), chronic bronchitis and liver atrophy.

The goal of defining precisely, in measurable terms, every possible health effect that may occur in the workplace as a result of chemical exposures cannot realistically be ac-complished. This does not negate the need for employees to be informed of such effects and protected from them.

Appendix B, which is also mandatory, outlines the principles and procedures of hazard assessment.

For purposes of this section, any chemicals which meet any of the following defini-tions, as determined by the criteria set forth in Appendix B are health hazards:

1. *Carcinogen:* A chemical is considered to be a carcinogen if:

    (a) It has been evaluated by the International Agency for Research on Cancer (IARC), and found to be a carcinogen or potential carcinogen; or

    (b) It is listed as a carcinogen or potential carcinogen in the *Annual Report on Carcinogens* published by the National Toxicology Program (NTP) (latest edition); or,

    (c) It is regulated by OSHA as a carcinogen.

2. *Corrosive:* A chemical that causes visible destruction of, or irreversible altera-tions in, living tissue by chemical action at the site of contact. For example, a chemical is considered to be corrosive if, when tested on the intact skin of albino rabbits by the method described by the U.S. Department of Transportation in Ap-pendix A to 49 CFR Part 173, it destroys or changes irreversibly the structure of the tissue at the site of contact following an exposure period of four hours. This term shall not refer to action on inanimate surfaces.

3. *Highly toxic:* A chemical falling within any of the following categories:

    (a) A chemical that has a median lethal dose ($LD_{50}$) of 50 milligrams or less per kilogram of body weight when administered orally to albino rats weigh-ing between 200 and 300 grams each.

    (b) A chemical that has a median lethal dose ($LD_{50}$) of 200 milligrams or less per kilogram of body weight when administered by continuous contact for

24 hours (or less if death occurs within 24 hours) with the bare skin of albino rabbits weighing between two and three kilograms each.

(c) A chemical that has a median lethal concentration ($LC_{50}$) in air of 200 parts per million by volume or less of gas or vapor, or 2 milligrams per liter or less of mist, fume, or dust, when administered by continuous inhalation for one hour (or less if death occurs within one hour) to albino rats weighing between 200 and 300 grams each.

4. *Irritant:* A chemical, which is not corrosive, but which causes a reversible inflammatory effect on living tissue by chemical action at the site of contact. A chemical is a skin irritant if, when tested on the intact skin of albino rabbits by the methods of 16 CFR 1500.41 for four hours exposure or by other appropriate techniques, it results in an empirical score of five or more. A chemical is an eye irritant if so determined under the procedure listed in 16 CFR 1500.42 or other appropriate techniques.

5. *Sensitizer:* A chemical that causes a substantial proportion of exposed people or animals to develop an allergic reaction in normal tissue after repeated exposure to the chemical.

6. *Toxic.* A chemical falling within any of the following categories:

(a) A chemical that has a median lethal dose ($LD_{50}$) of more than 50 milligrams per kilogram but not more than 500 milligrams per kilogram of body weight when administered orally to albino rats weighing between 200 and 300 grams each.

(b) A chemical that has a median lethal dose ($LD_{50}$) of more than 200 milligrams per kilogram but not more than 1,000 milligrams per kilogram of body weight when administered by continuous contact for 24 hours (or less if death occurs within 24 hours) with the bare skin of albino rabbits weighing between two and three kilograms each.

(c) A chemical that has a median lethal concentration (LC 50) in air of more than 200 parts per million but not more than 2,000 parts per million by volume of gas or vapor, or more than two milligrams per liter but not more than 20 milligrams per liter of mist, fume, or dust, when administered by continuous inhalation for one hour (or less if death occurs within one hour) to albino rats weighing between 200 and 300 grams each.

7. *Target organ effects.* The following is a target organ categorization of effects which may occur, including examples of signs and symptoms and chemicals which have been found to cause such effects. These examples are presented to illustrate the range and diversity of effects and hazards found in the workplace, and the broad scope employers must consider in this area, but are not intended to be all-inclusive.

| | | |
|---|---|---|
| (a) Hepatotoxins | Chemicals which produce liver damage | |
| Signs and symptoms | Jaundice, liver enlargement | |
| Chemicals | Carbon tetrachloride, nitrosamines | |
| (b) Nephrotoxins | Chemicals which produce kidney damage | |
| Signs and symptoms | Edema, proteinuria | |
| Chemicals | Halogenated hydrocarbons, uranium | |
| (c) Neurotoxins | Chemicals which produce their primary toxic effects on the nervous system | |
| Signs and symptoms | Narcosis, behavioral changes, decrease in motor functions | |
| Chemicals | Mercury, carbon disulfide | |

| (d) Agents which act on the blood or hematopoietic system | Decrease hemoglobin function, deprive the body tissues of oxygen |
| Signs and symptoms | Cyanosis, loss of consciousness |
| Chemicals | Carbon monoxide, cyanides |
| (e) Agents which damage the lung | Chemicals which irritate or damage the pulmonary tissue |
| Signs and symptoms | Cough, tightness in chest, shortness of breath |
| Chemicals | Silica, asbestos |
| (f) Reproductive toxins | Chemicals which affect the reproductive capabilities including chromosomal damage (mutations) and effects on fetuses (teratogenesis) |
| Signs and symptoms | Birth defects, sterility |
| Chemicals | Lead, DBCP |
| (g) Cutaneous hazards | Chemicals which affect the dermal layer of the body |
| Signs and symptoms | Defatting of the skin, rashes, irritation |
| Chemicals | Ketones, chlorinated compounds |
| (h) Eye hazards | Chemicals which affect the eye or visual capacity |
| Signs and symptoms | Conjunctivitis, corneal damage |
| Chemicals | Organic solvents, acids |

## APPENDIX B TO s1910.1200: HAZARD DETERMINATION (Mandatory)

The quality of a hazard communication program is largely dependent upon the adequacy and accuracy of the hazard determination. The hazard determination requirement of this standard is performance-oriented. Chemical manufacturers, importers, and employers evaluating chemicals are not required to follow any specific methods for determining hazards, but they must be able to demonstrate that they have adequately ascertained the hazards of the chemicals produced or imported in accordance with the criteria set forth in this Appendix.

Hazard evaluation is a process which relies heavily on the professional judgment of the evaluator, particularly in the area of chronic hazards. The performance-orientation of the hazard determination does not diminish the duty of the chemical manufacturer, importer or employer to conduct a thorough evaluation, examining all relevant data and producing a scientifically defensible evaluation. For purposes of this standard, the following criteria shall be used in making hazard determinations that meet the requirements of this standard.

1. *Carcinogenicity:* As described in paragraph (d)(4) and Appendix A of this section, a determination by the National Toxicology Program, the International Agency for Research on Cancer, or OSHA that a chemical is a carcinogen or potential carcinogen will be considered conclusive evidence for purposes of this section.

2. *Human data:* Where available, epidemiological studies and case reports of adverse health effects shall be considered in the evaluation.

3. *Animal data:* Human evidence of health effects in exposed populations is generally not available for the majority of chemicals produced or used in the workplace. Therefore, the available results of toxicological testing in animal populations shall be used to predict the health effects that may be experienced by exposed workers. In particular, the definitions of certain acute hazards refer to specific animal testing results (see Appendix A).

4. *Adequacy and reporting of data:* The results of any studies which are designed and conducted according to established scientific principles, and which report

statistically significant conclusions regarding the health effects of a chemical, shall be a sufficient basis for a hazard determination and reported on any material safety data sheet. The chemical manufacturer, importer, or employer may also report the results of other scientifically valid studies which tend to refute the findings of hazard.

## APPENDIX C TO s1910.1200: INFORMATION SOURCES (Advisory)

The following is a list of available data sources which the chemical manufacturer, importer, or employer may wish to consult to evaluate the hazards of chemicals he produces or imports:

- Any information in his own company files such as toxicity testing results or illness experience of company employees.

- Any information obtained from the supplier of the chemical, such as material safety data sheets or product safety bulletins.

- Any pertinent information obtained from the following source list (latest editions should be used):

*Condensed Chemical Dictionary*
    Van Nostrand Reinhold Co.
    135 West 50th Street
    New York, NY 10020

*The Merck Index: An Encyclopedia of Chemicals and Drugs*
    Merck and Company, Inc.
    126 E. Lincoln Avenue
    Rahway, NJ 07065

*IARC Monographs on the Evaluation of the Carcinogenic Risk of Chemicals to Man*
    Geneva: World Health Organization
    International Agency for Research on Cancer, 1972-1977
    (Multivolume work)
    49 Sheridan Street
    Albany, New York

*Industrial Hygiene and Toxicology, by F. A. Patty*
    John Wiley & Sons, Inc.
    New York, NY
    (Five volumes)

*Clinical Toxicology of Commercial Products*
    Gleason, Gosselin and Hodge

*Casarett and Doull's*
*Toxicology: The Basic Science of Poisons,*
    Doull, Klaassen, and Amdur
    Macmillan Publishing Co., Inc.
    New York, NY

*Industrial Toxicology, by Alice Hamilton and Harriet L. Hardy*
    Publishing Sciences Group, Inc.
    Acton, MA

*Toxicology of the Eye, by W. Morton Grant*
Charles C. Thomas
301-327 East Lawrence Avenue
Springfield, IL

*Recognition of Health Hazards in Industry*
William A. Burgess
John Wiley and Sons
605 Third Avenue
New York, NY 10158

*Chemical Hazards of the Workplace*
Nick H. Proctor and James P. Hughes
J. P. Lipincott Company
6 Winchester Terrace
New York, NY 10022

*Handbook of Chemistry and Physics*
Chemical Rubber Company
18901 Cranwood Parkway
Cleveland, OH 44128

*Threshold Limit Values for Chemical Substances and Physical Agents in the Workroom Environment with Intended Changes*
American Conference of Governmental Industrial Hygienists
6500 Glenway Avenue, Bldg. D-5
Cincinnati, OH 45211

NOTE: the following documents are on sale by the Superintendent of Documents, U.S. Government Printing Office, Washington, D.C. 20402

*Occupational Health Guidelines*
NIOSH/OSHA (NIOSH Pub. No. 81-123)

*NIOSH/OSHA Pocket Guide to Chemical Hazards*
NIOSH Pub. No. 78-210

*Registry of Toxic Effects of Chemical Substances*
U.S. Department of Health and Human Services
Public Health Service
Center for Disease Control
National Institute for Occupational Safety and Health
(NIOSH Pub. No. 80-102)

*The Industrial Environment—Its Evaluation and Control*
U.S. Department of Health and Human Services
Public Health Service
Center for Disease Control
National Institute for Occupational Safety and Health
(NIOSH Pub. No. 74-117)

## Miscellaneous Documents

National Institute for Occupational Safety and Health

1. Criteria for a recommended standard . . .
    Occupational Exposure to "_____"

2. Special Hazard Reviews

3. Occupational Hazard Assessment

4. Current Intelligence Bulletins

## Bibliographic Data Bases

| Service Provider | File Name |
|---|---|
| Bibliographic Retrieval Services (BRS) Corporation Park, Bldg. 702 Scotia, New York 12302 | AGRICOLA BIOSIS PREVIEWS CA CONDENSATES CA SEARCH DRUG INFORMATION MEDLARS MEDOC NTIS POLLUTION ABSTRACTS SCIENCE CITATION INDEX SSIE |
| Lockheed—DIALOG Lockheed Missiles & Space Company, Inc. P.O. Box 44481 San Francisco, CA 94144 | AGRICOLA BIOSIS PREV. 1972-PRESENT BIOSIS PREV. 1969-71 CA CONDENSATES 1970-71 CA SEARCH 1972-76 CA SEARCH 1977-PRESENT CHEMNAME CONFERENCE PAPERS INDEX FOOD SCIENCE & TECH. ABSTR. FOODS ADLIBRA INTL. PHARMACEUTICAL ABSTR. NTIS POLLUTION ABSTRACTS SCISEARCH 1978-PRESENT SCISEARCH 1974-77 SSIE CURRENT RESEARCH |
| SDC—ORBIT SDC Search Service Department No. 2230 Pasadena, CA 91051 | AGRICOLA BIOCODES BIOSIS/BIO6973 CAS6771/CAS7276 CAS77 CHEMDEX CONFERENCE ENVIROLINE LABORDOC NTIS POLLUTION SSIE |
| Chemical Information Systems (CIS) Chemical Information Systems, Inc. 7215 Yorke Road Baltimore, MD 21212 | Structure & Nomenclature Search System Acute Toxicity (RTECS) Clinical Toxicology of Commercial Products Oil and Hazardous Materials Technical Assistance Data System |
| National Library of Medicine Department of Health and Human Services Public Health Service National Institutes of Health Bethesda, MD 20209 | Toxicology Data Bank (TDB) MEDLINE TOXLINE CANCERLIT RTECS |

# Acronyms

| | |
|---|---|
| ACGIH | American Conference of Government Industrial Hygienists |
| ACS | American Chemical Society |
| ADR | European Agreement Concerning the International Carriage of Dangerous Goods by Road |
| ANPR | Advanced Notice of Proposed Rulemaking |
| ANSI | American National Standards Institute |
| ASTM | American Society for Testing and Materials |
| BIOS | Biological Information Service |
| BRS | Bibliographic Retrieval Services |
| CAS | Chemical Abstracts Service |
| CEFIC | European Council of Commercial Manufacturers |
| CFR/C.F.R. | Code of Federal Regulations |
| CIM | International Convention Concerning the Carriage of Goods by Rail |
| CMA | Chemical Manufacturers Association |
| CPR | Cardio-Pulmonary Resuscitation |
| CPSA | Consumer Product Safety Act |
| CPSC | Consumer Product Safety Commission |
| CSIN | Chemical Substances Information Network |
| DHEW/HEW | Department of Health, Education and Welfare |
| DOD | Department of Defense |
| DOT | Department of Transportation |
| EDF | Environmental Defense Fund |

| | |
|---|---|
| EEC | Council of the European Economic Community |
| EP | Extraction Procedure |
| EPA | Environmental Protection Agency |
| EUP | Environmental Use Permit |
| FDA | Food and Drug Administration |
| FFDCA | Federal Food, Drug and Cosmetic Act |
| FHSA | Federal Hazardous Substances Act |
| FIFRA | Federal Insecticide, Fungicide and Rodenticide Act |
| FOIA | Freedom of Information Act |
| FR/Fed. Reg. | Federal Register |
| FTC | Federal Trade Commission |
| GPO | Government Printing Office |
| HMAC | Hazardous Materials Advisory Council |
| HMIS | Hazardous Materials Identification System |
| IAEA | International Atomic Energy Authority |
| IATA | International Air Transport Association |
| ICAO | International Civil Aviation Organization |
| ICC | Interstate Commerce Commission/International Chamber of Commerce |
| IMCO | Intergovernmental Maritime Consultative Organization |
| IMDG | International Dangerous Goods Code |
| IMO | International Maritime Organization |
| IRLG | Interagency Regulatory Liaison Group |
| LAPI | Labels and Precautionary Information |
| MSDS | Material Safety Data Sheet |
| NACE | National Association of Corrosion Engineers |
| NFPA | National Fire Protection Association |
| NIOSH | National Institute for Occupational Safety and Health |
| NLM | National Library of Medicine |
| NOS/n.o.s. | Not Otherwise Specified |
| NPCA | National Paint and Coating Association |
| NTIS | National Technical Information Service |
| OCTI | Central Office for International Railway Transport |
| OECD | Organization for Economic Cooperation and Development |
| OMB | Office of Management and Budget |
| OPTS | Office of Pesticide and Toxic Substances |

| | |
|---|---|
| ORM | Other Regulated Materials |
| OSHA | Occupational Safety and Health Administration |
| PCB | Polychlorinated biphenyls |
| PEL | Permissible Exposure Level |
| RAR | Restricted Articles Regulations |
| RCRA | Resource Conservation and Recovery Act |
| RID | International Regulations Concerning the Carriage of Dangerous Goods by Rail |
| RPAR | Rebuttable Presumption Against Registration |
| RQ | Reportable Quantity |
| RTECS | Registry of Toxic Effects of Chemical Substances List |
| SDS | Systems Development Corp. |
| SIC | Standard Industrial Classification |
| SOLAS | International Conference on Safety of Life at Sea |
| SSIE | Smithsonian Science Information Exchange |
| TSCA | Toxic Substances Control Act |
| TSDF | Treatment Storage and Disposal Facility |
| UNEP | United Nations Environment Program |
| USDA | U.S. Department of Agriculture |

# Index

# D.O.T. PLACARDS

**DANGEROUS**  **FLAMMABLE** 3  **FLAMMABLE GAS** 2  **COMBUSTIBLE**

**ORGANIC PEROXIDE** 5  **NON-FLAMMABLE GAS** 2  **CORROSIVE** 8  **POISON** 6

**OXIDIZER** 5  **RADIOACTIVE** 7  **EXPLOSIVES A** 1  **EXPLOSIVES B** 1

**POISON GAS** 2  **FLAMMABLE SOLID**  **FLAMMABLE SOLID**  **OXYGEN** 2

**CHLORINE** 2  **FUEL OIL**  **GASOLINE** 3  **BLASTING AGENTS**

**RADIOACTIVE** 7  **EXPLOSIVES A** 1  **POISON GAS** 2  **EMPTY**

DANGER

AVOID ACCIDENTS

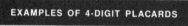

**EXAMPLES OF 4-DIGIT PLACARDS**

**1541**

**AVOID ACCIDENTS**
DO NOT REMOVE THIS DOME COVER
WHILE GAS PRESSURE EXISTS IN TANK
KEEP LIGHTED LANTERNS AWAY

 **1090** 3    **1072** 5    **1203** 3   **1005** 2